PURE MATHEMATICS

Textbook

ANN WHITEHOUSE BSc

HOLBORN COLLEGE COURSES

SPECIALIST DIPLOMAS IN LAW & BUSINESS
Validated by the University of Oxford Delegacy of Local Examinations at degree level.

9 month course.
Diplomas in: Contract Law • Commercial Law • Company Law • Revenue Law • European Community Law • Criminal Law • Evidence • Constitutional Law • English Legal System • Land Law • Organisational Theory • Economics • Accounting and Business Finance • Computer Systems & Information Technology • Maths for Economists • Statistics

Entry: Evidence of sufficient academic or work experience to study at degree level.

FUNDAMENTALS OF BRITISH BUSINESS
Examined internally by Holborn College

To familiarise European and other overseas business studies students with UK business practice. Courses are tailor-made for groups of students on a College to College basis. Courses run for ten to seventeen weeks and permit part-time relevant work experience.

Past courses have included: Aspects of Organisational Behaviour • Marketing • Statistics • Business English & Communication Skills • Economics • International Trade • Company Law

LAW
Examined externally by the University of London.

Three year course.
LLB Law.
Entry: 2 'A' levels grade E and 3 'O' levels.

ACCOUNTING & MANAGEMENT DEGREES
Examined externally by the University of London.

Three year course.
BSc (Econ) Accounting • BSc (Econ) Management Studies • BSc (Econ) Economics & Management Studies

Entry: 2 'A' levels grade E and 3 'O' levels to include Maths and English.

DIPLOMA IN ECONOMICS.
Examined externally by the University of London.

One year full time, two year part time courses.
Completion of the Diploma gives exemption from the first year of the BSc (Econ) Degree programmes reducing them to two years.

Entry: Mathematics and English 'O' level equivalent. Minimum age 18.

THE COMMON PROFESSIONAL EXAMINATION
Examined externally by Wolverhampton Polytechnic.

9 month, 6 month and short revision courses.

Entry: Acceptance by the Professional Body.

THE BAR EXAMINATION
Examined by the Council of Legal Education for non-UK practitioners.

9 month course.

Entry: Acceptance by the Professional Body.

THE SOLICITORS' FINAL
Examined by the Law Society.

6 month re-sit and short revision courses.
Entry: Acceptance by the Professional Body.

THE INSTITUTE OF LEGAL EXECUTIVES FINAL PART 2
Examined by the Institute of Legal Executives.

9 month course and short revision courses.

Entry: Acceptance by the Professional Body.

A & AS LEVEL COURSES
Examined by various UK Boards.

9 month course and short revision courses
Subjects offered: Law • Constitutional Law • Economics • Accounting • Business Studies • Mathematics Pure and Applied • Mathematics and Statistics • Sociology • Government and Politics

Entry: 3 'O' levels.

FULL-TIME, PART-TIME, REVISION & DISTANCE LEARNING

PURE MATHEMATICS

CONTENTS

1 ESSENTIAL ALGEBRA

1.1	Revision of basic algebra techniques	1
1.2	The quadratic function	10
1.3	General algebraic functions	23
1.4	Indices and logarithms	33

2 POWER SERIES AND EXPANSIONS

2.1	Series	41
2.2	Binomial theorem	47
2.3	Log and exponential functions	53
2.4	Expansions	62

3 TRIGONOMETRY

3.1	Basic trigonometry	70
3.2	Angle formulae and applications	80
3.3	Trigonometrical equations	88
3.4	Solutions of triangles	103

4 DIFFERENTIAL CALCULUS

4.1	Introduction to differentiation	114
4.2	Further differentiation	124
4.3	Other functions	141
4.4	Applications of differentiation	152

5 INTEGRAL CALCULUS

5.1	Introduction to integration	166
5.2	Further integration	181
5.3	Area under a curve	190
5.4	Further applications of integration	206

6 LINE, CIRCLE, PARABOLA

6.1	Graphical work	214
6.2	Introduction to the straight line	225
6.3	Loci	240
6.4	Circle and parabola	256

7 APPROXIMATIONS AND SERIES

7.1	Solutions of equations	272
7.2	Hyperbolic functions	284
7.3	Further work on series	302

8 CONICS

8.1	Curve sketching	321
8.2	Polar coordinates	334
8.3	Ellipse	346
8.4	Hyperbola	360

9 COMPLEX NUMBERS AND VECTORS

9.1	Complex numbers	377
9.2	Further theory of complex numbers	395
9.3	Vectors	416

PURE MATHEMATICS — CONTENTS

10 THREE-DIMENSIONAL GEOMETRY AND VECTORS

10.1	Three-dimensional coordinate geometry	435
10.2	Further three-dimensional coordinate geometry	452
10.3	Vectors in three dimensions	460
10.4	Further vector work	476

11 FURTHER CALCULUS; DIFFERENTIAL EQUATIONS

11.1	Differentiation and integration	494
11.2	Applications of calculus	518
11.3	Differential equations	534

12 SET THEORY AND PROBABILITY

12.1	Set theory	553
12.2	Functions	561
12.3	Permutations and combinations	571
12.4	Probability	583

PURE MATHEMATICS — ESSENTIAL ALGEBRA

1 ESSENTIAL ALGEBRA

   1.1   Revision of basic algebra techniques
   1.2   The quadratic function
   1.3   General algebraic functions
   1.4   Indices and logarithms

1.1 Revision of basic algebra techniques

a)   *The four rules*

The four rules are the four basic operations of addition, subtraction, multiplication and division. Taking each of these in turn:

i)   *Addition and subtraction*

Only *like* terms may be added or subtracted, ie x's may be added to x's and y's may be added to y's but x's may *not* be added to y's, and similarly for subtraction.

Example 1

Simplify:
(1) $7x - 5x + 3x$
(2) $5a + 6b + 4a - 3b$
(3) $2pq + 3pq - 2p - 3q$
(4) $3a^2b + 4ab^2 - 6a^2b - 3ab^2$

Solution
(1) $7x - 5x + 3x = (7 - 5 + 3)x = 5x$
(2) $5a + 6b + 4a - 3b = 5a + 4a + 6b - 3b$
$$= (5 + 4)a + (6 - 3)b = 9a + 3b$$
(3) $2pq + 3pq - 2p - 3q = (2 + 3)pq - 2p - 3q = 5pq - 2p - 3q$
(4) $3a^2b + 4ab^2 - 6a^2b - 3ab^2 = 3a^2b - 6a^2b + 4ab^2 - 3ab^2$
$$= (3 - 6)a^2b + (4 - 3)ab^2 = -3a^2b + ab^2$$

---

ii)   *Multiplication*

Any terms may be multiplied together as long as the usual rules of sign and number are obeyed, eg $(6a) \times (5b) = 6 \times 5 \times a \times b = 30ab$.

It must also be remembered that a number or a sign outside a bracket is applied to each of the terms inside the bracket, eg $-3(p + q) = -3p - 3q$.

When two brackets are multiplied together each term in the first bracket is multiplied by each term in the second bracket.

eg $(a + b)(c + d) = ac + ad + bc + bd$

Example 2

Find the products of the following:

(1) $(4r) \times (-3s)$

ESSENTIAL ALGEBRA — PURE MATHEMATICS

(2)   (a) x (-2b) x (3c) x (-4d)

(3) $-2(2x - 5)$

(4) $-3y(3x - 4)$

(5) $(5r - 2s)(3r + s)$

(6) $(x + 2y - z)(5 - 2x)$

Solution

(1) $(4r) \times (-3s) = (4 \times -3) \times (r \times s) = -12rs$

(2)   (a) x (-2b) x (3c) x (-4d) = $(-2 \times 3 \times -4) \times (a \times b \times c \times d) = 24abcd$

(3) $-2(2x - 5) = (-2 \times 2x) + (-2 \times -5) = -4x + 10$

(4) $-3y(3x - 4) = (-3y \times 3x) + (-3y \times -4) = -9xy + 12y$

(5) $(5r - 2s)(3r + s) = (5r \times 3r) + (5r \times s) + (-2s \times 3r) + (-2s \times s)$

$$= 15r^2 + 5rs - 6rs - 2s^2$$

$$= 15r^2 - rs - 2s^2$$

(6) $(x + 2y - z)(5 - 2x) = (x \times 5) + (x \times -2x) + (2y \times 5) + (2y \times -2x) + (-z \times 5) + (-z \times -2x)$

$$= 5x - 2x^2 + 10y - 4xy - 5z + 2xz$$

This will not simplify as all the terms are different (un-like).

---

iii)   *Division*

When dividing simple algebraic expressions it is often possible to cancel terms between the numerator and the denominator, eg (10ab) ÷ (2a) can be written as:

$$\frac{10ab}{2a} = \frac{\cancel{10}^5 \times \cancel{a} \times b}{\cancel{2} \times \cancel{a}} = 5b$$

More complicated expressions can be divided using a technique known as algebraic long division, eg $(2a^2 + 8a + 6) \div (a + 1)$. The working is set out in a very similar way to that used for numerical long division:

ie $\boxed{a} + 1 \,\overline{\big)\, \boxed{2a^2} + 8a + 6}$

        $\boxed{a}$ divides into $\boxed{2a^2}$ 2a times ie $2a \times a = 2a^2$

∴

$$\begin{array}{r} 2a \\ \boxed{a} + 1 \,\overline{\big)\, 2a^2 + 8a + 6} \\ \underline{\boxed{2a^2 + 2a}} \downarrow \\ \boxed{6a} + 6 \end{array}$$

PURE MATHEMATICS — ESSENTIAL ALGEBRA

a + 1 has been multiplied by 2a giving $\boxed{2a^2 + 2a}$ which is subtracted from $2a^2 + 8a + 6$ leaving $6a + 6$.

Repeating the process $\boxed{a}$ divides into $\boxed{6a}$ 6 times ie 6 x a = 6a.

∴

$$
\begin{array}{r}
2a + 6 \\
a+1 \overline{\smash{\big)}\, 2a^2 + 8a + 6} \\
2a^2 + 2a \\
\hline
6a + 6 \\
\underline{6a + 6} \\
\cdots\cdots
\end{array}
$$

a + 1 has been multiplied by 6 giving $\boxed{6a + 6}$ which has been subtracted from $6a + 6$ leaving zero remainder.

Example 3

Divide (1) (8abc) by (3ac)

(2) ($7a^2b^2$) by (2ab)

(3) ($-10xy^2$) by (5z)

(4) ($6x^2 - 7x - 20$) by ($2x - 5$)

(5) ($x^2 - 2xy + y^2$) by ($x - y$)

Solution

(1) $(8abc) \div (3ac) = \dfrac{8 \times \cancel{a} \times b \times \cancel{c}}{3 \times \cancel{a} \times \cancel{c}} = \dfrac{8b}{3}$

(2) $(7a^2b^2) \div (2ab) = \dfrac{7 \times \cancel{a} \times a \times \cancel{b} \times b}{2 \times \cancel{a} \times \cancel{b}} = \dfrac{7ab}{2}$

(3) $(-10xy^2) \div (5z) = \dfrac{-\cancel{10}^{\,2} \times x \times x \times y \times y}{\cancel{5} \times y} = \dfrac{-2xy^2}{z}$

(4)
$$
\begin{array}{r}
3x + 4 \\
2x-5 \overline{\smash{\big)}\, 6x^2 - 7x - 20} \\
6x^2 - 15x \\
\hline
8x - 20 \\
\underline{8x - 20} \\
\end{array}
$$

ESSENTIAL ALGEBRA — PURE MATHEMATICS

(5)
$$\begin{array}{r} x - y \\ x - y \overline{\smash{\big)}\, x^2 - 2xy + y^2} \\ \underline{x^2 - xy} \\ -xy + y^2 \\ \underline{-xy + y^2} \end{array}$$

b) Factorisation

A factor is a common part of two or more algebraic terms making up an algebraic expression, eg $3x + 3y$ is an expression containing two terms $3x$ and $3y$ and both these terms contain the common factor 3.

$\therefore \quad 3x + 3y = \boxed{3 \text{ x}}\, x + \boxed{3 \text{ x}}\, y = 3(x + y)$ and $3(x + y)$ is the factorised form of the original expression $3x + 3y$

Some expressions can be very complicated to factorise and the factors are not always as easy to find as in the above example.

Example 4

Factorise:
(1) $2x + 6$
(2) $5a + 10b + 15c$
(3) $6x^2 - 4x$
(4) $ab^3 - a^2b$
(5) $9r^3y - 6r^2y^2 + 3ry^3$

Solution

(1) $2x + 6 = \boxed{2 \text{ x}}\, x + \boxed{2 \text{ x}}\, 3 = 2(x + 3)$

(2) $5a - 10b + 15c = \boxed{5x}\, a - \boxed{5x}\, 2b + \boxed{5x}\, 3c = 5(a - 2b + 3c)$

(3) $6x^2 - 4x = 3 \text{ x } 2 \text{ x } x \text{ x } x - 2 \text{ x } 2 \text{ x } x$

$ = \boxed{2 \text{ x x x}}\, 3x - \boxed{2 \text{ x x x}}\, 2 = 2x(3x - 2)$

(4) $ab^3 - a^2b = a \text{ x } b \text{ x } b \text{ x } b - a \text{ x } a \text{ x } b$

$ = \boxed{a \text{ x } b \text{ x}}\, b^2 - \boxed{a \text{ x } b \text{ x}}\, a = ab(b^2 - a)$

(5) $9r^3y - 6r^2y^2 + 3ry^3 = 3 \text{ x } 3 \text{ x } r \text{ x } r \text{ x } r \text{ x } y - 3 \text{ x } 2 \text{ x } r \text{ x } r \text{ x } y \text{ x } y + 3 \text{ x } r \text{ x } y \text{ x } y \text{ x } y$

$ = \boxed{3 \text{ x } r \text{ x } y \text{ x}}\, 3r^2 - \boxed{3 \text{ x } r \text{ x } y \text{ x}}\, 2ry + \boxed{3 \text{ x } r \text{ x } y \text{ x}}\, y^2$

$ = 3ry(3r^2 - 2ry + y^2)$

PURE MATHEMATICS   ESSENTIAL ALGEBRA

c)   *Solutions of linear equations*

A linear equation only contains terms of the first power, ie terms in x or terms in y but not terms in x^2 or y^2 or xy. The usual method of solving these equations is to collect together all the terms in the variable on the left hand side of the equation and all the other terms on the right hand side of the equation. Remember terms can be moved from one side of the equation to the other as long as the sign of the term is changed.

eg $6x - 2 = 5x - 10$ moving 5x and -2 gives:

∴ $6x - 5x = -10 + 2$

∴ $x = -8$

Example 5

Solve the equations:

(1) $4(x - 5) = 7 - 5(3 - 2x)$

(2) $\dfrac{x}{5} - \dfrac{x}{3} = 2$

(3) $\dfrac{2x}{15} - \dfrac{(x-6)}{12} - \dfrac{3x}{20} = \dfrac{3}{2}$

(4) $\dfrac{3}{(x-1)} = \dfrac{2}{(x-5)}$

Solutions

(1) $4(x - 5) = 7 - 5(3 - 2x)$ removing brackets:

       $4x - 20 = 7 - 15 + 10x$ moving 10x and -20:

       $4x - 10x = 7 - 15 + 20$

       $-6x = 12$ dividing both sides by -6:

       $\dfrac{-6x}{-6} = \dfrac{12}{-6}$

       $x = -2$

---

(2) $\dfrac{x}{5} - \dfrac{x}{3} = 2$ multiplying both sides by 15 (the smallest number that cancels with 5 *and* 3):

       $\left(\cancel{15}^{3} \times \dfrac{x}{\cancel{5}}\right) - \left(\cancel{15}^{5} \times \dfrac{x}{\cancel{3}}\right) = 15 \times 2$

       $3x - 5x = 30$

       $-2x = 30$ dividing both sides by -2:

       $\dfrac{-2x}{-2} = \dfrac{30}{-2}$

       $x = -15$

(3) $\dfrac{2x}{15} - \dfrac{(x-6)}{12} - \dfrac{3x}{20} = \dfrac{3}{2}$

Multiplying both sides by 60 (the smallest number that will cancel with 15, 12, 20 and 2):

$$\left\{60^4 \times \dfrac{2x}{\cancel{15}}\right\} - \left\{60^5 \times \dfrac{(x-6)}{\cancel{12}}\right\} - \left\{60^3 \times \dfrac{3x}{\cancel{20}}\right\} = 60^{30} \times \dfrac{3}{\cancel{2}}$$

$8x - 5x + 30 - 9x = 90$

$8x - 5x - 9x = 90 - 30$

$-6x = 60$ dividing by -6:

$\dfrac{-6x}{-6} = \dfrac{60}{-6}$

$x = -10$

(4) $\dfrac{3}{(x-1)} = \dfrac{2}{(x-5)}$ Multiplying both sides by $(x-1)(x-5)$
                                        (To cancel with $(x-1)$ and $(x-5)$.)

$(x-1)(x-5)\dfrac{3}{(x-1)} = (x-1)(x-5)\dfrac{2}{(x-5)}$

$3(x-5) = 2(x-1)$

$3x - 15 = 2x - 2$

$3x - 2x = -2 + 15$

$x = 13$

d) *Solutions of simultaneous linear equations*

These are equations in two variables such as x and y where the highest power of either variable is one.

ie   equations such as $2x + 3y = 11$
                                   $4x - y = 1$

These can be solved algebraically using either of the following methods

i)   By elimination of one of the variables.

$2x + 3y = 11$ ---- (1) $2x + 3y = 11$
$4x - y = 1$ ---- (2) × 3 $12x - 3y = 3$ } adding to eliminate y

                                                      $14x = 14$

                                       $\therefore \;\; x = \dfrac{14}{14} = 1$

Replacing x = 1 in equation (1) $2(1) + 3y = 11$

                                        $\therefore \;\; 3y = 11 - 2 = 9$

PURE MATHEMATICS   ESSENTIAL ALGEBRA

$$\therefore \quad y = \frac{9}{3} = 3$$

So the solutions are $x = 1$ and $y = 3$.

ii)   By substitution from one equation into the other.

$$2x + 3y = 11 \quad \text{---} \quad (1)$$
$$4x - y = 1 \quad \text{---} \quad (2) \quad \therefore 4x = 1 + y \text{ and } 4x - 1 = y$$

Replacing $y = 4x - 1$ in equation (1):

$$2x + 3(4x - 1) = 11$$
$$\therefore \quad 2x + 12x - 3 = 11$$
$$\therefore \quad 14x = 11 + 3 = 14$$
$$x = \frac{14}{14} = 1 \text{ (again!)}$$

and replacing $x = 1$ in $4x - 1 = y$ gives $y = 3$ (again!).

It should be realised that either of these methods can be used when solving simultaneous linear equations.

Example 6

Solve the following equations by different methods:

(1) $2x - 5y = 11$

        $4x + y = -11$

(2) $5x = 6y + 11$

        $3x - 8y = 0$

Solution

(1)   Using the method of elimination

$$2x - 5y = 11 \quad \text{---} \quad (1) \qquad 2x - 5y = 11$$
$$4x + y = -11 \quad \text{---} \quad (2) \times 5 \qquad 20x + 5y = -55$$

} adding to eliminate y

$$22x = -44$$
$$\therefore \quad x = \frac{-44}{22} = -2$$

Replacing $x = -2$ in equation (1) $2(-2) - 5y = 11$

$$-4 - 5y = 11$$
$$\therefore \quad -5y = 11 + 4 = 15$$
$$\therefore \quad y = \frac{15}{-5} = -3$$

So the solutions are $x = -2$, $y = -3$.

ESSENTIAL ALGEBRA   PURE MATHEMATICS

(2)   Using the method of substitution:

$5x = 6y + 11$ ----- (1)

$3x - 8y = 0$ ----- (2) $3x = 8y$ and $x = \dfrac{8y}{3}$

Replacing $x = \dfrac{8y}{3}$ in equation (1) $5\left(\dfrac{8y}{3}\right) = 6y + 11$

$\therefore \quad \dfrac{40y}{3} = 6y + 11$ multiplying by 3

$40y = 18y + 33$

$40y - 18y = 33$

$22y = 33$

$\therefore \quad y = \dfrac{33}{22} = \dfrac{3}{2}$

and replacing $y = \dfrac{3}{2}$ in $x = \dfrac{8y}{3}$ gives $x = \dfrac{8 \times \dfrac{3}{2}}{3}$

$x = \dfrac{12}{3} = 4$

So the solutions are $x = 4$, $y = \dfrac{3}{2}$ or 1.5

Pairs of simultaneous equations can also be solved graphically by plotting each equation as a straight line graph and finding the point of intersection of the two lines.

eg $2x + 3y = 11$
                         } these equations have already been solved giving $x = 1$ and $y = 3$
      $4x - y = 1$

Before plotting the graphs of these two straight lines, it is necessary to re-arrange both equations into the form $y = x$'s.

     $2x + 3y = 11$ and $4x - y = 1$

$\therefore \quad 3y = -2x + 11 \quad \therefore \quad -y = -4x + 1$ changing signs

$\therefore \quad y = \dfrac{-2x + 11}{3} \quad \therefore \quad y = 4x - 1$

x	0	1	2	3	4
-2x	0	-2	-4	-6	-8
+11	11	11	11	11	11
- 2x + 11	11	9	7	5	3
$y = \dfrac{-2x + 11}{3}$	3.67	3	2.33	1.67	1
4x	0	4	8	12	16
-1	-1	-1	-1	-1	-1
y = 4x - 1	-1	3	7	11	15

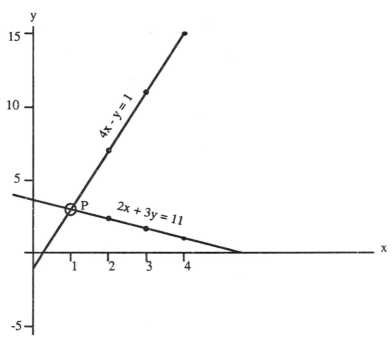

The point of intersecton P, is (1,3) i.e. x=1 when y=3

---

e) *Inequalities and the modulus sign*

In mathematics quantities are not always equal to one another, there are times when one quantity is greater than (or less than) another quantity.

eg   7 is greater than 5 but 3 is less than 5. These can be written more simply as 7 > 5 but 3 < 5.

Just as equations can be solved by applying set methods, so inequalities can be solved in a very similar way.

eg   if x + 3 > 7 then x > 7 - 3 ie x > 4.

The rules that govern the manipulation of inequalities are:

(a)   Any number may be added to or subtracted from each side of an inequality

        eg   6 > -2 then 6 + 4 > -2 + 4 ie 10 > 2

        or   12 > 3 then 12 - 6 > 3 - 6 ie 6 > -3

(b)   Both sides of an inequality may be multiplied by or divided by a positive number.

        eg   6 > -2 then 6 x 4 > -2 x 4 ie 24 > -8

        or   12 > 3 then 12 ÷ 6 > 3 ÷ 6 ie 2 > 0.5

(c)   Both sides of an inequality may be multiplied by or divided by a negative number as long as the inequality sign is reversed.

        eg   6 > -2 then 6 x -4 < -2 x - 4 ie -24 < 8

        or   12 > 3 then 12 ÷ -6 < 3 ÷ -6 ie -2 < -0.5

It is worth remembering that if a quantity, y say, is positive then it is greater than zero ie y > 0 or if it is negative it is less than zero y < 0.

ESSENTIAL ALGEBRA — PURE MATHEMATICS

The modulus of x (written as $|x|$ and usually called mod x) means:

$|x| = x$ when $x = 0$ or $x > 0$ ie $x \geq 0$

$|x| = -x$ when $x < 0$

eg $\quad |2x + 1| = 2x + 1$ when $2x + 1 \geq 0$ ie $2x \geq -1$ and $x \geq -\frac{1}{2}$

but $\quad |2x + 1| = -(2x + 1)$ when $2x + 1 < 0$ ie $2x < -1$ and $x < -\frac{1}{2}$

Example 7

Draw the graph of $y = |x|$

Solution

$y = x$ for $|x| = x$ when $x \geq 0$

$y = -x$ for $|x| = -x$ when $x < 0$

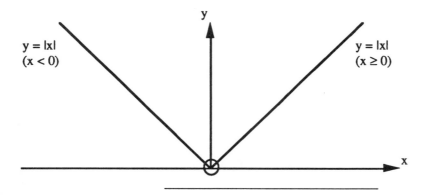

1.2 The quadratic function

a) Solutions of quadratic equations

These are equations where the highest power of the variable is two, ie equations such as $2x^2 + 7x - 15 = 0$, or $3x^2 + 10xy - 8y^2 = 0$.

There are two methods of solving this type of equation which should already be familiar.

i) by factorisation;

ii) by using the formula $x = \dfrac{-b \pm \sqrt{b^2 - 4ac}}{2a}$ where $ax^2 + bx + c = 0$

Example 8

Solve the following quadratic equations:

(1) $\quad 2x^2 + 7x - 15 = 0$

(2) $\quad 3x^2 + 10xy - 8y^2 = 0$

(3) $\quad 2x^2 - 5x + 1 = 0$

(4) $\quad x^2 + 6x + 10 = 0$

Solution

(1) This equation will factorise:

$$2x^2 + 7x - 15 = 0 \quad \text{becomes} \quad (2x - 3)(x + 5) = 0$$

so either $2x - 3 = 0$ or $x + 5 = 0$

$\therefore 2x = 3$ $\therefore x = -5$

$$x = \frac{3}{2} = 1.5$$

So the solutions are $x = 1.5$ or -5.

(2) This equation will also factorise:

$$3x^2 + 10xy - 8y^2 = 0$$

becomes $(3x - 2y)(x + 4y) = 0$

so either $3x - 2y = 0$ or $x + 4y = 0$

$\therefore 3x = 2y$ $\therefore x = -4y$

$$x = \frac{2y}{3}$$

So the solutions are $x = \frac{2y}{3}$ or $-4y$

(3) This equation does not factorise so the formula must be used:

$$2x^2 - 5x + 1 = 0 \quad \ldots \quad a = 2, b = -5, c = 1$$

$$x = \frac{-b \pm \sqrt{b^2 - 4ac}}{2a} \quad = \frac{-(-5) \pm \sqrt{(-5)^2 - 4 \cdot 2 \cdot 1}}{2 \cdot 2}$$

$$= \frac{5 \pm \sqrt{25 - 8}}{4} \quad = \frac{5 \pm \sqrt{17}}{4}$$

$\therefore$ either $x = \dfrac{5 + \sqrt{17}}{4}$ or $x = \dfrac{5 - \sqrt{17}}{4}$

$x = \dfrac{5 + 4.123}{4}$ or $x = \dfrac{5 - 4.123}{4}$

$x = 2.28$ or $x = 0.22$

So the solutions are $x = 2.28$ or 0.22 (to 3 significant figures)

ESSENTIAL ALGEBRA   PURE MATHEMATICS

(4)   This equation does not factorise so the formula must be used again:

$$x^2 + 6x + 10 = 0 \qquad \therefore a = 1, b = 6, c = 10$$

$$x = \frac{-b \pm \sqrt{b^2 - 4ac}}{2a} = \frac{-6 \pm \sqrt{6^2 - 4.1.10}}{2.1}$$

$$= \frac{-6 \pm \sqrt{36 - 40}}{2} = \frac{-6 \pm \sqrt{-4}}{2}$$

However, it is impossible to square root a negative number so this equation has no *real* solutions.

Following on from the last part of this example are some very important results relating to the solution of quadratic equations.

They are if:

$b^2 - 4ac > 0$ ie $b^2 > 4ac$, the solutions are real and different

$b^2 - 4ac = 0$ ie $b^2 = 4ac$, the solutions are real and equal

$b^2 - 4ac < 0$ ie $b^2 < 4ac$, there are no real solutions)

---

b)   *Solutions of simultaneous equations - 1 linear, 1 quadratic*

These are equations in two variables such as

$$\left. \begin{array}{l} x - y - 6 = 0 \\ y^2 = 8x \end{array} \right\} \begin{array}{l} \text{- linear equation} \\ \text{- quadratic equation} \end{array}$$

The usual method of solving these equations is to substitute from the linear equation into the quadratic equation, ie

$$x - y - 6 = 0$$

$$\therefore \quad x = y + 6$$

Replacing for x in the quadratic equation gives:

$$y^2 = 8(y + 6)$$

$$y^2 = 8y + 48$$

$$\therefore \quad y^2 - 8y - 48 = 0$$

$$(y + 4)(y - 12) = 0$$

either $y + 4 = 0$ or $y - 12 = 0$

$\therefore$ $y = -4$ $\therefore y = 12$

and $x = y + 6 = -4 + 6$ and $x = y + 6 = 12 + 6$

$\therefore$ $x = 2$ $\therefore x = 18$

So there are two pairs of solutions: $x = 2, y = -4$ and $x = 18, y = 12$.

PURE MATHEMATICS   ESSENTIAL ALGEBRA

Example 9

Solve these simultaneous equations:

$$y^2 = 16x$$
$$y = x - 5$$

Solution

Substituting $y = x - 5$ in $y^2 = 16x$ gives

$$(x - 5)^2 = 16x$$
$$x^2 - 10x + 25 = 16x$$
$$x^2 - 10x - 16x + 25 = 0$$
$$x^2 - 26x + 25 = 0$$
$$(x - 25)(x - 1) = 0$$

either $x - 25 = 0$ or $x - 1 = 0$

∴ $x = 25$ ∴ $x = 1$

and $y = x - 5 = 25 - 5$ and $y = x - 5 = 1 - 5$

∴ $y = 20$ ∴ $y = -4$

So there are two pairs of solutions: $x = 25, y = 20$ and
                                              $x = 1, \ y = -4$.

c)   *Graphical solutions*

i)   *Quadratic equations*

eg $2x^2 + 7x - 15 = 0$, which has already been solved giving $x = -5$ and 1.5.

To solve a quadratic equation graphically it is necessary to plot the graph of $y = 2x^2 + 7x - 15$ and then take the points where the curve crosses the x axis, ie the points where $y = 0$ and hence $2x^2 + 7x - 15 = 0$

x	-6	-5	-4	-3	-2	-1	0	1	2	3
x^2	36	25	16	9	4	1	0	1	4	9
$2x^2$	72	50	32	18	8	2	0	2	8	18
$7x$	-42	-35	-28	-21	-14	-7	0	7	14	21
-15	-15	-15	-15	-15	-15	-15	-15	-15	-15	-15
$y = 2x^2 - 7x - 15$	15	0	-11	-18	-21	-20	-15	-6	7	24

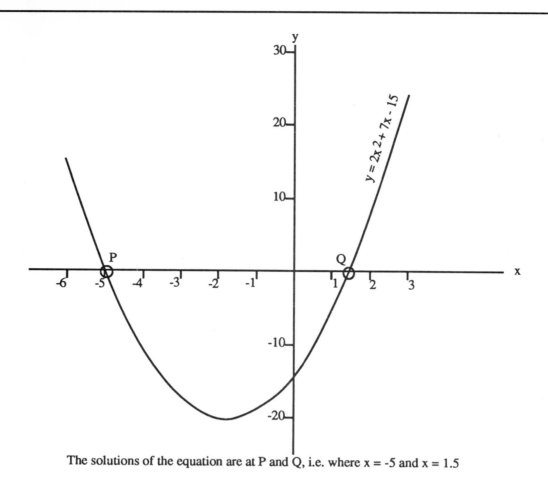

The solutions of the equation are at P and Q, i.e. where x = -5 and x = 1.5

ii) *Simultaneous equations - 1 linear, 1 quadratic*

eg x - y - 6 = 0 } as before these have already been solved
    }   giving x = 2, y = -4 and x = 18, y = 12
    $y^2 = 8x$

These equations also need to be re-arranged into the form y = x's.

x - y - 6 = 0 $y^2 = 8x$

- y = -x + 6 changing signs $y = \pm \sqrt{8x}$

y = x - 6

(Note: When square rooting a number the result can be positive or negative. This means that for every value of x there are two values of y either $+ \sqrt{8x}$ or $- \sqrt{8x}$.)

x	0	4	8	12	16	20
8x	0	32	64	96	128	160
$y = \pm \sqrt{8x}$	0	5.66 / -5.66	8 / -8	9.80 / -9.80	11.31 / -11.31	12.65 / -12.65

x	0	4	8	12	16	20
-6	-6	-6	-6	-6	-6	-6
y = x - 6	-6	-2	2	6	10	14

PURE MATHEMATICS   ESSENTIAL ALGEBRA

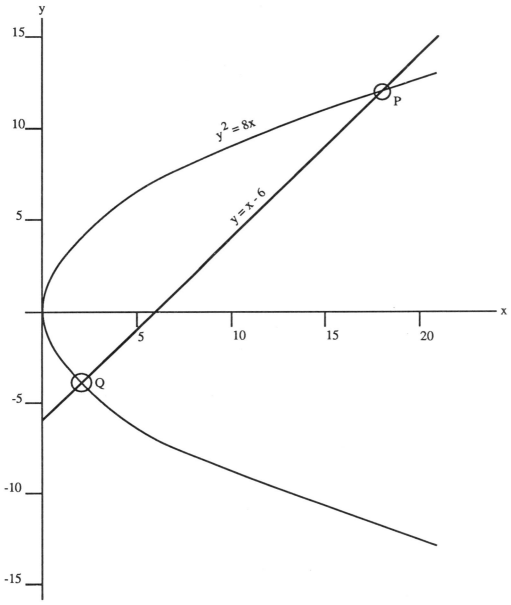

Points of intersection P (18,12), i.e. $x = 18$, $y = 12$
Q (2,-4), i.e. $x = 2$, $y = -4$

Example 10

(a)   Find the set of real values of x for which $x^2 - 9x + 20$ is negative.

(b)   Find the set of values of k for which $x^2 + kx + 9$ is positive for all real values of x.

Solution

(a)   Drawing a sketch graph of $y = x^2 - 9x + 20$ can be very helpful in answering this type of question. It is not necessary to plot many points accurately - just several key points such as where the curve crosses the axes.

On the y axis $x = 0$, ∴ $y = 0 - (9 \times 0) + 20 = 20$ ie coordinates (0, 20) -----   (A)

On the x axis $y = 0$, ∴ $x^2 - 9x + 20 = 0$

$(x - 5)(x - 4) = 0$

either $x = 5$ ie coordinates $(5, 0)$ ----- (C)

or $x = 4$ ie coordinates $(4, 0)$ ----- (B)

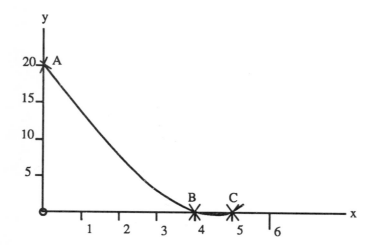

So $x^2 - 9x + 20 < 0$ between $x = 4$ and $x = 5$ which can be written as $4 < x < 5$.

(b) Let $y = x^2 + kx + 9$, if $x^2 + kx + 9$ is positive then $x^2 + kx + 9 > 0$

∴ $y > 0$. This means that the graph of $y = x^2 + kx + 9$ must be completely above the x-axis as shown in this sketch.

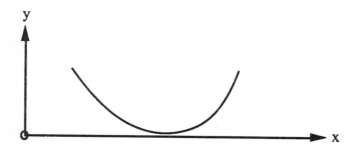

Since the graph does not cross the x axis the quadratic equation $x^2 + kx + 9 = 0$ does not have any real solutions, which means '$b^2 < 4ac$' where $a = 1$, $b = k$ and $c = 9$.

$b^2 < 4ac$ becomes $k^2 < 4.1.9$
$k^2 < 36$

This inequality is valid for values of k between -6 and +6, ie $-6 < k < 6$.

PURE MATHEMATICS   ESSENTIAL ALGEBRA

d)   *Roots and coefficients*

The solutions of an equation are also called the roots of the equation. Sometimes it is useful to obtain information about the roots of an equation without actually solving the equation.

If a quadratic equation such as $ax^2 + bx + c = 0$ has roots α and β then the equation can be written as:

$$(x - \alpha)(x - \beta) = 0$$

Multiplying out the brackets gives

$$x^2 - \alpha x - \beta x + \alpha\beta = 0$$

$$x^2 - (\alpha + \beta)x + \alpha\beta = 0 \quad\text{-----}\quad (1)$$

But α and β are also the roots of the equation

$$ax^2 + bx + c = 0$$

Dividing through by a gives

$$x^2 + \frac{b}{a}x + \frac{c}{a} = 0 \quad\text{-----}\quad (2)$$

Equations (1) and (2) are the same equation written out in a different form so it follows that

$$-(\alpha + \beta) = \frac{b}{a} \text{ or } \alpha + \beta = -\frac{b}{a} \quad \text{the sum of the roots}$$

$$\text{and } \alpha\beta = \frac{c}{a} \quad \text{the product of the roots}$$

Example 11

The equation $2x^2 + 3x - 4 = 0$ has roots α and β. Find the values of:

(1) $\alpha^2 + \beta^2$

(2) $\frac{1}{\alpha} + \frac{1}{\beta}$

(3) $(\alpha + 1)(\beta + 1)$

(4) $\frac{\beta}{\alpha} + \frac{\alpha}{\beta}$

Solution

Since the given equation has roots α and β it follows that $\alpha + \beta = -\frac{3}{2}$ and $\alpha\beta = -\frac{4}{2} = -2$

Using $a = 2, b = 3, c = -4$

(1) $\alpha^2 + \beta^2$ has to be re-written using $\alpha + \beta$ and $\alpha\beta$ only

$$\alpha^2 + \beta^2 = \alpha^2 + \beta^2 + 2\alpha\beta - 2\alpha\beta$$

$$\alpha^2 + \beta^2 = (\alpha + \beta)^2 - 2\alpha\beta$$

$$\alpha^2 + \beta^2 = \left(-\frac{3}{2}\right)^2 - 2(-2)$$

ESSENTIAL ALGEBRA   PURE MATHEMATICS

$$\therefore \quad \alpha^2 + \beta^2 = \frac{9}{4} + 4 = \frac{25}{4}$$

(2) $\quad \dfrac{1}{\alpha} + \dfrac{1}{\beta} = \dfrac{\beta + \alpha}{\alpha\beta} = \dfrac{\alpha + \beta}{\alpha\beta}$

$$\dfrac{1}{\alpha} + \dfrac{1}{\beta} = \dfrac{-\frac{3}{2}}{-2}$$

$$\therefore \quad \dfrac{1}{\alpha} + \dfrac{1}{\beta} = \dfrac{3}{4}$$

(3) $\quad (\alpha + 1)(\beta + 1) = \alpha\beta + \alpha + \beta + 1 = \alpha\beta + (\alpha + \beta) + 1$

$$(\alpha + 1)(\beta + 1) = -2 + \left(-\dfrac{3}{2}\right) + 1$$

$$\therefore \quad (\alpha + 1)(\beta + 1) = -\dfrac{5}{2}$$

(4) $\quad \dfrac{\beta}{\alpha} + \dfrac{\alpha}{\beta} = \dfrac{\beta^2 + \alpha^2}{\alpha\beta}$

$$\dfrac{\beta}{\alpha} + \dfrac{\alpha}{\beta} = \dfrac{\frac{25}{4}}{-2}$$

$$\therefore \quad \dfrac{\beta}{\alpha} + \dfrac{\alpha}{\beta} = \dfrac{-25}{8}$$

e) *Completing the square, greatest and least values*

$x^2 + 6x + 9$ is a perfect square because it factorises into brackets as $(x + 3)(x + 3)$, ie $x^2 + 6x + 9 = (x + 3)^2$. However, to make $x^2 + 10x$ into a perfect square it would be necessary to add 25 to the expression giving $x^2 + 10x + 25 = (x + 5)(x + 5) = (x + 5)^2$.

This method of completing the square gives an alternative method of solving a quadratic equation as will be seen from the next example.

Example 12

Solve the equation $x^2 - 4x - 2 = 0$ by completing the square.

Solution

As written $x^2 - 4x - 2$ is not a perfect square so it is necessary to rearrange the equation to obtain the required form.

$x^2 - 4x - 2 = 0$
$x^2 - 4x = 2$

By adding 4 to $x^2 - 4x$ a perfect square can be obtained because $x^2 - 4x + 4 = (x - 2)^2$. However, with equations, the balance must always be maintained so if 4 is added to one side it must also be added to the other side of the equation.

∴ $x^2 - 4x + 4 = 2 + 4$

∴ $(x - 2)^2 = 6$

Square rooting both sides:

$x - 2 = \pm \sqrt{6}$

∴ $x = 2 \pm \sqrt{6}$

When completing the square it is easier if the coefficient of the squared term is 1.

Example 13

Solve $3x^2 + 4x + 1 = 0$ by completing the square

Solution

$3x^2 + 4x + 1 = 0$

Divide by 3

$$\frac{3x^2}{3} + \frac{4x}{3} + \frac{1}{3} = 0$$

$$x^2 + \frac{4x}{3} = -\frac{1}{3}$$

To complete the square $\frac{4}{9}$ must be added to both sides

$$x^2 + \frac{4x}{3} + \frac{4}{9} = -\frac{1}{3} + \frac{4}{9}$$

$$\left(x + \frac{2}{3}\right)^2 = \frac{-3 + 4}{9}$$

$$\left(x + \frac{2}{3}\right)^2 = \frac{1}{9} \text{ square rooting both sides:}$$

$$x + \frac{2}{3} = \pm \sqrt{\frac{1}{9}} = \pm \frac{1}{3}$$

∴ $x = -\frac{2}{3} \pm \frac{1}{3}$

∴ either $x = -\frac{2}{3} + \frac{1}{3} = -\frac{1}{3}$

  or $x = -\frac{2}{3} - \frac{1}{3} = -1$

∴ $x = -\frac{1}{3}$ or -1

ESSENTIAL ALGEBRA   PURE MATHEMATICS

In general when completing the square for equation

$$ax^2 + bx + c = 0 \qquad \text{ie} \qquad ax^2 + bx = -c$$

divide by a:
$$x^2 + \left(\frac{b}{a}\right)x = -\frac{c}{a}$$

halve coefficient of x:
$$\left(\frac{b}{2a}\right)$$

square and add to both sides:

$$x^2 + \left(\frac{b}{a}\right)x + \left(\frac{b}{2a}\right)^2 = -\frac{c}{a} + \left(\frac{b}{2a}\right)^2$$

Perfect square
$$\left(x + \frac{b}{2a}\right)^2 = -\frac{c}{a} + \left(\frac{b}{2a}\right)^2$$

$$= \frac{b^2 - 4ac}{4a^2}$$

If this is now continued further:

$$x + \frac{b}{2a} = \frac{\pm\sqrt{b^2 - 4ac}}{2a}$$

$$\therefore \quad x = -\frac{b}{2a} \pm \frac{\sqrt{b^2 - 4ac}}{2a}$$

which is the formula for solving quadratic equations.

Check that this is true for examples 12 and 13.

The same results would have been obtained for examples 12 and 13 if the formula had been applied in 12 and if 13 had been factorised. So completing the square is a third alternative method for solving quadratic equations. However it is especially useful if the greatest (or least) value of an expression is required instead.

Example 14

Show that $3x^2 + 10x + 9$ cannot be negative and find its least value.

Solution

Note this is *not* an equation but a quadratic expression; so it is necessary to adopt a slightly different approach as this is not equal to zero.

$$3x^2 + 10x + 9 = 3\left(x^2 + \frac{10x}{3}\right) + 9$$

To make $x^2 + \frac{10x}{3}$ into a perfect square $\frac{25}{9}$ must be added to it. However, so as not to alter the basic function, $\frac{25}{9}$ must also be subtracted because $+\frac{25}{9} - \frac{25}{9} = 0$ and does not alter the value of the expression.

$$\therefore \quad 3\left(x^2 + \frac{10x}{3}\right) + 9 = 3\left(x^2 + \frac{10x}{3} + \frac{25}{9} - \frac{25}{9}\right) + 9$$

PURE MATHEMATICS   ESSENTIAL ALGEBRA

$$= 3\left(x^2 + \frac{10x}{3} + \frac{25}{9}\right) - 3\left(\frac{25}{9}\right) + 9$$

$$= 3\left(x + \frac{5}{3}\right)^2 - \frac{25}{3} + 9$$

∴ $\qquad 3x^2 + 10x + 9 = 3\left(x + \frac{5}{3}\right)^2 + \frac{2}{3}$

Since $\left(x + \frac{5}{3}\right)^2$ is a square it cannot be negative at the very least it will be zero $\left(\text{when } x = -\frac{5}{3}\right)$. It follows that $3x^2 + 10x + 9$ cannot be less than $\frac{2}{3}$, therefore it cannot be negative and its least value is $\frac{2}{3}$.

f)   *Example 'A' level question*

If the equation $x^2 - qx + r = 0$ has roots $\alpha + 2$, $\beta + 1$ where α and β are the roots of the equation $2x^2 - bx + c = 0$ and $\alpha \geq \beta$, find q and r in terms of b and c. In the case $\alpha = \beta$, show $q^2 = 4r + 1$.

Solution

$\qquad \alpha$ and β are roots of $2x^2 - bx + c = 0$

∴ $\qquad \alpha + \beta = -\frac{(-b)}{2} = \frac{b}{2} \ldots$ (1) $\qquad$ and $\qquad \alpha\beta = \frac{c}{2} \ldots$ (2)

$\qquad \alpha + 2$ and $\beta + 1$ are roots of $x^2 - qx + r = 0$

∴ $\qquad (\alpha + 2) + (\beta + 1) = \frac{-(-q)}{1} = q \qquad$ and $\qquad (\alpha + 2)(\beta + 1) = r$

∴ $\qquad \alpha + \beta + 3 = q \ldots$ (3) $\qquad\qquad \alpha\beta + \alpha + 2\beta + 2 = 4 \ldots$ (4)

Replacing for $\alpha + \beta$ from (1) into (3) gives:

$\qquad \frac{b}{2} + 3 \quad = q$ (multiply by 2)

∴ $\qquad b + 6 \quad = 2q$

∴ $\qquad \frac{1}{2}(b + 6) = q$

Returning to equation (4)

$\qquad \alpha\beta + \alpha + \beta + \beta + 2 = 4$

Replacing for $\alpha\beta$ from (2) and $\alpha + \beta$ from (1)

$\qquad \frac{c}{2} + \frac{b}{2} + \beta + 2 = 4$

However it is still necessary to replace for β. Using equations (1) and (2) it is possible to eliminate α and find an expression for β.

∴ $\qquad \alpha = \frac{b}{2} - \beta \qquad \left(\frac{b}{2} - \beta\right)\beta = \frac{c}{2}$

∴ $\qquad \frac{b\beta}{2} - \beta^2 = \frac{c}{2}$ multiplying by 2:

ESSENTIAL ALGEBRA   PURE MATHEMATICS

$\therefore \quad b\beta - 2\beta^2 = c$

Rearranging $2\beta^2 - b\beta + c = 0$

This will not factorise therefore the formula

$$x = \frac{-b \pm \sqrt{b^2 - 4ac}}{2} \text{ will be used with 'a' = 2, 'b' = -b, 'c' = c.}$$

$$\beta = \frac{-(-b) \pm \sqrt{(-b)^2 - 4.2.c}}{2.2}$$

$$= \frac{b \pm \sqrt{b^2 - 8c}}{4}$$

$$\beta = \frac{b + \sqrt{b^2 - 8c}}{4} \text{ or } \frac{b - \sqrt{b^2 - 8c}}{4}$$

These two values for β seem a little confusing but they are just the roots of the equation $2x^2 - bx + c = 0$ and so they are really the values of α and β, therefore since $\alpha \geq \beta$,

$$\alpha = \frac{b + \sqrt{b^2 - 8c}}{4} \quad \text{and} \quad \beta = \frac{b - \sqrt{b^2 - 8c}}{4}$$

$$\frac{c}{2} + \frac{b}{2} + \frac{b - \sqrt{b^2 - 8c}}{4} + 2 = r$$

$\therefore \quad \dfrac{c}{2} + \dfrac{b}{2} + \dfrac{b}{4} - \dfrac{\sqrt{b^2 - 8c}}{4} + 2 = r$

$\therefore \quad \dfrac{c}{2} + \dfrac{3b}{4} - \dfrac{\sqrt{b^2 - 8c}}{4} + 2 = r$

If $\alpha = \beta$ then

$$\frac{b + \sqrt{b^2 - 8c}}{4} = \frac{b - \sqrt{b^2 - 8c}}{4}$$

$\therefore \quad \dfrac{b}{4} + \dfrac{\sqrt{b^2 - 8c}}{4} = \dfrac{b}{4} - \dfrac{\sqrt{b^2 - 8c}}{4}$

$$\frac{b}{4} + \frac{\sqrt{b^2 - 8c}}{4} - \frac{b}{4} + \frac{\sqrt{b^2 - 8c}}{4} = 0$$

$\therefore \quad \dfrac{2\sqrt{b^2 - 8c}}{4} = 0$

$\therefore \quad b^2 = 8c$

But $q = \dfrac{1}{2}(b + 6)$ squaring both sides

$\therefore \quad q^2 = \left\{\dfrac{1}{2}(b + 6)\right\}^2 = \dfrac{1}{4}(b^2 + 12b + 36)$

$\therefore \quad q^2 = \dfrac{b^2}{4} + 3b + 9$

PURE MATHEMATICS   ESSENTIAL ALGEBRA

But $\quad 4r = 4\left(\dfrac{c}{2} + \dfrac{3b}{4} - \dfrac{\sqrt{b^2 - 8c}}{4} + 2\right)$

$\quad\quad\quad\quad = 2c + 3b + 8 \ \text{ using } b^2 = 8c$

$\quad\quad\quad\quad = 2\left(\dfrac{b^2}{8}\right) + 3b + 8$

$\quad\quad\quad\quad = \dfrac{b^2}{4} + 3b + 8$

$\therefore \quad q^2 = 4r + 1$

1.3   General algebraic functions

a)   *Function notation*

An algebraic expression containing terms in x and higher powers (such as x^2, x^3, etc) is called a function of x, and this is abbreviated to f(x). f(a) means the value of the function when x = a.

Example 15

Find the values of f(0), f(1), f(-2) when $f(x) = 7x^3 - 5x^2 + 4x - 3$

Solution

$\quad\quad\quad\quad f(0) \quad = 7(0)^3 - 5(0)^2 + 4(0) - 3$

$\therefore \quad\quad f(0) \quad = 0 - 0 + 0 - 3$

$\therefore \quad\quad f(0) \quad = -3$

$\quad\quad\quad\quad f(1) \quad = 7(1)^3 - 5(1)^2 + 4(1) - 3$

$\therefore \quad\quad f(1) \quad = 7 - 5 + 4 - 3$

$\therefore \quad\quad f(1) \quad = 3$

$\quad\quad\quad\quad f(-2) \quad = 7(-2)^3 - 5(-2)^2 + 4(-2) - 3$

$\therefore \quad\quad f(-2) \quad = -56 - 20 - 8 - 3$

$\therefore \quad\quad f(-2) \quad = -87$

ESSENTIAL ALGEBRA PURE MATHEMATICS

b) *Remainder theorem*

When $x^3 - 3x^2 + 6x + 5$ is divided algebraically by $x - 2$ there is a remainder of 13.

$$
\begin{array}{r}
x^2 - x - 4 \\
x-2 \overline{\smash{)}\, x^3 - 3x^2 + 6x + 5} \\
\underline{x^3 - 2x^2} \\
-x^2 + 6x \\
\underline{-x^2 + 2x} \\
4x + 5 \\
\underline{4x - 8} \\
13 \text{ remainder}
\end{array}
$$

This can also be written as

$x^3 - 3x^2 + 6x + 5 = (x - 2)(x^2 - x + 4) + 13$

However, there is a quicker and easier way of calculating the remainder and that is to evaluate f(2).

 f(x) = $x^3 - 3x^2 + 6x + 5$

So f(2) = $2^3 - 3(2)^2 + 6(2) + 5$

∴ f(2) = 8 - 12 + 12 + 5

∴ f(2) = 13 (remainder)

The value 2 has been chosen for x because it is the value of the variable that will make the divisor (x - 2) equal to zero, ie

 x - 2 = 0 when x = 2

Example 16

Find the value of a if $x^3 - 3x^2 + ax + 5$ has a remainder of 17 when divided by x - 3.

Solution

 f(x) = $x^3 - 3x^2 + ax + 5$

The divisor is x - 3 and so x - 3 = 0 gives x = 3, hence evaluating f(3) will give a value of the remainder in terms of a.

 f(3) = $(3)^3 - 3(3)^2 + a(3) + 5$

∴ f(3) = 27 - 27 + 3a + 5

∴ f(3) = 3a + 5 (remainder)

but remainder = 17

∴ 3a + 5 = 17

∴ 3a = 17 - 5

∴ 3a = 12

∴ $\underline{a = 4}$

PURE MATHEMATICS ESSENTIAL ALGEBRA

c) Factor theorem

If a function f(x) is divided by (x - a) resulting in a zero remainder, ie f(a) = 0 then (x - a) is called a factor of the function.

Example 17

Show that $(2x - 1)$ is a factor of the function $2x^3 + 3x^2 - 32x + 15$ and find the other factors.

Solution

$$f(x) = 2x^3 + 3x^2 - 32x + 15$$

putting $2x - 1 = 0$ gives $x = \frac{1}{2}$

$$f\left(\frac{1}{2}\right) = 2\left(\frac{1}{2}\right)^3 + 3\left(\frac{1}{2}\right)^2 - 32\left(\frac{1}{2}\right) + 15$$

$\therefore \quad f\left(\frac{1}{2}\right) = \frac{1}{4} + \frac{3}{4} - 16 + 15$

$\therefore \quad f\left(\frac{1}{2}\right) = 0$

so $(2x - 1)$ is a factor of $2x^3 + 3x^2 - 32x + 15$

The other factors can be determined in either one of two ways, ie either

i) *Algebraic long division*

```
                x² + 2x - 15
         ┌─────────────────────
  2x - 1 │ 2x³ + 3x² - 32x + 15
           2x³ -  x²
           ─────────
                4x² - 32x
                4x² -  2x
                ─────────
                    - 30x + 15
                    - 30x + 15
                    ──────────
                         ...
```

$\therefore \quad 2x^3 + 3x^2 - 32x + 15 = (2x - 1)(x^2 + 2x - 15)$

And $x^2 + 2x - 15$ can be factorised as $(x + 5)(x - 3)$

$\therefore \quad 2x^3 + 3x^2 - 32x + 15 = (2x - 1)(x + 5)(x - 3)$

ii) *Applying the factor theorem*

Other values of x are tried in f(x) until the ones are found that will make f(x) = 0, eg

try $\quad x = 1 \quad f(1) = 2\,(1^3) + 3\,(1^2) - 32\,(1) + 15$

$\therefore \quad\quad\quad f(1) = 2 + 3 - 32 + 15$

$\therefore \quad\quad\quad f(1) = -12$

So $f(1) \neq 0$ ... $x - 1$ is *not* a factor.

This method of trial and error can be rather tedious as only the values of $x = -5$ and $x = 3$ will make $f(x) = 0$.

To check $\quad f(-5) = 2(-5)^3 + 3(-5)^2 - 32(-5) + 15$

$\therefore \quad f(-5) = -250 + 75 + 160 + 15$

$\therefore \quad f(-5) = 0 \therefore x + 5$ is a factor

and $\quad f(3) = 2(3)^3 + 3(3)^2 - 32(3) + 15$

$\therefore \quad f(3) = 54 + 27 - 96 + 15$

$\therefore \quad f(3) = 0 \therefore x - 3$ is a factor

d) *Partial fractions*

When an algebraic fraction is split up into several component fractions, it is said to be in partial fractions, eg

$$\frac{5}{x^2 + x - 6} = \frac{1}{x - 2} - \frac{1}{x + 3}$$

This process has two main applications - in series expansions (such as the binomial) and in integration. The rules for splitting an algebraic function into partial fractions will be detailed.

Rule 1

The numerator (top line) must always be at least one degree less than the denominator (bottom line) before an expression can be partial fractioned. If this is not the case, then the expression must first be divided out algebraically. Therefore,

$$\frac{x^3 + 2x^2 - 2x + 2}{(x - 1)(x + 3)} = \frac{x^3 + 2x^2 - 2x + 2}{x^2 + 2x - 3}$$

is a 'top heavy' or 'improper' fraction because the highest power in the numerator is cubed, ie x^3, and the highest power in the denominator is squared, ie x^2. So, dividing out:

$$\begin{array}{r} x \\ x^2 + 2x - 3 \overline{\smash{\big)}\, x^3 + 2x^2 - 2x + 2} \\ \underline{x^3 + 2x^2 - 3x} \\ x + 2 \text{ remainder} \end{array}$$

$\therefore \quad \dfrac{x^3 + 2x^2 - 2x + 2}{x^2 + 2x - 3} = x + \dfrac{x + 2}{(x - 1)(x + 3)}$

Example 18

Divide the following expressions algebraically:

(a) $\quad \dfrac{3x^2 - 2x - 7}{(x - 2)(x + 1)}$

(b) $\quad \dfrac{x^3 - x^2 - 4x + 1}{x^2 - 4}$

Solution

(a) $\dfrac{3x^2 - 2x - 7}{(x - 2)(x + 1)} = \dfrac{3x^2 - 2x - 7}{x^2 - x - 2}$

$$\begin{array}{r} 3 \\ x^2 - x - 2 \overline{\smash{)}\, 3x^2 - 2x - 7} \\ 3x^2 - 3x - 6 \\ \hline x - 1 \text{ remainder} \end{array}$$

$\therefore \quad \dfrac{3x^2 - 2x - 7}{(x - 2)(x + 1)} = 3 + \dfrac{x - 1}{(x - 2)(x + 1)}$

(b) $\dfrac{x^3 - x^2 - 4x + 1}{x^2 - 4}$

$$\begin{array}{r} x - 1 \\ x^2 - 4 \overline{\smash{)}\, x^3 - x^2 - 4x + 1} \\ x^3 - 4x \\ \hline -x^2 + 1 \\ -x^2 + 4 \\ \hline -3 \text{ remainder} \end{array}$$

$\therefore \quad \dfrac{x^3 - x^2 - 4x + 1}{x^2 - 4} = x - 1 - \dfrac{3}{x^2 - 4}$

Rule 2

This is the method of obtaining partial fractions when the denominator contains only linear terms such as $x + 3$, $2x - 1$, etc.

Considering $\dfrac{6}{(x + 3)(x - 3)}$. Firstly, it is not an improper fraction so it is not necessary to divide out and, secondly, the denominator contains only linear terms: $x + 3$ and $x - 3$.

It is *assumed* that this expression can be split into two fractions

$$\dfrac{A}{x + 3} + \dfrac{B}{x - 3}$$

where A and B are constants whose values need to be found to make the process complete.

Therefore: $\dfrac{6}{(x + 3)(x - 3)} \equiv \dfrac{A}{x + 3} + \dfrac{B}{x - 3}$

Multiplying through each term by $(x + 3)(x - 3)$ gives

$\quad 6 \equiv A(x - 3) + B(x + 3)$

In order to find A and B values of x are chosen to make each bracket zero in turn, ie

try $\quad x = 3 \quad 6 = A(0) + B(3 + 3)$

$\qquad\qquad\quad 6 = 6B$

$\qquad \therefore \quad 1 = B$

ESSENTIAL ALGEBRA   PURE MATHEMATICS

try $\quad x = -3 \quad 6 = A(-3 - 3) + B(0)$

$\qquad\qquad\qquad 6 = -6A$

$\qquad\therefore\qquad -1 = A$

So $\quad \dfrac{6}{(x + 3)(x - 3)} \equiv \dfrac{-1}{x + 3} + \dfrac{1}{x - 3}$

This can be checked by adding the partial fractions together and (hopefully) obtaining the original function.

(*Note*: $\equiv$ means identically equal to.)

Example 19

Express in partial fractions $\dfrac{x - 1}{3x^2 - 11x + 10}$

Solution

The denominator does not look as though it contains linear terms but it factorises. Therefore:

$$3x^2 - 11x + 10 = (3x - 5)(x - 2)$$

$\therefore\quad \dfrac{x - 1}{3x^2 - 11x + 10} = \dfrac{x - 1}{(3x - 5)(x - 2)}$

Assuming partial fractions then:

$$\dfrac{x - 1}{(3x - 5)(x - 2)} \equiv \dfrac{A}{3x - 5} + \dfrac{B}{x - 2}$$

Multiplying through by $(3x - 5)(x - 2)$ gives:

$\qquad x - 1 \equiv A(x - 2) + B(3x - 5)$

let $x = 2 \qquad 2 - 1 = A(0) + B(3(2) - 5)$

$\qquad\therefore\qquad 1 = B$

let $x = \dfrac{5}{3} \qquad \dfrac{5}{3} - 1 = A\left(\dfrac{5}{3} - 2\right) + B(0)$

$\qquad\qquad \dfrac{2}{3} = A\left(-\dfrac{1}{3}\right)$

$\qquad\therefore\qquad -2 = A$

So $\quad \dfrac{x - 1}{3x^2 - 11x + 10} = -\dfrac{2}{3x - 5} + \dfrac{1}{x - 2}$

Rule 3

This is the method of obtaining partial fractions when the denominator contains a quadratic factor which will *not* factorise (unlike the last example when $3x^2 - 11x + 10$ factorised to give $(3x - 5)(x - 2)$ - the product of two linear factors). The denominator will also contain one or more linear factors.

Considering $\dfrac{6 - x}{(1 - x)(4 + x^2)}$. It is not an improper fraction so there is no need to divide out, and the denominator contains a linear factor $(1 - x)$ and a quadratic factor $(4 + x^2)$ which does not factorise.

PURE MATHEMATICS   ESSENTIAL ALGEBRA

Having a quadratic factor in the denominator means that the corresponding numerator is of the form Bx + C. Therefore

$$\frac{6-x}{(1-x)(4+x^2)} \equiv \frac{A}{1-x} + \frac{Bx+C}{4+x^2}$$

Multiplying through by $(1-x)(4+x^2)$ gives:

$$6 - x = A(4 + x^2) + (Bx + C)(1 - x)$$

let $x = 1$ $6 - 1 = A(4 + 1^2) + (B(1) + C)(0)$

$$5 = A(5)$$

$\therefore$ $1 = A$

There is no value of x that will make $(4 + x^2)$ equal to zero as $x^2 = -4$ is impossible. So any other values of x may be used as follows:

let $x = 0$ $6 - 0 = 1(4 + 0) + (B(0) + C)(1 - 0)$

$$6 = 4 + C$$

$\therefore$ $2 = C$

let $x = 2$ $6 - 2 = 1(4 + 2^2) + (B(2) + 2)(1 - 2)$

$$4 = 8 - 2B - 2$$

$\therefore$ $-2 = -2B$

$\therefore$ $1 = B$

$$\frac{6-x}{(1-x)(4+x^2)} \equiv \frac{1}{1-x} + \frac{x+2}{x^2+4}$$

Example 20

Express in partial fractions $\dfrac{4}{(x+1)(2x^2+x+3)}$

Solution

At first it looks as though $2x^2 + x + 3$ will factorise into two linear factors - but it does not. Therefore:

$$\frac{4}{(x+1)(2x^2+x+3)} \equiv \frac{A}{x+1} + \frac{Bx+C}{2x^2+x+3}$$

$\therefore$ $4 \equiv A(2x^2 + x + 3) + (Bx + C)(x + 1)$

Let $x = -1$ $4 = A(2(-1)^2 + (-1) + 3) + (B(-1) + C)(0)$

$$4 = A(4)$$

$\therefore$ $1 = A$

Let $x = 0$ $4 = 1(2(0)^2 + 0 + 3) + (B(0) + (C))(0 + 1)$

$$4 = 3 + C$$

$\therefore$ $1 = C$

Let x = 1 $4 = 1(2(1)^2 + 1 + 3) + (B(1) + 1)(1 + 1)$

$\qquad\qquad 4 = 6 + 2B + 2$

∴ $\qquad -4 = 2B$

$\qquad\qquad -2 = B$

∴ $\dfrac{4}{(x + 1)(2x^2 + x + 3)} \equiv \dfrac{1}{x + 1} + \dfrac{1 - 2x}{2x^2 + x + 3}$

Rule 4

This is the method of obtaining partial fractions when the denominator contains a repeated factor (linear or quadratic).

Considering $\dfrac{2x^2 - 5x + 7}{(x - 2)(x - 1)^2}$. Since this is not an improper fraction it can be expressed directly in partial fractions as:

$$\dfrac{2x^2 - 5x + 7}{(x - 2)(x - 1)^2} \equiv \dfrac{A}{x - 2} + \dfrac{B}{x - 1} + \dfrac{C}{(x - 1)^2}$$

ie it is necessary to allow for two terms, one with denominator of $x - 1$ and the other with $(x - 1)^2$.

Multiplying through by $(x - 2)(x - 1)^2$ gives

$$2x^2 - 5x + 7 \equiv A(x - 1)^2 + B(x - 1)(x - 2) + C(x - 2)$$

let x = 1 $2(1)^2 - 5(1) + 7 = A(0) + B(0)(1 - 2) + C(1 - 2)$

$\qquad\qquad 4 = -C$

$\qquad\qquad -4 = C$

let x = 2 $2(2)^2 - 5(2) + 7 = A(2 - 1)^2 + B(2 - 1)(0) - 4(0)$

$\qquad\qquad 5 = A$

let x = 0 $2(0)^2 - 5(0) + 7 = 5(0 - 1)^2 + B(0 - 1)(0 - 2) - 4(0 - 2)$

$\qquad\qquad 7 = 5 + 2B + 8$

$\qquad\qquad -6 = 2B$

$\qquad\qquad -3 = B$

∴ $\dfrac{2x^2 - 5x + 7}{(x - 2)(x - 1)^2} \equiv \dfrac{5}{x - 2} - \dfrac{3}{x - 1} - \dfrac{4}{(x - 1)^2}$

Example 21

Express $\dfrac{3x - 8}{(x + 1)(x - 2)^2}$ in partial fractions

Solution

Let $\dfrac{3x-8}{(x+1)(x-2)^2} = \dfrac{A}{(x+1)} + \dfrac{B}{(x-2)} + \dfrac{C}{(x-2)^2}$

Multiplying through by $(x+1)(x-2)^2$ gives:

$$3x - 8 = A(x-2)^2 + B(x+1)(x-2) + C(x+1)$$

Let $x = -1$ $3(-1) - 8 = A(-1-2)^2 + B(0)(-1-2) + C(0)$

$\qquad -11 = 9A$

$\therefore \quad -\dfrac{11}{9} = A$

Let $x = 2$ $3(2) - 8 = -\dfrac{11}{9}(0) + B(2+1)(0) + C(2+1)$

$\qquad -2 = 3C$

$\therefore \quad -\dfrac{2}{3} = C$

Let $x = 0$ $3(0) - 8 = -\dfrac{11}{9}(0-2)^2 + B(0+1)(0+2) - \dfrac{2}{3}(0+1)$

$\qquad -8 = -\dfrac{44}{9} - 2B - \dfrac{2}{3}$

$\therefore \quad 2B = -\dfrac{44}{9} - \dfrac{2}{3} + 8$

$\qquad = \dfrac{-44 - 6 + 72}{9}$

$\qquad = \dfrac{22}{9}$

$\therefore \quad B = \dfrac{22}{18} = \dfrac{11}{9}$

$\therefore \quad \dfrac{3x-8}{(x+1)(x-2)^2} = \dfrac{-11}{9(x+1)} + \dfrac{11}{9(x-2)} - \dfrac{2}{3(x-2)^2}$

---

e)   *Example* ('A' level question)

Show that $x = 2$ is a root of the equation $x^3 + 3x - 14 = 0$.

Given that the other roots are α and β show that $\alpha + \beta = -2$ and find the value of $\alpha\beta$.

Find the equation with numerical coefficients whose roots are:

(a) $\alpha + 3$ and $\beta + 3$

(b) $5, \alpha + 3$ and $\beta + 3$

ESSENTIAL ALGEBRA   PURE MATHEMATICS

Solution

Replacing x with 2 in the given equation gives:

$$f(2) = 2^3 + 3(2) - 14$$

$$f(2) = 8 + 6 - 14$$

$$\therefore f(2) = 0$$

ie $x = 2$ satisfies the equation and is a root of the equation. It follows that $(x - 2)$ is a factor of $x^3 + 3x - 14$ and the other factors can be found by division:

$$
\begin{array}{r}
x^2 + 2x + 7 \\
x - 2 \overline{\smash{)}x^3 + 3x - 14} \\
\underline{x^3 - 2x^2} \\
2x^2 + 3x \\
\underline{2x^2 - 4x} \\
7x - 14 \\
\underline{7x - 14} \\
\cdots
\end{array}
$$

So $x^3 + 3x - 14 = 0$ becomes

$$(x - 2)(x^2 + 2x + 7) = 0$$

and the roots of $x^2 + 2x + 7 = 0$ are α and β

$$\therefore \quad \alpha + \beta = \frac{-b}{a} = \frac{-2}{1} = -2$$

and $\quad \alpha\beta = \frac{c}{a} = \frac{7}{1} = 7$

(a) Assuming that the equation with roots $\alpha + 3$ and $\beta + 3$ is of the form $Ax^2 + Bx + C = 0$, then the sum of the roots $(\alpha + 3 + \beta + 3) = \frac{-B}{A}$

$$\therefore \quad \alpha + \beta + 6 = \frac{-B}{A}$$

Using $\alpha + \beta = -2$ then $\quad -2 + 6 = \frac{-B}{A}$

$$4 = \frac{-B}{A}$$

The product of the roots $(\alpha + 3)(\beta + 3) = \frac{C}{A}$

$$\alpha\beta + 3\beta + 3\alpha + 9 = \frac{C}{A}$$

$$\alpha\beta + 3(\alpha + \beta) + 9 = \frac{C}{A}$$

PURE MATHEMATICS ESSENTIAL ALGEBRA

Using $\alpha + \beta = -2$, $\alpha\beta = 7$ then $7 + 3(-2) + 9 = \dfrac{C}{A}$

$$10 = \dfrac{C}{A}$$

So it follows that $A = 1$, $B = -4$, $C = 10$. ∴ equation becomes $x^2 - 4x + 10 = 0$.

(b) If the equation has an additional root of $x = 5$, then $(x - 5)$ is also a factor of the equation, ie as well as $(x^2 - 4x + 10)$.

So the equation is	$(x - 5)(x^2 - 4x + 10)$	$= 0$
multiplying out	$x^3 - 4x^2 + 10x - 5x^2 + 20x - 50$	$= 0$
rearranging	$x^3 - 9x^2 + 30x - 50$	$= 0$

1.4 Indices and logarithms

a) *Laws of indices*

There are three basic rules of indices. These are:

$$a^m \times a^n = a^{m+n}$$
$$a^m \div a^n = a^{m-n}$$
$$(a^m)^n = a^{mn}$$

Example 22

Use the laws of indices to simplify the following expressions:

(1) $\quad 4^{3n} \times 2^{2n} \div 8^n$

(2) $\quad 3^{n+1} \times 9^n \div 27^{\frac{2n}{3}}$

Solution

(1) $\quad 4^{3n} \times 2^{2n} \div 8^n$ - this expression cannot be simplified until all the terms are expressed in a common factor 2; ie

$4 = 2^2 \qquad \therefore 4^{3n} = (2^2)^{3n} = 2^{6n}$

$8 = 2^3 \qquad \therefore 8^n = (2^3)^n = 2^{3n}$

$\therefore \quad 4^{3n} \times 2^{2n} \div 8^n = 2^{6n} \times 2^{2n} \div 2^{3n}$

$\qquad = 2^{6n + 2n - 3n} = 2^{5n}$

(2) $\quad 3^{n+1} \times 9^n \div 27^{\frac{2n}{3}}$ the common factor will be 3 this time.

$9 = 3^2 \qquad \therefore 9^n = (3^2)^n = 3^{2n}$

ESSENTIAL ALGEBRA　　　　　　　　PURE MATHEMATICS

$27 = 3^3 \qquad \therefore 27^{\frac{2n}{3}} = (3^3)^{\frac{2n}{3}} = 3^{2n}$

$\therefore 3^{n+1} \times 9^n + 27^{\frac{2}{3n}} = 3^{n+1} \times 3^{2n} + 3^{2n}$

$\qquad\qquad\qquad\qquad\qquad = 3^{n+1+2n-2n} = 3^{n+1}$

b) *Fractional and negative indices*

i) *Fractional indices*

Consider $3^{\frac{1}{2}}$, then by the law of indices

$$3^{\frac{1}{2}} \times 3^{\frac{1}{2}} = 3^{\frac{1}{2} + \frac{1}{2}} = 3^1$$

but $\quad 3^{\frac{1}{2}} \times 3^{\frac{1}{2}} = \left(3^{\frac{1}{2}}\right)^2 \qquad \therefore \left(3^{\frac{1}{2}}\right)^2 = 3$

$\therefore$ square rooting $\qquad 3^{\frac{1}{2}} = \sqrt{3}$

Similarly for $5^{\frac{1}{3}}$, say

$$5^{\frac{1}{3}} \times 5^{\frac{1}{3}} \times 5^{\frac{1}{3}} = 5^{\frac{1}{3} + \frac{1}{3} + \frac{1}{3}} = 5^1$$

but $\quad 5^{\frac{1}{3}} \times 5^{\frac{1}{3}} \times 5^{\frac{1}{3}} \left(5^{\frac{1}{3}}\right)^3 \quad \therefore \left(5^{\frac{1}{3}}\right)^3 = 5$

$\therefore$ cube rooting $\qquad 5^{\frac{1}{3}} = \sqrt[3]{5}$

So in general terms it follows that

$$\boxed{a^{\frac{1}{n}} = \sqrt[n]{a}}$$

Also $5^{\frac{2}{3}} \qquad = 5^{\frac{1}{3}} \times 5^{\frac{1}{3}}$

$\qquad\qquad\qquad = \left(5^{\frac{1}{3}}\right)^2$

$\qquad\qquad\qquad = (\sqrt[3]{5})^2$ or $\sqrt[3]{5}\,^2$

Similarly $6^{\frac{3}{2}} \quad = 6^{\frac{1}{2}} \times 6^{\frac{1}{2}} \times 6^{\frac{1}{2}}$

$\qquad\qquad\qquad = \left(6^{\frac{1}{2}}\right)^3$

$\qquad\qquad\qquad = (\sqrt{6})^3$ or $\sqrt{6}\,^3$

So in general terms it follows that

$$\boxed{a^{m/n} = \left(\sqrt[n]{a}\right)^m \text{ or } \sqrt[n]{a}\,^m}$$

PURE MATHEMATICS   ESSENTIAL ALGEBRA

Example 23

Find the values of:

(1) $25^{\frac{1}{2}}$

(2) $(-8)^{\frac{1}{3}}$

(3) $27^{\frac{2}{3}}$

(4) $\left(\frac{16}{81}\right)^{\frac{1}{4}}$

Solution

(1) $25^{\frac{1}{2}} = \sqrt{25} = 5 \quad \because 5^2 = 25$

(2) $(-8)^{\frac{1}{3}} = \sqrt[3]{-8} = -2 \quad \dots (-2)^3 = -8$

(3) $27^{\frac{2}{3}} = \sqrt[3]{27^2} = 3^2 = 9$

(4) $\left(\frac{16}{81}\right)^{\frac{1}{4}} = \sqrt[4]{\frac{16}{81}} = \frac{\sqrt[4]{16}}{\sqrt[4]{81}} = \frac{2}{3}$

---

ii) *Zero index*

$4^3 \div 4^3 = 4^{3-3} = 4^0$

but $\quad 4^3 \div 4^3 = 64 \div 64 = 1$

$\therefore \quad 4^0 = 1$

Similarly $\quad 6^n \div 6^n = 6^{n-n} = 6^0$

but $\quad 6^n \div 6^n = 1$

$\therefore \quad 6^0 = 1$

So in general terms it follows that:

$$\boxed{a^n \div a^n = a^0 = 1}$$

iii) *Negative indices*

In order to understand the meaning of a negative index it is again necessary to apply the basic laws of indices to a particular problem.

Consider $\quad 5^0 \div 5^2 = 5^{0-2} = 5^{-2}$

but $\quad 5^0 \div 5^2 = \frac{5^0}{5^2} = \frac{1}{5^2}$

ESSENTIAL ALGEBRA   PURE MATHEMATICS

$\therefore \qquad 5^{-2} = \dfrac{1}{5^2}$

Similarly $\qquad 3^0 \div 3^n = 3^{0-n} = 3^{-n}$

but $\qquad 3^0 \div 3^n = \dfrac{3^0}{3^n} = \dfrac{1}{3^n}$

$\therefore \qquad 3^{-n} = \dfrac{1}{3^n}$

So in general terms it follows that:

$$\boxed{a^{-n} = \dfrac{1}{a^n}}$$

Example 24

Find the values of:

(1) $\quad 3^{-4}$

(2) $\quad \left(\dfrac{2}{3}\right)^{-2}$

(3) $\quad 4^{\frac{-1}{2}}$

(4) $\quad 81^{\frac{-3}{4}}$

Solution

(1) $\quad 3^{-4} = \dfrac{1}{3^4} = \dfrac{1}{81}$

(2) $\quad \left(\dfrac{2}{3}\right)^{-2} = \dfrac{1}{\left(\frac{2}{3}\right)^2} = \dfrac{1}{\left(\frac{4}{9}\right)} = \dfrac{9}{4}$

(3) $\quad 4^{\frac{-1}{2}} = \dfrac{1}{4^{\frac{1}{2}}} = \dfrac{1}{\sqrt{4}} = \dfrac{1}{2}$

(4) $\quad 81^{\frac{-3}{4}} = \dfrac{1}{81^{\frac{3}{4}}} = \dfrac{1}{\sqrt[4]{81^3}} = \dfrac{1}{3^3} = \dfrac{1}{27}$

c) *Logarithms*

Before calculators became so commonplace many calculations involving multiplication and division of decimals were done with logarithmic tables. It is now necessary to study the basic theory of logarithms and to establish the rules that govern their use.

From logarithmic tables it is found that the logarithm of 3 is 0.4771. These tables are given in base 10 which means

$\qquad \log 3 \quad = \quad 0.4771$ or more correctly

PURE MATHEMATICS — ESSENTIAL ALGEBRA

if $\quad \log_{10} 3 = 0.4771$

then $\quad 3 = 10^{0.4771}$

Similarly $\log_{10} 8 = 0.9031$

then $\quad 8 = 10^{0.9031}$

So it follows that a logarithm is an index and that bases other than 10 can be used.

For example $\quad 8 = 2^3 \qquad\qquad 100 = 10^2$

Using base 2 $\therefore \quad \log_2 8 = 3 \qquad$ Using base 10 $\therefore \quad \log_{10} 100 = 2$

$\qquad\qquad 25 = 5^2 \qquad\qquad 81 = 3^4$

Using base 5 $\therefore \quad \log_5 25 = 2 \qquad$ Using base 3 $\therefore \quad \log_3 81 = 4$

$\qquad\qquad 1 = 4^0 \qquad\qquad 1 = a^0$

Using base 4 $\therefore \quad \log_4 1 = 0 \qquad$ Using base a $\therefore \quad \log_a 1 = 0$

$\qquad\qquad 2 = \sqrt{4} \qquad\qquad 9 = \sqrt[3]{27^2}$

$\qquad\qquad 2 = 4^{\frac{1}{2}} \qquad\qquad 9 = 27^{\frac{2}{3}}$

Using base 4 $\therefore \quad \log_4 2 = \frac{1}{2} \qquad$ Using base 27 $\therefore \quad \log_{27} 9 = \frac{2}{3}$

In general terms:

$$\boxed{\text{if} \quad a = c^x \qquad \text{then} \quad \log_c a = x}$$

Example 25

Write the following in index notation:

(1) $\quad \log_2 32 = 5$

(2) $\quad \log_3 \left(\frac{1}{9}\right) = -2$

(3) $\quad \log_3 a = b$

(4) $\quad \log_a 8 = c$

Solution

(1) $\quad \log_2 32 = 5$

$\qquad \therefore \quad 32 = 2^5$

(2) $\quad \log_3 \left(\frac{1}{9}\right) = -2$

ESSENTIAL ALGEBRA   PURE MATHEMATICS

$$\therefore \quad \frac{1}{9} = 3^{-2} = \frac{1}{3^2}$$

(3) $\log_3 a = b$

$$\therefore \quad a = 3b$$

(4) $\log_a 8 = c$

$$\therefore \quad 8 = a^c$$

Example 26

Evaluate the following:

(1) $\log_2 64$

(2) $\log_7 1$

(3) $\log_a a^2$

(4) $\log_{27} 3$

Solution

(1) $64 = 2^6$

$$\therefore \quad \log_2 64 = 6$$

(2) $1 = 7^0$

$$\therefore \quad \log_7 1 = 0$$

(3) $a^2 = a^2$

$$\therefore \quad \log_a a^2 = 2$$

(4) $3 = \sqrt[3]{27}$

$3 = 27^{\frac{1}{3}}$

$$\therefore \quad \log_{27} 3 = \frac{1}{3}$$

d)   *Laws of logarithms*

(a)   When two numbers are multiplied the logarithms are added. This can be shown as follows:

$$\text{If } a = c^x \text{ then } \log_c a = x$$
$$\text{and } b = c^y \text{ then } \log_c b = y$$
$$ab = c^x \cdot c^y$$
$$\therefore \quad ab = c^{x+y}$$

$$\therefore \quad \log_c(ab) = x + y$$

$$\boxed{\therefore \log_c(ab) = \log_c a + \log_c b}$$

(b) When two numbers are divided the logarithms are subtracted. This can be shown as follows:

$$a + b = c^x + c^y$$

$$\therefore \quad a + b = c^{x-y}$$

$$\therefore \quad \log_c(a + b) = x - y$$

$$\boxed{\therefore \log_c(a + b) = \log_c a - \log_c b}$$

(c) This law of logarithms relates to powers of numbers:

$$\text{If} \quad a = c^x$$

$$\text{then} \quad a^n = (c^x)^n$$

$$a^n = c^{nx}$$

$$\therefore \quad \log_c a^n = nx$$

$$\boxed{\therefore \log_c a^n = n.\log_c a}$$

(d) The last law of logarithms is known as 'change of base'.

$$\text{If} \quad x = \log_a b$$

$$\text{then} \quad a^x = b$$

Now taking logs to base c of both sides

$$\log_c a^x = \log_c b$$

$$\therefore \quad x \log_c a = \log_c b \text{ but } x = \log_a b$$

$$\therefore \quad \log_a b \cdot \log_c a = \log_c b$$

$$\boxed{\therefore \log_a b = \frac{\log_c b}{\log_c a}}$$

e) *Example* ('A' level question)

Simplify:

(1) $20 \times 8^{2n} - 5 \times 4^{3n+1}$

(2) $(\log_2 5) \times (\log_5 8)$

Solution

(1) It is first necessary to find a common factor, ie

$$20 = 5 \times 4 = 5 \times 2^2,\ 8 = 2^3,\ 4 = 2^2$$

ESSENTIAL ALGEBRA   PURE MATHEMATICS

$\therefore \quad 20 \times 8^{2n} - 5 \times 4^{3n+1} = (5 \times 2^2) \times (2^3)^{2n} - 5 \times (2^2)^{3n+1}$

$\qquad\qquad\qquad\qquad\quad = 5 \times 2^2 \times 2^{6n} - 5 \times 2^{6n+2}$

$\qquad\qquad\qquad\qquad\quad = 5 \times 2^{6n+2} - 5 \times 2^{6n+2}$

$\therefore \quad 20 \times 8^{2n} - 5 \times 4^{3n+1} = 0$

(2) It is necessary to use a change of base either into base 5 or into base 2. Whichever change of base is used the answer will be the same. To prove this both methods will be shown.

Either:

Using a change of base to base 2

$\therefore \quad \log_5 8 = \dfrac{\log_2 8}{\log_2 5}$

$\therefore \quad (\log_2 5) \times (\log_5 8) = (\log_2 5) \times \left(\dfrac{\log_2 8}{\log_2 5}\right)$

$\qquad\qquad\qquad\qquad\quad = \log_2 8$

But $\quad 8 = 2^3$

$\therefore \quad \log_2 8 = \log_2 (2^3) = 3\log_2 2$

but $\quad \log_2 2 = 1 \because 2 = 2^1$

$\therefore \quad \log_2 8 = 3$

$\therefore \quad (\log_2 5) \times (\log_5 8) = 3$

———————————————

Or:

Using a change of base to base 5

$\therefore \quad \log_2 5 = \dfrac{\log_5 5}{\log_5 2} = \dfrac{1}{\log_5 2}$

$\therefore \quad (\log_2 5) \times (\log_5 8) = \dfrac{1}{(\log_5 2)} \times (\log_5 8)$

But $\quad 8 = 2^3$

$\therefore \quad \log_5 8 = \log_5 (2^3) = 3\log_5 2$

$\therefore \quad (\log_2 5) \times (\log_5 8) = \dfrac{1}{(\log_5 2)} \times (3\log_5 2)$

$\therefore \quad (\log_2 5) \times (\log_5 8) = 3$

———————————————

PURE MATHEMATICS — POWER SERIES AND EXPANSIONS

2 POWER SERIES AND EXPANSIONS

2.1 Series
2.2 Binomial theorem
2.3 Log and exponential functions
2.4 Expansions

2.1 Series

a) *Sequences*

A sequence is a set of numbers written down in a definite order, and it is as such that subsequent members of the set can be obtained by applying a simple rule.

Example 1

Find the next two terms of these sequences:

(1)   1, 3, 5, 7, ...

(2)   1, 2, 4, 8, ...

(3)   1, 4, 9, 16, ...

(4)   4, 2, 0, -2, ...

(5)   1, 2, 6, 24, 120, ...

Solution

It is quite easy to find more terms in each sequence by studying the number patterns. Therefore:

(1)   These are odd numbers - 1, 3, 5, 7, 9, 11, ...

(2)   The numbers are being doubled - 1, 2, 4, 8, 16, 32, ...

(3)   These are square numbers - 1, 4, 9, 16, 25, 36, ...

(4)   2 is being subtracted each time - 4, 2, 0, -2, -4, -6, ...

(5)   The numbers are being multiplied by 2, then 3, then 4, then 5, etc., ie 1 x 2 = 2, 2 x 3 = 6, 6 x 4 = 24, 24 x 5 = 120, so the next two terms are 120 x 6 = 720, 720 x 7 = 5,040, ie 1, 2, 6, 24, 120, 720, 5,040, ...

If the terms of a sequence are added together to form a sum, then the expression is called a series.

A finite series ends after a given number of terms:

$$1 + 2 + 3 + 4 \ldots + 100$$

An infinite series has no last term:

$$1 + \frac{1}{2} + \frac{1}{4} + \frac{1}{8} \ldots$$

b) *Arithmetical progressions*

An arithmetical progression (AP) is a series in which every term is obtained from the preceding term by adding to it a certain number, called the common difference (d), eg

POWER SERIES AND EXPANSIONS — PURE MATHEMATICS

$$1 + 6 + 11 + 16 + 21 + \ldots + 101 \quad \text{----} \quad d = 5$$

$$50 + 48 + 46 + 44 + 42 + \ldots + 2 \quad \text{----} \quad d = -2$$

It is possible to obtain a formula for summing an AP:

If a = first term of an AP
 d = common difference
 L = last term
 n = number of terms
 S_n = sum of the first n terms

Then $S_n = a + (a + d) + \ldots + (L - d) + L$

also $S_n = L + (L - d) + \ldots + (a + d) + a$ reversing the order of the terms

Adding these two expressions for S_n together gives:

$$S_n + S_n = (a + L) + (a + L) + \ldots + (a + L) + (a + L)$$

$$\therefore 2S_n = n(a + L)$$

$$\therefore \boxed{S_n = \frac{n}{2}(a + L)}$$

But $L = a + (n - 1)d$ so S_n can be written as:

$$S_n = \frac{n}{2}(a + a + (n - 1)d)$$

$$\therefore \boxed{S_n = \frac{n}{2}(2a + (n - 1)d)}$$

Example 2

In an arithmetical progression the thirteenth term is 27 and the seventh term is three times the second term.

Find the first term, the common difference and the sum of the first ten terms.

Solution

Thirteenth term: $a + 12d = 27$ ---- (1)

Seventh term: $a + 6d$
Second term: $a + d$ } $a + 6d = 3(a + d)$

$\therefore \quad a + 6d = 3a + 3d$

$\therefore \quad 6d - 3d = 3a - a$

$\therefore \quad 3d = 2a$ ---- (2)

Multiplying (2) × 4: $12d = 8a$

Replacing 12d in (1): $a + 8a = 27$

$9a = 27$
$a = 3$

PURE MATHEMATICS — POWER SERIES AND EXPANSIONS

Replacing a = 3 in (2) $3d = 2 \times 3$

∴ $3d = 6$

∴ $d = 2$

So a = 3, d = 2 and n = 10 and using the formula:

$$S_n = \frac{n}{2}(2a + (n-1)d)$$

$$S_{10} = \frac{10}{2}(2 \times 3 + (10-1) \times 2)$$

∴ $S_{10} = 5(6 + 18)$

∴ $S_{10} = 5(24)$

∴ $S_{10} = 120$

c) *Geometrical progressions*

A geometrical progression (GP) is a series in which every term is obtained from the preceding term by multiplying it by a certain number, called the common ratio (r).

$1 + 2 + 4 + 8 + \ldots + 512$ ---- $r = 2$

$27 + 9 + 3 + 1 + \ldots + \frac{1}{729}$ ---- $r = \frac{1}{3}$

It is possible to obtain a formula for summing a GP.

If a = first term
     r = common ratio
     ar^{n-1} = last term
     n = number of terms
     S_n = sum of the first n terms

Then $S_n = a + ar + \ldots + ar^{n-2} + ar^{n-1}$

and $r.S_n = ar + ar^2 + \ldots + ar^{n-1} + ar^n$

Subtracting these two expressions for S_n gives

$$S_n - r.S_n = a - ar^n$$

∴ $S_n(1 - r) = a(1 - r^n)$

∴ $\boxed{S_n = \frac{a(1 - r^n)}{(1 - r)}}$

Example 3

In a geometrical progression the sum of the second and third terms is 6, and the sum of the third and fourth terms is -12. Find the first term and the common ratio.

POWER SERIES AND EXPANSIONS — PURE MATHEMATICS

Solution

Second term: ar

Third term: ar^2 $\quad \}\ ar + ar^2 = 6 \qquad \text{---------- (1)}$

Fourth term: ar^3 $\quad \}\ ar^2 + ar^3 = -12 \qquad \text{---------- (2)}$

Using (1) $\qquad ar(1 + r) = 6$

Using (2) $\qquad ar^2(1 + r) = -12$

Dividing (1) by (2) $\qquad \dfrac{ar(1 + r)}{ar^2(1 + r)} = \dfrac{6}{-12}$

$\qquad \therefore \quad \dfrac{1}{r} = \dfrac{1}{-2}$

$\qquad \therefore \quad r = -2$

Replacing $r = -2$ in (1) $\quad a(-2) + a(4) = 6$

$\qquad \therefore \quad -2a + 4a = 6$

$\qquad \therefore \quad 2a = 6$

$\qquad \therefore \quad a = 3$

d) *Infinite geometrical progressions*

This is a GP with an infinite number of terms.

$\qquad -3 + 6 - 12 + 24 - 48 + \ldots \qquad (r = -2)$

$\qquad \dfrac{3}{10} + \dfrac{3}{100} + \dfrac{3}{1000} + \dfrac{3}{10000} + \dfrac{3}{100000} + \ldots \qquad \left(r = \dfrac{1}{10}\right)$

In certain circumstances it is possible to sum an infinite GP

Considering again $S_n = \dfrac{a(1 - r^n)}{(1 - r)} = \dfrac{a}{(1 - r)} - \dfrac{ar^n}{(1 - r)}$

If $r > +1$ or < -1 then r^n will become larger and larger as n increases to infinity, but if $r < +1$ or > -1, ie $-1 < r < +1$, r^n will become closer and closer to zero as n increases to infinity.

$\therefore \quad \boxed{\text{as } n \to \infty,\ r^n \to 0 \text{ and } S_\infty = \dfrac{a}{1 - r} \quad \text{as } -1 < r < +1}$

Example 4

The sum to infinity of a geometrical progression with a positive common ratio is 9 and the sum of the first two terms is 5. Find the first four terms of the progression.

Solution

Sum to infinity: $\qquad S_\infty = \dfrac{a}{1 - r} = 9 \qquad \text{---------- (1)}$

PURE MATHEMATICS   POWER SERIES AND EXPANSIONS

First term: a

Second term: ar

$\}\ a + ar = 5$ ---------- (2)

Using (1) $a = 9(1 - r)$

Using (2) $a(1 + r) = 5$

Replacing a from (1) into (2) $9(1 - r)(1 + r) = 5$

$$9(1 + r - r - r^2) = 5$$

$$9 - 9r^2 = 5$$

Changing the signs $9r^2 - 4 = 0$

$(3r - 2)(3r + 2) = 0$ using difference of squares

either $3r - 2 = 0$ or $3r + 2 = 0$

$\therefore$ $3r = 2$ $\therefore$ $3r = -2$

$\therefore$ $r = \frac{2}{3}$ $\therefore$ $r = \frac{-2}{3}$

However, r must be positive so $r = \frac{2}{3}$

Replacing $r = \frac{2}{3}$ in (1) $a = 9\left(1 - \frac{2}{3}\right)$

$\therefore$ $a = 9\left(\frac{1}{3}\right)$

$\therefore$ $a = 3$

So the first four terms are:

a,	ar,	ar^2,	ar^3
3,	$3 \times \frac{2}{3}\ (=2)$,	$3 \times \left(\frac{2}{3}\right)^2\ \left(=\frac{4}{3}\right)$,	$3 \times \left(\frac{2}{3}\right)^3\ \left(=\frac{8}{9}\right)$
3	2	$\frac{4}{3}$	$\frac{8}{9}$

e)   *Example* ('A' level question)

The first, second and third terms of an arithmetic series are a, b and c respectively.

Prove that the sum of the first ten terms can be expressed as $\frac{5}{2}(9c - 5a)$.

These numbers a, b and c are also the first, third and fourth terms respectively of a geometric series. Prove that $(2b - c)c^2 = b^3$.

POWER SERIES AND EXPANSIONS PURE MATHEMATICS

Arithmetic series: First term: a

Second term: $b = a + d$ ---------- (1)

Third term: $c = a + 2d$ ---------- (2)

Using $S_n = \frac{n}{2}(2a + (n - 1)d)$

$$S_{10} = \frac{10}{2}(2a + (10 - 1)d)$$

$\therefore \quad S_{10} = 5(2a + 9d)$

However, it is necessary to find an alternative expression for d in terms of a and c.

Using (2) $c - a = 2d$

$$\frac{c - a}{2} = d$$

$\therefore \quad S_{10} = 5\left(2a + 9\left(\frac{c - a}{2}\right)\right)$

$\therefore \quad S_{10} = 5\left(\frac{4a + 9(c - a)}{2}\right)$

$\therefore \quad S_{10} = \frac{5}{2}(4a + 9c - 9a)$

$\therefore \quad S_{10} = \frac{5}{2}(9c - 5a)$

Geometric series: First term: a

Third term: $b = ar^2$ ---------- (3)

Fourth term: $c = ar^3$ ---------- (4)

Using (3) $r^2 = \frac{b}{a}$

Using (4) ÷ (3) $r = \frac{c}{b}$ and $r^2 = \left(\frac{c}{b}\right)^2$

$\therefore \quad \left(\frac{c}{b}\right)^2 = \frac{b}{a}$

$\frac{c^2}{b^2} = \frac{b}{a}$

$\therefore \quad ac^2 = b^3$

However, it is necessary to find an alternative expression for a in terms of b and c.

Using (1) $b - a = d$

PURE MATHEMATICS   POWER SERIES AND EXPANSIONS

Using (2) $\dfrac{c-a}{2} = d$

$\therefore \quad b - a = \dfrac{c-a}{2}$

$\therefore \quad 2(b-a) = c - a$

$\therefore \quad 2b - 2a = c - a$

$\therefore \quad 2b - c = 2a - a$

$\therefore \quad 2b - c = a$

Replacing a with $2b - c$ ... $(2b-c)c^2 = b^3$

2.2 Binomial theorem

a) *Pascal's triangle*

The expansion of $(a+b)^2$ is very easy to obtain, ie

$$(a+b)^2 = (a+b)(a+b)$$
$$= a^2 + ab + ab + b^2$$

$\therefore \quad (a+b)^2 = a^2 + 2ab + b^2$

Similarly for $(a+b)^3 = (a+b)(a+b)(a+b)$

$$= (a+b)(a^2 + 2ab + b^2)$$
$$= a^3 + 2a^2b + ab^2 + a^2b + 2ab^2 + b^3$$

$\therefore \quad (a+b)^3 = a^3 + 3a^2b + 3ab^2 + b^3$

Repeating this process would lead to

$$(a+b)^4 = a^4 + 4a^3b + 6a^2b^2 + 4ab^3 + b^4$$

It is clearly very tedious to continue working out powers of $(a + b)$ in this way and a shorter method needs to be found.

One such method is to construct the rows of Pascal's triangle. These form a table of coefficients for the powers of a and b.

```
                    1
                 1     1
              1     2     1          power of 2
           1     3     3     1       power of 3
        1     4     6     4     1    power of 4
     1     5    10    10     5     1
```

Each line is obtained from the preceding line by addition. Therefore the next line is formed as follows:

POWER SERIES AND EXPANSIONS   PURE MATHEMATICS

```
Using    1   5   10   10   5    1      power of 5
          \+/ \+/ \+/  \+/ \+/
Gives    1   6   15   20   15   6   1  power of 6
```

So the expansion of $(a + b)^6$ is obtained using this last line as follows:

$$(a + b)^6 = 1a^6 + 6a^5b + 15a^4b^2 + 20a^3b^3 + 15a^2b^4 + 6ab^5 + 1b^6$$

$\therefore \quad (a + b)^6 = a^6 + 6a^5b + 15a^4b^2 + 20a^3b^3 + 15a^2b^4 + 6ab^5 + b^6$

Example 5

Expand $(3x - y)^4$.

Solution

Using Pascal's triangle the coefficients of the terms will be 1 4 6 4 1.

$$(3x - y)^4 = 1(3x)^4 + 4(3x)^3(-y) + 6(3x)^2(-y)^2 + 4(3x)(-y)^3 + 1(-y)^4$$

$\therefore \quad (3x - y)^4 = 81x^4 + 4(27x^3)(-y) + 6(9x^2)y^2 + 4(3x)(-y^3) + y^4$

$\therefore \quad (3x - y)^4 = 81x^4 - 108x^3y + 54x^2y^2 - 12xy^3 + y^4$

---

b)   *Binomial expansion for a positive index*

For even larger powers of $(a + b)$ it would be easier to have a formula to use, since only the first few terms are usually required and not the complete expansion.

$$\boxed{(a + b)^n = a^n + na^{n-1}b + \frac{n(n-1)}{2!}a^{n-2}b^2 + \frac{n(n-1)(n-2)}{3!}a^{n-3}b^3 + \ldots + b^n}$$

This is the binomial expansion for all positive integers where

   2! called 2 factorial = 2 x 1   = 2

   3! called 3 factorial = 3 x 2 x 1   = 6

   4! called 4 factorial = 4 x 3 x 2 x 1   = 24

   5! called 5 factorial = 5 x 4 x 3 x 2 x 1 = 120   etc.

It can be shown that this expansion is valid by replacing n with given values, eg

$n = 2 \quad (a + b)^2 = a^2 + 2.a.b + b^2$

$n = 3 \quad (a + b)^3 = a^3 + 3a^2b + \frac{3.2}{2!}ab^2 + b^3$

$\qquad\qquad\qquad = a^3 + 3a^2b + 3ab^2 + b^3$

$n = 4 \quad (a + b)^4 = a^4 + 4a^3b + \frac{4.3}{2!}a^2b^2 + \frac{4.3.2}{3!}ab^3 + b^4$

$\qquad\qquad\qquad = a^4 + 4a^3b + 6a^2b^2 + 4ab^3 + b^4$

$n = 5 \quad (a + b)^5 = a^5 + 5a^4b + \frac{5.4}{2!}a^3b^2 + \frac{5.4.3}{3!}a^2b^3 + \frac{5.4.3.2}{4!}ab^4 + b^5$

$\qquad\qquad\qquad = a^5 + 5a^4b + 10a^3b^2 + 10a^2b^3 + 5ab^4 + b^5 \qquad$ etc.

PURE MATHEMATICS — POWER SERIES AND EXPANSIONS

Example 6

Write down the first four terms in the expansion of $\left(2 + \frac{x}{2}\right)^8$.

Solution

$$\left(2 + \frac{x}{2}\right)^8 = 2^8 + 8.2^7\left(\frac{x}{2}\right) + \frac{8.7}{2!}.2^6.\left(\frac{x}{2}\right)^2 + \frac{8.7.6}{3!}.2^5\left(\frac{x}{2}\right)^3 + \ldots$$

$$= 256 + 512x + 448x^2 + 224x^3 + \ldots$$

Sometimes only one or two particular terms of the expansion are required and these need not be the first terms. On such occasions it is often easier to use an alternative form of the binomial expansion.

$$(a + b)^n = a^n + \binom{n}{1}a^{n-1}b + \binom{n}{2}a^{n-2}b^2 + \ldots + \binom{n}{r}a^{n-r}b^r + \ldots + b^n$$

where $\binom{n}{1} = \dfrac{n!}{(n-1)!1!} = n$ as before

$\binom{n}{2} = \dfrac{n!}{(n-2)!2!} = \dfrac{n.(n-1)}{2!}$ as before

$\binom{n}{3} = \dfrac{n!}{(n-3)(!3!)} = \dfrac{n(n-1)(n-2)}{3!}$ as before

and $\binom{n}{r} = \dfrac{n!}{(n-r)(!r!)}$ is the coefficient of any general term ($0 \leq r \leq n$)

Example 7

For $(2x - 3)^7$ write down the term in x^5.

Solution

Using general term = $\binom{n}{r}a^{n-r}b^r$ with $a = 2x$, $b = (-3)$, $n = 7$ and $r = 2$.

Required term $= \binom{7}{2}(2x)^5(-3)^2$

$= \dfrac{7!}{(7-2)!2!} \cdot (32x^5) \cdot 9$

$= \dfrac{7!}{5!2!} \cdot (288x^5)$

$= 21.(288x^5)$

$= 6048x^5$

c) *Binomial expansion for any index*

When n is a positive integer there is always a finite number of terms, for example:

If $n = 2$ then $(a + b)^2$ has 3 terms

POWER SERIES AND EXPANSIONS — PURE MATHEMATICS

n = 3   then $(a + b)^3$ has 4 terms

n = 4   then $(a + b)^4$ has 5 terms   etc.

So the number of terms is always one more than the power of n.

However, this is not true when n is other than a positive integer, ie when n has a negative or fractional value.

In this section the binomial expansion will be used in a slightly different form. Consider

$$(1 + x)^n = 1 + nx + \frac{n(n-1)}{2!}x^2 + \frac{n(n-1)(n-2)}{3!}x^3 + \ldots$$

If n is other than a positive integer this series will continue indefinitely but as long as x lies between +1 and -1 (ie $-1 < x < +1$) the expansion is valid.

Note: Fractional and negative powers are used in the following examples. Any student who is not familiar with these may need to study section 1.4 (b) pages 35-37 first.

Example 8

Expand $\dfrac{1}{(2+x)^2}$ as far as the term in x^3 and state the range of values of x for which the expansion is valid.

Solution

$\dfrac{1}{(2+x)^2}$ can be written as $(2 + x)^{-2}$. Before commencing the expansion it is necessary to take out a factor of 2 so that the inside of the bracket is in the form (1 + *).

$$(2 + x) = 2\left(1 + \frac{x}{2}\right)$$

$$\therefore \quad (2+x)^{-2} = 2^{-2}\left(1 + \frac{x}{2}\right)^{-2} = \frac{1}{4}\left(1 + \frac{x}{2}\right)^{-2}$$

$$\frac{1}{4}\left(1 + \frac{x}{2}\right)^{-2} = \frac{1}{4}\left(1 + (-2)\left(\frac{x}{2}\right) + \frac{(-2)(-3)}{2!}\left(\frac{x}{2}\right)^2 + \frac{(-2)(-3)(-4)}{3!}\left(\frac{x}{2}\right)^3 + \ldots\right)$$

$$= \frac{1}{4}\left(1 - 2 \cdot \frac{x}{2} + \frac{6}{2} \cdot \frac{x^2}{4} - \frac{24}{6} \cdot \frac{x^3}{8} + \ldots\right)$$

$$\therefore \quad \frac{1}{(2+x)^2} = \frac{1}{4}\left(1 - x + \frac{3x^2}{4} - \frac{x^3}{2} + \ldots\right)$$

This is valid for the following range of values:

$$-1 < \frac{x}{2} < +1$$

ie $-2 < x < +2$

PURE MATHEMATICS   POWER SERIES AND EXPANSIONS

Example 9

Expand $\sqrt[3]{(3+x)}$ as far as the term in x^3 and state the range of values of x for which the expansion is valid.

Solution

$\sqrt[3]{(3+x)}$ can be written as $(3+x)^{\frac{1}{3}}$. Taking out a factor of 3 gives:

$$(3+x) = 3\left(1 + \frac{x}{3}\right) \quad \therefore \quad (3+x)^{\frac{1}{3}} = 3^{\frac{1}{3}}\left(1 + \frac{x}{3}\right)^{\frac{1}{3}}$$

$$= 3^{\frac{1}{3}}\left(1 + \left(\frac{1}{3}\right)\left(\frac{x}{3}\right) + \frac{\left(\frac{1}{3}\right)\left(-\frac{2}{3}\right)}{2!}\left(\frac{x}{3}\right)^2 + \frac{\left(\frac{1}{3}\right)\left(-\frac{2}{3}\right)\left(-\frac{5}{3}\right)}{3!}\left(\frac{x}{3}\right)^3 + \ldots\right)$$

$$= 3^{\frac{1}{3}}\left(1 + \frac{1}{3}\cdot\frac{x}{3} - \frac{2}{9.2}\cdot\frac{x^2}{9} + \frac{10}{27.6}\cdot\frac{x^3}{27} - \ldots\right)$$

$$\therefore \quad \sqrt[3]{(3+x)} = \sqrt[3]{3}\left(1 + \frac{x}{9} - \frac{x^2}{81} + \frac{5x^3}{2187} - \ldots\right)$$

This is valid for the following range of values:

$$-1 < \frac{x}{3} < +1 \quad \text{ie} \quad -3 < x < +3$$

---

d)   *Approximate calculations using binomial expansion*

The binomial expansion can be used to obtain approximate values for numerical calculations such as $(1.005)^{10}$, $\sqrt{9.09}$, $\frac{1}{(10.04)^2}$, etc.

Example 10

Obtain the first four terms in the expansion of $\left(1 + \frac{x}{2}\right)^{10}$ and hence find the value of 1.005^{10} correct to four decimal places.

Solution

$$\left(1 + \frac{x}{2}\right)^{10} = 1 + 10\cdot\frac{x}{2} + \frac{10.9}{2!}\left(\frac{x}{2}\right)^2 + \frac{10.9.8}{3!}\left(\frac{x}{2}\right)^3 + \ldots$$

$$= 1 + 10\cdot\frac{x}{2} + \frac{90}{2}\cdot\frac{x^2}{4} + \frac{720}{6}\cdot\frac{x^3}{8} + \ldots$$

$$\therefore \quad \left(1 + \frac{x}{2}\right)^{10} = 1 + 5x + \frac{45x^2}{4} + 15x^3 + \ldots$$

If $\quad 1 + \frac{x}{2} = 1.005$

then $\quad \frac{x}{2} = 1.005 - 1 = 0.005$

$\therefore \quad x = 0.005 \times 2 = 0.01$

POWER SERIES AND EXPANSIONS — PURE MATHEMATICS

So replacing x with 0.01 in the above expansion gives an approximate value for 1.005^{10}, ie

$$(1.005)^{10} = \left(1 + \frac{0.01}{2}\right)^{10} = 1 + 5\,(0.01) + \frac{45\,(0.01)^2}{4} + 15\,(0.01)^3$$

$$= 1 + 0.05 + 0.001125 + 0.000015$$

$$= 1.051140$$

$\therefore \quad (1.005)^{10} = 1.0511$ (to 4 decimal places)

---

e) *Example* ('A' level question)

Expand $\sqrt{4 - x}$ as a series in ascending powers of x up to and including the term in x^2.

If terms in x^n, $n \geq 3$, can be neglected find the quadratic approximation to $\sqrt{\dfrac{4 - x}{1 - 2x}}$. State the range of values of x for which this approximation is valid.

Solution

$$\sqrt{4 - x} = (4 - x)^{\frac{1}{2}} \text{ but factorising gives}$$

$$(4 - x) = 4\left(1 - \frac{x}{4}\right)$$

$$\therefore \quad (4 - x)^{\frac{1}{2}} = 4^{\frac{1}{2}}\left(1 - \frac{x}{4}\right)^{\frac{1}{2}} = 2.\left(1 - \frac{x}{4}\right)^{\frac{1}{2}}$$

Expanding

$$\therefore \quad 2\left(1 - \frac{x}{4}\right)^{\frac{1}{2}} = 2\left[1 + \left(\frac{1}{2}\right)\left(-\frac{x}{4}\right) + \frac{\left(\frac{1}{2}\right)\left(-\frac{1}{2}\right)}{2!}\left(-\frac{x}{4}\right)^2 + \ldots\right]$$

$$= 2\left(1 - \frac{1}{2}\cdot\frac{x}{4} - \frac{1}{4.2}\cdot\frac{x^2}{16} + \ldots\right)$$

$$= 2\left(1 - \frac{x}{8} - \frac{x^2}{128} + \ldots\right)$$

$$\therefore \quad \sqrt{4 - x} = 2 - \frac{x}{4} - \frac{x^2}{64} + \ldots$$

$$\sqrt{\frac{4 - x}{1 - 2x}} = \frac{(4 - x)^{\frac{1}{2}}}{(1 - 2x)^{\frac{1}{2}}} = (4 - x)^{\frac{1}{2}}.(1 - 2x)^{-\frac{1}{2}}$$

Expanding $(1 - 2x)^{-\frac{1}{2}} = 1 + \left(-\frac{1}{2}\right).(-2x) + \dfrac{\left(\frac{-1}{2}\right)\left(\frac{-3}{2}\right)}{2!}(-2x)^2 + \ldots$

$$= 1 + \frac{1}{2}.2x + \frac{3}{4.2}.4x^2 + \ldots$$

52

PURE MATHEMATICS — POWER SERIES AND EXPANSIONS

$$= 1 + x + \frac{3x^2}{2} + \ldots$$

$$\sqrt{\frac{4-x}{1-2x}} = (4-x)^{\frac{1}{2}} \cdot (1-2x)^{-\frac{1}{2}}$$

$$= \left(2 - \frac{x}{4} - \frac{x^2}{64}\right)\left(1 + x + \frac{3x^2}{2}\right)$$

$$= 2 + 2x + 3x^2 - \frac{x}{4} - \frac{x^2}{4} - \frac{x^2}{64} \text{ ignoring higher order terms}$$

$$= 2 + \frac{8x - x}{4} + \frac{192x^2 - 16x^2 - x^2}{64}$$

$$\sqrt{\frac{4-x}{1-2x}} = 2 + \frac{7x}{4} + \frac{175x^2}{64}$$

Range of values:

$$\sqrt{4-x} = 2\left(1 - \frac{x}{4}\right)^{\frac{1}{2}} \text{ is valid for } -1 < \frac{x}{4} < +1 \qquad \text{ie } -4 < x < +4$$

and $\quad \dfrac{1}{\sqrt{1-2x}} = (1-2x)^{-\frac{1}{2}}$ is valid for $-1 < 2x < +1 \qquad$ ie $-\frac{1}{2} < x < +\frac{1}{2}$

So $\quad \sqrt{\dfrac{4-x}{1-2x}}$ is valid for $\quad -\frac{1}{2} < x < +\frac{1}{2}$

because only in this range are *both* expansions valid.

2.3 Log and exponential functions

a) *Exponential functions*

Functions in which the variable is in the index are called exponential functions. These are functions such as:

$$2^x, \quad 10^{\sin x}, \quad 3^{\frac{1}{x}}$$

Example 11

Draw the graph of $y = 2^x$ for values of x from -3 to $+2$.

Solution

Drawing up a table of values gives:

x	-3	-2	-1	0	1	2
2^x	2^{-3}	2^{-2}	2^{-1}	2^0	2^1	2^2
y	$\frac{1}{8}$	$\frac{1}{4}$	$\frac{1}{2}$	1	2	4

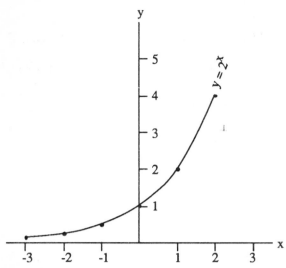

Note: As x takes larger and larger negative values so 2^x gets smaller and smaller, ie as $x \to -\infty$, $2^x \to 0$ and so the curve approaches the x axis but does not meet it.

b) *The exponential function*

A very important exponential function is $y = e^x$ where e is a constant (value 2.718828). This function and other similar functions such as $y = e^{x^2}$, $y = e^{-3x}$, $y = e^{\cos x}$ will frequently appear in later work.

Example 12

Draw graphs of $y = e^x$ and $y = e^{-x}$ on the same axes, taking x from -3 to +3.

Solution

Drawing up the table of values gives:

x	-3	-2	-1	0	1	2	3
e^x	e^{-3}	e^{-2}	e^{-1}	e^0	e^1	e^2	e^3
y	0.050	0.135	0.368	1	2.718	7.389	20.08

x	-3	-2	-1	0	1	2	3
e^{-x}	e^3	e^2	e^1	e^0	e^{-1}	e^{-2}	e^{-3}
y	20.08	7.389	2.718	1	0.368	0.135	0.050

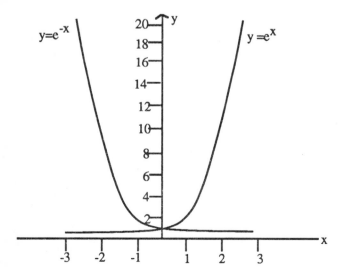

PURE MATHEMATICS — POWER SERIES AND EXPANSIONS

c) *Logarithmic functions*

As has already been stated a logarithm is an index, and this will now be considered in more detail.

Logarithms to the base 10 (as found in books of logarithmic and other tables) are called *common* logarithms. However, logs are often given to the base 'e' and these are called *natural* logarithms.

Logarithmic functions are functions of a variable (say, x) such as $\log_e(1+x)$, $\log_e(3\sin^2 x)$, $\log_e(x^2 - 4)$. These functions still obey all the laws of logarithms as outlined in chapter 1 pages 37-41.

Example 13

Express the following as a single logarithmic function:

(1) $\quad 3\log_e x + 3 - \log_e 3x$

(2) $\quad \frac{1}{2}\log_e(1+y) + \frac{1}{2}\log_e(1-y) + \log_e k$

Solution

(1) Using the laws of logarithms it follows that $\quad 3\log_e x = \log_e x^3$

and so $\quad 3\log_e x + 3 - \log_e 3x = \log_e x^3 - \log_e 3x + 3$

$$= \log_e(x^3 + 3x) + 3$$

$$= \log_e\left(\frac{x^3}{3x}\right) + 3$$

However, it is necessary to write 3 in terms of $\log_e$. So since $\log_e e = 1$

it follows that $\quad 3\log_e e = 3$

$\therefore \quad \log_e e^3 = 3$

Using this now in the above expression gives

$$3\log_e x + 3 - \log_e 3x = \log_e\left(\frac{x^2}{3}\right) + \log_e e^3$$

$$= \log_e\left(\frac{x^2 e^3}{3}\right)$$

(2) Again applying the laws of logarithms it follows that:

$$\frac{1}{2}\log_e(1+y) = \log_e(1+y)^{\frac{1}{2}} = \log_e\sqrt{1+y}$$

$$\frac{1}{2}\log_e(1-y) = \log_e(1-y)^{\frac{1}{2}} = \log_e\sqrt{1-y}$$

$\therefore \quad \frac{1}{2}\log_e(1+y) + \frac{1}{2}\log_e(1-y) + \log_e k = \log_e\sqrt{1+y} + \log_e\sqrt{1-y} + \log_e k$

$$= \log_e\left(\sqrt{1+y} \cdot \sqrt{1-y} \cdot k\right)$$

$$= \log_e\left(k \cdot \sqrt{(1+y)(1-y)}\right)$$

POWER SERIES AND EXPANSIONS   PURE MATHEMATICS

$$\therefore \quad \tfrac{1}{2}\log_e(1+y) + \tfrac{1}{2}\log_e(1-y) + \log_e k = \log_e\left(k.\sqrt{1-y^2}\right)$$

Example 14

Express as the sum or difference of logarithmic functions:

(1) $\log_e(3\sin^2 x)$

(2) $\log_e(x^2 - 4)$

Solution

(1) $\log_e(3\sin^2 x)$ $= \log_e 3 + \log_e \sin^2 x$

$\qquad\qquad\qquad\qquad = \log_e 3 + 2\log_e \sin x$

(2) $\log_e(x^2 - 4)$ $= \log_e[(x-2)(x+2)]$

$\qquad\qquad\qquad\qquad = \log_e(x-2) + \log_e(x+2)$

Example 15

Draw the graph of $y = \log_e x$ for values of x from 0 to +5.

Solution

Drawing up a table of values gives:

x	0	1	2	3	4	5
$\log_e x$	$-\infty$	0	0.693	1.099	1.386	1.609

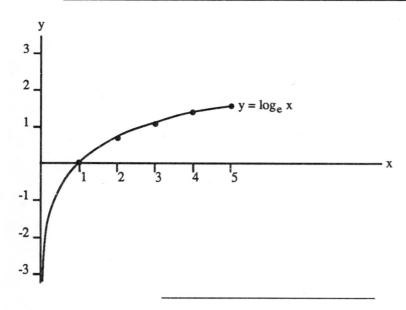

d)   *Equations involving these functions*

It is often necessary to solve equations involving exponential and/or logarithmic functions. These may be functions of one variable or two variables; in the latter case this usually results in simultaneous equations.

The following examples will help to explain the method of solving such equations.

PURE MATHEMATICS — POWER SERIES AND EXPANSIONS

Example 16

Solve the following equations:

(1) $\quad 2^x \cdot 2^{x+1} = 10$

(2) $\quad \log_{10} x + \log_{10} y = 1, \quad x + y = 11$

Solution

(1) $\qquad 2^x \cdot 2^{x+1} = 10 \qquad$ Adding indices gives:

$\qquad\qquad 2^{x+(x+1)} = 10$

$\therefore \qquad 2^{2x+1} = 10 \qquad$ Taking logarithms to base 10 gives:

$\qquad \log_{10} 2^{2x+1} = \log_{10} 10$

$\therefore \quad (2x+1)\log_{10} 2 = \log_{10} 10$ but $\log_{10} 2 = 0.301$ and $\log_{10} 10 = 1$

$\therefore \quad (2x+1)\, 0.301 = 1$

$\therefore \quad 0.602x + 0.301 = 1$

$\therefore \quad 0.602x = 1 - 0.301 = 0.699$

$\therefore \quad x = \dfrac{0.699}{0.602} = 1.161$

(2) $\quad \log_{10} x + \log_{10} y = 1$

$\therefore \quad \log_{10}(xy) = 1$

$\qquad xy = 10^1 = 10 \qquad \therefore xy = 10 \qquad$ ---------- (1)

$\qquad x + y = 11 \qquad \therefore x = 11 - y \qquad$ ---------- (2)

Replacing for x from equation (2) into (1)

$\therefore \qquad xy = 10$ becomes

$\qquad (11 - y)y = 10$

$\qquad 11y - y^2 = 10$

$\qquad 11y - y^2 - 10 = 0 \quad$ changing signs

$\therefore \quad y^2 - 11y + 10 = 0$

$\therefore \quad (y - 10)(y - 1) = 0$

either $\quad y - 10 = 0 \qquad\qquad$ or $\quad y - 1 = 0$

$\therefore \qquad y = 10 \qquad\qquad\qquad \therefore \quad y = 1$

and $\quad x = 11 - 10 = 1 \qquad$ and $\quad x = 11 - 1 = 10$

So the solutions are $x = 1, y = 10$ or $x = 10, y = 1$.

POWER SERIES AND EXPANSIONS — PURE MATHEMATICS

The following examples use the same methods but are more complicated to solve.

Example 17

Given that $\log_2(x - 5y + 4) = 0$ and $\log_2(x + 1) - 1 = 2\log_2 y$ find the values of x and y.

Solution

Using the laws of logarithms:

$\log_2(x - 5y + 4) = 0$ gives $(x - 5y + 4) = 2^0$

$\therefore \quad x - 5y + 4 = 1$

$\therefore \quad x - 5y + 3 = 0$ ——— (1)

$\log_2(x + 1) - 1 = 2\log_2 y$ but $2\log_2 y = \log_2 y^2$

$\therefore \quad \log_2(x + 1) - \log_2 y^2 = 1$

$\therefore \quad \log_2[(x+1) + y^2] = 1$ gives $\dfrac{x+1}{y^2} = 2^1 = 2$

$\therefore \quad x + 1 = 2y^2$

$\therefore \quad x = 2y^2 - 1$ ——— (2)

Replacing for x from equation (2) into equation (1):

$x - 5y + 3 = 0$ becomes

$(2y^2 - 1) - 5y + 3 = 0$

$2y^2 - 5y + 2 = 0$

$(2y - 1)(y - 2) = 0$

either $2y - 1 = 0$ or $y - 2 = 0$

$\therefore \quad 2y = 1$ $\therefore \quad y = 2$

and $y = \dfrac{1}{2}$

and using equation (2) to find x:

$x = 2\left(\dfrac{1}{2}\right)^2 - 1$ or $x = 2(2)^2 - 1$

$x = \dfrac{1}{2} - 1 = -\dfrac{1}{2}$ $= 8 - 1 = 7$

So the solutions of the two equations are $x = -\dfrac{1}{2}$, $y = \dfrac{1}{2}$ and $x = 7$, $y = 2$.

PURE MATHEMATICS POWER SERIES AND EXPANSIONS

Example 18

Given that $\log_2 x + 2\log_4 y = 4$ show that $xy = 16$. Hence solve for x and y the simultaneous equations $\log_{10}(x+y) = 1$ and $\log_2 x + 2\log_4 y = 4$.

Solution

Since the logarithmic functions are given to different bases (ie base 2 and base 4) it is necessary to change the base of either one of the functions.

$$\therefore \quad \log_2 x = \frac{\log_4 x}{\log_4 2} \quad \text{but } \log_4 2 = \log_4 (4)^{\frac{1}{2}} = \frac{1}{2}\log_4 4 = \frac{1}{2}$$

$$= \frac{\log_4 x}{\frac{1}{2}}$$

$$\therefore \quad \log_2 x = 2\log_4 x$$

So $\log_2 x + 2\log_4 y = 4$ becomes $2\log_4 x + 2\log_4 y = 4$

$$\therefore \quad 2(\log_4 x + \log_4 y) = 4$$

$$\therefore \quad 2\log_4(xy) = 4$$

$$\therefore \quad \log_4(xy) = 2$$

$$\therefore \quad xy = 4^2$$

$$\therefore \quad xy = 16$$

If $\log_{10}(x+y) = 1$ then $x + y = 10^1 = 10$

So the equations to be solved are:

$$x + y = 10 \quad \text{---------} \quad (1)$$

and $\quad xy = 16 \quad \text{---------} \quad (2)$ and so $y = \frac{16}{x}$

Replacing $y = \frac{16}{x}$ from equation (2) into equation (1)

$$x + y = 10$$

$$x + \frac{16}{x} = 10$$

$$\therefore \quad x^2 + 16 = 10x$$

$$\therefore \quad x^2 - 10x + 16 = 0$$

$$(x - 2)(x - 8) = 0$$

either $\quad x - 2 = 0 \quad$ or $\quad x - 8 = 0$

$\therefore \quad x = 2 \quad\quad\quad \therefore \quad x = 8$

POWER SERIES AND EXPANSIONS PURE MATHEMATICS

and $\quad y = \dfrac{16}{x} = \dfrac{16}{2}\quad$ and $\quad y = \dfrac{16}{x} = \dfrac{16}{8}$

$\therefore \quad y = 8 \qquad\qquad \therefore \quad y = 2$

So the solutions of the two equations are $x = 2, y = 8$ and $x = 8, y = 2$.

Example 19

Find x and y given that $e^x + 3e^y = 3$ and $e^{2x} - 9e^{2y} = 6$, expressing each answer as a logarithm to base e.

Solution

$$e^x + 3e^y = 3 \quad \text{gives } e^x = 3 - 3e^y$$

but an expression for e^{2x} is required and so squaring both sides:

$$(e^x)^2 = (3 - 3e^y)^2$$

$$\therefore \quad e^{2x} = 9 - 18e^y + 9e^{2y}$$

Replacing this in the other equation:

$$(9 - 18e^y + 9e^{2y}) - 9e^{2y} = 6$$

$$\therefore \quad 9 - 18e^y + 9e^{2y} - 9e^{2y} = 6$$

$$9 - 18e^y = 6$$

$$-18e^y = 6 - 9 = -3$$

$$e^y = \dfrac{-3}{-18} = \dfrac{1}{6}$$

Taking logs of both sides: $\log_e e^y = \log_e\left(\dfrac{1}{6}\right)$

$\therefore \quad y\log_e e = \log_e 1 - \log_e 6$

but $\quad \log_e e = 1 \quad$ and $\quad \log_e 1 = 0$

$\therefore \quad y = -\log_e 6$

Replacing e^y by $\dfrac{1}{6}$ to find the value of x:

$$\therefore \quad e^x + 3 \times \dfrac{1}{6} = 3$$

$$e^x = 3 - \dfrac{1}{2} = \dfrac{5}{2}$$

Taking logs at both sides: $\log_e e^x = \log_e\left(\dfrac{5}{2}\right)$

$$x \log_e e = \log_e\left(\dfrac{5}{2}\right)$$

PURE MATHEMATICS   POWER SERIES AND EXPANSIONS

$$x = \log_e\left(\frac{5}{2}\right)$$

So the solution of the two equations is $x = \log_e\left(\frac{5}{2}\right)$, $y = -\log_e 6$

From these examples it can be seen that there is no one method of solving equations involving logarithmic and/or exponential functions. Each question needs to be studied carefully and the basic rules applied.

e)   *Example* ('A' level question)

Given that $\log_{10} 2 = p$ and $\log_{10} 3 = q$, solve for x in terms of p and q

(1) $\log_3 x = \log_x 2$

(2) $(6^{3-4x})(4^{x+4}) = 2$

Solution

(1) $\log_3 x = \log_x 2$ Using a change of base into base 10

$$\log_3 x = \frac{\log_{10} x}{\log_{10} 3} = \frac{\log_{10} x}{q}$$

$$\log_x 2 = \frac{\log_{10} 2}{\log_{10} x} = \frac{p}{\log_{10} x}$$

Since $\log_3 x = \log_x 2$ it follows that $\dfrac{\log_{10} x}{q} = \dfrac{p}{\log_{10} x}$

$\therefore \quad \log_{10} x \cdot \log_{10} x = pq$

$\quad\quad (\log_{10} x)^2 = pq$

$\therefore \quad \log_{10} x = \sqrt{pq}$

$\therefore \quad x = 10^{\sqrt{pq}}$

(2) $(6^{3-4x})(4^{x+4}) = 2$

Taking logarithms to base 10 on both sides gives:

$$\log_{10}[(6^{3-4x})(4^{x+4})] = \log_{10} 2$$

$$\log_{10}(6^{3-4x}) + \log_{10}(4^{x+4}) = p$$

$$(3-4x)\log_{10} 6 + (x+4)\log_{10} 4 = p$$

Since $6 = 2 \times 3$ it follows that $\log_{10} 6 = \log_{10}(2 \times 3)$

$\therefore \quad \log_{10} 6 = \log_{10} 2 + \log_{10} 3$

$$\log_{10} 6 = p + q$$

Also $4 = 2^2$ $\therefore$ $\log_{10} 4 = \log_{10}(2^2) = 2\log_{10} 2$

$\therefore \quad \log_{10} 4 = 2p$

Replacing for $\log_{10} 6$ and $\log_{10} 4$ in the above equation gives:

$$(3 - 4x)(p + q) + (x + 4)2p = p$$

$$3p - 4px + 3q - 4qx + 2px + 8p = p$$

$$-2px - 4qx + 11p + 3q = p$$

$\therefore \qquad -2(p + 2q)x = p - 11p - 3q$

$\therefore \qquad -2(p + 2q)x = -10p - 3q$

$\therefore \qquad x = \dfrac{-(10p + 3q)}{-2(p + 2q)}$

$\therefore \qquad x = \dfrac{(10p + 3q)}{2(p + 2q)}$

2.4 Expansions

a) *Exponential series*

In section 2.2 the binomial expansion was introduced for expanding functions such as $(a + bx)^n$ as a series of terms in powers of x. It is possible to expand other functions in a very similar way. At this stage these series expansions will be stated and examples will be given on using the expansions. Later it will be possible to deduce these results.

Taking the exponential series first, this can be expanded as:

$$e^x = 1 + x + \frac{x^2}{2!} + \frac{x^3}{3!} + \dots$$

and this expansion is valid for all values of x.

Example 20

Expand the following functions as far as the fourth non-zero term:

(1) $\quad e^{x^3}$

(2) $\quad (1 + x)e^x$

(3) $\quad (1 + e^x)(1 + e^{2x})$

PURE MATHEMATICS   POWER SERIES AND EXPANSIONS

Solution

(1) e^{x^3}: Using the expansion for e^x but replacing x with x^3.

$$\therefore \quad e^{x^3} = 1 + x^3 + \frac{(x^3)^2}{2!} + \frac{(x^3)^3}{3!} + \ldots$$

$$= 1 + x^3 + \frac{x^6}{2!} + \frac{x^9}{3!} + \ldots$$

---

(2) $(1 + x)e^x$: Using the expansion for e^x and then multiplying through by $(1 + x)$.

$$\therefore \quad (1 + x)e^x = (1 + x)(1 + x + \frac{x^2}{2!} + \frac{x^3}{3!} + \ldots)$$

$$= 1 + x + \frac{x^2}{2!} + \frac{x^3}{3!} + x + x^2 + \frac{x^3}{2!} + \frac{x^4}{3!} \text{ (ignore)} \ldots$$

$$= 1 + 2x + \frac{3x^2}{1} + \frac{4x^3}{6} + \ldots$$

$$\therefore \quad (1 + x)e^x = 1 + 2x + \frac{3x^2}{2} + \frac{2x^3}{3} + \ldots$$

---

(3) $(1 + e^x)(1 + e^{2x})$: Before using the expansion it is best to multiply out the brackets.

$$(1 + e^x)(1 + e^{2x}) = 1 + e^x + e^{2x} + e^x \cdot e^{2x}$$

$$= 1 + e^x + e^{2x} + 3^{3x}$$

$$e^x = 1 + x + \frac{x^2}{2!} + \frac{x^3}{3!} + \ldots \quad = 1 + x + \frac{x^2}{2} + \frac{x^3}{6} + \ldots$$

$$e^{2x} = 1 + 2x + \frac{(2x)^2}{2!} + \frac{(2x)^3}{3!} + \ldots \quad = 1 + 2x + \frac{4x^2}{2} + \frac{8x^3}{6} + \ldots$$

$$e^{3x} = 1 + 3x + \frac{(3x)^2}{2!} + \frac{(3x)^3}{3!} + \ldots \quad = 1 + 3x + \frac{9x^2}{2} + \frac{27x^3}{6} + \ldots$$

Now collecting together like terms gives:

$$1 + e^x + e^{2x} + e^{3x} = 1 + (1 + 1 + 1) + (x + 2x + 3x) + \left(\frac{x^2}{2} + \frac{4x^2}{2} + \frac{9x^2}{2}\right)$$

$$+ \left(\frac{x^3}{6} + \frac{8x^3}{6} + \frac{27x^3}{6}\right) + \ldots$$

$$\therefore \quad (1 + e^x)(1 + e^{2x}) = 4 + 6x + 7x^2 + 6x^3 + \ldots$$

---

b)   *Logarithmic series*

It is necessary to give a series expansion for $\log_e (1 + x)$ and not for $\log_e x$ as no expansion exists for this function.

$$\boxed{\log_e (1 + x) = x - \frac{x^2}{2} + \frac{x^3}{3} - \frac{x^4}{4} + \ldots}$$

POWER SERIES AND EXPANSIONS PURE MATHEMATICS

However, this expansion is only valid for values of x in the range $-1 < x \leq +1$.

If x is replaced with -x the following expansion is found:

$$\log_e(1-x) = -x - \frac{x^2}{2} - \frac{x^3}{3} - \frac{x^4}{4} \ldots$$

This expansion is valid in the range $-1 \leq x < +1$.

Example 21

Expand the following functions as far as the fourth term, and state the range of values of x for which the expansion is valid:

(1) $\log_e(2 - 5x)$

(2) $\log_e\left(\frac{3+x}{3-x}\right)$

(3) $\log_e(1 - x^2)$

Solution

(1) $\log_e(2 - 5x)$: It is first necessary to factorise '2' from the bracket so that the inside of the bracket starts with 1, ie

$$(2 - 5x) = 2\left(1 - \frac{5x}{2}\right)$$

$$\therefore \log_e(2 - 5x) = \log_e 2\left(1 - \frac{5x}{2}\right) = \log_e 2 + \log_e\left(1 - \frac{5x}{2}\right)$$

$$\log_e\left(1 - \frac{5x}{2}\right) = -\left(\frac{5x}{2}\right) - \frac{\left(\frac{5x}{2}\right)^2}{2} - \frac{\left(\frac{5x}{2}\right)^3}{3} - \ldots$$

$$= -\frac{5x}{2} - \frac{25x^2}{8} - \frac{125x^3}{24} - \ldots$$

$$\therefore \log_e(2 - 5x) = \log_e 2 + \left(-\frac{5x}{2} - \frac{25x^2}{8} - \frac{125x^3}{24} - \ldots\right)$$

$$= \log_e 2 - \frac{5x}{2} - \frac{25x^2}{8} - \frac{125x^3}{24}$$

This is valid for $-1 \leq \frac{5x}{2} < +1$

Multiplying by $\frac{2}{5}$ gives $-\frac{2}{5} \leq x < \frac{2}{5}$

PURE MATHEMATICS   POWER SERIES AND EXPANSIONS

(2) $\log_e\left(\dfrac{3+x}{3-x}\right)$: First it is necessary to factorise 3 from each bracket.

$$\therefore \left(\dfrac{3+x}{3-x}\right) = \dfrac{3\left(1+\dfrac{x}{3}\right)}{3\left(1-\dfrac{x}{3}\right)}$$

$$\log_e\left(\dfrac{3+x}{3-x}\right) = \log_e\left(\dfrac{1+\dfrac{x}{3}}{1-\dfrac{x}{3}}\right) = \log_e\left(1+\dfrac{x}{3}\right) - \log_e\left(1-\dfrac{x}{3}\right)$$

But: $\log\left(1+\dfrac{x}{3}\right) = \dfrac{x}{3} - \dfrac{\left(\dfrac{x}{3}\right)^2}{2} + \dfrac{\left(\dfrac{x}{3}\right)^3}{3} - \dfrac{\left(\dfrac{x}{3}\right)^4}{4} + \ldots$

$$= \dfrac{x}{3} - \dfrac{x^2}{18} + \dfrac{x^3}{81} - \dfrac{x^4}{324} + \ldots$$

and $\log_e\left(1-\dfrac{x}{3}\right) = -\dfrac{x}{3} - \dfrac{x^2}{18} - \dfrac{x^3}{81} - \dfrac{x^4}{324} - \ldots$

$\therefore \log_e\left(1+\dfrac{x}{3}\right) - \log_e\left(1-\dfrac{x}{3}\right) = \left(\dfrac{x}{3} - \dfrac{x^2}{18} + \dfrac{x^3}{81} - \dfrac{x^4}{324} + \ldots\right) - \left(-\dfrac{x}{3} - \dfrac{x^2}{18} - \dfrac{x^3}{81} - \dfrac{x^4}{324} - \ldots\right)$

$$= \dfrac{x}{3} + \dfrac{x}{3} - \dfrac{x^2}{18} + \dfrac{x^2}{18} + \dfrac{x^3}{81} + \dfrac{x^3}{81} - \dfrac{x^4}{324} + \dfrac{x^4}{324} \ldots$$

$\therefore \log_e\left(\dfrac{3+x}{3-x}\right) = \dfrac{2x}{3} + \dfrac{2x^3}{81} + \ldots$

$\log_e(3+x)$ is valid for $-1 < \dfrac{x}{3} \le +1$, ie $-3 < x \le +3$.

$\log_e(3-x)$ is valid for $-1 \le \dfrac{x}{3} < +1$, ie $-3 \le x < +3$.

So, to satisfy both inequalities it follows that the range must be $-3 < x < +3$.

---

(3) $\log_e(1-x^2)$: This time it is necessary to factorise the function of x itself, ie

$(1-x^2) = (1-x)(1+x)$

$\log_e(1-x^2) = \log_e[(1-x)(1+x)] = \log_e(1-x) + \log_e(1+x)$

But $\log_e(1-x) = -x - \dfrac{x^2}{2} - \dfrac{x^3}{3} - \dfrac{x^4}{4} - \ldots$

and $\log_e(1+x) = x - \dfrac{x^2}{2} + \dfrac{x^3}{3} - \dfrac{x^4}{4} \ldots$

$\therefore \log_e(1-x) + \log_e(1+x) = \left(-x - \dfrac{x^2}{2} - \dfrac{x^3}{3} - \dfrac{x^4}{4} - \ldots\right) + \left(x - \dfrac{x^2}{2} + \dfrac{x^3}{3} - \dfrac{x^4}{4} \ldots\right)$

$$= -x + x - \dfrac{x^2}{2} - \dfrac{x^2}{2} - \dfrac{x^3}{3} + \dfrac{x^3}{3} - \dfrac{x^4}{4} - \dfrac{x^4}{4} + \ldots$$

POWER SERIES AND EXPANSIONS — PURE MATHEMATICS

$$= -\frac{2x^2}{2} - \frac{2x^4}{4} - \ldots$$

$$\therefore \quad \log_e(1 - x^2) = -x^2 - \frac{x^4}{2} - \ldots$$

$\log_e(1 + x)$ is valid for $-1 < x \leq +1$

$\log_e(1 - x)$ is valid for $-1 \leq x < +1$

So to satisfy both inequalities it follows that the range must be $-1 < x < +1$.

c) *Trigonometric series*

It is possible to expand both sin x and cos x as a series:

$$\sin x = x - \frac{x^3}{3!} + \frac{x^5}{5!} - \frac{x^7}{7!} + \ldots$$

$$\cos x = 1 - \frac{x^2}{2!} + \frac{x^4}{4!} - \frac{x^6}{6!} + \ldots$$

However the angles must be expressed in radians not degrees. These are explained in detail in the next chapter.

These expansions are valid for all values of x.

Example 22

Expand the following functions as far as the fourth term:

(1) $\cos x^2$

(2) $\sin \frac{x}{2}$

Solution

(1) $\cos x^2$: Replacing x with x^2 in the expansion of cos x gives

$$\cos x^2 = 1 - \frac{(x^2)^2}{2!} + \frac{(x^2)^4}{4!} - \frac{(x^2)^6}{6!} + \ldots$$

$$\therefore \quad \cos x^2 = 1 - \frac{x^4}{2!} + \frac{x^8}{4!} - \frac{x^{12}}{6!} + \ldots$$

(2) $\sin \frac{x}{2}$: Replacing x with $\frac{x}{2}$ in the expansion of sin x gives:

$$\sin \frac{x}{2} = \frac{x}{2} - \frac{\left(\frac{x}{2}\right)^3}{3!} + \frac{\left(\frac{x}{2}\right)^5}{5!} - \frac{\left(\frac{x}{2}\right)^7}{7!} + \ldots$$

$$\therefore \quad \sin \frac{x}{2} = \frac{x}{2} - \frac{x^3}{48} + \frac{x^5}{3840} + \frac{x^7}{645120} + \ldots$$

PURE MATHEMATICS — POWER SERIES AND EXPANSIONS

d) *Approximate calculations using these expansions*

Just as the binomial expansion can be used to calculate approximate numerical values so too can the other expansions given in this section. In particular the exponential and logarithmic series have many such applications.

Example 23

Using the result that:

$$\log_e \sqrt{\frac{n}{n-1}} = \frac{1}{2n-1} + \frac{1}{3(2n-1)^3} + \frac{1}{5(2n-1)^5} + \dots$$

calculate an approximate value for $\log_e \sqrt{1.5}$ to 4 decimal places.

Solution

If $\sqrt{\frac{n}{n-1}} = \sqrt{1.5} = \sqrt{\frac{3}{2}}$ it follows that n must be replaced by 3 in the given series, and therefore $2n-1$ replaced by 5.

$$\therefore \quad \log_e \sqrt{1.5} = \frac{1}{5} + \frac{1}{3 \cdot 5^3} + \frac{1}{5 \cdot 5^5} + \dots$$

$$= 0.2 + 0.00267 + 0.00006$$

$$= 0.20273$$

$\therefore \quad \log_e \sqrt{1.5} = 0.2027$ (to 4 decimal places).

Example 24

Use the expansion for e^x to find an approximate value for e to 4 decimal places.

Solution

$$e^x = 1 + x + \frac{x^2}{2!} + \frac{x^3}{3!} + \frac{x^4}{4!} + \frac{x^5}{5!} + \frac{x^6}{6!} + \frac{x^7}{7!} + \frac{x^8}{8!} + \dots$$

Replacing x with 1 will give a series for e^1 (or e) which can be summed to give an approximate value for e.

$$\therefore \quad e^1 = 1 + 1 + \frac{1^2}{2} + \frac{1^3}{6} + \frac{1^4}{24} + \frac{1^5}{120} + \frac{1^6}{720} + \frac{1^7}{5040} + \frac{1^8}{40320}$$

$$= 1 + 1 + 0.5 + 0.16667 + 0.04167 + 0.00833 + 0.00139 + 0.00020 + 0.00002$$

$$= 2.71828$$

$\therefore \quad e = 2.7183$ (to 4 decimal places).

e) *Example ('A' level question)*

Write down in ascending powers of y the first four terms in the series for e^y. By taking $e^y = 2^x$ show that:

$$2^x = 1 + x \log_e 2 + x^2 \frac{(\log_e 2)^2}{2} + x^3 \frac{(\log_e 2)^3}{6} + \dots$$

POWER SERIES AND EXPANSIONS   PURE MATHEMATICS

Given that $2^{3x} + 5(2^x) = 6 + Ax + Bx^2 + Cx^3 + ...$ for all real x, find the values of the constants A, B and C in terms of $\log_e 2$.

Solution

i) $\quad e^y = 1 + y + \dfrac{y^2}{2!} + \dfrac{y^3}{3!} + ...$

If $e^y = 2^x$: taking logs to base e of both sides

$\log_e e^y = \log_e 2^x \therefore y \log_e e = x \log_e 2$ but $\log_e e = 1$

$\therefore \quad y = x \log_e 2$; replacing in the above expression

gives $\quad e^y = 1 + x \log_e 2 + \dfrac{(x\log_e 2)^2}{2} + \dfrac{(x\log_e 2)^3}{6} + ...$

$\therefore \quad 2^x = 1 + x\log_e 2 + x^2 \cdot \dfrac{(\log_e 2)^2}{2} + x^3 \cdot \dfrac{(\log_e 2)^3}{6} + ...$

ii) $\quad 2^{3x} + 5(2^x) = 6 + Ax + Bx^2 + Cx^3 + ...$

It is first necessary to find a series expansion for 2^{3x}, using the above expansion for 2^x.

$\therefore \quad 2^{3x} = 1 + 3x\log_e 2 + \dfrac{(3x\log_e 2)^2}{2!} + \dfrac{(3x\log_e 2)^3}{6} + ...$

$\therefore \quad 2^{3x} = 1 + x.3\log_e 2 + x^2 \cdot \dfrac{9(\log_e 2)^2}{2} + x^3 \cdot \dfrac{27(\log_e 2)^3}{6} + ...$

and $\quad 5(2^x) = 5(1 + x.\log_e 2 + x^2 \cdot \dfrac{(\log_e 2)^2}{2} + x^3 \cdot \dfrac{(\log_e 2)^3}{6} + ...$

$= 5 + x.5\log_e 2 + x^2 \cdot \dfrac{5(\log_e 2)^2}{2} + x^3 \cdot \dfrac{5(\log_e 2)^3}{6} + ...$

Adding together like terms gives:

$2^{3x} + 5(2^x) = 1 + 5 + x3\log_e 2 + x5\log_e 2 + x^2 \cdot \dfrac{9(\log_e 2)^2}{2} + x^2 \cdot \dfrac{5(\log_e 2)^2}{2}$
$\qquad + x^3 \cdot \dfrac{27(\log_e 2)^3}{6} + x^3 \cdot \dfrac{5(\log_e 2)^3}{6}$

$= 6 + x(3\log_e 2 + 5\log_e 2) + x^2 \left(\dfrac{9(\log_e 2)^2}{2} + \dfrac{5(\log_e 2)^2}{2} \right) + x^3 \left(\dfrac{27(\log_e 2)^3}{6} + \dfrac{5(\log_e 2)^3}{6} \right) + ...$

$\therefore \quad 2^{3x} + 5(2^x) = 6 + x.8\log_e 2 + x^2.7(\log_e 2)^2 + x^3.\dfrac{16}{3}(\log_e 2)^3$

PURE MATHEMATICS — POWER SERIES AND EXPANSIONS

Comparing this result with

$$2^{3x} + 5(2^x) = 6 + Ax + Bx^2 + Cx^3 + \ldots$$

gives $\quad Ax = x \cdot 8\log_e 2 \qquad \therefore \quad A = 8\log_e 2$

$\quad\quad\quad Bx^2 = x^2 7(\log_e 2)^2 \qquad \therefore \quad B = 7(\log_e 2)^2$

$\quad\quad\quad Cx^3 = x^3 \dfrac{16}{3}(\log_e 2)^3 \qquad \therefore \quad C = \dfrac{16}{3}(\log_e 2)^3$

TRIGONOMETRY   PURE MATHEMATICS

3   **TRIGONOMETRY**

     3.1   Basic trigonometry
     3.2   Angle formulae and applications
     3.3   Trigonometrical equations
     3.4   Solutions of triangles

3.1   Basic trigonometry

a)   *Sine, cosine, tangent*

Most people will have met the definitions of these ratios in earlier work. However, they have been included at this stage to give a complete introduction to trigonometry.

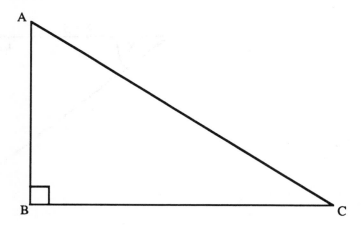

It must be stressed that these definitions only apply to right-angled triangles such as ABC above.

$$\text{Sine} = \frac{\text{opposite}}{\text{hypotenuse}} = \frac{O}{H}$$

$$\text{Cosine} = \frac{\text{adjacent}}{\text{hypotenuse}} = \frac{A}{H}$$

$$\text{Tangent} = \frac{\text{opposite}}{\text{adjacent}} = \frac{O}{A}$$

$\therefore \quad \sin A = \dfrac{BC}{AC} \qquad \cos A = \dfrac{AB}{AC} \qquad \tan A = \dfrac{BC}{AB}$

and $\quad \sin C = \dfrac{AB}{AC} \qquad \cos C = \dfrac{BC}{AC} \qquad \tan C = \dfrac{AB}{BC}$

Another result to note is that $\dfrac{\sin A}{\cos A} = \dfrac{\frac{BC}{AC}}{\frac{AB}{AC}} = \dfrac{BC}{AB} = \tan A$

and also $\dfrac{\sin C}{\cos C} = \dfrac{\frac{AB}{AC}}{\frac{BC}{AC}} = \dfrac{AB}{BC} = \tan C$

PURE MATHEMATICS   TRIGONOMETRY

Example 1

In each of the following triangles find the angle X:

(1) 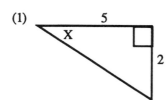   (2) 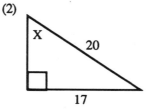   (3)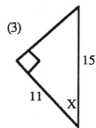

Solution

(1)

Using $\tan = \dfrac{O}{A}$

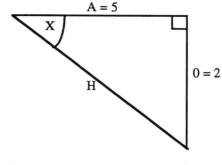

then $\tan X = \dfrac{2}{5} = 0.4$

∴ $X = 21.8°$ or $21° 48'$

This can also be written as $X = \tan^{-1} 0.4$

ie $21.8°$ is the angle whose tangent is 0.4.

(2)

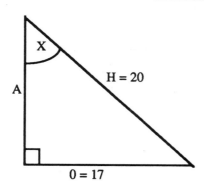

Using $\sin = \dfrac{O}{H}$

then $\sin x = \dfrac{17}{20} = 0.85$

∴ $X = 58.2°$ or $58° 18'$

This can also be written as $X = \sin^{-1} 0.85$

ie $58.2°$ is the angle whose sine is 0.85.

(3)

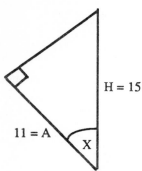

Using $\cos = \dfrac{A}{H}$

then $\cos X = \dfrac{11}{15} = 0.7333$

$\therefore \quad X = 42.8°$ or $42°\ 50'$

This can also be written as $X = \cos^{-1} 0.7333$

ie $42.8°$ is the angle whose cosine is 0.7333.

b) *Angles of any size*

For angles between 0° and 90° sine, cosine and tangent are all positive. However, for angles between 90° and 360° the rules of sign are best summarised in a diagram. It is assumed that angles are measured in an anti-clockwise direction from 0°.

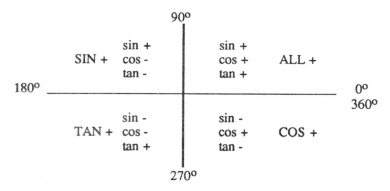

Another assumption is that angles are always measured from the horizontal. This will be best explained by means of a numerical example.

Example 2

Find the sin, cos and tan for each of the following angles:

(1) 75°

(2) 110°

(3) 225°

(4) 303°

Solution

(1)

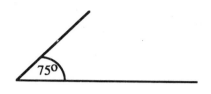

$\sin 75° = 0.9659 \qquad \cos 75° = 0.2588 \qquad \tan 75° = 3.7321$

(2)

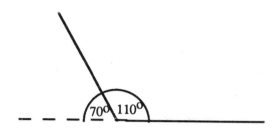

$\sin 110° = \sin 70° \qquad \cos 110° = -\cos 70° \qquad \tan 110° = -\tan 70°$
$\qquad\quad = 0.9397 \qquad\qquad\quad = -0.3420 \qquad\qquad\quad = -2.7475$

$(\because 180° - 110° = 70°)$

(3)

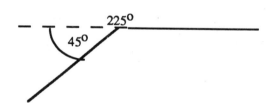

$\sin 225° = -\sin 45° \qquad \cos 225° = -\cos 45° \qquad \tan 225° = \tan 45°$
$\qquad\quad = -0.7071 \qquad\qquad\quad = -0.7071 \qquad\qquad\quad = 1$

$(\because 225° - 180° = 45°)$

(4)

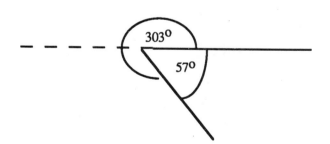

$\sin 303° = -\sin 57° \qquad \cos 303° = \cos 57° \qquad \tan 303° = -\tan 57°$
$\qquad\quad = -0.8387 \qquad\qquad\quad = 0.5446 \qquad\qquad\quad = -1.5399$

$(\because 360° - 303° = 57°)$

TRIGONOMETRY   PURE MATHEMATICS

c)   *Graphs*

It is useful to know what the graphs of sin x°, cos x° and tan x° look like. This will be particularly useful later when the methods of solving trigonometrical equations are explained as these can involve drawing graphs of trigonometrical functions.

(a) $y = \sin x°$

Taking values of x at 30° intervals gives:

x°	0	30	60	90	120	150	180	210	240	270	300	330	360°
y = sin x	0	0.5	0.866	0.5	0.866	0.5	0	-0.5	-0.866	-1	-0.866	-0.5	0

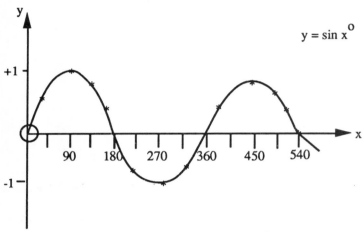

The graph repeats this pattern every 360°

(b) $y = \cos x°$

Taking values of x as before

x°	0	30	60	90	120	150	180	210	240	270	300	330	360°
y = cos x	1	0.866	0.5	0	-0.5	-0.866	-1	-0.866	-0.5	0	0.5	0.866	1

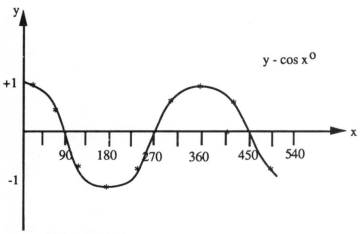

This graph also repeats every 360°

PURE MATHEMATICS — TRIGONOMETRY

(c) $y = \tan x°$

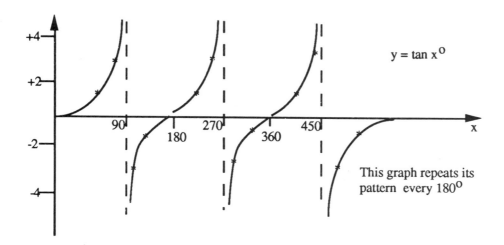

$x°$	0	30	60	90	120	150	180	210	240	270	300	330	360°
$y = \tan x$	0	0.577	1.732	∞	-1.732	-0.577	0	0.577	1.732	∞	-1.732	-0.577	0

This graph repeats its pattern every 180°

d) *Application of Pythagoras theorem*

(a)   Sometimes the trigonometrical ratios are required for 30°, 45°, 60° without using tables or calculators.

The method of obtaining these results is detailed below and the actual results themselves should be memorised for future use.

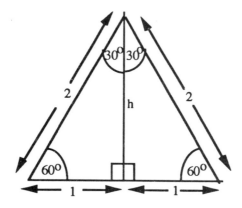

This is an equilateral triangle with sides of length 2 units (and angles of 60° of course). The height, h, of the triangle has also been drawn in.

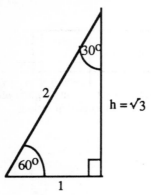

By applying Pythagoras theorem to this triangle it follows that $h = \sqrt{3}$, ie

$$1^2 + h^2 = 2^2$$

$\therefore \quad 1 + h^2 = 4$

$\therefore \quad h^2 = 4 - 1 = 3$

$\therefore \quad h = \sqrt{3}$

Therefore, from the diagram it follows that:

$\sin 60° = \dfrac{\sqrt{3}}{2}$	$\cos 60° = \dfrac{1}{2}$	$\tan 60° = \sqrt{3}$
and $\sin 30° = \dfrac{1}{2}$	$\cos 30° = \dfrac{\sqrt{3}}{2}$	$\tan 30° = \dfrac{1}{\sqrt{3}}$

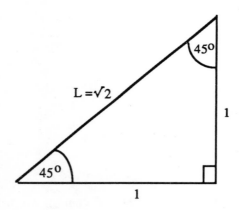

This is a right angled isosceles triangle with two sides of length 1 unit. By applying Pythagoras theorem again it follows that the length of the hypotenuse $L = \sqrt{2}$, ie

$$1^2 + 1^2 = L^2$$

$\therefore \quad 1 + 1 = L^2$

$\therefore \quad 2 = L^2$

$\therefore \quad \sqrt{2} = L$

PURE MATHEMATICS — TRIGONOMETRY

Therefore, from the diagram it follows that:

| $\sin 45° = \dfrac{1}{\sqrt{2}}$ | $\cos 45° = \dfrac{1}{\sqrt{2}}$ | $\tan 45° = 1$ |

(b) Some further *very* important results arise from applying Pythagoras theorem to any right angled triangle. These results provide relationships between certain of the trigonometrical ratios.

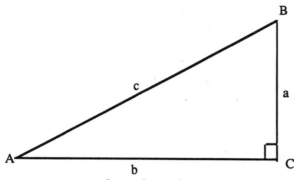

By Pythagoras: $a^2 + b^2 = c^2$

Also $\sin A = \dfrac{a}{c}$ $\cos A = \dfrac{b}{c}$ $\tan A = \dfrac{a}{b}$

There are three other trigonometrical ratios which will now be introduced. They are cosec, sec and cot where:

| $\operatorname{cosec} A = \dfrac{1}{\sin A} = \dfrac{c}{a}$ $\sec A = \dfrac{1}{\cos A} = \dfrac{c}{b}$ $\cot A = \dfrac{1}{\tan A} = \dfrac{b}{a}$ |

Taking $a^2 + b^2 = c^2$ and dividing through by c^2

gives $\dfrac{a^2}{c^2} + \dfrac{b^2}{c^2} = \dfrac{c^2}{c^2}$

∴ $\left(\dfrac{a}{c}\right)^2 + \left(\dfrac{b}{c}\right)^2 = 1$ and this becomes

$(\sin A)^2 + (\cos A)^2 = 1$ which is usually written as

| $\sin^2 A + \cos^2 A = 1$ |

Taking $a^2 + b^2 = c^2$ and dividing through by a^2

gives $\dfrac{a^2}{a^2} + \dfrac{b^2}{a^2} = \dfrac{c^2}{a^2}$

∴ $1 + \left(\dfrac{b}{a}\right)^2 = \left(\dfrac{c}{a}\right)^2$ and this becomes

$1 + (\cot A)^2 = (\operatorname{cosec} A)^2$ which is usually written as

| $1 + \cot^2 A = \operatorname{cosec}^2 A$ |

TRIGONOMETRY   PURE MATHEMATICS

Finally, taking $a^2 + b^2 = c^2$ and dividing through by b^2

$$\frac{a^2}{b^2} + \frac{b^2}{b^2} = \frac{c^2}{b^2}$$

$$\therefore \quad \left(\frac{a}{b}\right)^2 + 1 = \left(\frac{c}{b}\right)^2$$

$(\tan A)^2 + 1 = (\sec A)^2$ which is usually written as

$$\boxed{\tan^2 A + 1 = \sec^2 A}$$

It cannot be stressed too much how very important these three identities are: THEY MUST BE LEARNED.

e) *Example* ('A' level question)

A man walks North. When he is at a point A he sees a pole on a bearing of 40°. After walking 200 m he is at the point B from which the bearing of the pole is 70°. Find, to the nearest metre, the distance of the pole from:

(1) the man's path;

(2) the mid-point of AB.

Solution

Before attempting the calculations it is essential to draw a diagram containing all the above information.

(1)

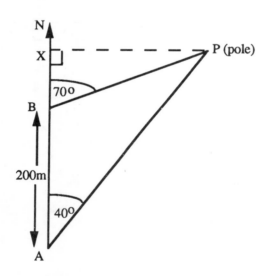

The distance of the pole from the man's path is the perpendicular distance PX as shown on the diagram. Using △APX gives:

$$\tan 40° = \frac{PX}{AX} = \frac{PX}{AB + BX} = \frac{PX}{200 + BX}; \text{ but } \tan 40° = 0.8391$$

$$\therefore \quad 0.8391 = \frac{PX}{200 + BX}; \text{ multiplying both sides by } (200 + BX)$$

$$\therefore \quad 0.8291 (200 + BX) = PX; \text{ multiplying through the bracket}$$

$$\therefore \quad 167.82 + 0.8391 BX = PX$$

PURE MATHEMATICS   TRIGONOMETRY

Since this equation contains 2 unknowns - PX and BX - it is necessary to find another relationship between these lengths.

Using $\triangle$BPX gives

$$\tan 70° = \frac{PX}{BX}; \text{ but } \tan 70° = 2.7475$$

$$\therefore \quad 2.7475 = \frac{PX}{BX}; \text{ multiplying both sides by BX}$$

$$\therefore \quad 2.7475 BX = PX; \text{ dividing both sides by 2.7475}$$

$$\therefore \quad BX = \frac{PX}{2.7475}$$

This can now be used to replace BX in the first equation.

$$167.82 + 0.8391 \left(\frac{PX}{2.7475}\right) = PX$$

$$167.82 + 0.3054 PX = PX$$

$$\therefore \quad 167.82 = PX - 0.3054 PX$$

$$\therefore \quad 167.82 = PX (1 - 0.3054)$$

$$\therefore \quad 167.82 = PX (0.6946)$$

$$\therefore \quad \frac{167.82}{0.6946} = PX$$

$$242 = PX$$

So the required distance is 242 m to the nearest metre.

---

(2)

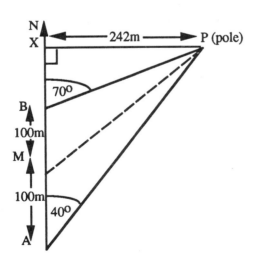

The distance to be calculated this time is PM. If the distance BX is calculated then Pythagoras theorem can be used to find PM.

$$BX = \frac{PX}{2.7475} = \frac{242}{2.7475} = 88 \text{ m}$$

∴   MX = MB + BX

∴   MX = 100 + 88 = 188 m

Using Pythagoras gives:

$PM^2 = 188^2 + 242^2$

$= 35344 + 58564$

$= 93908$

∴ $PM = \sqrt{93908} = 306$

Therefore, the required distance is 306 m to the nearest metre.

(*Note*: There are alternative ways of arriving at these answers, by using different trigonometrical ratios.)

3.2 Angle formulae and applications

a)   *Compound angle formulae*

The following formulae are very important and will be used in solving many problems that will arise in later work. It is essential that they are learnt and practised many times. No attempt is made to prove these formulae which are known as the compound angle formulae.

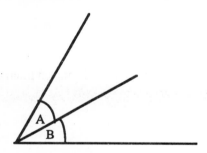

| sin (A + B) = sin A . cos B + cos A . sin B |
| cos (A + B) = cos A . cos B - sin A . sin B |

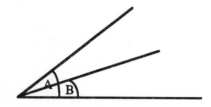

| sin (A - B) = sin A . cos B - cos A . sin B |
| cos (A - B) = cos A . cos B + sin A . sin B |

The formulae for tan (A + B) and tan (A - B) are derived from the above using $\tan = \frac{\sin}{\cos}$

∴ $\tan(A + B) = \dfrac{\sin(A + B)}{\cos(A + B)} = \dfrac{\sin A . \cos B + \cos A . \sin B}{\cos A . \cos B - \sin A . \sin B}$

Dividing every term by cos A.cos B gives:

$$\tan(A+B) = \frac{\frac{\sin A.\cos B}{\cos A.\cos B} + \frac{\cos A.\sin B}{\cos A.\cos B}}{\frac{\cos A.\cos B}{\cos A.\cos B} - \frac{\sin A.\sin B}{\cos A.\cos B}}$$

$$= \frac{\frac{\sin A}{\cos A} + \frac{\sin B}{\cos B}}{1 - \frac{\sin A}{\cos A} \cdot \frac{\sin B}{\cos B}}$$

$$\therefore \boxed{\tan(A+B) = \frac{\tan A + \tan B}{1 - \tan A.\tan B}} \quad \text{Similarly}$$

$$\boxed{\tan(A-B) = \frac{\tan A - \tan B}{1 + \tan A.\tan B}}$$

Example 3

If $\sin A = \frac{3}{5}$ and $\sin B = \frac{5}{13}$ where A and B are acute angles, find, without using tables, the values of:

(1) $\sin(A+B)$

(2) $\cos(A+B)$

(3) $\tan(A+B)$

Solution

If $\sin A = \frac{3}{5}$ then cos A and tan A can be found by applying Pythagoras theorem to a right angled triangle as follows:

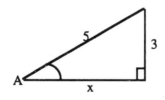

$3^2 + x^2 = 5^2$

$9 + x^2 = 25$

$x^2 = 25 - 9$

$x^2 = 16$

$\therefore x = \sqrt{16} = 4$

$\therefore$ If $\sin A = \frac{3}{5}$, $\cos A = \frac{4}{5}$, $\tan A = \frac{3}{4}$

TRIGONOMETRY — PURE MATHEMATICS

Similarly

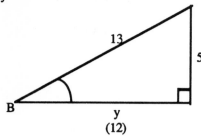

$5^2 + y^2 = 13^2$

$25 + y^2 = 169$

$y^2 = 169 - 25$

$y^2 = 144$

$y = \sqrt{144} = 12$

$\therefore$ If $\sin B = \dfrac{5}{13}$, $\cos B = \dfrac{12}{13}$, $\tan B = \dfrac{5}{12}$

(1) $\sin(A+B) = \sin A \cdot \cos B + \cos A \cdot \sin B$

$$= \dfrac{3}{5} \cdot \dfrac{12}{13} + \dfrac{4}{5} \cdot \dfrac{5}{13} = \dfrac{36}{65} + \dfrac{20}{65}$$

$\therefore \sin(A+B) = \dfrac{56}{65}$

---

(2) $\cos(A+B) = \cos A \cdot \cos B - \sin A \cdot \sin B$

$$= \dfrac{4}{5} \cdot \dfrac{12}{13} - \dfrac{3}{5} \cdot \dfrac{5}{13} = \dfrac{48}{65} - \dfrac{15}{65}$$

$\therefore \cos(A+B) = \dfrac{33}{65}$

---

(3) $\tan(A+B) = \dfrac{\tan A + \tan B}{1 - \tan A \cdot \tan B}$

$$= \dfrac{\dfrac{3}{4} + \dfrac{5}{12}}{1 - \dfrac{3}{4} \cdot \dfrac{5}{12}} = \dfrac{\dfrac{9+5}{12}}{\dfrac{48-15}{48}}$$

$$= \dfrac{\dfrac{14}{12}}{\dfrac{33}{48}} = \dfrac{\dfrac{56}{48}}{\dfrac{33}{48}}$$

$\therefore \tan(A+B) = \dfrac{56}{33}$

PURE MATHEMATICS   TRIGONOMETRY

b)   *Double angle formulae*

This is a special case which occurs when angles A and B are equal.

$\therefore$ If $A = B$ then $A + B = 2A$ and $A - B = 0$

$\therefore$ $\sin(A + A) = \sin A . \cos A + \cos A . \sin A$

$$\boxed{\therefore \sin 2A = 2 \sin A . \cos A}$$

$\cos(A + A) = \cos A . \cos A - \sin A . \sin A$

$\therefore$ $\cos 2A = \cos^2 A - \sin^2 A$

However, this can be written in two alternative ways using $\sin^2 A + \cos^2 A = 1$, which gives $\sin^2 A = 1 - \cos^2 A$.

$\cos 2A = \cos^2 A - \sin^2 A$

$\therefore$ $\cos 2A = \cos^2 A - (1 - \cos^2 A)$

             $= \cos^2 A - 1 + \cos^2 A$

$\therefore$ $\cos 2A = 2\cos^2 A - 1$

Using $\sin^2 A + \cos^2 A = 1$ (again!) giving $\cos^2 A = 1 - \sin^2 A$

$\cos 2A = \cos^2 A - \sin^2 A$

       $= (1 - \sin^2 A) - \sin^2 A$

$\therefore$ $\cos 2A = 1 - 2\sin^2 A$

So the following three formulae go together:

$$\boxed{\begin{array}{l} \cos 2A = \cos^2 A - \sin^2 A \\ \cos 2A = 2\cos^2 A - 1 \\ \cos 2A = 1 - 2\sin^2 A \end{array}}$$

Finally $\tan(A + A) = \dfrac{\tan A + \tan A}{1 - \tan A . \tan A}$

$$\boxed{\therefore \tan 2A = \frac{2\tan A}{1 - \tan^2 A}}$$

Example 4

Prove the following identities:

(1) $\sin 3A = 3\sin A - 4\sin^3 A$

(2) $\cos 3A = 4\cos^3 A - 3\cos A$

(3)   If $2A + B = 45°$ show that $\tan B = \dfrac{1 - 2\tan A - \tan^2 A}{1 + 2\tan A - \tan^2 A}$

TRIGONOMETRY   PURE MATHEMATICS

Solution

(1) $\sin 3A = \sin(2A + A)$

$= \sin 2A \cdot \cos A + \cos 2A \cdot \sin A$

$= (2\sin A \cdot \cos A)\cos A + (1 - 2\sin^2 A)\sin A$

$= 2\sin A \cdot \cos^2 A + \sin A - 2\sin^3 A$

$= 2\sin A(1 - \sin^2 A) + \sin A - 2\sin^3 A$

$= 2\sin A - 2\sin^3 A + \sin A - 2\sin^3 A$

$\therefore \sin 3A = 3\sin A - 4\sin^3 A$

---

(2) $\cos 3A = \cos(2A + A)$

$= \cos 2A \cdot \cos A - \sin 2A \cdot \sin A$

$= (2\cos^2 A - 1)\cos A - (2\sin A \cdot \cos A)\sin A$

$= 2\cos^3 A - \cos A - 2\sin^2 A \cdot \cos A$

$= 2\cos^3 A - \cos A - 2(1 - \cos^2 A)\cos A$

$= 2\cos^3 A - \cos A - 2\cos A + 2\cos^3 A$

$\therefore \cos 3A = 4\cos^3 A - 3\cos A$

---

(3) If $2A + B = 45°$

then $2A = 45 - B$

and $\tan 2A = \tan(45 - B)$

$\therefore \dfrac{2\tan A}{1 - \tan^2 A} = \dfrac{\tan 45 - \tan B}{1 + \tan 45 \cdot \tan B}$ but $\tan 45° = 1$

$\therefore \dfrac{2\tan A}{1 - \tan^2 A} = \dfrac{1 - \tan B}{1 + \tan B}$ cross multiplying

$2\tan A(1 + \tan B) = (1 - \tan^2 A)(1 - \tan B)$

$2\tan A + 2\tan A \cdot \tan B = 1 - \tan B - \tan^2 A + \tan^2 A \cdot \tan B$

$\therefore 2\tan A \cdot \tan B + \tan B - \tan^2 A \cdot \tan B = 1 - \tan^2 A - 2\tan A$

$\therefore \tan B(2\tan A + 1 - \tan^2 A) = 1 - 2\tan A - \tan^2 A$

$\therefore \tan B = \dfrac{1 - 2\tan A - \tan^2 A}{1 + 2\tan A - \tan^2 A}$

---

c) *Factor formulae*

Factors are often used in algebra to solve equations or to simplify an expression. Similarly, in trigonometry it is useful to be able to factorise a sum or difference of two terms.

PURE MATHEMATICS TRIGONOMETRY

The method of obtaining the factor formulae is given below but, as ever, it is the final results that must be committed to memory.

$$\text{If} \quad \sin(A + B) = \sin A \cos B + \cos A \sin B$$

$$\text{and} \quad \sin(A - B) = \sin A \cos B - \cos A \sin B$$

then adding these two expressions gives

$$\sin(A + B) + \sin(A - B) = 2\sin A \cos B$$

and subtracting the same two expressions gives

$$\sin(A + B) - \sin(A - B) = 2\cos A \sin B$$

$$\text{If} \quad \cos(A + B) = \cos A \cos B - \sin A \sin B$$

$$\text{and} \quad \cos(A - B) = \cos A \cos B + \sin A \sin B$$

then adding these two expressions gives

$$\cos(A + B) + \cos(A - B) = 2\cos A \cos B$$

and subtracting the same two expressions gives

$$\cos(A + B) - \cos(A - B) = -2\sin A \sin B$$

This, however, is not the easiest way to remember the factor formulae. They are easier to remember if the left hand side is written in the form sin P + sin Q, ie as the sum of two sines.

$$\therefore \quad \text{Let } P = A + B \quad\quad P + Q = 2A \quad\quad P - Q = 2B$$
$$\text{and } Q = A - B \quad \therefore \quad \frac{P + Q}{2} = A \quad \text{and} \quad \frac{P - Q}{2} = B$$

So the factor formulae are

$$\boxed{\begin{aligned} \sin P + \sin Q &= 2\sin\left(\frac{P + Q}{2}\right)\cos\left(\frac{P - Q}{2}\right) \\ \sin P - \sin Q &= 2\cos\left(\frac{P + Q}{2}\right)\sin\left(\frac{P - Q}{2}\right) \\ \cos P + \cos Q &= 2\cos\left(\frac{P + Q}{2}\right)\cos\left(\frac{P - Q}{2}\right) \\ \cos P - \cos Q &= -2\sin\left(\frac{P + Q}{2}\right)\sin\left(\frac{P - Q}{2}\right) \end{aligned}}$$

One of the most important applications of these formulae is in helping to solve trigonometrical equations. These will be considered in the next section. They are also used to simplify expressions and to prove identities as will be seen in the next example.

Example 5

Express the following in factors:

(1) $\sin 5x + \sin 3x$

(2) $\sin 4x - \sin 2x$

(3) $1 + \sin 2x$

TRIGONOMETRY   PURE MATHEMATICS

Solution

(1) $\sin 5x + \sin 3x = 2\sin\left(\dfrac{5x + 3x}{2}\right) \cos\left(\dfrac{5x - 3x}{2}\right)$

                         $= 2\sin 4x \cos x$

(2) $\sin 4x - \sin 2x = 2\cos\left(\dfrac{4x + 2x}{2}\right) \sin\left(\dfrac{4x - 2x}{2}\right)$

                         $= 2\cos 3x \sin x$

(3) $1 + \sin 2x$: at first it seems impossible to factorise this as there is only one trigonometrical term. However, using $1 = \sin 90°$

     $\therefore$ $1 + \sin 2x = \sin 90 + \sin 2x$

                        $= 2\sin\left(\dfrac{90 + 2x}{2}\right) \cos\left(\dfrac{90 - 2x}{2}\right)$

                        $= 2\sin(45 + x) \cos(45 - x)$

Now for a harder example on proving identities.

Example 6

Prove the following identities:

(1) $\sin x + \sin 2x + \sin 3x = \sin 2x(2\cos x + 1)$

(2) $\cos x + 2\cos 3x + \cos 5x = 4\cos^2 x \cos 3x$

Solution

(1) $\sin x + \sin 2x + \sin 3x$: Since the right hand side already contains a term in $\sin 2x$ it would seem most sensible to factorise $\sin x$ and $\sin 3x$.

     $\therefore$ $\sin x + \sin 2x + \sin 3x = \sin 2x + 2\sin\left(\dfrac{3x + x}{2}\right) \cos\left(\dfrac{3x - x}{2}\right)$

                                   $= \sin 2x + 2\sin 2x \cos x$

     $\therefore$ $\sin x + \sin 2x + \sin 3x = \sin 2x(1 + 2\cos x)$

(2) $\cos x + 2\cos 3x + \cos 5x$: Since the right hand side already contains $\cos 3x$ it would seem best to start by factorising $\cos x$ and $\cos 5x$.

     $\therefore$ $\cos x + 2\cos 3x + \cos 5x = 2\cos\left(\dfrac{x + 5x}{2}\right) \cos\left(\dfrac{5x - x}{2}\right) + 2\cos 3x$

                               $= 2\cos 3x \cos 2x + 2\cos 3x$

                               $= 2\cos 3x (\cos 2x + 1)$

                               $= 2\cos 3x ((2\cos^2 x - 1) + 1)$

                               $= 2\cos 3x (2\cos^2 x - 1 + 1)$

PURE MATHEMATICS   TRIGONOMETRY

$\therefore \quad \cos x + 2\cos 3x + \cos 5x = 4\cos 3x \cos^2 x$

d)   *Example* ('A' level question)

If $\tan(x + y) = a$ and $\tan(x - y) = b$, express $\dfrac{\sin 2x + \sin 2y}{\sin 2x - \sin 2y}$ in terms of a and b.

Show that $\tan 2y = \dfrac{a - b}{1 + ab}$ and by using this result obtain an expression for $\tan(x + 3y)$ in terms of a and b.

Solution

$$\frac{\sin 2x + \sin 2y}{\sin 2x - \sin 2y} = \frac{2 \sin\left(\frac{2x + 2y}{2}\right) \cos\left(\frac{2x - 2y}{2}\right)}{2 \cos\left(\frac{2x + 2y}{2}\right) \sin\left(\frac{2x - 2y}{2}\right)}$$

$$= \frac{\sin(x + y)}{\cos(x + y)} \cdot \frac{\cos(x - y)}{\sin(x - y)}$$

$$= \tan(x + y) \cdot \cot(x - y)$$

$$= a \cdot \frac{1}{b}$$

$\therefore \quad \dfrac{\sin 2x + \sin 2y}{\sin 2x - \sin 2y} = \dfrac{a}{b}$

$2y = (x + y) - (x - y)$

$\therefore \quad \tan 2y = \tan((x + y) - (x - y))$

$$= \frac{\tan(x + y) - \tan(x - y)}{1 + \tan(x + y)\tan(x - y)}$$

$\therefore \quad \tan 2y = \dfrac{a - b}{1 + ab}$

$x + 3y = (x + y) + 2y$

$\therefore \quad \tan(x + 3y) = \tan[(x + y) + 2y]$

$$= \frac{\tan(x + y) + \tan 2y}{1 - \tan(x + y) \tan 2y}$$

$$= \frac{a + \dfrac{a - b}{1 + ab}}{1 - a \cdot \dfrac{(a - b)}{1 + ab}}$$

Multiplying through every term by $(1 + ab)$ gives

$$\tan(x + 3y) = \frac{a(1 + ab) + (a - b)}{1 \cdot (1 + ab) - a \cdot (a - b)}$$

$$= \frac{a + a^2 b + a - b}{1 + ab - a^2 + ab}$$

TRIGONOMETRY PURE MATHEMATICS

∴ $\tan(x + 3y) = \dfrac{2a - b + a^2b}{1 - a^2 + 2ab}$

3.3 Trigonometrical equations

a) *Solution by factorisation*

Just as an algebraic equation (such as $2x^2 - 3x = 0$) can be solved by removing a common factor from the terms ($x(2x - 3) = 0$) so can a trigonometrical equation, as will be seen from the following examples.

Example 7

Solve the following equations for values of θ from $0°$ to $360°$.

(1) $2\sin^2\theta + \sin\theta = 0$

(2) $\tan\theta + 2\cot\theta = 3$

(3) $3\cos^2\theta = 2\sin\theta \cos\theta$

Solution

(1) $2\sin^2\theta + \sin\theta = 0$

∴ $\sin\theta(2\sin\theta + 1) = 0$

either $\sin\theta = 0$ or $2\sin\theta + 1 = 0$

∴ $\theta = 0°, 180°, 360°$ ∴ $2\sin\theta = -1$

∴ $\sin\theta = -\dfrac{1}{2} = -0.5$ (use quadrants 3 and 4)

∴ $\theta = 210°, 330°$

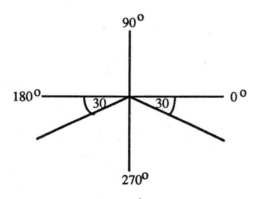

Therefore, solutions are $0°, 180°, 210°, 330°, 360°$

---

(2) $\tan\theta + 2\cot\theta = 3$

As the equation is written it does not look possible to find any common factors but $\cot\theta = \dfrac{1}{\tan\theta}$

∴ $\tan\theta + 2 \cdot \dfrac{1}{\tan\theta} = 3$ Multiplying through by $\tan\theta$

$\tan\theta.\tan\theta + 2.\dfrac{1}{\tan\theta}.\tan\theta = 3.\tan\theta$

$\therefore \quad \tan^2\theta - 3\tan\theta + 2 = 0$

This is a quadratic equation in $\tan\theta$

$(\tan\theta - 2)(\tan\theta - 1) = 0$

either $\quad \tan\theta - 2 = 0 \qquad\qquad$ or $\qquad \tan\theta - 1 = 0$

$\qquad\quad \tan\theta = 2 \qquad\qquad\qquad\qquad \tan\theta = 1$

$\therefore \qquad \theta = 63°26', 243°26' \qquad$ or $\qquad \theta = 45°, 225°$

Use quadrants 1 and 3 both times.

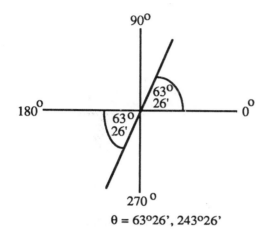

$\theta = 63°26', 243°26' \qquad\qquad\qquad \theta = 45°, 225°$

Therefore, solutions are $45°, 63°26', 225°, 243°26'$

---

(3) $\qquad\qquad\qquad 3\cos^2\theta = 2\sin\theta\cos\theta$

$\therefore \quad 3\cos^2\theta - 2\sin\theta\cos\theta = 0$

$\quad \cos\theta(3\cos\theta - 2\sin\theta) = 0$

either $\quad \cos\theta = 0 \qquad\qquad$ or $\qquad 3\cos\theta - 2\sin\theta = 0 \;(\div \cos\theta)$

$\qquad \theta = 90°, 270° \qquad\qquad\qquad \dfrac{3\cos\theta}{\cos\theta} - \dfrac{2\sin\theta}{\cos\theta} = 0$

$\qquad\qquad\qquad\qquad\qquad\qquad\qquad 3 - 2\tan\theta = 0$

$\therefore \qquad\qquad\qquad\qquad\qquad\qquad \tan\theta = \dfrac{3}{2} = 1.5$

TRIGONOMETRY   PURE MATHEMATICS

Use quadrants 1 and 3

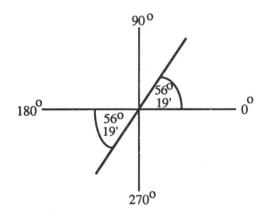

$\theta = 56°19', 236°19'$

Therefore, solutions are $56°19', 90°, 236°19', 270°$

As was seen in (2) of the last example it is sometimes necessary to rewrite part of an equation before it can be solved. In particular, it may be necessary to use one of the formulae derived in section d).

$$\sin^2\theta + \cos^2\theta = 1$$

$$\tan^2\theta + 1 = \sec^2\theta$$

$$1 + \cot^2\theta = \csc^2\theta$$

For example, the equation $2\cos^2\theta + \sin\theta = 1$ is a mixture of sines and cosines and impossible to solve as it stands, but by replacing $\cos^2\theta$ as $1 - \sin^2\theta$ a quadratic equation in $\sin\theta$ is formed which hopefully can be solved.

The next examples will all involve substitutions of this type.

Example 8

Solve the following equations for values of θ from 0 to $360°$:

(1) $2\cos^2\theta + \sin\theta = 1$

(2) $5\sec\theta - 2\sec^2\theta = \tan^2\theta - 1$

(3) $4\cot^2\theta + 39 = 24\csc\theta$

Solution

(1) $2\cos^2\theta + \sin\theta = 1$ but $\cos^2\theta = 1 - \sin^2\theta$

∴ $2(1 - \sin^2\theta) + \sin\theta = 1$

           $2 - 2\sin^2\theta + \sin\theta = 1$

∴ $-2\sin^2\theta + \sin\theta + 2 - 1 = 0$ changing signs gives

           $2\sin^2\theta - \sin\theta - 1 = 0$

           $(2\sin\theta + 1)(\sin\theta - 1) = 0$

either $2\sin\theta + 1 = 0$ or $\sin\theta - 1 = 0$

$$2\sin\theta = -1 \quad\quad \text{or} \quad\quad \sin\theta = 1$$

$$\sin\theta = -\frac{1}{2} \quad\quad\quad\quad \theta = 90°$$

Use quadrants 3 and 4

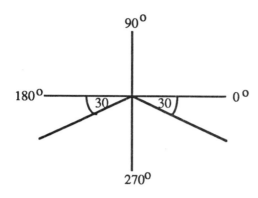

$\theta = 210°, 330°$

Therefore, solutions are $\theta = 90°, 210°, 330°$

---

(2) $\quad\quad 5\sec\theta - 2\sec^2\theta = \tan^2\theta - 1 \quad\quad$ but $\tan^2\theta = \sec^2\theta - 1$

$\quad\quad\quad\quad 5\sec\theta - 2\sec^2\theta = (\sec^2\theta - 1) - 1$

$\quad\quad\quad\quad 5\sec\theta - 2\sec^2\theta = \sec^2\theta - 2$

∴ $\quad 5\sec\theta - 2\sec^2\theta - \sec^2\theta + 2 = 0$

∴ $\quad -3\sec^2\theta + 5\sec\theta + 2 = 0 \quad\quad$ changing signs

$\quad\quad 3\sec^2\theta - 5\sec\theta - 2 = 0$

$\quad\quad (3\sec\theta + 1)(\sec\theta - 2) = 0$

either $\quad 3\sec\theta + 1 = 0 \quad\quad$ or $\quad \sec\theta - 2 = 0$

$\quad\quad\quad 3\sec\theta = -1 \quad\quad\quad\quad \sec\theta = 2$

∴ $\quad\quad \sec\theta = -\frac{1}{3} \quad\quad$ ∴ $\quad \frac{1}{\cos\theta} = 2$

∴ $\quad\quad \frac{1}{\cos\theta} = -\frac{1}{3} \quad\quad$ ∴ $\quad \cos\theta = \frac{1}{2}$

∴ $\quad\quad \cos\theta = -3$ (impossible) $\quad\quad$ Use quadrants 1 and 4

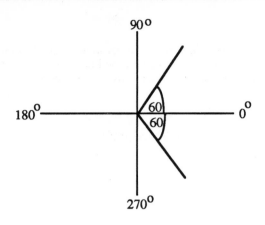

Therefore, solutions are θ = 60°, 300°

(3) $4\cot^2\theta + 39 = 24\csc\theta$ but $\cot^2\theta = \csc^2\theta - 1$

$4(\csc^2\theta - 1) + 39 = 24\csc\theta$

$4\csc^2\theta - 4 + 39 = 24\csc\theta$

$4\csc^2\theta - 24\csc\theta + 35 = 0$

$(2\csc\theta - 7)(2\csc\theta - 5) = 0$

either $2\csc\theta - 7 = 0$ or $2\csc\theta - 5 = 0$

∴ $2\csc\theta = 7$ ∴ $2\csc\theta = 5$

∴ $\csc\theta = \frac{7}{2}$ ∴ $\csc\theta = \frac{5}{2}$

∴ $\frac{1}{\sin\theta} = \frac{7}{2}$ ∴ $\frac{1}{\sin\theta} = \frac{5}{2}$

∴ $\sin\theta = \frac{2}{7} = 0.2857$ ∴ $\sin\theta = \frac{2}{5} = 0.4$

Use quadrants 1 and 2 for both.

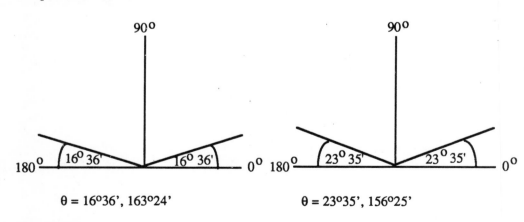

θ = 16°36', 163°24'   θ = 23°35', 156°25'

Therefore, solutions are θ = 16°36', 23°35', 156°25', 163°24'

b) *Solution by angle formulae*

To solve the following equations it will be necessary to use one or more of the formulae given in the last section for compound angles, double angles and factors. Having decided which formula is required the method of solution is very similar to that outlined in the previous examples in this section.

Example 9

Solve the following equations for values of θ between $0°$ and $360°$:

(1) $4\cos(\theta - 60°) = 5\cos(\theta - 30°)$

(2) $\tan 2\theta + \tan\theta = 0$

(3) $\sin\theta + \sin 2\theta + \sin 3\theta + \sin 4\theta = 0$

Solution

(1) $4\cos(\theta - 60°) = 5\cos(\theta - 30°)$

Using the expansion for $\cos(A - B)$

$4(\cos\theta \cdot \cos 60 + \sin\theta \cdot \sin 60) = 5(\cos\theta \cdot \cos 30 + \sin\theta \cdot \sin 30)$

but $\cos 60 = \frac{1}{2}$, $\sin 60 = \frac{\sqrt{3}}{2}$, $\cos 30 = \frac{\sqrt{3}}{2}$, $\sin 30 = \frac{1}{2}$

$\therefore \quad 4 \cdot \cos\theta \cdot \frac{1}{2} + 4 \cdot \sin\theta \cdot \frac{\sqrt{3}}{2} = 5 \cdot \cos\theta \cdot \frac{\sqrt{3}}{2} + 5 \cdot \sin\theta \cdot \frac{1}{2}$

$\therefore \quad 4\cos\theta + 4\sqrt{3}\sin\theta = 5\sqrt{3}\cos\theta + 5\sin\theta$

$4\sqrt{3}\sin\theta - 5\sin\theta = 5\sqrt{3}\cos\theta - 4\cos\theta$

$(4\sqrt{3} - 5)\sin\theta = (5\sqrt{3} - 4)\cos\theta$

$\therefore \quad \dfrac{\sin\theta}{\cos\theta} = \dfrac{(5\sqrt{3} - 4)}{(4\sqrt{3} - 5)} = \dfrac{4.66}{1.928}$

$\therefore \quad \tan\theta = 2.417$

Use the quadrants 1 and 3.

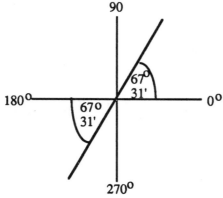

Therefore, solutions are $\theta = 67°31'$, $247°31'$

TRIGONOMETRY   PURE MATHEMATICS

(2) $\tan 2\theta + \tan\theta = 0$

$\therefore \quad \dfrac{2\tan\theta}{1 - \tan^2\theta} + \tan\theta = 0$

$\dfrac{2\tan\theta + \tan\theta(1 - \tan^2\theta)}{1 - \tan^2\theta} = 0$

$\therefore \quad 2\tan\theta + \tan\theta - \tan^3\theta = 0$

$\therefore \quad 3\tan\theta - \tan^3\theta = 0$

$\tan\theta(3 - \tan^2\theta) = 0$

either $\tan\theta = 0$ or $3 - \tan^2\theta = 0$

$\theta = 0, 180°, 360°$ $\therefore \tan^2\theta = 3$

$\therefore \tan\theta = \pm\sqrt{3}$

Use all four quadrants.

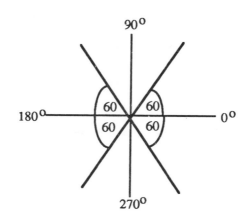

$\theta = 60°, 120°, 240°, 300°$

Therefore, solutions are $\theta = 0°, 60°, 120°, 180°, 240°, 300°, 360°$

---

(3) $\sin\theta + \sin 2\theta + \sin 3\theta + \sin 4\theta = 0$

The problem here is knowing which pairs of terms to take in order to apply the factor formula for the sum of two sines.

If $\sin\theta$ is paired with $\sin 4\theta$ and $\sin 2\theta$ paired with $\sin 3\theta$, then a common factor will appear.

$\sin\theta + \sin 4\theta + \sin 2\theta + \sin 3\theta = 0$

$2\sin\left(\dfrac{\theta + 4\theta}{2}\right) \cdot \cos\left(\dfrac{4\theta - \theta}{2}\right) + 2\sin\left(\dfrac{3\theta + 2\theta}{2}\right) \cdot \cos\left(\dfrac{3\theta - 2\theta}{2}\right) = 0$

$2\sin\left(\dfrac{5\theta}{2}\right) \cdot \cos\left(\dfrac{3\theta}{2}\right) + 2\sin\left(\dfrac{5\theta}{2}\right) \cdot \cos\left(\dfrac{\theta}{2}\right) = 0$

$2\sin\left(\dfrac{5\theta}{2}\right) \cdot \left(\cos\left(\dfrac{3\theta}{2}\right) + \cos\left(\dfrac{\theta}{2}\right)\right) = 0$

PURE MATHEMATICS   TRIGONOMETRY

Now the formula for the sum of two cosines can be used.

$$2\sin\left(\frac{5\theta}{2}\right)\left(2\cos\left(\frac{\frac{3\theta}{2}+\frac{\theta}{2}}{2}\right)\cos\left(\frac{\frac{3\theta}{2}-\frac{\theta}{2}}{2}\right)\right) = 0$$

$$2\sin\left(\frac{5\theta}{2}\right) 2\cos\theta \cdot \cos\left(\frac{\theta}{2}\right) = 0$$

$$\therefore \quad 4\sin\left(\frac{5\theta}{2}\right) \cdot \cos\theta \cdot \cos\left(\frac{\theta}{2}\right) = 0$$

either $\sin\left(\frac{5\theta}{2}\right) = 0$ or $\cos\theta = 0$ or $\cos\left(\frac{\theta}{2}\right) = 0$

$\theta = 90°, 270°$ $\frac{\theta}{2} = 90°, 270°$

$\frac{5\theta}{2} = 0°, 180°, 360°$ $\theta = 2 \times (90°, 270°)$

$\theta = \frac{2}{5} \times (0°, 180°, 360°)$ $\theta = 180°, 540°$

$= 0°, 72°, 144°$ (540° is outside range)

Therefore solutions are $\theta = 0°, 72°, 90°, 144°, 180°, 270°$

c)   *Solution using $R\cos(\theta \pm \alpha)$, $R\sin(\theta \pm \alpha)$*

This method is used to solve equations of the type $a\cos\theta + b\sin\theta = c$, eg $3\cos\theta + 4\sin\theta = 2$. None of the methods used so far would help in solving this type of equation.

Start by assuming that $3\cos\theta + 4\sin\theta$ can be expressed in the form $R\cos(\theta - \alpha)$.

$\therefore \quad 3\cos\theta + 4\sin\theta = R(\cos\theta \cos\alpha + \sin\theta \sin\alpha)$

$3 = R\cos\alpha$

$4 = R\sin\alpha$ } ... $\tan\alpha = \dfrac{R\sin\alpha}{R\cos\alpha} = \dfrac{4}{3}$

$\alpha = 53°08'$

and $3^2 + 4^2 = R^2\cos^2\alpha + R^2\sin^2\alpha$

$\therefore \quad 9 + 16 = R^2(\cos^2\alpha + \sin^2\alpha)$

$\therefore \quad 25 = R^2$

$\therefore \quad 5 = R$

Therefore $3\cos\theta + 4\sin\theta = 2$ becomes

$5\cos(\theta - 53°08') = 2$

$\cos(\theta - 53°08') = \dfrac{2}{5} = 0.4$

TRIGONOMETRY   PURE MATHEMATICS

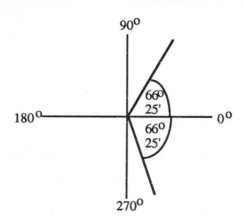

$\therefore \quad \theta - 53°08' = 66°25', 293°35'$

$\theta = \quad 66°25' \quad$ and $\quad 293°35'$
$ 53°08' + \phantom{\quad\text{and}\quad} 53°08' +$

$ 119°33' \phantom{\quad\text{and}\quad\quad} 346°43'$

These solutions could equally well have been found using $R\sin(\theta + \alpha)$:
R will have the same value but α will be different.

Example 10

Express $4\sin 2x + 3\cos 2x$ in the form $R\sin(2x + \alpha)$ where $R > 0$ and α is an acute angle.

Hence find the values of x between $0°$ and $360°$ for which $4\sin 2x + 3\cos 2x = 2.49$.

Solution

$\quad 4\sin 2x + 3\cos 2x \equiv R\sin(2x + \alpha)$

$\quad 4\sin 2x + 3\cos 2x \equiv R(\sin 2x \cos\alpha + \cos 2x \sin\alpha)$

$\therefore \quad 4 = R\cos\alpha$

$ \} \therefore \quad \tan\alpha = \dfrac{R\sin\alpha}{R\cos\alpha} = \dfrac{3}{4}$

$\quad 3 = R\sin\alpha$

$\therefore \quad \alpha = 36°52' \quad$ or $\quad 216°52'$ (not allowed)

and $\quad 4^2 + 3^2 = R^2\cos^2\alpha + R^2\sin^2\alpha$

$\quad 16 + 9 = R^2(\cos^2\alpha + \sin^2\alpha)$

$\quad 25 = R^2$

$\quad 5 = R$

$\therefore \quad 4\sin 2x + 3\cos 2x = 5\sin(2x + 36°52')$

since $\quad 4\sin 2x + 3\cos 2x = 2.49$ it follows that

$\quad 5\sin(2x + 36°52') = 2.49$

$\therefore \quad \sin(2x + 36°52') = \dfrac{2.49}{5} = 0.498$

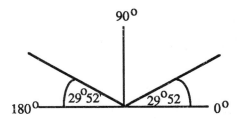

Since x can take values between 0° and 360° it follows that 2x can take values between 0° and 720°.

∴ 2x + 36°52' = 29°52',   150°08'   389°52'   510°08'

∴   2x   = 29°52'-   150°08'- 389°52'- 510°08'-
                   36°52'   36°52'   36°52'   36°52'
                   ──────   ──────   ──────   ──────
                   -7°0'   113°16'   353°0'   473°16'
                   ──────   ──────   ──────   ──────

∴   x   = -3°30',   56°38'   176°30'   236°38'

A negative angle such as -3°30' has been measured clockwise from zero instead of anti-clockwise. Since x takes values from 0° to 360° (measured anti-clockwise) x = -3°30' should be written as

x = 360° - 3°30'

x = 356°30'

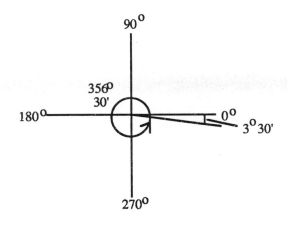

Therefore, solutions are x = 56°38', 176°30', 236°38', 356°30'.

TRIGONOMETRY   PURE MATHEMATICS

d) *Radians*

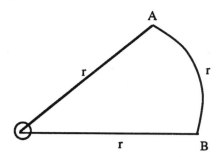

An arc of a circle equal in length to the radius (r) subtends an angle of 1 radian at the centre of the circle.

∴   AÔB = 1 radian or 1 rad

The total circumference of a circle is of length $2\pi r$ and so subtends an angle of 2π radians at the centre of the circle.

However, the angle at the centre of a circle is 360°.

∴ 2π radians $\equiv 360°$

∴ π radians $\equiv 180°$

So   1 radian $= \dfrac{180°}{\pi} = 57°17'$

and $1° = \dfrac{\pi}{180}$ radians

∴ $x° = \dfrac{\pi}{180}(x)$ radians $= \theta$ rads

If   AÔB $= x°$

then arc APB $= \dfrac{x}{360} \times 2\pi r$

$= \left(\dfrac{x}{180} \times \pi\right) r$

∴   arc APB $= \theta r$

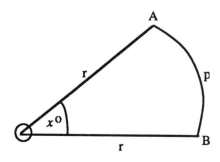

AÔB $=$ $x°$ or θ rads.

Also area sector AOB $= \dfrac{x}{360} \times \pi r^2$

$$= \frac{1}{2}\left(\frac{x}{180} \times \pi r\right)r^2$$

∴ area AOB $= \frac{1}{2}\theta r^2$

Therefore, the important results are:

π radians	$\equiv 180°$
arc length	$= r\theta$
area of sector	$= \frac{1}{2}r^2\theta$

Example 11

Convert the following angles into radians:

(1)   90°

(2)   45°

(3)   60°

(4)   120°

(5)   270°

(6)   300°

(7)   15°

(8)   30°

Solution

Using $x° \equiv \frac{\pi}{180°}(x°)$ rads

$x°$ =	15	30	45	60	90	120	270	300
θ_{rads} =	$\frac{\pi}{12}$	$\frac{\pi}{6}$	$\frac{\pi}{4}$	$\frac{\pi}{3}$	$\frac{\pi}{2}$	$\frac{2\pi}{3}$	$\frac{3\pi}{2}$	$\frac{5\pi}{3}$

e)   *Graphical solutions of equations*

Equations can be solved graphically whether the angles are in degrees or radians, as the following examples will show.

Example 12

Plot the graphs of $y = \sin 2x°$ and $y = \cos 3x°$ on the same axes for values of x from 0° to 90°. Find from the graph the root of the equation $\sin 2x° = \cos 3x°$ which lies in this range.

Solution

Taking values of x at 15° intervals, say:

TRIGONOMETRY   PURE MATHEMATICS

x	0	15	30	45	60	75	90
2x	0	30	60	90	120	150	180
y = sin2x	0	0.5	0.866	1	0.866	0.5	0

x	0	15	30	45	60	75	90
3x	0	45	90	135	180	225	270
y = cos3x	1	0.707	0	-0.707	-1	-0.707	0

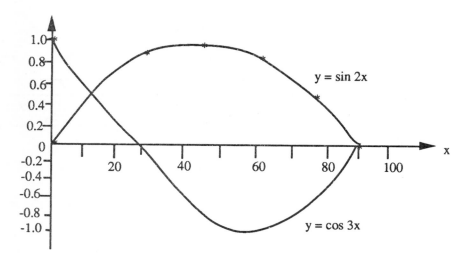

The root of the equation sin2x = cos3x is by the intersection of the graphs ie 18°.

Example 13

A chord AB divides a circle, of radius a and centre O, into 2 segments. The perimeter of the minor segment is 4a, and the angle AOB is 2θ radians.

Show that $\sin\theta = 2 - \theta$

Find graphically the value of θ to 2 significant figures.

Solution

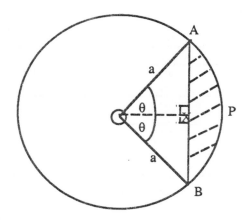

Perimeter of minor segment = arc APB + AB

Using arc length = radius x sector angle in radians

$$APB = a.2\theta = 2a\theta$$

Using $\triangle AOX$ gives $\sin\theta = \dfrac{AX}{a}$

PURE MATHEMATICS   TRIGONOMETRY

∴ $AX = a \sin\theta$

and $AB = 2a \sin\theta$

So $APB + AB = 4a$

        $2a\theta + 2a\sin\theta = 4a$ dividing by $2a$

        $\theta + \sin\theta = 2$

∴ $\sin\theta = 2 - \theta$

Taking θ from O to 2 radians, say at intervals of 0.2 radians, and drawing graphs of $y = \sin\theta$ and $y = 2 - \theta$:

θ	0	0.2	0.4	0.6	0.8	1.0	1.2	1.4	1.6	1.8	2.0
$y = \sin\theta$	0	0.199	0.389	0.565	0.717	0.841	0.932	0.985	1.0	0.974	0.909
$y = 2 - \theta$	2	1.8	1.6	1.4	1.2	1.0	0.8	0.6	0.4	0.2	0

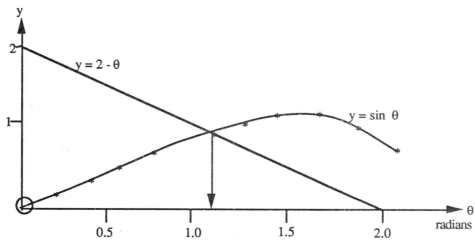

So the value of θ that satisfies $\sin\theta = 2 - \theta$ is 1.1 radians, ie the value of θ at the point of intersection of the graphs $y = \sin\theta$ and $y = 2 - \theta$.

---

f)   *Example* ('A' level questions)

Solve the simultaneous equations

$$\cos A + \cos B = \frac{2k}{3}$$

$$\cos A . \cos B = \frac{-k^2}{3}$$

to find $\cos A$ and $\cos B$ in terms of k.

Find the range of values of k for which A and B exist.

Solution

When solving simultaneous equations it is often easiest to replace from one equation into the other.

Taking $\cos A + \cos B = \frac{2k}{3}$ ∴ $\cos A = \frac{2k}{3} - \cos B$

TRIGONOMETRY — PURE MATHEMATICS

So equation $\cos A \cdot \cos B = \dfrac{-k^2}{3}$ becomes

$$\left(\dfrac{2k}{3} - \cos B\right) \cdot \cos B = \dfrac{-k^2}{3}$$

$\therefore \quad \dfrac{2k}{3}\cos B - \cos^2 B = \dfrac{-k^2}{3}$

$\therefore \quad -\cos^2 B + \dfrac{2k}{3}\cos B + \dfrac{k^2}{3} = 0 \quad$ changing signs

$\quad \cos^2 B - \dfrac{2k}{3}\cos B - \dfrac{k^2}{3} = 0 \quad$ multiplying by 3

$\quad 3\cos^2 B - 2k\cos B - k^2 = 0$

$\quad (3\cos B + k)(\cos B - k) = 0$

$\therefore \quad$ either $3\cos B + k = 0 \quad$ or $\quad \cos B - k = 0$

$\therefore \quad\quad 3\cos B = -k \quad\quad\quad\quad \cos B = k$

$\therefore \quad\quad \cos B = \dfrac{-k}{3}$

but $\quad\quad \cos A = \dfrac{2k}{3} - \cos B \quad\quad \cos A = \dfrac{2k}{3} - \cos B$

$\quad\quad\quad\quad = \dfrac{2k}{3} - \left(-\dfrac{k}{3}\right) \quad\quad = \dfrac{2k}{3} - k$

$\quad\quad\quad\quad = \dfrac{2k}{3} + \dfrac{k}{3} \quad\quad\quad\quad = \dfrac{2k - 3k}{3}$

$\therefore \quad\quad \cos A = \dfrac{3k}{3} = k \quad\quad\quad \cos A = \dfrac{-k}{3}$

So the solutions are $\cos A = k$, $\cos B = -\dfrac{k}{3}$ and $\cos A = -\dfrac{k}{3}$, $\cos B = k$.

The range of values for cosine is from -1 to $+1$, so the range of values for k is also from -1 to $+1$, ie $-1 \leq k \leq +1$.

Note it can be shown that when an angle of θ radians is small $\sin\theta \approx \theta$ and $\cos\theta \approx 1 - \dfrac{\theta^2}{2}$

Example if $\theta = 0.1$ radians then using a calculator

$\quad \sin\theta = 0.0998 \approx 0.1$

$\quad \tan\theta = 0.1003 \approx 0.1$

$\quad \cos\theta = 0.9950 = 1 - \dfrac{(0.1)^2}{2} = 0.995$

These relations can be useful if a question involves such small angles.

PURE MATHEMATICS — TRIGONOMETRY

3.4 Solutions of triangles

a) *Sine formula*

For many people this will be a revision topic.

In any triangle ABC

$$\frac{a}{\sin A} = \frac{b}{\sin B} = \frac{c}{\sin C} = 2R$$

where R is the radius of the circumcircle, ie the circle passing through the vertices of the triangle.

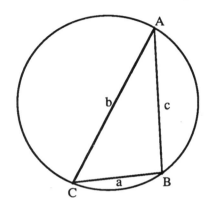

Example 13

Solve the following triangle:

$$A = 36°, \ b = 2.37, \ C = 49°$$

Solution

If $A = 36°$ and $C = 49°$ then $B = 180 - (36 + 49) = 180 - 85 = 95°$

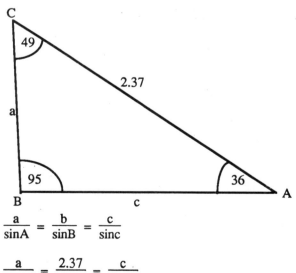

$$\frac{a}{\sin A} = \frac{b}{\sin B} = \frac{c}{\sin c}$$

$$\therefore \quad \frac{a}{\sin 36} = \frac{2.37}{\sin 95} = \frac{c}{\sin 49}$$

To find a $\quad \dfrac{a}{\sin 36} = \dfrac{2.37}{\sin 95}$

$$\therefore \quad a = \frac{2.37 \times \sin 36}{\sin 95} = 1.398$$

$$\therefore \quad a = 1.40 \text{ (to 3 significant figures)}$$

To find c
$$\frac{c}{\sin 49} = \frac{2.37}{\sin 95}$$

$$\therefore \quad c = \frac{2.37 \times \sin 49}{\sin 95} = 1.795$$

$$\therefore \quad c = 1.80 \text{ (to 3 significant figures)}$$

Therefore, the triangle is completely solved knowing

$$B = 95°, \quad a = 1.40, \quad c = 1.80$$

b) *Cosine formula*

Again a revision topic for many people.

In any triangle ABC:

$$a^2 = b^2 + c^2 - 2bc \cos A$$
$$\therefore \quad b^2 = a^2 + c^2 - 2ac \cos B$$
$$\text{and} \quad c^2 = a^2 + b^2 - 2ab \cos C$$

The formula is used in this form if a side is to be found. However, if an angle is required, the formula has to be arranged to give:

$$\cos A = \frac{b^2 + c^2 - a^2}{2bc}$$

$$\cos B = \frac{a^2 + c^2 - b^2}{2ac}$$

$$\cos C = \frac{a^2 + b^2 - c^2}{2ab}$$

Example 14

In the following triangle calculate angle C

$$a = 8, \quad b = 10, \quad c = 15$$

Solution

$$\cos C = \frac{a^2 + b^2 - c^2}{2ab} = \frac{8^2 + 10^2 - 15^2}{2 \times 8 \times 10} = \frac{64 + 100 - 225}{160}$$

$$\cos C = \frac{-61}{160} = -0.3812$$

The negative sign denotes the fact that C is an obtuse angle.

$$\therefore \quad C = 180 = 67°36'$$

$$\therefore \quad C = 112°24'$$

Example 15

In the following triangle calculate side b:

$$a = 17, \hat{B} = 120°, c = 63.$$

Solution

$$\begin{aligned}
b^2 &= a^2 + c^2 - 2ac \cos B \\
&= 17^2 + 63^2 - 2 \times 17 \times 63 \cos 120° \\
&= 289 + 3969 - 2142 \cos 120°
\end{aligned}$$

since $\hat{B}$ is obtuse $\cos 120° = - \cos (180 - 120)$

$$= - \cos 60 = - 0.5$$

$$\begin{aligned}
b^2 &= 4258 - 2142 (- 0.5) \\
&= 4258 + 1071 \\
&= 5329 \\
b &= \sqrt{5329} = 73.00
\end{aligned}$$

c)   *Area formulae*

There are three formulae that can be used to find the area of the triangle:

i)

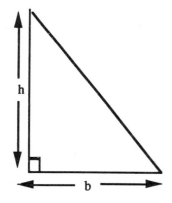 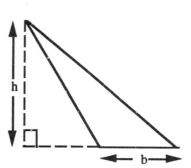

For all these triangles, where the base and height are known, the formula to use is:

$$\boxed{A = \tfrac{1}{2} b h}$$

TRIGONOMETRY   PURE MATHEMATICS

ii)

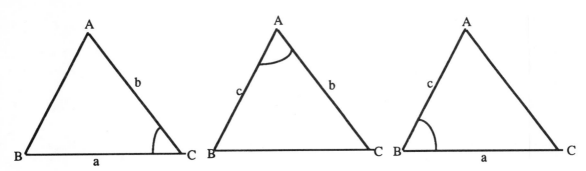

For these triangles, where 2 sides and the angle between are known, the formula to use is:

| Area = $\frac{1}{2}$ ab sinC | Area = $\frac{1}{2}$ bc sinA | Area = $\frac{1}{2}$ ac sin b |

iii)

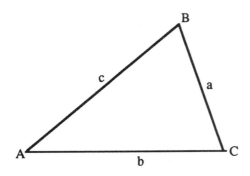

For this triangle, where all three sides are known, the formula to use is:

$$A = \sqrt{s(s-a)(s-b)(s-c)} \text{ where } s = \frac{1}{2}(a+b+c)$$

Example 16

The perimeter of a triangle is 42 cm, one side is of length 12 cm and the area is $21\sqrt{5}$ cm². Find the lengths of the other sides and show that the cosine of the smallest angle is $\frac{11}{16}$.

Solution

Assume that the triangle is ABC as shown.

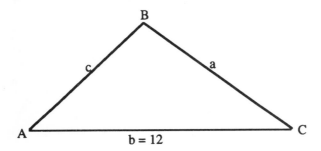

Assume   b = 12

$a + b + c = 42$

$\therefore a + c = 30$

$\therefore a = 30 - c$

Using the fact that the area $= 21\sqrt{5}$

but area $= \sqrt{s(s-a)(s-b)(s-c)}$

where $s = \frac{1}{2}(42) = 21$

$\therefore \sqrt{21(21-a)(21-12)(21-c)} = 21\sqrt{15}$ Squaring both sides

$\therefore 21(21-a)(9)(21-c) = 21^2 \times 15$

but $a = 30 - c$

$\therefore 21(21-(30-c))(9)(21-c) = 21^2 \times 15$ Dividing by 21

$(21 - 30 + c)(9)(21 - c) = 21 \times 15$

$9(-9 + c)(21 - c) = 315$ Dividing by 9

$(-9 + c)(21 - c) = 35$

$-189 + 9c + 21c - c^2 - 35 = 0$ Changing signs

$c^2 - 30c + 224 = 0$

$(c - 14)(c - 16) = 0$

either $c - 14 = 0$ or $c - 16 = 0$

$\therefore c = 14$ $\therefore c = 16$

and $a = 16$ and $a = 14$

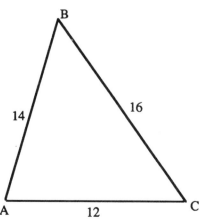

The smallest angle will be opposite the smallest side, ie $\hat{B}$ opposite b = 12

$\cos B = \dfrac{a^2 + c^2 - b^2}{2ac} = \dfrac{16^2 + 14^2 - 12^2}{2 \times 16 \times 14}$

TRIGONOMETRY   PURE MATHEMATICS

$$= \frac{256 + 196 - 144}{448}$$

$$= \frac{308}{448} = \frac{11}{16} \text{ dividing by 28}$$

∴   cosine of smallest angle is $\frac{11}{16}$

---

d)   *3-D problems*

Three dimensional problems require special care and it is often necessary to draw several diagrams to find the triangles that are to be used in answering different parts of the question.

Having found the correct triangles, it is a matter of applying the basic trigonometrical ratio and/or Pythagoras theorem if the triangles are right angled, or otherwise sine and cosine formulae.

The following examples will help to explain the approach that needs to be adopted to these problems.

Example 17

A plane is inclined at angle α to the horizontal and a line PQ on the plane makes an acute angle β with PR which is a line of greatest slope on the plane. Show that the inclination θ of PQ to the horizontal is given by $\sin\theta = \sin\alpha \cos\beta$.

Show that the angle φ between the vertical plane through PQ and the vertical plane through PR is given by $\cos\phi \cos\theta = \cos\alpha \cos\beta$.

Solution

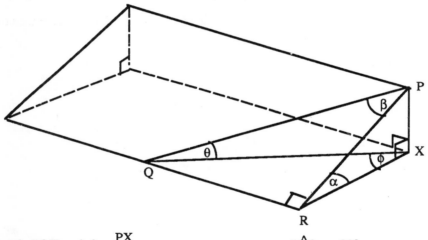

From triangle PQX $\sin\theta = \frac{PX}{PQ}$ as $P\hat{X}Q = 90°$

From triangle PRX $\sin\alpha = \frac{PX}{PR}$ ∴ $PR\sin\alpha = PX$ as $P\hat{X}R = 90°$

From triangle PQR $\cos\beta = \frac{PR}{PQ}$ ∴ $\frac{PR}{\cos\beta} = PQ$ as $P\hat{R}Q = 90°$

$$\sin\theta = \frac{PX}{PQ} = \frac{PR\sin\alpha}{\left(\frac{PR}{\cos\beta}\right)}$$

108

PURE MATHEMATICS TRIGONOMETRY

$$= \frac{PR\sin\alpha \cos\beta}{PR}$$

$\therefore \quad \sin\theta = \sin\alpha \cos\beta$

Let angle $Q\hat{X}R) = \phi$ and $X\hat{R}Q = 90°$

From triangle QRX $\cos\phi = \dfrac{XR}{XQ}$

From triangle PQX $\cos\theta = \dfrac{QX}{QP}$

From triangle PRX $\cos\alpha = \dfrac{RX}{RP} \quad \therefore \; RP\cos\alpha = RX$

From triangle PQR $\cos\beta = \dfrac{PR}{PQ} \qquad \dfrac{PR}{\cos\beta} = PQ$

$\cos\phi \cdot \cos\theta = \dfrac{XR}{XQ} \cdot \dfrac{QX}{QP} = \dfrac{XR}{QP}$

$$= \frac{RP\cos\alpha}{\left(\dfrac{PR}{\cos\beta}\right)} = \frac{RP\cos\alpha \cos\beta}{PR}$$

$\therefore \quad \cos\phi \cdot \cos\theta = \cos\alpha \cos\beta$

Example 18

A tetrahedron has a horizontal equilateral triangular base of side 6 cm and the sloping edges are each of length 4 cm.

(1) Find the height of the tetrahedron.

(2) Find the inclination of a sloping edge to the horizontal.

(3) Show that the sloping faces are inclined to one another at angle with a cosine of $\dfrac{-1}{7}$

Solution

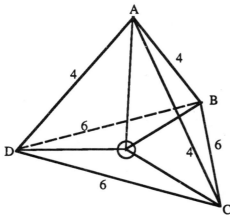

AB = AC = AD = 4 cm

BC = CD = DB = 6 cm

109

TRIGONOMETRY   PURE MATHEMATICS

(1)   The vertex, A, of the tetrahedron will be directly above the 'centre', O, of the triangular base BCD.

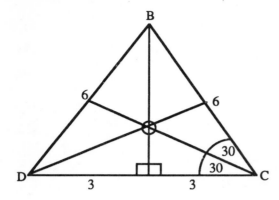

To find OC:

$\cos 30 = \dfrac{3}{OC}$

$\therefore \quad OC = \dfrac{3}{\cos 30} = \dfrac{3}{\frac{\sqrt{3}}{2}}$

$\therefore \quad OC = \dfrac{6}{\sqrt{3}}$

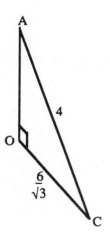

To find AO:

$AO^2 + \left(\dfrac{6}{\sqrt{3}}\right)^2 = 4^2$

$AO^2 + \dfrac{36}{3} = 16$

$\therefore \quad AO^2 = 16 - 12 = 4$

$\therefore \quad AO = 2 \text{ cm}$

PURE MATHEMATICS — TRIGONOMETRY

(2) The inclination of a sloping edge to the horizontal is $A\hat{C}O = A\hat{B}O = A\hat{D}O$

To find $A\hat{C}O$:

$$\sin A\hat{C}O = \frac{AO}{AC} = \frac{2}{4} = \frac{1}{2}$$

$$\therefore A\hat{C}O = 30°$$

(3)

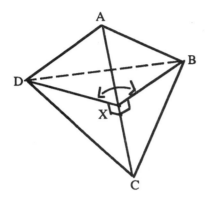

The sloping faces are inclined to one another at an angle given by $D\hat{X}B$ where X is the point on AC such that $D\hat{X}C = 90°$ and $B\hat{X}C = 90°$

To find $A\hat{C}D$, thence DX and hence $D\hat{X}B$

Consider face ACD, draw AY perpendicular to DC

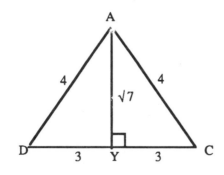
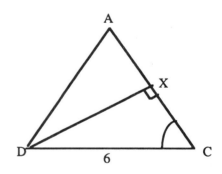

$$AY^2 + 3^2 = 4^2$$

$$AY^2 + 9 = 16$$

$$\therefore AY^2 = 16 - 9$$

$$\therefore AY = \sqrt{7}$$

$$\sin A\hat{C}D = \frac{AY}{AC} = \frac{DX}{CD}$$

TRIGONOMETRY — PURE MATHEMATICS

$$\therefore \quad \frac{\sqrt{7}}{4} = \frac{DX}{6}$$

$$\therefore \quad DX = \frac{6\sqrt{7}}{4} = \frac{3\sqrt{7}}{2}$$

$$\therefore \quad \cos D\hat{X}B = \frac{DX^2 + BX^2 - DB^2}{2 \cdot DX \cdot BX}$$

$$= \frac{\left(\frac{3\sqrt{7}}{2}\right)^2 + \left(\frac{3\sqrt{7}}{2}\right)^2 - 6^2}{2\left(\frac{3\sqrt{7}}{2}\right)\left(\frac{3\sqrt{7}}{2}\right)}$$

$$= \frac{\frac{9 \times 7}{4} + \frac{9 \times 7}{4} - 36}{2 \times \frac{9 \times 7}{4}} \quad \text{Multiplying by 4}$$

$$= \frac{63 + 63 - (36 \times 4)}{126}$$

$$= \frac{126 - 144}{126} = \frac{-18}{126}$$

$$\therefore \quad \cos D\hat{X}B = \frac{-1}{7}$$

---

e) *Example* ('A' level question)

Given that $\sin^{-1}x$, $\cos^{-1}x$ and $\sin^{-1}(1 - x)$ are acute angles:

(1) prove that $\sin[\sin^{-1}x - \cos^{-1}x] = 2x^2 - 1$

(2) solve the equation $\sin^{-1}x - \cos^{-1}x = \sin^{-1}(1 - x)$

Solution

(1) Let $\sin^{-1}x = a \quad \therefore \quad x = \sin a$

$\cos^{-1}x = b \quad \therefore \quad x = \cos b$

$\sin^{-1}(1 - x) = c \quad \therefore \quad 1 - x = \sin c$

$\sin[\sin^{-1}x - \cos^{-1}x] = \sin[a - b]$

$\qquad = \sin a \cdot \cos b - \cos a \cdot \sin b$

If $\sin a = x$, then using $\sin^2 a + \cos^2 a = 1$

$$x^2 + \cos^2 a = 1$$

$$\therefore \quad \cos^2 a = 1 - x^2$$

$$\therefore \quad \cos a = \sqrt{1 - x^2}$$

PURE MATHEMATICS — TRIGONOMETRY

Similarly $\cos b = x$ $\therefore$ $\sin b = \sqrt{1 - x^2}$

$$\sin(a - b) = \sin a \cos b - \cos a \sin b$$
$$= x.x - (\sqrt{1 - x^2})(\sqrt{1 - x^2})$$
$$= x^2 - (1 - x^2)$$
$$= x^2 - 1 + x^2$$

$\therefore \sin[\sin^{-1}x - \cos^{-1}x] = 2x^2 - 1$

---

(2) $\sin^{-1}x - \cos^{-1}x = \sin^{-1}(1 - x)$

$a - b = c$

$\therefore \sin(a - b) = \sin c$

$\therefore 2x^2 - 1 = 1 - x$

$\therefore 2x^2 + x - 2 = 0$

Using formula to solve quadratic equation

$$\therefore x = \frac{-1 \pm \sqrt{1^2 - 4 \times 2 \times (-2)}}{2 \times 2}$$

$$= \frac{-1 \pm \sqrt{1 + 16}}{4}$$

$$= \frac{-1 \pm \sqrt{17}}{4}$$

either $x = \dfrac{-1 - \sqrt{17}}{4}$ or $x = \dfrac{-1 + \sqrt{17}}{4}$

Since $x = \sin a$ it has limits of ± 1, ie $-1 \leq x \leq +1$.

$\therefore x \neq \dfrac{-1 - \sqrt{17}}{4}$

So $x = \dfrac{-1 + \sqrt{17}}{4} = \dfrac{1}{4}(\sqrt{17} - 1)$

DIFFERENTIAL CALCULUS — PURE MATHEMATICS

4 DIFFERENTIAL CALCULUS

    4.1   Introduction to differentiation
    4.2   Further differentiation
    4.3   Other functions
    4.4   Applications of differentiation

4.1 Introduction to differentiation

a)   *Gradients*

The gradient of a straight line is constant, ie it remains the same all the way along the line. However, it is not true for curves where the gradient is continuously changing and therefore is more difficult to calculate.

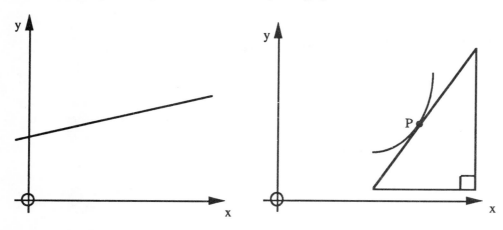

The gradient at any point P on a curve is the gradient of the tangent to the curve at that point. (A tangent is a straight line which just touches a curve at one point only.)

If an approximate value of the gradient is all that is required, then the tangent can be drawn in 'by eye' and the gradient of the line found in the usual way. However, this method will not be very accurate and so another, more precise, way is needed.

b)   *Gradient functions*

In order to deduce the general rule for obtaining gradient functions of x, two examples will be explained from first principles.

Consider $y = x^2$

Consider any point P on the curve $y = x^2$ with coordinates $x = a$, $y = a^2$, ie P is the point (a, a^2).

Q is another point on the same curve with coordinates $x = a + h$, $y = (a + h)^2 = a^2 + 2ah + h^2$, ie Q is the point $(a + h,\ a^2 + 2ah + h^2)$.

PURE MATHEMATICS — DIFFERENTIAL CALCULUS

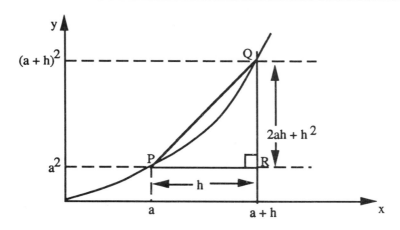

The line joining PQ is a chord and the gradient of $PQ = \dfrac{QR}{RP} = \dfrac{(a^2 + 2ah + h^2) - a^2}{(a + h) - a} = \dfrac{2ah + h^2}{h}$

$\therefore$ gradient of chord PQ $= 2a + h$

If Q moves along the curve until it reaches point P, then the chord PQ will have become a tangent to the curve at P, and h will equal zero. So the gradient of the chord PQ will have become the gradient of the tangent to the curve at P.

$\therefore$ gradient of tangent $= 2a$

In general terms if P is any point (x, y) on the curve $y = x^2$, the gradient at that point will be $2x$. This is known as the gradient function of the curve $y = x^2$.

Consider $y = x^3$

As before, considering two points P and Q on the curve $y = x^3$ where P has coordinates (a, a^3) and Q has coordinates $[(a + h), (a + h)^3] = [(a + h), (a^3 + 3a^2h + 3ah^2 + h^3)]$.

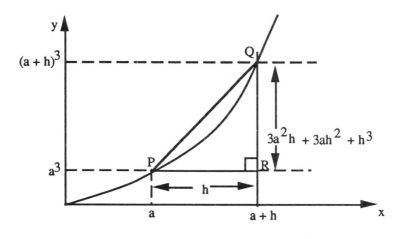

Gradient of chord PQ $= \dfrac{QR}{RP} = \dfrac{(a^3 + 3a^2h + 3ah^2 + h^3) - a^3}{(a + h) - a}$

$= \dfrac{3a^2h + 3ah^2 + h^3}{h}$

$\therefore$ gradient of chord PQ $= 3a^2 + 3ah + h^2$

As Q moves along the curve until it reaches P, h will become zero and the gradient of tangent $= 3a^2$

DIFFERENTIAL CALCULUS — PURE MATHEMATICS

Therefore, in general terms, if P is any pont (x, y) on $y = x^3$, the gradient at that point will be $3x^2$. This is known as the gradient function of the curve $y = x^3$.

These two results suggest a rule for finding the gradient function of any power of x.

> If $y = x^n$ then gradient function $= nx^{n-1}$,
> ie multiply by the index (n) and reduce the power of x by 1 to (n - 1)

c) *Differentiation of algebraic functions*

The process of finding the gradient function of a curve $y = f(x)$ is known as differentiation, and the gradient function itself is more commonly called the derived function, $\frac{dy}{dx}$ and is called (dy by dx).

Therefore, $\frac{dy}{dx}$ means that the original function has been differentiated with respect to x in order to find the gradient function, eg

$$\text{If} \quad y = x^2 \quad \text{then} \frac{dy}{dx} = 2x$$

$$y = x^3 \quad \text{then} \frac{dy}{dx} = 3x^2$$

$$y = x^4 \quad \text{then} \frac{dy}{dx} = 4x^3$$

There are some other important rules that need to be learnt. These are:

(a) > If $y = k$ (a constant) then $\frac{dy}{dx} = 0$

(b) > If $y = kx^n$ then $\frac{dy}{dx} = k.nx^{n-1}$

(c) The above result is true for values of n that are positive or negative, and for n as an integer or a fraction.

Example 1

Differentiate the following functions of *x* by applying the basic rule

ie   If $y = kx^n$ then $\frac{dy}{dx} = k.nx^{n-1}$

(1) $y = x^2$ $\therefore$ $\frac{dy}{dx} = 2x^{2-1} = 2x^1 = 2x$

(2) $y = 3x^7$ $\therefore$ $\frac{dy}{dx} = 3.7.x^{7-1} = 21x^6$

(3) $y = 5x = 5x^1$ $\therefore$ $\frac{dy}{dx} = 5.1.x^{1-1} = 5x^0 = 5$ (as $x^0 = 1$)

(4) $y = 2x^4$ $\therefore$ $\frac{dy}{dx} = 2.4.x^{4-1} = 8x^3$

(5) $y = 4$ $\therefore$ $\frac{dy}{dx} = 0$ $\therefore$ y is equal to a constant $\therefore$ no *x*'s

PURE MATHEMATICS   DIFFERENTIAL CALCULUS

(6) $y = \dfrac{2}{x^3} = 2.x^{-3}$ $\therefore$ $\dfrac{dy}{dx} = 2.(-3)x^{-3-1} = -6x^{-4} = -\dfrac{6}{x^4}$

(7) $y = \dfrac{-3}{x^2} = -3x^{-2}$ $\therefore$ $\dfrac{dy}{dx} = -3(-2)x^{-2-1} = 6x^{-3} = \dfrac{6}{x^3}$

(8) $y = \dfrac{2}{5.x} = \dfrac{2}{5}x^{-1}$ $\therefore$ $\dfrac{dy}{dx} = \dfrac{2}{5}(-1)x^{-1-1} = \dfrac{-2}{5}x^{-2} = -\dfrac{2}{5x^2}$

(9) $y = \dfrac{1}{3x^3} = \dfrac{1}{3}x^{-3}$ $\therefore$ $\dfrac{dy}{dx} = \dfrac{1}{3}(-3)x^{-3-1} = -1.x^{-4} = -\dfrac{1}{x^4}$

(10) $y = \dfrac{-1}{x^4} = -1.x^{-4}$ $\therefore$ $\dfrac{dy}{dx} = -1(-4)x^{-4-1} = 4x^{-5} = \dfrac{4}{x^5}$

(11) $y = x^{\frac{1}{3}}$ $\therefore$ $\dfrac{dy}{dx} = \dfrac{1}{3}x^{\frac{1}{3}-1} = \dfrac{1}{3}x^{-\frac{2}{3}} = \dfrac{1}{3x^{\frac{2}{3}}}$

(12) $y = 2x^{\frac{3}{4}}$ $\therefore$ $\dfrac{dy}{dx} = 2.\dfrac{3}{4}x^{\frac{3}{4}-1} = \dfrac{3}{2}x^{-\frac{1}{4}} = \dfrac{3}{2x^{\frac{1}{4}}}$

(13) $y = \sqrt{4x} = 2x^{\frac{1}{2}}$ $\therefore$ $\dfrac{dy}{dx} = 2.\dfrac{1}{2}x^{\frac{1}{2}-1} = 1.x^{-\frac{1}{2}} = \dfrac{1}{\sqrt{x}}$

(14) $y = 5\sqrt{x^3} = 5x^{\frac{3}{2}}$ $\therefore$ $\dfrac{dy}{dx} = 5.\dfrac{3}{2}x^{\frac{3}{2}-1} = \dfrac{15}{2}x^{\frac{1}{2}} = \dfrac{15}{2}\sqrt{x}$

(15) $y = \dfrac{1}{\sqrt{x}} = x^{-\frac{1}{2}}$ $\therefore$ $\dfrac{dy}{dx} = -\dfrac{1}{2}x^{-\frac{1}{2}-1} = -\dfrac{1}{2}x^{-\frac{3}{2}} = \dfrac{-1}{2\sqrt{x^3}}$

It is more usual for a function of x to be composed of several individual terms, in which case each term is differentiated separately.

$\therefore$ If $y = x^3 - 2x^2 + 3x$

then $\dfrac{dy}{dx} = 3x^2 - 2.2x + 3.1x^0 = 3x^2 - 4x + 3$

Example 2

Differentiate the following functions of x:

(1) $4x^4 - 3x^2 + 5$

(2) $ax^2 + bx + c$

(3) $5x^2(x - 4)$

(4) $(x + 3)(2x - 1)$

(5) $\dfrac{10x^5 + 3x^4}{2x^2}$

DIFFERENTIAL CALCULUS — PURE MATHEMATICS

Solution

(1) Let $y = 4x^4 - 3x^2 + 5$

$\therefore \dfrac{dy}{dx} = 4.4x^3 - 3.2x + 0$

$\phantom{\therefore \dfrac{dy}{dx}} = \underline{16x^3 - 6x}$

(2) Let $y = ax^2 + bx + c$

$\therefore \dfrac{dy}{dx} = a.2x + b.1x^0 + 0$

$\phantom{\therefore \dfrac{dy}{dx}} = \underline{2ax + b}$

(3) Let $y = 5x^2(x - 4)$, terms must first be separated

$\therefore y = 5x^3 - 20x^2$, now it is possible to differentiate

$\therefore \dfrac{dy}{dx} = 5.3x^2 - 20.2x$

$\phantom{\therefore \dfrac{dy}{dx}} = \underline{15x^2 - 40x}$

(4) Let $y = (x + 3)(2x - 1)$, multiplying out the brackets

$y = 2x^2 + 6x - x - 3$

$y = 2x^2 + 5x - 3$, now differentiating

$\therefore \dfrac{dy}{dx} = 2.2x + 5.1x^0 + 0$

$\phantom{\therefore \dfrac{dy}{dx}} = \underline{4x + 5}$

(5) Let $y = \dfrac{10x^5 + 3x^4}{2x^2}$, separating out gives

$\phantom{\text{Let } y} = \dfrac{10x^5}{2x^2} + \dfrac{3x^4}{2x^2} = 5x^{5-2} + \dfrac{3}{2}x^{4-2}$

$\phantom{\text{Let } y} = 5x^3 + \dfrac{3}{2}x^2$, now differentiating

$\therefore \dfrac{dy}{dx} = 5.3x^2 + \dfrac{3}{2}.2x$

$\phantom{\therefore \dfrac{dy}{dx}} = \underline{15x^2 + 3x}$

PURE MATHEMATICS — DIFFERENTIAL CALCULUS

d) Gradients of tangents and normals

Differentiation is the accurate way of finding the gradient of a tangent to a curve at a particular point, P, on the curve.

A normal to a curve at the same point P is the straight line through P at right angles to the tangent at P.

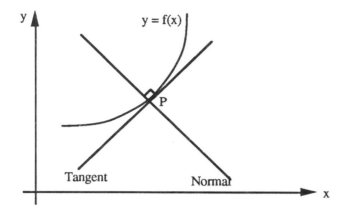

An important fact (which will be derived later) concerning lines at right angles to each other such as tangents and normals is that the product of their gradients is -1:

If m = gradient of tangent

and m' = gradient of normal, then $m \cdot m' = -1$ ∴ $m' = -\dfrac{1}{m}$

Therefore, in order to find the gradient at a particular point on the curve the equation of the curve is differentiated with respect to x and the appropriate numerical value substituted for x. Having found the gradient of the tangent, the gradient of the normal at the same point is found using $m' = -\dfrac{1}{m}$.

Example 3

Calculate the gradients of the tangents and normals to the following curves at the points given:

(1) $y = x^2 - 2x + 1$ at $(2, 1)$

(2) $y = 3x^2 - 2x^3$ at $(-2, 28)$

(3) $y = (x + 2)(x - 4)$ at $(3, -5)$

(4) $y = (4x - 5)^2$ at $(\tfrac{1}{2}, 9)$

Solution

Throughout this solution m has been used to denote the gradient of the tangent and m' the gradient of the normal.

(1) $y = x^2 - 2x + 1$

∴ $\dfrac{dy}{dx} = 2x - 2$ so at $(2, 1)$ $m = 2(2) - 2 = 2$ and $m' = \dfrac{-1}{m} = -\dfrac{1}{2}$

(2) $y = 3x^2 - 2x^3$

∴ $\dfrac{dy}{dx} = 6x - 6x^2$ so at $(-2, 28)$ $m = 6(-2) - 6(-2)^2 = -36$ and $m' = \dfrac{-1}{-36} = \dfrac{1}{36}$

DIFFERENTIAL CALCULUS   PURE MATHEMATICS

(3) $y = (x + 2)(x - 4) = x^2 + 2x - 4x - 8 = x^2 - 2x - 8$

$\therefore \dfrac{dy}{dx} = 2x - 2$ so at $(3, -5)$ $m = 2(3) - 2 = 4$ and $m' = -\dfrac{1}{4}$

(4) $y = (4x - 5)^2 = (4x - 5)(4x - 5) = 16x^2 - 20x - 20x + 25 = 16x^2 - 40x + 25$

$\therefore \dfrac{dy}{dx} = 32x - 40$ so at $\left(\dfrac{1}{2}, 9\right)$ $m = 32\left(\dfrac{1}{2}\right) - 40 = -24$ and $m' = \dfrac{-1}{-24} = \dfrac{1}{24}$

An alternative problem arises when the gradient of the tangent is known and it is the coordinates of the point with that gradient that are required. This is in fact the reverse of the above example - m is known and the corresponding point(s) (x, y) are to be found.

Example 4

Find the coordinates of the points on the following curves at which the gradient has the given value:

(1) $y = x^2$ and $m = 8$

(2) $y = x^3 - 6x^2 + 4$ and $m = -12$

(3) $y = x^2 - 3x + 1$ and $m = 0$

(4) $y = x^2 - x^3$ and $m = -1$

Solution

(1) If $y = x^2$

then $\dfrac{dy}{dx} = 2x$ $\therefore$ $m = 2x = 8$

$\therefore$ $2x = 8$

and $x = 4$

To find the y coordinate $x = 4$ must be replaced in the equation of the curve.

$\therefore$ $y = x^2 = 4^2$

   $y = 16$

The gradient is 8 at the point (4, 16).

---

(2) If $y = x^3 - 6x^2 + 4$

then $\dfrac{dy}{dx} = 3x^2 - 12x$ $\therefore$ $m = 3x^2 - 12x = -12$

$\therefore$ $3x^2 - 12x + 12 = 0$ dividing by 3

$\therefore$ $x^2 - 4x + 4 = 0$

$\therefore$ $(x - 2)^2 = 0$

$\therefore$ $x - 2 = 0$ and $x = 2$

PURE MATHEMATICS — DIFFERENTIAL CALCULUS

Replacing to find y:

$$y = x^3 - 6x^2 + 4$$
$$= 2^3 - 6(2)^2 + 4$$
$$= 8 - 24 + 4 = -12$$

The gradient is -12 at the point (2, -12).

---

(1) If $y = x^2 - 3x + 1$

then $\dfrac{dy}{dx} = 2x - 3$ ∴ $m = 2x - 3 = 0$

$$2x = 3$$
$$x = \tfrac{3}{2} = 1.5$$

Replacing to find y:

$$y = x^2 - 3x + 1$$
$$= \left(\tfrac{3}{2}\right)^2 - 3\left(\tfrac{3}{2}\right) + 1$$
$$= 2.25 - 4.5 + 1 = -1.25$$

The gradient is 0 at the point (1.5, -1.25).

---

(4) If $y = x^2 - x^3$

then $\dfrac{dy}{dx} = 2x - 3x^2$ ∴ $m = 2x - 3x^2 = -1$

$$2x - 3x^2 + 1 = 0 \quad \text{changing signs}$$
$$3x^2 - 2x - 1 = 0$$
$$(3x + 1)(x - 1) = 0$$

either $3x + 1 = 0$ or $x - 1 = 0$

$$x = -\tfrac{1}{3} \qquad x = 1$$

Because there are two values of x it follows that there are two points on the curve with a gradient of -1.

Replacing to find y:

$$y = x^2 - x^3 \qquad\qquad y = x^2 - x^3$$
$$= \left(-\tfrac{1}{3}\right)^2 - \left(-\tfrac{1}{3}\right)^3 \qquad = 1^2 - 1^3 = 0$$
$$= \tfrac{1}{9} + \tfrac{1}{27}$$

DIFFERENTIAL CALCULUS — PURE MATHEMATICS

$$= \frac{3+1}{27} = \frac{4}{27}$$

So the points are $\left(-\frac{1}{3}, \frac{4}{27}\right)$ and $(1, 0)$.

---

e) *Harder examples*

These are not 'A' level questions but are harder than the previous examples and require more thought.

Example 5

Find the coordinates of the points of intersection of the line $x - 3y = 0$ with the curve $y = x(1 - x^2)$. If these points are in order P, O, Q prove that the tangents to the curve at P and Q are parallel, and that the tangent at O is perpendicular to them. Draw a sketch of the curve and its tangents.

Solution

Solving $x - 3y = 0$ and $y = x(1 - x^2)$ for their points of intersection gives:

$$x = 3y$$

$\therefore \quad y = \frac{x}{3}$ replacing this in the other equation

$y = x(1 - x^2)$ becomes

$\frac{x}{3} = x(1 - x^2)$ multiplying by 3

$x = 3x(1 - x^2)$

$x = 3x - 3x^3$

$3x^3 - 2x = 0$

$x(3x^2 - 2) = 0$

Either $x = 0$ or $3x^2 - 2 = 0$

$\therefore \quad 3x^2 = 2$ dividing by 3

$\therefore \quad x^2 = \frac{2}{3}$ square rooting

$\therefore \quad x = \pm\sqrt{\frac{2}{3}}$

Replacing to find the corresponding values of y

$y = \frac{0}{3}$ or $\quad y = \frac{\pm\sqrt{\frac{2}{3}}}{3}$

$y = 0 \quad\quad\quad\quad = \pm\frac{1}{3}\sqrt{\frac{2}{3}}$

So the points P, O, Q in order are: $P\left(-\sqrt{\frac{2}{3}}, -\frac{1}{3}\sqrt{\frac{2}{3}}\right)$ $O(0,0)$ $Q\left(\sqrt{\frac{2}{3}}, \frac{1}{3}\sqrt{\frac{2}{3}}\right)$

To find the gradients of the tangents at these three points it is first necessary to differentiate the equation of the curve.

$$y = x(1 - x^2)$$

$$y = x - x^3$$

$$\therefore \quad \frac{dy}{dx} = 1 - 3x^2$$

At P gradient $= 1 - 3\left(-\sqrt{\frac{2}{3}}\right)^2$ $\therefore$ $m_P = 1 - 3\left(\frac{2}{3}\right) = -1$

At O gradient $= 1 - 3(0)^2$ $\therefore$ $m_O = 1 - 0 = 1$

At Q gradient $= 1 - 3\left(\sqrt{\frac{2}{3}}\right)^2$ $\therefore$ $m_Q = 1 - 3 \cdot \left(\frac{2}{3}\right) = -1$

If lines are parallel then they have the same gradient:

$m_P = -1$
$m_Q = -1$ } so these tangents are parallel

If lines are perpendicular then the product of the gradients is -1:

$m_P \cdot m_O = -1 \cdot (1) = -1$
$m_Q \cdot m_O = -1 \cdot (1) = -1$ } so these lines are perpendicular

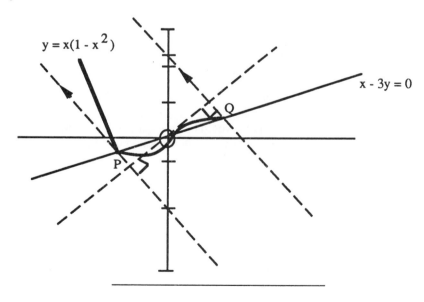

Example 6

Show that the curve $y = (x - 2)(x - 3)(x - 4)$ cuts the x axis at $P(2, 0)$, $Q(3, 0)$, $R(4, 0)$. Prove that the tangents at P and R are parallel.

DIFFERENTIAL CALCULUS   PURE MATHEMATICS

Solution

On the x axis $\quad y = 0$. Therefore putting:

$$(x - 2)(x - 3)(x - 4) = 0$$

either $(x - 2) = 0 \quad$ or $(x - 3) = 0 \quad$ or $(x - 4) = 0$

$\therefore \quad x = 2 \qquad \therefore \quad x = 3 \qquad \therefore \quad x = 4$

$\therefore \quad$ P(2, 0) $\qquad \therefore \quad$ Q(3, 0) $\qquad \therefore \quad$ R(4, 0)

In order to differentiate $y = (x - 2)(x - 3)(x - 4)$ the brackets must be removed:

$$y = (x - 2)(x^2 - 7x + 12)$$
$$= x^3 - 7x^2 + 12x - 2x^2 + 14x - 24$$

$\therefore \quad y = x^3 - 9x^2 + 26x - 24$

So $\quad \dfrac{dy}{dx} = 3x^2 - 18x + 26$

At P, gradient $= 3(2)^2 - 18(2) + 26 = 2$

$\therefore \quad m_P = 2$

At R, gradient $= 3(4)^2 - 18(4) + 26 = 2$

$\therefore \quad m_R = 2$

$m_P = m_R$ and so the tangents at P and R are parallel.

4.2 Further differentiation

a) *Function of a function*

Functions such as $(x + 3)^2$ have already been differentiated by first multiplying out the brackets to obtain separate terms, which were differentiated individually. Functions such as $(2x^2 + 6)^7$ could be expanded using the binomial theorem and then differentiated but this would be very tedious. An alternative method of dealing with these functions would be helpful, and this is known as the 'function of a function' method.

If y is a function of t and t is a function of x, then the following result can be used:

$$\boxed{\dfrac{dy}{dx} = \dfrac{dy}{dt} \cdot \dfrac{dt}{dx}}$$

So, considering the function given above

If $\quad y = (2x^2 + 6)^7 \quad$ then this can be written as

$\quad y = t^7 \qquad$ where $t = 2x^2 + 6$

$\therefore \quad \dfrac{dy}{dt} = 7t^6 \qquad$ and $\dfrac{dt}{dx} = 4x$

PURE MATHEMATICS — DIFFERENTIAL CALCULUS

$$\therefore \quad \frac{dy}{dx} = \frac{dy}{dt} \cdot \frac{dt}{dx}$$

$$= 7t^6 \cdot 4x$$

Since the function was given in terms of x originally, the answer should be given in the same terms, ie

$$\frac{dy}{dx} = 7.(2x^2 + 6)^6 \cdot 4x = 28x(2x^2 + 6)^6$$

This method is obviously very useful where brackets and powers are involved.

Example 7

Differentiate the following functions of x:

(1) $\quad (3x^2 + 5)^3$

(2) $\quad (7x^2 - 4)^{\frac{1}{3}}$

(3) $\quad (6x^3 - 4x)^{-2}$

(4) $\quad (3x^2 - 5x)^{-\frac{2}{3}}$

(5) $\quad \dfrac{1}{\sqrt{3x + 2}}$

Solution

(1) Let $y = (3x^2 + 5)^3$ and $t = 3x^2 + 5$

then $y = t^3$

$\therefore \dfrac{dy}{dt} = 3t^2$ and $\dfrac{dt}{dx} = 6x$

so $\dfrac{dy}{dx} = \dfrac{dy}{dt} \cdot \dfrac{dt}{dx}$

$= 3t^2 \cdot 6x = 3(3x^2 + 5)^2 \cdot 6x$

$\therefore \dfrac{dy}{dx} = 18x(3x^2 + 5)^2$

---

(2) Let $y = (7x^2 - 4)^{\frac{1}{3}}$ and $t = 7x^2 - 4$

then $y = t^{\frac{1}{3}}$

$\therefore \dfrac{dy}{dt} = \dfrac{1}{3}t^{-\frac{2}{3}}$ and $\dfrac{dt}{dx} = 14x$

so $\dfrac{dy}{dx} = \dfrac{dy}{dt} \cdot \dfrac{dt}{dx}$

$$= \tfrac{1}{3}t^{-\tfrac{2}{3}}.14x = \tfrac{1}{3}(7x^2 - 4)^{-\tfrac{2}{3}}.14x$$

$$\therefore \quad \frac{dy}{dx} = \frac{14x}{3}(7x^2 - 4)^{-\tfrac{2}{3}}$$

---

(3) Let $y = (6x^3 - 4x)^{-2}$ and $t = 6x^3 - 4x$

then $y = t^{-2}$

$\therefore \quad \dfrac{dy}{dt} = -2t^{-3}$ and $\dfrac{dt}{dx} = 18x^2 - 4$

so $\dfrac{dy}{dx} = \dfrac{dy}{dt} \cdot \dfrac{dt}{dx}$

$$= -2t^{-3}.(18x^2 - 4) = -2(6x^3 - 4x)^{-3}.(18x^2 - 4)$$

$\therefore \quad \dfrac{dy}{dx} = -2(18x^2 - 4)(6x^3 - 4x)^{-3}$

---

(4) Let $y = (3x^2 - 5x)^{-\tfrac{2}{3}}$ and $t = 3x^2 - 5x$

then $y = t^{-\tfrac{2}{3}}$

$\therefore \quad \dfrac{dy}{dt} = -\tfrac{2}{3}t^{-\tfrac{5}{3}}$ and $\dfrac{dt}{dx} = 6x - 5$

so $\dfrac{dy}{dx} = \dfrac{dy}{dt} \cdot \dfrac{dt}{dx}$

$$= -\tfrac{2}{3}t^{-\tfrac{5}{3}}(6x - 5) = -\tfrac{2}{3}(3x^2 - 5x)^{-\tfrac{5}{3}}(6x - 5)$$

$\therefore \quad \dfrac{dy}{dx} = -\tfrac{2}{3}(6x - 5)(3x^2 - 5x)^{-\tfrac{5}{3}}$

---

(5) Let $y = \dfrac{1}{\sqrt{3x + 2}} = \dfrac{1}{(3x + 2)^{\tfrac{1}{2}}} = (3x + 2)^{-\tfrac{1}{2}}$ and $t = 3x + 2$

$\therefore \quad y = t^{-\tfrac{1}{2}}$

$\dfrac{dy}{dt} = \tfrac{1}{2}t^{-\tfrac{3}{2}}$ and $\dfrac{dt}{dx} = 3$

PURE MATHEMATICS — DIFFERENTIAL CALCULUS

so $\quad \dfrac{dy}{dx} = \dfrac{dy}{dt} \cdot \dfrac{dt}{dx}$

$\quad = -\dfrac{1}{2} t^{-\frac{3}{2}} \cdot 3 \qquad = -\dfrac{1}{2}(3x+2)^{-\frac{3}{2}}(3)$

$\quad = -\dfrac{3}{2}(3x+2)^{-\frac{3}{2}} \qquad = -\dfrac{-3}{2(3x+2)^{\frac{3}{2}}}$

$\therefore \quad \dfrac{dy}{dx} = \dfrac{-3}{2\sqrt{(3x+2)^3}}$

This method is also very useful when roots and reciprocals are involved inside the brackets.

Example 8

Differentiate the following functions:

(1) $\quad \dfrac{-1}{(1+\sqrt{x})^2}$

(2) $\quad \left(1 - \dfrac{1}{x}\right)^3$

(3) $\quad (3\sqrt{x} - 2x)^3$

(4) $\quad \sqrt[3]{(1 - \sqrt{x})}$

Solution

(1) Let $\quad t = 1 + \sqrt{x} \quad = \quad 1 + x^{\frac{1}{2}}$

$\quad y = -\dfrac{1}{(1+\sqrt{x})^2} = -(1+\sqrt{x})^{-2}$

$\therefore \quad y = -t^{-2}$

$\therefore \quad \dfrac{dy}{dt} = -(-2)t^{-3} \quad$ and $\quad \dfrac{dt}{dx} = \dfrac{1}{2} x^{-\frac{1}{2}}$

so $\quad \dfrac{dy}{dx} = \dfrac{dy}{dt} \cdot \dfrac{dt}{dx}$

$\quad = 2t^{-3}\left(\dfrac{1}{2} x^{-\frac{1}{2}}\right) = 2(1+\sqrt{x})^{-3}\left(\dfrac{1}{2} x^{-\frac{1}{2}}\right)$

$\quad = (1+\sqrt{x})^{-3} x^{-\frac{1}{2}}$

$\therefore \quad \dfrac{dy}{dx} = \dfrac{1}{(1+\sqrt{x})^3 \sqrt{x}}$

(2) Let $y = \left(1 - \frac{1}{x}\right)^3$ and $t = 1 - \frac{1}{x}$

$\therefore \quad y = t^3 \qquad t = 1 - x^{-1}$

$\therefore \quad \frac{dy}{dt} = 3t^2 \qquad \frac{dt}{dx} = -(-1)x^{-2} = \frac{1}{x^2}$

so $\quad \frac{dy}{dx} = \frac{dy}{dt} \cdot \frac{dt}{dx}$

$\qquad = 3t^2 \cdot \frac{1}{x^2} = 3\left(1 - \frac{1}{x}\right)^2 \cdot \frac{1}{x^2}$

$\therefore \quad \frac{dy}{dx} = \frac{3}{x^2}\left(1 - \frac{1}{x}\right)^2$

---

(3) Let $y = (3\sqrt{x} - 2x)^3$ and $t = (3\sqrt{x} - 2x) \qquad t = 3x^{\frac{1}{2}} - 2x$

$\therefore \quad y = t^3$

so $\quad \frac{dy}{dt} = 3t^2 \qquad$ and $\quad \frac{dt}{dx} = 3\left(\frac{1}{2}x^{-\frac{1}{2}}\right)(-2)$

$\therefore \quad \frac{dy}{dx} = \frac{dy}{dt} \cdot \frac{dt}{dx}$

$\qquad = 3t^2\left(\frac{3}{2}x^{-\frac{1}{2}} - 2\right) = 3(3x^{\frac{1}{2}} - 2x)^2\left(\frac{3}{2x^{\frac{1}{2}}} - 2\right)$

$\therefore \quad \frac{dy}{dx} = 3(3\sqrt{x} - 2x)^2\left(\frac{3}{2\sqrt{x}} - 2\right)$

---

(4) Let $y = \sqrt[3]{(1 - \sqrt{x})}$ and $t = (1 - \sqrt{x}) \qquad t = 1 - x^{\frac{1}{2}}$

$\therefore \quad y = \sqrt[3]{t} = t^{\frac{1}{3}}$

so $\quad \frac{dy}{dt} = \frac{1}{3}t^{-\frac{2}{3}} \qquad$ and $\quad \frac{dt}{dx} = -\frac{1}{2}x^{-\frac{1}{2}}$

$\therefore \quad \frac{dy}{dx} = \frac{dy}{dt} \cdot \frac{dt}{dx}$

$\qquad = \frac{1}{3}t^{-\frac{2}{3}}\left(-\frac{1}{2}x^{-\frac{1}{2}}\right) = \frac{1}{3}\left(1 - x^{\frac{1}{2}}\right)^{-\frac{2}{3}}\left(-\frac{1}{2}x^{-\frac{1}{2}}\right)$

$\qquad = \frac{1}{3}(1 - \sqrt{x})^{-\frac{2}{3}}\left(-\frac{1}{2x^{\frac{1}{2}}}\right)$

$$= - \frac{1}{6(\sqrt[3]{1 - \sqrt{x}})^2 (\sqrt{x})}$$

By now it may have become clear that the 'function of a function' method of differentiation consists of two distinct steps:

(a) differentiating the power of the function;

(b) differentiating the function itself;

ie if $y = (f(x))^n$

then $\frac{dy}{dx} = \underset{(a)}{[n(f(x))^{n-1}]} . \underset{(b)}{[\text{derivative of } f(x)]}$

so if $y = (2x^2 + 6)^7$

$\frac{dy}{dx} = 7(2x^2 + 6)^6 .(4x)$

$= 28x(2x^2 + 6)^6$

This shortened method of dealing with this process will be used from now on.

b) *Products and quotients*

Some very complicated functions of x have just been differentiated. However, the previous method cannot be used if the problem involves a product of two separate functions of x such as $(x^2 + 1)(x^3 + 1)$.

If two functions of x eg u and v, are multiplied, they can be differentiated using the following formula:

$$\boxed{y = u.v. \qquad \frac{dy}{dx} = u.\frac{dv}{dx} + v.\frac{du}{dx}}$$

Using the above formula,

if $y = (x^2 + 1)(x^3 + 1)$

then $u = x^2 + 1$ and $v = x^3 + 1$

$\frac{du}{dx} = 2x \qquad \frac{dv}{dx} = 3x^2$

Using $\frac{dy}{dx} = u.\frac{dv}{dx} + v.\frac{du}{dx}$

$= (x^2 + 1)(3x^2) + (x^3 + 1)(2x)$

$= 3x^4 + 3x^2 + 2x^4 + 2x$

$\therefore \frac{dy}{dx} = 5x^4 + 3x^2 + 2x$

DIFFERENTIAL CALCULUS PURE MATHEMATICS

This particular example could have been solved by first multiplying the brackets and then using term by term differentiation. Just to show that the same answer would have resulted:

$$y = (x^2 + 1)(x^3 + 1)$$
$$= x^5 + x^2 + x^3 + 1$$
$$= x^5 + x^3 + x^2 + 1$$
$$\therefore \frac{dy}{dx} = 5x^4 + 3x^2 + 2x$$

This second solution may seem quicker and easier but there are many occasions when the individual functions cannot be multiplied out and then the product formula must be used. This will become obvious in the following example.

Example 9

Differentiate the following functions:

(1) $x^2\sqrt{1 + x^2}$

(2) $\sqrt{x + 2}.\sqrt{x + 3}$

(3) $\sqrt{x + 1}\,\sqrt{(x + 2)^3}$

Solution

(1) If $y = x^2\sqrt{1 + x^2}$ then $u = x^2$ $v = (1 + x^2)^{\frac{1}{2}}$

ie $y = x^2(1 + x^2)^{\frac{1}{2}}$ $\frac{du}{dx} = 2x$ $\frac{dv}{dx} = \frac{1}{2}(1 + x^2)^{-\frac{1}{2}}(2x)$

using $\frac{dy}{dx} = u.\frac{dv}{dx} + v.\frac{du}{dx}$

$$= x^2 . \frac{1}{2}(1 + x^2)^{-\frac{1}{2}}(2x) + (1 + x^2)^{\frac{1}{2}}(2x)$$

$$\frac{dy}{dx} = \frac{x^3}{\sqrt{1 + x^2}} + 2x\sqrt{(1 + x^2)}$$

It is often necessary to arrange these terms to obtain the answer in the neatest and most concise form. Therefore, continuing:

$$\frac{dy}{dx} = \frac{x^3 + 2x(\sqrt{1 + x^2}).(\sqrt{1 + x^2})}{\sqrt{(1 + x^2)}}$$

$$= \frac{x^3 + 2x(1 + x^2)}{\sqrt{1 + x^2}}$$

$$= \frac{x^3 + 2x + 2x^3}{\sqrt{1 + x^2}} = \frac{2x + 3x^3}{\sqrt{1 + x^2}}$$

$$\therefore \frac{dy}{dx} = \frac{x(2 + 3x^2)}{\sqrt{1 + x^2}}$$

PURE MATHEMATICS   DIFFERENTIAL CALCULUS

(2)   If $y = \sqrt{x+2} \; \sqrt{x+3}$

     ie $y = (x+2)^{\frac{1}{2}} (x+3)^{\frac{1}{2}} \quad$ then $u = (x+2)^{\frac{1}{2}} \quad$ and $\quad v = (x+3)^{\frac{1}{2}}$

$$\frac{du}{dx} = \frac{1}{2}(x+2)^{-\frac{1}{2}} \quad \frac{dv}{dx} = \frac{1}{2}(x+3)^{-\frac{1}{2}}$$

Using $\quad \dfrac{dy}{dx} = u \cdot \dfrac{dv}{dx} + v \cdot \dfrac{du}{dx}$

$$= (x+2)^{\frac{1}{2}} \frac{1}{2}(x+3)^{-\frac{1}{2}} + (x+3)^{\frac{1}{2}} \frac{1}{2}(x+2)^{-\frac{1}{2}}$$

$$= \frac{1}{2} \frac{(x+2)^{\frac{1}{2}}}{(x+3)^{\frac{1}{2}}} + \frac{1}{2} \frac{(x+3)^{\frac{1}{2}}}{(x+2)^{\frac{1}{2}}}$$

$\therefore \quad \dfrac{dy}{dx} = \dfrac{1}{2}\sqrt{\dfrac{x+2}{x+3}} + \dfrac{1}{2}\sqrt{\dfrac{x+3}{x+2}}$

Rearranging to obtain a more concise form of the answer:

$$\frac{dy}{dx} = \frac{\sqrt{x+2}\sqrt{x+2} + \sqrt{x+3}\sqrt{x+3}}{2\sqrt{x+3}\sqrt{x+2}}$$

$$= \frac{x+2+x+3}{2\sqrt{x+3}\sqrt{x+2}}$$

$\therefore \quad \dfrac{dy}{dx} = \dfrac{2x+5}{2\sqrt{x+3}\sqrt{x+2}}$

---

(3)   If $y = \sqrt{x+1} \; \sqrt{(x+2)^3}$

         $y = (x+1)^{\frac{1}{2}} (x+2)^{\frac{3}{2}} \quad$ then $u = (x+1)^{\frac{1}{2}} \quad v = (x+2)^{\frac{3}{2}}$

$$\frac{du}{dx} = \frac{1}{2}(x+1)^{-\frac{1}{2}} \quad \frac{dv}{dx} = \frac{3}{2}(x+2)^{\frac{1}{2}}$$

Using $\quad \dfrac{dy}{dx} = u \cdot \dfrac{dv}{dx} + v \cdot \dfrac{du}{dx}$

$$= (x+1)^{\frac{1}{2}} \frac{3}{2}(x+2)^{\frac{1}{2}} + (x+2)^{\frac{3}{2}} \frac{1}{2}(x+1)^{-\frac{1}{2}}$$

$\therefore \quad \dfrac{dy}{dx} = \dfrac{3}{2}\sqrt{x+1}\sqrt{x+2} + \dfrac{1}{2}\dfrac{\sqrt{(x+2)^3}}{\sqrt{x+1}}$

DIFFERENTIAL CALCULUS PURE MATHEMATICS

Rearranging as before gives:

$$\frac{dy}{dx} = \frac{3\sqrt{x+1}\sqrt{x+2}\sqrt{x+1} + \sqrt{(x+2)^3}}{2\sqrt{x+1}}$$

$$= \frac{3(x+1)\sqrt{x+2} + \sqrt{(x+2)^3}}{2\sqrt{x+1}}$$

$$= \frac{3(x+1)\sqrt{x+2} + (x+2)\sqrt{x+2}}{2\sqrt{x+1}}$$

$$= \frac{\sqrt{x+2}\,(3x+3+x+2)}{2\sqrt{x+1}}$$

$$\therefore \quad \frac{dy}{dx} = \frac{(4x+5)\sqrt{x+2}}{2\sqrt{x+1}}$$

Now that there is a formula for differentiating a product it would be useful to have an equivalent formula for differentiating a quotient, ie

$$\boxed{\; y = \frac{u}{v} \qquad \frac{dy}{dx} = \frac{v\cdot\frac{du}{dx} - u\cdot\frac{dv}{dx}}{v^2} \;}$$

Example 10

Differentiate the following functions:

(1) $\dfrac{1 - x^2}{1 + x^2}$

(2) $\dfrac{x^2}{\sqrt{1 + x^2}}$

(3) $\dfrac{1 - \sqrt{x}}{1 + \sqrt{x}}$

Solution

(1) If $y = \dfrac{1 - x^2}{1 + x^2}$ $u = 1 - x^2$ and $v = 1 + x^2$

$\dfrac{du}{dx} = -2x$ $\dfrac{dv}{dx} = 2x$

Using $\dfrac{dy}{dx} = \dfrac{v\cdot\frac{du}{dx} - u\cdot\frac{dv}{dx}}{v^2}$

$$= \frac{(1+x^2)(-2x) - (1-x^2)(2x)}{(1+x^2)^2}$$

$$= \frac{-2x - 2x^3 - 2x + 2x^3}{(1+x^2)^2}$$

PURE MATHEMATICS — DIFFERENTIAL CALCULUS

$$\therefore \quad \frac{dy}{dx} = \frac{4x}{(1+x^2)^2}$$

---

(2) If $y = \dfrac{x^2}{\sqrt{1+x^2}}$ $u = x^2$ and $v = (1+x^2)^{\frac{1}{2}}$

$$\frac{du}{dx} = 2x \qquad \frac{dv}{dx} = \frac{1}{2}(1+x^2)^{-\frac{1}{2}} 2x$$

Using $\dfrac{dy}{dx} = \dfrac{v \cdot \frac{du}{dx} - u \cdot \frac{dv}{dx}}{v^2}$

$$= \frac{(1+x^2)^{\frac{1}{2}} 2x - x^2 (1+x^2)^{-\frac{1}{2}} x}{\left((1+x^2)^{\frac{1}{2}}\right)^2}$$

$$\therefore \quad \frac{dy}{dx} = \frac{2x\sqrt{1+x^2} - \dfrac{x^3}{\sqrt{1+x^2}}}{1+x^2}$$

In order to obtain a more concise form of answer the numerator will be considered.

Numerator $= 2x\sqrt{1+x^2} - \dfrac{x^3}{\sqrt{1+x^2}}$

$$= \frac{2x\sqrt{1+x^2}(1+x^2) - x^3}{\sqrt{1+x^2}} = \frac{2x(1+x^2) - x^3}{\sqrt{1+x^2}}$$

$$= \frac{2x + 2x^3 - x^3}{\sqrt{1+x^2}} = \frac{x(2+x^2)}{\sqrt{1+x^2}}$$

$$\therefore \quad \frac{dy}{dx} = \frac{\dfrac{x(2+x^2)}{\sqrt{1+x^2}}}{1+x^2} = \frac{x(2+x^2)}{(\sqrt{1+x^2})(1+x^2)}$$

$$= \frac{x(2+x^2)}{\sqrt{(1+x^2)^3}}$$

---

(3) $y = \dfrac{1 - \sqrt{x}}{1 + \sqrt{x}}$

$y = \dfrac{1 - x^{\frac{1}{2}}}{1 + x^{\frac{1}{2}}}$ $u = 1 - x^{\frac{1}{2}}$ $v = 1 + x^{\frac{1}{2}}$

$$\frac{du}{dx} = -\frac{1}{2}x^{-\frac{1}{2}} \qquad \frac{dv}{dx} = \frac{1}{2}x^{-\frac{1}{2}}$$

DIFFERENTIAL CALCULUS — PURE MATHEMATICS

$$\text{Using } \frac{dy}{dx} = \frac{v \cdot \frac{du}{dx} - u \cdot \frac{dv}{dx}}{v^2}$$

$$= \frac{\left(1 + x^{\frac{1}{2}}\right)\left(-\frac{1}{2} x^{-\frac{1}{2}}\right) - \left(1 - x^{\frac{1}{2}}\right)\left(\frac{1}{2} x^{-\frac{1}{2}}\right)}{\left(1 + x^{\frac{1}{2}}\right)^2}$$

$$= \frac{\frac{-(1 + \sqrt{x})}{2\sqrt{x}} - \frac{(1 - \sqrt{x})}{2\sqrt{x}}}{(1 + \sqrt{x})^2}$$

$$= \frac{-(1 + \sqrt{x}) - (1 - \sqrt{x})}{2\sqrt{x}\,(1 + \sqrt{x})^2}$$

$$= \frac{-1 - \sqrt{x} - 1 + \sqrt{x}}{2\sqrt{x}\,(1 + \sqrt{x})^2} = \frac{-2}{2\sqrt{x}\,(1 + \sqrt{x})^2}$$

$$\therefore \frac{dy}{dx} = -\frac{1}{\sqrt{x}\,(1 + \sqrt{x})^2}$$

c)   *Implicit functions*

All the functions that have been considered so far have been *explicit* functions of x, ie with y on one side of the equation and all the terms in x on the other side.

If, however, y is given *implicitly* it means that y cannot be expressed in terms of x, eg the equation of a curve is:

$$x^2 + 2xy - 2y^2 + x = 2$$

There is no way that x and y can be separated onto different sides of the equation, as can be seen from this attempt, ie

$$2xy - 2y^2 = 2 - x - x^2$$

$$y(2x - 2y) = 2 - x - x^2$$

$$\therefore y = \frac{2 - x - x^2}{2x - 2y}$$

Therefore another method is required for differentiating functions of this type, ie implicit functions. The whole expression is differentiated with respect to x and powers of y are differentiated as a 'function of a function' as y is a function of x.

The above example will now be differentiated implicitly to explain the process.

$$x^2 + 2xy - 2y^2 + x = 2$$

$\therefore$ differentiating gives:

$$\frac{d(x^2)}{dx} + \frac{d}{dx}(2xy) - \frac{d}{dx}(2y^2) + \frac{d}{dx}(x) = \frac{d}{dx}(2)$$

PURE MATHEMATICS — DIFFERENTIAL CALCULUS

Considering each term in turn:

(1) $\frac{d}{dx}(x^2) = 2x$

(2) $\frac{d}{dx}(2xy)$ $2xy$ is a product.

∴   if $u = 2x$ and $v = y$

then $\frac{du}{dx} = 2$ $\frac{dv}{dx} = 1 \cdot \frac{dy}{dx} = \frac{dy}{dx}$

∴ $\frac{d}{dx}(2xy) = 2x \cdot \frac{dy}{dx} + y \cdot 2 = 2x\frac{dy}{dx} + 2y$

(3) $\frac{d}{dx}(2y^2) = 4y \cdot \frac{dy}{dx}$ ie $2y^2$ is differentiated with respect to y giving 4y and then y is differentiated with respect to x giving $\frac{dy}{dx}$

(4) $\frac{d}{dx}(x) = 1$

(5) $\frac{d}{dx}(2) = 0$

∴ $x^2 + 2xy - 2y^2 + x = 2$ becomes after differentiation

$(2x) + \left(2x\frac{dy}{dx} + 2y\right) - \left(4y\frac{dy}{dx}\right) + (1) = 0$

$2x + 2x\frac{dy}{dx} + 2y - 4y\frac{dy}{dx} + 1 = 0$

$2x\frac{dy}{dx} - 4y\frac{dy}{dx} = -2x - 2y - 1$

$\frac{dy}{dx}(2x - 4y) = -(2x + 2y + 1)$

∴ $\frac{dy}{dx} = \frac{-(2x + 2y + 1)}{(2x - 4y)}$

Example 11

Differentiate the following functions:

(1) $x^2 + 2xy + y^2 = 3$

(2) $x^2 - 3xy + y^2 - 2y + 4x = 0$

(3) $x^2 + 3xy - y^2 = 3$

DIFFERENTIAL CALCULUS PURE MATHEMATICS

Solution

(1) $x^2 + 2xy + y^2 = 3$ differentiating implicitly

$$\frac{d}{dx}(x^2) + \frac{d}{dx}(2xy) + \frac{d}{dx}(y^2) = \frac{d}{dx}(3)$$

$$(2x) + \left(2x\frac{dy}{dx} + 2y\right) + \left(2y\frac{dy}{dx}\right) = 0$$

$\therefore \quad 2x\frac{dy}{dx} + 2y\frac{dy}{dx} = -2x - 2y$

$\frac{dy}{dx} 2(x+y) = -2(x+y)$

$\therefore \quad \frac{dy}{dx} = \frac{-2(x+y)}{2(x+y)} = -1$

---

(2) $x^2 - 3xy + y^2 - 2y + 4x = 0$

$$\frac{d}{dx}(x^2) - \frac{d}{dx}(3xy) + \frac{d}{dx}(y^2) - \frac{d}{dx}(2y) + \frac{d}{dx}(4x) = 0$$

$$(2x) - \left(3x\frac{dy}{dx} + 3y\right) + \left(2y\frac{dy}{dx}\right) - \left(2\frac{dy}{dx}\right) + (4) = 0$$

$2x - 3x\frac{dy}{dx} - 3y + 2y\frac{dy}{dx} - 2\frac{dy}{dx} + 4 = 0$

$\frac{dy}{dx}(-3x + 2y - 2) = -2x + 3y - 4$

$\therefore \quad \frac{dy}{dx} = \frac{-2x + 3y - 4}{-3x + 2y - 2}$

---

(3) $x^2 + 3xy - y^2 = 3$

$$\frac{d}{dx}(x^2) + \frac{d}{dx}(3xy) - \frac{d}{dx}(y^2) = \frac{d}{dx}(3)$$

$$(2x) + \left(3x\frac{dy}{dx} + 3y\right) - \left(2y\frac{dy}{dx}\right) = 0$$

$3x\frac{dy}{dx} - 2y\frac{dy}{dx} = -2x - 3y$

$\frac{dy}{dx}(3x - 2y) = -2x - 3y$

$\therefore \quad \frac{dy}{dx} = \frac{-2x - 3y}{3x - 2y}$

PURE MATHEMATICS — DIFFERENTIAL CALCULUS

d) *Parameters*

On occasions x and y are given in terms of another variable, say t,

ie $\quad x =$ function of t

and $\quad y =$ function of t

This variable is known as a parameter, eg

$$x = t^3 + t^2 \quad \text{and} \quad y = t^2 + t$$

In order to find $\dfrac{dy}{dx}$ it is necessary to find $\dfrac{dx}{dt}$ and $\dfrac{dy}{dt}$ and then use:

$$\boxed{\dfrac{dy}{dx} = \dfrac{dy}{dt} \cdot \dfrac{dt}{dx} \quad \text{where} \quad \dfrac{dt}{dx} = \dfrac{1}{\dfrac{dx}{dt}}}$$

Therefore, for the above example:

$$x = t^3 + t^2 \qquad y = t^2 + t$$

$$\dfrac{dx}{dt} = 3t^2 + 2t \qquad \dfrac{dy}{dt} = 2t + 1$$

Using $\quad \dfrac{dy}{dx} = \dfrac{dy}{dt} \cdot \dfrac{dt}{dx}$

$$= (2t + 1) \cdot \dfrac{1}{t(3t + 2)}$$

$$\therefore \quad \dfrac{dy}{dx} = \dfrac{2t + 1}{t(3t + 2)}$$

Example 12

Differentiate the following to find $\dfrac{dy}{dx}$:

(1) $\quad x = (t + 1)^2, \; y = (t^2 - 1)$

(2) $\quad x = \dfrac{t}{1 - t}, \; y = \dfrac{t^2}{1 - t}$

(3) $\quad x = \dfrac{2t}{t + 2}, \; y = \dfrac{3t}{t + 3}$

Solution

(1) $\qquad x = (t + 1)^2 \qquad\qquad y = t^2 - 1$

$\qquad \therefore \quad \dfrac{dx}{dt} = 2(t + 1) \qquad \dfrac{dy}{dt} = 2t$

$\qquad$ Using $\quad \dfrac{dy}{dx} = \dfrac{dy}{dt} \cdot \dfrac{dt}{dx}$

$\qquad\qquad = 2t \cdot \dfrac{1}{2(t + 1)}$

DIFFERENTIAL CALCULUS — PURE MATHEMATICS

$$\therefore \quad \frac{dy}{dx} = \frac{t}{t+1}$$

(2) $x = \dfrac{t}{1-t}$, $y = \dfrac{t^2}{1-t}$ these both have to be differentiated as quotients.

$$\frac{dx}{dt} = \frac{(1-t)(1) - (t)(-1)}{(1-t)^2} \qquad \frac{dy}{dt} = \frac{(1-t)(2t) - (t^2)(-1)}{(1-t)^2}$$

$$= \frac{1 - t + t}{(1-t)^2} \qquad\qquad\qquad = \frac{2t - 2t^2 + t^2}{(1-t)^2}$$

$$= \frac{1}{(1-t)^2} \qquad\qquad\qquad = \frac{2t - t^2}{(1-t)^2} \qquad = \frac{t(2-t)}{(1-t)^2}$$

Using $\dfrac{dy}{dx} = \dfrac{dy}{dt} \cdot \dfrac{dt}{dx}$

$$= \frac{t(2-t)}{(1-t)^2} \cdot \frac{1}{\dfrac{1}{(1-t)^2}}$$

$$= \frac{t(2-t)}{(1-t)^2} \cdot (1-t)^2$$

$$\therefore \quad \frac{dy}{dx} = t(2-t)$$

(3) $x = \dfrac{2t}{t+2}$ \qquad $y = \dfrac{3t}{t+3}$ \quad both quotients again!

$$\frac{dx}{dt} = \frac{(t+2)(2) - (2t)(1)}{(t+2)^2} \qquad \frac{dy}{dt} = \frac{(t+3)(3) - (3t)(1)}{(t+3)^2}$$

$$= \frac{2t + 4 - 2t}{(t+2)^2} \qquad\qquad\qquad = \frac{3t + 9 - 3t}{(t+3)^2}$$

$$= \frac{4}{(t+2)^2} \qquad\qquad\qquad\qquad = \frac{9}{(t+3)^2}$$

Using $\dfrac{dy}{dx} = \dfrac{dy}{dt} \cdot \dfrac{dt}{dx}$

$$= \frac{9}{(t+3)^2} \cdot \frac{1}{\left[\dfrac{4}{(t+2)^2}\right]}$$

$$= \frac{9}{(t+3)^2} \cdot \frac{(t+2)^2}{4}$$

$$\therefore \quad \frac{dy}{dx} = \frac{9(t+2)^2}{4(t+3)^2}$$

f) *Example* ('A' level question)

i) Find, in its simplest form, $\dfrac{dy}{dx}$ when:

$$y = \frac{\sqrt{1+x^3}}{x^3}$$

ii) Given that $x = \dfrac{t}{1+t^2}$ and $y = \dfrac{1}{1+t^2}$ prove that

$$(1-t^2)\frac{dy}{dx} + 2t = 0$$

Solution

i) $y = \dfrac{\sqrt{1+x^3}}{x^3} = \dfrac{(1+x^3)^{\frac{1}{2}}}{x^3} = \dfrac{u}{v}$ This is a quotient form.

$\therefore \quad u = (1+x^3)^{\frac{1}{2}} \qquad\qquad v = x^3$

$\therefore \quad \dfrac{du}{dx} = \dfrac{1}{2}(1+x^3)^{-\frac{1}{2}}(3x^2) \quad\therefore\quad \dfrac{dv}{dx} = 3x^2$

Using $\dfrac{dy}{dx} = \dfrac{v \cdot \dfrac{du}{dx} - u \cdot \dfrac{dv}{dx}}{v^2}$

$$\frac{dy}{dx} = \frac{x^3 \cdot \frac{1}{2}(1+x^3)^{-\frac{1}{2}}(3x^2) - (1+x^3)^{\frac{1}{2}}(3x^2)}{(x^3)^2}$$

$$= \frac{\frac{3}{2}x^5(1+x^3)^{-\frac{1}{2}} - 3x^2(1+x^3)^{\frac{1}{2}}}{x^6}$$

In order to rearrange this formula the numerator will be considered first.

Numerator $= \dfrac{3}{2} \dfrac{x^5}{(1+x^3)^{\frac{1}{2}}} - 3x^2(1+x^3)^{\frac{1}{2}}$

$$= \frac{3x^5 - 3x^2(1+x^3)^{\frac{1}{2}} \cdot 2(1+x^3)^{\frac{1}{2}}}{2(1+x^3)^{\frac{1}{2}}}$$

$$= \frac{3x^5 - 6x^2(1+x^3)}{2(1+x^3)^{\frac{1}{2}}}$$

$$= \frac{3x^5 - 6x^2 - 6x^5}{2(1+x^3)^{\frac{1}{2}}}$$

DIFFERENTIAL CALCULUS   PURE MATHEMATICS

$$= \frac{-6x^2 - 3x^5}{2(1+x^3)^{\frac{1}{2}}}$$

$$= \frac{-3x^2(2+x^3)}{2(1+x^3)^{\frac{1}{2}}}$$

$$\therefore \quad \frac{dy}{dx} = \frac{-3x^2(2+x^3)}{2(1+x^3)^{\frac{1}{2}} \cdot x^6} = \frac{-3(2+x^3)}{2x^4(1+x^3)^{\frac{1}{2}}}$$

ii) $x = \dfrac{t}{1+t^2} \left(= \dfrac{u}{v} \right)$ $y = \dfrac{1}{1+t^2} = (1+t^2)^{-1}$

$$\therefore \quad \frac{dx}{dt} = \frac{v \cdot \frac{du}{dt} - u \cdot \frac{dv}{dt}}{v^2} \qquad \frac{dy}{dt} = -1(1+t^2)^{-2} \cdot 2t$$

$$\frac{dx}{dt} = \frac{(1+t^2) \cdot 1 - t \cdot (2t)}{(1+t^2)^2} \qquad \frac{dy}{dt} = \frac{-2t}{(1+t^2)^2}$$

$$= \frac{1 + t^2 - 2t^2}{(1+t^2)^2}$$

$$= \frac{1 - t^2}{(1+t^2)^2}$$

Using $\dfrac{dy}{dx} = \dfrac{dy}{dt} \cdot \dfrac{dt}{dx}$

$$= \frac{-2t}{(1+t^2)^2} \cdot \frac{1}{\left[\frac{1-t^2}{(1+t^2)^2}\right]} = \frac{-2t}{(1+t^2)^2} \cdot \frac{(1+t^2)^2}{(1-t^2)}$$

$$\therefore \quad \frac{dy}{dx} = \frac{-2t}{1-t^2}$$

$$(1-t^2) \cdot \frac{dy}{dx} = -2t$$

$(1-t^2) \cdot \dfrac{dy}{dx} + 2t = 0$ as required

PURE MATHEMATICS — DIFFERENTIAL CALCULUS

4.3 Other functions

a) *Trigonometric functions*

Trigonometric functions can be differentiated with the following results:

$$\text{If } y = \sin x \text{ then } \frac{dy}{dx} = \cos x$$
$$y = \cos x \text{ then } \frac{dy}{dx} = -\sin x$$

Note: The angle x is given in radians not degrees.

All the rules already given for algebraic differentiation in the preceeding two sections apply to these trigonometric functions, in particular the 'function of a function' rule.

So for differentiating $\quad y = 3\sin(5x - 1) \quad$ let $t = 5x - 1$

$$\therefore \quad y = 3\sin t$$

$$\therefore \quad \frac{dy}{dt} = 3\cos t \quad \text{and} \quad \frac{dt}{dx} = 5$$

Using $\quad \dfrac{dy}{dx} = \dfrac{dy}{dt} \cdot \dfrac{dt}{dx}$

$$= (3\cos t)\,5 = 15\cos t$$

$$\therefore \quad \frac{dy}{dx} = 15\cos(5x - 1)$$

or for a function such as $\quad y = \cos^3 x \quad$ let $t = \cos x$

$$\therefore \quad y = t^3$$

$$\therefore \quad \frac{dy}{dt} = 3t^2 \quad \text{and} \quad \frac{dt}{dx} = -\sin x$$

Using $\quad \dfrac{dy}{dx} = \dfrac{dy}{dt} \cdot \dfrac{dt}{dx}$

$$= 3t^2 \cdot (-\sin x)$$

$$= 3\cos^2 x \cdot (-\sin x) = -3\cos^2 x \cdot \sin x$$

The following example will show that the product and quotient formulae also apply.

Example 13

Differentiate the following functions:

(1) $y = -2\sin^3 3x$

(2) $y = \sqrt{\sin 2x}$

(3) $y = x^2 \sin x$

(4) $y = \dfrac{x^2}{\cos x}$

(5) $y = \tan x$

DIFFERENTIAL CALCULUS　　　　PURE MATHEMATICS

Solution

(1)　　$y = -2\sin^3 3x$

$\dfrac{dy}{dx} = -2 \cdot (3\sin^2 3x)(3\cos 3x)$

　　　$= -18\sin^2 3x \cos 3x$

The 'function of a function' rule has been used without actually introducing t. First the power was differentiated ($3\sin^2 3x$) and then $\sin 3x$ was differentiated ($3\cos 3x$).

---

(2)　　$y = \sqrt{\sin 2x} = \sin^{\frac{1}{2}} 2x$

$\therefore \quad \dfrac{dy}{dx} = \left(\dfrac{1}{2}\sin^{-\frac{1}{2}} 2x\right)(2\cos(2x))$

　　　　$= \sin^{-\frac{1}{2}}(2x) \cos(2x)$

$\therefore \quad \dfrac{dy}{dx} = \dfrac{\cos 2x}{\sqrt{\sin 2x}}$

---

(3)　　$y = x^2 \sin x$　　　- a product form

　　　$y = u.v.$　　　$u = x^2$　and　$v = \sin x$

　　　　　$\dfrac{du}{dx} = 2x$　　　　$\dfrac{dv}{dx} = \cos x$

Using　　$\dfrac{dy}{dx} = u.\dfrac{dv}{dx} + v.\dfrac{du}{dx}$

　　　　　　$= x^2.\cos x + 2x \sin x$

$\therefore \quad \dfrac{dy}{dx} = x(x\cos x + 2\sin x)$

---

(4)　　$y = \dfrac{x^2}{\cos x}$　　　a quotient form

　　　$y = \dfrac{u}{v}$　　　$u = x^2$　　and　　$v = \cos x$

　　　　　$\dfrac{du}{dx} = 2x$　　　　$\dfrac{dv}{dx} = -\sin x$

Using　　$\dfrac{dy}{dx} = \dfrac{v.\dfrac{du}{dx} - u.\dfrac{du}{dx}}{v^2}$

　　　　　　$= \dfrac{(\cos x)(2x) - x^2(-\sin x)}{(\cos x)^2}$

$$= \frac{2x \cos x + x^2 \sin x}{\cos^2 x}$$

$$\therefore \quad \frac{dy}{dx} = \frac{x(2\cos x + x \sin x)}{\cos^2 x}$$

(5) $\quad y = \tan x = \dfrac{\sin x}{\cos x} \quad$ - a quotient form

$$y = \frac{u}{v} \qquad u = \sin x \quad \text{and} \quad v = \cos x$$

$$\frac{du}{dx} = \cos x \qquad \frac{dv}{dx} = -\sin x$$

Using $\quad \dfrac{dy}{dx} = \dfrac{v \cdot \frac{du}{dx} - u \cdot \frac{dv}{dx}}{v^2}$

$$= \frac{(\cos x)(\cos x) - (\sin x)(-\sin x)}{(\cos x)^2}$$

$$= \frac{\cos^2 x + \sin^2 x}{\cos^2 x} = \frac{1}{\cos^2 x}$$

$$\therefore \quad \frac{dy}{dx} = \sec^2 x$$

This gives another useful result:

$$\boxed{\text{If } y = \tan x \quad \frac{dy}{dx} = \sec^2 x}$$

b) *Inverse trigonometric functions*

These are functions of the type $\sin^{-1}x$, $\cos^{-1}x$, $\tan^{-1}x$ and, whilst formulae can be quoted for differentiating these functions, the method of differentiation is very important.

Consider the function $\sin^{-1}x$

Let $\quad y = \sin^{-1}x$

$\therefore \quad \sin y = x \qquad$ differentiating both sides with respect to x gives

$$\frac{d}{dx}(\sin y) = \frac{d}{dx}(x)$$

$$\cos y \cdot \frac{dy}{dx} = 1$$

$\therefore \quad \dfrac{dy}{dx} = \dfrac{1}{\cos y} \quad$ but the answer needs to be given in terms of x

$$\sin^2 y + \cos^2 y = 1$$

$\therefore \quad x^2 + \cos^2 y = 1$

$\quad \cos^2 y = 1 - x^2$

DIFFERENTIAL CALCULUS — PURE MATHEMATICS

$$\cos y = \sqrt{1 - x^2}$$

So

$$\boxed{\frac{d}{dx}(\sin^{-1}x) = \frac{1}{\sqrt{1-x^2}}}$$

Results can be derived for $\cos^{-1}x$ and $\tan^{-1}x$ in a similar way. They are:

$$\boxed{\frac{d}{dx}(\cos^{-1}x) = \frac{-1}{\sqrt{1-x^2}}}$$

$$\boxed{\frac{d}{dx}(\tan^{-1}x) = \frac{1}{1+x^2}}$$

Example 14

Differentiate the following functions:

(1) $\cos^{-1}(2x - 1)$

(2) $\text{cosec}^{-1}(1 + x^2)$

(3) $\tan^{-1}\left(\frac{1}{x^2}\right)$

Solution

(1) Let $y = \cos^{-1}(2x - 1)$

$\therefore \cos y = 2x - 1$

$\frac{d}{dx}(\cos y) = \frac{d}{dx}(2x - 1)$

$-\sin y \cdot \frac{dy}{dx} = 2$

$\therefore \frac{dy}{dx} = \frac{2}{-\sin y}$ but $\cos^2 y + \sin^2 y = 1$

$(2x - 1)^2 + \sin^2 y = 1$

$4x^2 - 4x + 1 + \sin^2 y = 1$

$\sin^2 y = 1 - 4x^2 + 4x - 1$

$\sin^2 y = 4(x - x^2)$

$\sin y = 2(\sqrt{x - x^2})$

$\therefore \frac{dy}{dx} = \frac{2}{-2\sqrt{x - x^2}} = \frac{-1}{\sqrt{x(1 - x)}}$

PURE MATHEMATICS — DIFFERENTIAL CALCULUS

(2) Let $y = \operatorname{cosec}^{-1}(1 + x^2)$

$\therefore \operatorname{cosec} y = 1 + x^2$

$\dfrac{1}{\sin y} = 1 + x^2$

$\sin y = \dfrac{1}{1 + x^2} = (1 + x^2)^{-1}$

$\dfrac{d}{dx}(\sin y) = \dfrac{d}{dx}((1 + x^2)^{-1})$

$\cos y \dfrac{dy}{dx} = -1(1 + x^2)^{-2}(2x) = \dfrac{-2x}{(1 + x^2)^2}$

$\dfrac{dy}{dx} = \dfrac{-2x}{\cos y (1 + x^2)^2}$ and $\cos^2 y + \sin^2 y = 1$

$\cos^2 y + \left(\dfrac{1}{1 + x^2}\right)^2 = 1$

$\cos^2 y = 1 - \dfrac{1}{(1 + x^2)^2}$

$\cos^2 y = \dfrac{(1 + x^2)^2 - 1}{(1 + x^2)^2}$

$\cos^2 y = \dfrac{1 + 2x^2 + x^4 - 1}{(1 + x^2)^2}$

$= \dfrac{x^2(2 + x^2)}{(1 + x^2)^2}$

$\therefore \dfrac{dy}{dx} = \dfrac{-2x}{\dfrac{x}{(1 + x^2)} \cdot \sqrt{2 + x^2} \cdot (1 + x^2)^2}$ $\therefore \cos y = \dfrac{x}{(1 + x^2)}\sqrt{2 + x^2}$

$\therefore \dfrac{dy}{dx} = \dfrac{-2}{(1 + x^2)\sqrt{2 + x^2}}$

---

(3) Let $y = \tan^{-1}\left(\dfrac{1}{x^2}\right)$

$\therefore \tan y = \dfrac{1}{x^2} = x^{-2}$

$\therefore \dfrac{d}{dx}(\tan y) = \dfrac{d}{dx}(x^{-2})$

$\sec^2 y \dfrac{dy}{dx} = -2x^{-3} = \dfrac{-2}{x^3}$

$\therefore \dfrac{dy}{dx} = \dfrac{-2}{x^3 \sec^2 y}$ $\sec^2 y = 1 + \tan^2 y$

$= 1 + \dfrac{1}{(x^2)^2}$

DIFFERENTIAL CALCULUS PURE MATHEMATICS

$$\sec^2 y = 1 + \frac{1}{x^4}$$

$$\therefore \quad \frac{dy}{dx} = \frac{-2}{\left(\frac{x^4+1}{x^4}\right)x^3} \qquad \therefore \quad \sec^2 y = \frac{x^4+1}{x^4}$$

$$\therefore \quad \frac{dy}{dx} = \frac{-2}{\left(\frac{x^4+1}{x}\right)} = \frac{-2x}{x^4+1}$$

c) *Exponential functions*

$$\boxed{\text{If } y = e^x \text{ then } \frac{dy}{dx} = e^x}$$

Exponential functions are not always in such a simple form. They are more likely to be like:

$$e^{\sqrt{x^2+1}}$$

which again is a 'function of a function' of x.

Let $\quad y = e^{\sqrt{x^2+1}} \qquad\qquad$ and $t = (x^2+1)^{\frac{1}{2}}$

$\therefore \quad y = e^t$

$\therefore \quad \dfrac{dy}{dt} = e^t \qquad\qquad \dfrac{dt}{dx} = \dfrac{1}{2}(x^2+1)^{-\frac{1}{2}} 2x = \dfrac{x}{(x^2+1)^{\frac{1}{2}}}$

Using $\quad \dfrac{dy}{dx} = \dfrac{dy}{dt} \cdot \dfrac{dt}{dx}$

$$= e^t \cdot \frac{x}{(x^2+1)^{\frac{1}{2}}}$$

$$\therefore \quad \frac{dy}{dx} = \frac{\left(e^{\sqrt{x^2+1}}\right)x}{\sqrt{x^2+1}}$$

Example 15

Differentiate the following functions:

(1) $\quad e^{2x^2}$

(2) $\quad e^{\sin 2x}$

(3) $\quad e^x \sin x$

(4) $\quad x^2 e^x$

(5) $\quad e^{x^2} \csc x$

PURE MATHEMATICS — DIFFERENTIAL CALCULUS

Solution

(1) Let $y = e^{2x^2}$

$\therefore \quad \dfrac{dy}{dx} = 4x \cdot e^{2x^2}$

Again the 'function of a function' rule without introducing t. First the power was differentiated ($4x$) and then e^{2x^2} itself (e^{2x^2}).

---

(2) Let $y = e^{\sin 2x}$

$\therefore \quad \dfrac{dy}{dx} = (2\cos 2x)\, e^{\sin 2x}$

Again the power was differentiated first ($2\cos 2x$) and then $e^{\sin 2x}$ itself ($e^{\sin 2x}$).

---

(3) Let $y = e^{x \sin x}$

$\therefore \quad \dfrac{dy}{dx} = (x \cdot \cos x + \sin x) e^{x \sin x}$

The power is itself a product $x \cdot \sin x$ and so has been differentiated using the formula for u.v. ie ($x \cos x + \sin x$) and then $e^{x \sin x}$ has been differentiated ($e^{x \sin x}$).

---

(4) Let $y = x^2 e^x$ a product form

$y = u.v.$ with $u = x^2$ and $v = e^x$

$\therefore \quad \dfrac{dy}{dx} = x^2 \cdot e^x + e^x \cdot 2x$

$\therefore \quad \dfrac{dy}{dx} = x \cdot (x + 2) e^x$

---

(5) Let $y = e^{x^2} \operatorname{cosec} x = \dfrac{e^{x^2}}{\sin x}$ a quotient form

$y = \dfrac{u}{v}$ with $u = e^{x^2}$ and $v = \sin x$

$\therefore \quad \dfrac{dy}{dx} = \dfrac{\sin x (2x\, e^{x^2}) - e^{x^2} (\cos x)}{(\sin x)^2}$

$\therefore \quad \dfrac{dy}{dx} = \dfrac{(2x \sin x - \cos x) e^{x^2}}{\sin^2 x}$

However, the question was given in terms of cosec x.

$\therefore \quad \dfrac{dy}{dx} = \left(\dfrac{2x \sin x}{\sin^2 x} - \dfrac{\cos x}{\sin^2 x} \right) e^{x^2}$

$\qquad\qquad = \left(\dfrac{2x}{\sin x} - \dfrac{\cos x}{\sin x} \cdot \dfrac{1}{\sin x} \right) e^{x^2}$

DIFFERENTIAL CALCULUS   PURE MATHEMATICS

$$= (2x \operatorname{cosec} x - \cot x . \operatorname{cosec} x)e^{x^2}$$

$$\therefore \quad \frac{dy}{dx} = e^{x^2}(2x - \cot x) \operatorname{cosec} x$$

d) *Logarithmic functions*

In order to differentiate $\log_e x$ it is necessary to use the basic definition of a logarithm.

If $\quad y = \log_e x$

then $\quad e^y = x$

So $\quad \frac{d}{dx}(e^y) = \frac{d}{dx}(x)$

$$e^y \frac{dy}{dx} = 1$$

$$\therefore \quad \frac{dy}{dx} = \frac{1}{e^y} = \frac{1}{x}$$

$$\boxed{\text{So if } y = \log_e x \quad \text{then} \quad \frac{dy}{dx} = \frac{1}{x}}$$

Needless to say, logarithmic functions are usually more complicated than this, $\log_e(x^3 - 2)$ for instance, which is a 'function of a function' of x.

Let $\quad y = \log_e(x^3 - 2) \quad$ and $\quad t = x^3 - 2$

$\therefore \quad y = \log_e t$

$\therefore \quad \frac{dy}{dt} = \frac{1}{t} \quad$ and $\quad \frac{dt}{dx} = 3x^2$

$\therefore \quad \frac{dy}{dx} = \frac{dy}{dt} \cdot \frac{dt}{dx}$

$$= \frac{1}{t} \cdot 3x^2$$

$\therefore \quad \frac{dy}{dx} = \frac{3x^2}{(x^3 - 2)}$

Example 16

Differentiate the following functions:

(1) $\quad \log_e 3x^2$

(2) $\quad \log_e \cos x$

(3) $\quad \log_e \cos^3 2x$

(4) $\quad \log_e \frac{(x + 1)^2}{\sqrt{x - 1}}$

PURE MATHEMATICS — DIFFERENTIAL CALCULUS

(5) $\log_e(x\sqrt{x^2 - 1})$

(6) $x.\log_e x$

(7) $\dfrac{\log_e x}{x^2}$

Solution

(1) Let $y = \log_e 3x^2$

$$\frac{dy}{dx} = \frac{6x}{3x^2} = \frac{2}{x}$$

The 'function of a function' method has been used without introducing t. The function of x has been differentiated $(6x)$ and then the logarithm itself $\left(\dfrac{1}{3x^2}\right)$.

---

(2) Let $y = \log_e \cos x$

$$\frac{dy}{dx} = \frac{-\sin x}{\cos x} = -\tan x$$

As before, the function of x has been differentiated $(-\sin x)$ and then the logarithm $\left(\dfrac{1}{\cos x}\right)$

---

(3) Let $y = \log_e \cos^3 2x = \log_e (\cos 2x)^3 = 3 \log_e(\cos 2x)$

$$\therefore \quad \frac{dy}{dx} = 3\left(\frac{-2 \sin 2x}{\cos 2x}\right) = \frac{-6 \sin 2x}{\cos 2x}$$

$$= -6 \tan 2x$$

So, differentiating the function of x $(-2 \sin 2x)$ is followed by differentiating the logarithm itself $\left(\dfrac{1}{\cos 2x}\right)$

---

(4) Let $y = \log_e \dfrac{(x+1)^2}{\sqrt{x-1}} = \log_e(x+1)^2 - \log_e(x-1)^{\frac{1}{2}}$

$$y = 2\log_e(x+1) - \frac{1}{2}\log_e(x-1)$$

$$\therefore \quad \frac{dy}{dx} = 2\left(\frac{1}{x+1}\right) - \frac{1}{2}\left(\frac{1}{x-1}\right)$$

$$= \frac{2.2(x-1) - 1(x+1)}{(x+1)(2)(x-1)}$$

$$= \frac{4x - 4 - x - 1}{2(x+1)(x-1)}$$

$$\therefore \quad \frac{dy}{dx} = \frac{3x - 5}{2(x+1)(x-1)}$$

(5) Let $y = \log_e(x\sqrt{x^2-1}) = \log_e\left(x(x^2-1)^{\frac{1}{2}}\right)$

$$= \log_e x + \log_e(x^2-1)^{\frac{1}{2}}$$

$$= \log_e x + \frac{1}{2}\log_e(x^2-1)$$

$\therefore \dfrac{dy}{dx} = \dfrac{1}{x} + \dfrac{1}{2}\left(\dfrac{2x}{x^2-1}\right)$

$$= \frac{2(x^2-1) + 2x \cdot (x)}{(x)(2)(x^2-1)}$$

$$= \frac{2x^2 - 2 + 2x^2}{2x(x^2-1)} = \frac{4x^2 - 2}{2x(x^2-1)}$$

$\therefore \dfrac{dy}{dx} = \dfrac{2x^2 - 1}{x(x^2-1)}$

---

(6) Let $y = x \cdot \log_e x$ a product form

$y = u.v.$ with $u = x$ $v = \log_e x$

$\dfrac{du}{dx} = 1$ $\dfrac{dv}{dx} = \dfrac{1}{x}$

$\therefore \dfrac{dy}{dx} = x \cdot \dfrac{1}{x} + \log_e x \cdot 1$

$= 1 + \log_e x$

---

(7) Let $y = \dfrac{\log_e x}{x^2}$ a quotient form

$y = \dfrac{u}{v}$ with $u = \log_e x$ $v = x^2$

$\dfrac{du}{dx} = \dfrac{1}{x}$ $\dfrac{dv}{dx} = 2x$

$\therefore \dfrac{dy}{dx} = \dfrac{x^2\left(\dfrac{1}{x}\right) - \log_e x (2x)}{(x^2)^2}$

$$= \frac{x - 2x\log_e x}{x^4} = \frac{x(1 - 2\log_e x)}{x^4}$$

$\therefore \dfrac{dy}{dx} = \dfrac{1 - 2\log_e x}{x^3}$

PURE MATHEMATICS — DIFFERENTIAL CALCULUS

e) *Example* ('A' level question)

If $y = 1 - \cos 2t$ and $x = \sqrt{1 + t^2}$ show that

$$\frac{dy}{dx} = \frac{2\sqrt{1 + t^2} \cdot \sin 2t}{t}$$

Find the first three non-zero terms in the expansion, in ascending powers of t, of $\frac{dy}{dx}$

Solution

These are parametric equations, ie x and y are both given in terms of t.

$$y = 1 - \cos 2t \qquad\qquad x = \sqrt{1 + t^2} = (1 + t^2)^{\frac{1}{2}}$$

$$\frac{dy}{dt} = 0 - (-2 \sin 2t) \qquad\qquad \frac{dx}{dt} = \frac{1}{2}(1 + t^2)^{-\frac{1}{2}}(2t)$$

$$\qquad = 2 \sin 2t \qquad\qquad\qquad\qquad = \frac{t}{(1 + t^2)^{\frac{1}{2}}}$$

Using $\frac{dy}{dx} = \frac{dy}{dt} \cdot \frac{dt}{dx}$

$$= (2 \sin 2t)\left[\frac{1}{\left[\frac{t}{(1+t^2)^{\frac{1}{2}}}\right]}\right] \qquad = (2 \sin 2t)\frac{(1+t^2)^{\frac{1}{2}}}{t}$$

$$\therefore \quad \frac{dy}{dx} = \frac{2 \sin 2t \sqrt{1+t^2}}{t}$$

To obtain a series expansion, it is necessary to expand $\sqrt{1 + t^2}$ binomially and to write sin 2t as a power series.

$$(1 + t^2)^{\frac{1}{2}} = 1 + \left(\frac{1}{2}\right)t^2 + \frac{\left(\frac{1}{2}\right)\left(-\frac{1}{2}\right)}{2!}(t^2)^2 + \ldots$$

$$= 1 + \frac{t^2}{2} - \frac{t^4}{8}$$

$$\sin 2t = 2t - \frac{(2t)^3}{3!} + \frac{(2t)^5}{5!} - \ldots$$

$$= 2t - \frac{8t^3}{6} + \frac{32t^5}{120} - \ldots$$

$$= 2t - \frac{4t^3}{3} + \frac{4t^5}{15} - \ldots$$

$$(\sin 2t)(1 + t^2)^{\frac{1}{2}} = \left(2t - \frac{4t^3}{3} + \frac{4t^5}{15} - \ldots\right)\left(1 + \frac{t^2}{2} - \frac{t^4}{8} + \ldots\right)$$

$$= 2t.(1) + 2t\left(\frac{t^2}{2}\right) + 2t\left(-\frac{t^4}{8}\right) - \frac{4t^3}{3}.(1) - \frac{4t^3}{3}\left(\frac{t^2}{2}\right) + \frac{4t^5}{15}.(1)$$

$$= 2t + t^3 - \frac{t^5}{4} - \frac{4t^3}{3} - \frac{2t^5}{3} + \frac{4t^5}{15}$$

$$= 2t - \frac{t^3}{3} - \frac{39t^5}{60}$$

(*Note*: Higher order powers of x were ignored when the brackets were multiplied.)

$$\therefore \quad \frac{dy}{dx} = \frac{2}{t}\left(2t - \frac{t^3}{3} - \frac{39t^5}{60}\right)$$

$$= 4 - \frac{2t^2}{3} - \frac{13t^4}{10}$$

As can be seen from this question and solution the topics that have so far been met and treated separately do not exist in isolation when it comes to answering questions.

4.4 Applications of differentiation

a) *Equations of tangents and normals*

As has already been discussed differentiation is a means of finding the gradients of tangents and normals at specific points on a curve. The next step following from this will be to use the gradient to find the actual equation of the tangent (or normal) at that point. Since there are several methods of finding the equation of a straight line, depending on the information given, any further work on this topic will be delayed to a later chapter.

b) *Successive differentiation*

So far only the first derivative of a function has been considered. However, there are occasions when the first derivative is differentiated to give the second derivative and so on, eg

If $y = 3x^6$

then $\frac{dy}{dx} = 3.6x^5 = 18x^5$

and differentiating both sides of this expression gives

$$\frac{d}{dx}\left(\frac{dy}{dx}\right) = 18.5x^4$$

$$\therefore \quad \frac{d^2y}{dx^2} = 90x^4 \text{ is the second derivative}$$

and differentiating both sides of this expression again gives

$$\frac{d}{dx}\left(\frac{d^2y}{dx^2}\right) = 90.4x^3$$

$$\therefore \quad \frac{d^3y}{dx^3} = 360x^3$$

and so on...

PURE MATHEMATICS DIFFERENTIAL CALCULUS

Example 17

Find the second derivative of the following functions:

(1) $\quad y = \dfrac{x^2}{\sqrt{x+1}}$

(2) $\quad x = (t^2 - 1)^2, \quad y = t^3$

(3) $\quad x^2 + 3xy - y^2 = 3$ at the point $(1, 1)$

Solution

(1) $\quad y = \dfrac{x^2}{\sqrt{x+1}} = \dfrac{x^2}{(x+1)^{\frac{1}{2}}} \qquad$ a quotient form

$\qquad y = \dfrac{u}{v} \quad \therefore \quad u = x^2 \quad$ and $\quad v = (x+1)^{\frac{1}{2}}$

$\qquad \dfrac{du}{dx} = 2x \qquad \dfrac{dv}{dx} = \dfrac{1}{2}(x+1)^{-\frac{1}{2}}$

$\therefore \quad \dfrac{dy}{dx} = \dfrac{(x+1)^{\frac{1}{2}} \cdot 2x - x^2 \cdot \frac{1}{2}(x+1)^{-\frac{1}{2}}}{\left((x+1)^{\frac{1}{2}}\right)^2}$

$\qquad = \dfrac{2x(x+1)^{\frac{1}{2}} - \frac{x^2}{2}(x+1)^{-\frac{1}{2}}}{(x+1)}$

Simplifying the numerator:

$\qquad \text{Numerator} = 2x(x+1)^{\frac{1}{2}} - \dfrac{x^2}{2(x+1)^{\frac{1}{2}}}$

$\qquad = \dfrac{2x(x+1)^{\frac{1}{2}} 2(x+1)^{\frac{1}{2}} - x^2}{2(x+1)^{\frac{1}{2}}}$

$\qquad = \dfrac{4x(x+1) - x^2}{2(x+1)^{\frac{1}{2}}} = \dfrac{4x^2 + 4x - x^2}{2(x+1)^{\frac{1}{2}}}$

$\qquad = \dfrac{3x^2 + 4x}{2(x+1)^{\frac{1}{2}}}$

$\therefore \quad \dfrac{dy}{dx} = \dfrac{3x^2 + 4x}{2(x+1)^{\frac{1}{2}}(x+1)} = \dfrac{3x^2 + 4x}{2(x+1)^{\frac{3}{2}}}$

To obtain the second derivative this quotient has to be differentiated again.

DIFFERENTIAL CALCULUS — PURE MATHEMATICS

$$\frac{d}{dx}\left(\frac{dy}{dx}\right) = \frac{d}{dx}\left(\frac{3x^2 + 4x}{2(x+1)^{\frac{3}{2}}}\right) = \frac{d}{dx}\left(\frac{u}{v}\right)$$

$$u = 3x^2 + 4x \quad \text{and} \quad v = 2(x+1)^{\frac{3}{2}}$$

$$\therefore \quad \frac{du}{dx} = 6x + 4 \qquad \frac{dv}{dx} = 2 \cdot \frac{3}{2}(x+1)^{\frac{1}{2}} = 3(x+1)^{\frac{1}{2}}$$

So
$$\frac{d^2y}{dx^2} = \frac{2(x+1)^{\frac{3}{2}}(6x+4) - (3x^2+4x)(3)(x+1)^{\frac{1}{2}}}{\left(2(x+1)^{\frac{3}{2}}\right)^2}$$

$$= \frac{(x+1)^{\frac{1}{2}}[2(x+1)(6x+4) - 3(3x^2+4x)]}{4(x+1)^3}$$

$$= \frac{(x+1)^{\frac{1}{2}}[12x^2 + 8x + 12x + 8 - 9x^2 - 12x]}{4(x+1)^3}$$

$$= \frac{(x+1)^{\frac{1}{2}}[3x^2 + 8x + 8]}{4(x+1)^3}$$

$$\therefore \quad \frac{d^2y}{dx^2} = \frac{3x^2 + 8x + 8}{4(x+1)^{\frac{5}{2}}}$$

---

(2) A parametric equation this time:

$$x = (t^2 - 1)^2 \qquad y = t^3$$

$$\frac{dx}{dt} = 2(t^2 - 1)\cdot(2t) \qquad \frac{dy}{dt} = 3t^2$$

Using $\quad \dfrac{dy}{dx} = \dfrac{dy}{dt} \cdot \dfrac{dt}{dx}$

$$= 3t^2 \cdot \frac{1}{4t(t^2 - 1)}$$

$$\frac{dy}{dx} = \frac{3t}{4(t^2 - 1)}$$

In order to obtain the second derivative both sides will be differentiated with respect to x.

$$\frac{d}{dx}\left(\frac{dy}{dx}\right) = \frac{d}{dx}\left(\frac{3t}{4(t^2 - 1)}\right)$$

so $\quad \dfrac{d^2y}{dx^2} = \dfrac{d}{dt}\left(\dfrac{3t}{4(t^2-1)}\right) \cdot \dfrac{dt}{dx}$

$\dfrac{d}{dt}\left(\dfrac{3t}{4(t^2-1)}\right)$ is a quotient form with $u = 3t \quad v = 4(t^2 - 1)$

$$\therefore \quad \frac{du}{dt} = 3 \qquad \frac{dv}{dt} = 8t$$

So $\dfrac{d}{dt}\left(\dfrac{3t}{4(t^2-1)}\right) = \dfrac{4(t^2-1).3 - 3t.(8t)}{(4(t^2-1))^2}$

$= \dfrac{12t^2 - 12 - 24t^2}{16(t^2-1)^2}$

$= \dfrac{-12 - 12t^2}{16(t^2-1)^2} = \dfrac{-12(1+t^2)}{16(t^2-1)^2}$

$= \dfrac{-3(t^2+1)}{4(t^2-1)^2}$

$\therefore \dfrac{d^2y}{dx^2} = \dfrac{-3(t^2+1)}{4(t^2-1)^2} \cdot \dfrac{1}{4t(t^2-1)}$

$\therefore \dfrac{d^2y}{dx^2} = \dfrac{-3(t^2+1)}{16t(t^2-1)^3}$

(3) An implicit function of x and y:

$x^2 + 3xy - y^2 = 3$

$\therefore \dfrac{d}{dx}(x^2) + \dfrac{d}{dx}(3xy) - \dfrac{d}{dx}(y^2) = \dfrac{d}{dx}(3)$

$2x + \left(3x.\dfrac{dy}{dx} + y.3\right) - 2y\dfrac{dy}{dx} = 0$

$\dfrac{dy}{dx}(3x - 2y) = -2x - 3y$

$\dfrac{dy}{dx} = \dfrac{-(2x+3y)}{(3x-2y)}$

at the point (1, 1) $\dfrac{dy}{dx} = \dfrac{-(2+3)}{(3-2)} = -5$

Differentiating again $\dfrac{d}{dx}\left(\dfrac{dy}{dx}\right) = \dfrac{d}{dx}\left(\dfrac{-(2x+3y)}{(3x-2y)}\right)$ a quotient form

$u = -2x - 3y$ $\qquad v = 3x - 2y$

$\dfrac{du}{dx} = -2 - 3\dfrac{dy}{dx}$ $\qquad \dfrac{dv}{dx} = 3 - 2\dfrac{dy}{dx}$

$\therefore \dfrac{d^2y}{dx^2} = \dfrac{(3x-2y)(-2-3\dfrac{dy}{dx}) - (-2x-3y)\left(3 - 2\dfrac{dy}{dx}\right)}{(3x-2y)^2}$

Since a numerical answer is required no attempt will be made to simplify this but the values $x = 1$, $y = 1$, $\dfrac{dy}{dx} = -5$ will be substituted.

DIFFERENTIAL CALCULUS — PURE MATHEMATICS

$$\frac{d^2y}{dx^2} = \frac{(3-2)(-2-(3)(-5)) - (-2-3)(3-2(-5))}{(3-2)^2}$$

$$= \frac{1(-2+15) - (-5)(3+10)}{1^2}$$

$$= 13 - (-65) = 13 + 65$$

$$\therefore \quad \frac{d^2y}{dx^2} = 78$$

c) Stationary values

The gradient of a tangent to a curve is zero when the tangent is parallel to the x axis. If the gradient of the tangent is zero then:

$$\boxed{\frac{dy}{dx} = 0 \text{ and } y \text{ is said to have a stationary value}}$$

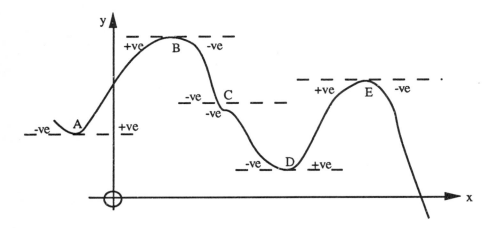

On this curve there are five stationary points:

A } are called minimum turning points and the gradient of the curve is changing
D } from negative to positive.

B } are called maximum turning points and the gradient of the curve is changing
E } from positive to negative.

C } is called a point of inflexion and the gradient is not changing in sign.

Therefore, stationary values can be found by obtaining $\frac{dy}{dx}$ from the equation of the curve and putting this equal to zero.

So, taking the curve $y = 3x^5 - 5x^3$ the stationary values can be found by differentiating the function of x:

$$\frac{dy}{dx} = 3.5x^4 - 5.3x^2$$

$$= 15x^4 - 15x^2$$

$$= 15x^2(x^2 - 1)$$

PURE MATHEMATICS — DIFFERENTIAL CALCULUS

and then equating $\frac{dy}{dx} = 0$ ∴ $15x^2(x^2 - 1) = 0$

either $15x^2 = 0$ or $x^2 - 1 = 0$

$x = 0$ $x^2 = 1$

$x = \pm 1$

There are three stationary points:

$x = 0$ $y = 3(0) - 5(0) = 0$ ∴ $(0, 0)$

$x = -1$ $y = 3.(-1)^5 - 5.(-1)^3 = -3 + 5 = 2$ ∴ $(-1, 2)$

$x = 1$ $y = 3.(1)^5 - 5(1)^3 = 3 - 5 = -2$ ∴ $(1, -2)$

However, the nature of these stationary points is important, ie maximum, minimum or inflexion.

This information can be determined graphically:

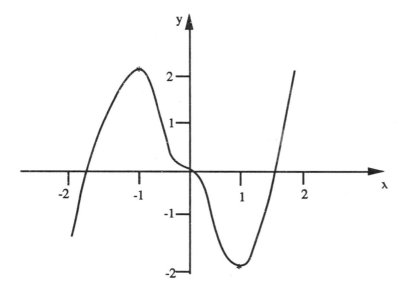

∴ $(-1, 2)$ is a maximum turning point

$(0, 0)$ is a point of inflexion

$(1, -2)$ is a minimum turning point

However, there is another quicker way of determining the nature of these stationary values and that is by considering the second derivative $\frac{d^2y}{dx^2}$, in particular its sign, ie positive, negative or zero.

$\frac{d^2y}{dx^2} = 15.4x^3 - 15.2x$

$= 60x^3 - 30x$

when $x = -1$ $\frac{d^2y}{dx^2} = 60(-1)^3 - 30(-1)$

$= -60 + 30$ ∴ negative

DIFFERENTIAL CALCULUS — PURE MATHEMATICS

when $\quad x = 0 \quad \dfrac{d^2y}{dx^2} = 60(0) - 30(0)$

$\qquad\qquad\qquad\qquad = 0 \qquad\qquad \therefore$ zero

when $\quad x = 1 \quad \dfrac{d^2y}{dx^2} = 60(1)^3 - 30(1)$

$\qquad\qquad\qquad\qquad = 60 - 30 \qquad \therefore$ positive

---

It is true in general terms that:

$\dfrac{d^2y}{dx^2} < 0 \;$ for maximum turning points

$\dfrac{d^2y}{dx^2} = 0 \;$ for points of inflexion

$\dfrac{d^2y}{dx^2} > 0 \;$ for minimum turning points

Example 18

Find the turning points on the curve

$$y = e^x \sin x \text{ for } 0 \leq x \leq 2\pi$$

Determine the nature of these turning points.

Solution

If $\quad y = e^x \sin x \quad$ product form

then $\quad y = u v \qquad u = e^x \qquad v = \sin x$

$\qquad\qquad\qquad\qquad \dfrac{du}{dx} = e^x \qquad \dfrac{dv}{dx} = \cos x$

$\therefore \quad \dfrac{dy}{dx} = e^x . \cos x + \sin x . e^x$

$\qquad\quad = e^x(\cos x + \sin x)$

For turning points $\dfrac{dy}{dx} = 0 \qquad \therefore e^x(\cos x + \sin x) = 0$

$\qquad$ either $\quad e^x = 0 \quad$ or $\qquad \cos x + \sin x = 0$

$\qquad$ impossible $\qquad\qquad\qquad \sin x = -\cos x$

$\qquad\qquad\qquad\qquad\qquad\qquad \dfrac{\sin x}{\cos x} = \dfrac{-\cos x}{\cos x}$

$\qquad \therefore \qquad\qquad\qquad\qquad \tan x = -1$

But tan is negative in the second and fourth quadrants.

PURE MATHEMATICS — DIFFERENTIAL CALCULUS

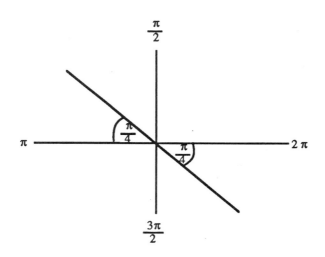

Note: Angles are given in radians and not degrees when differentiating trigonometric functions such as sin x, cos x.

$\therefore \quad x = \dfrac{3\pi}{4}$ radians or $\dfrac{7\pi}{4}$ radians (ie 45°, 225°)

$\quad\quad = 2.356$ radians or 5.498 radians

To find the corresponding y coordinates, these values of x are substituted in the equation $y = e^x \sin x$.

when $\quad x = \dfrac{3\pi}{4} \quad\quad y = e^{\frac{3\pi}{4}} \sin \dfrac{3\pi}{4} = 7.46$

when $\quad x = \dfrac{7\pi}{4} \quad\quad y = e^{\frac{7\pi}{4}} \sin \dfrac{7\pi}{4} = -172.6$

In order to determine the nature of these turning points $\dfrac{d^2y}{dx^2}$ is needed.

$\quad \dfrac{dy}{dx} = e^x(\cos x + \sin x) \quad\quad$ product form $\quad \dfrac{dy}{dx} = u\,v$

$\quad\quad u = e^x \quad\quad\quad\quad v = \cos x + \sin x$

$\quad\quad \dfrac{du}{dx} = e^x \quad\quad\quad \dfrac{dv}{dx} = -\sin x + \cos x$

$\therefore \quad \dfrac{d^2y}{dx^2} = e^x(-\sin x + \cos x) + (\cos x + \sin x)e^x$

$\quad\quad = -e^x \sin x + e^x \cos x + e^x \cos x + e^x \sin x$

$\quad\quad = 2e^x \cos x$

To determine the nature of the turning points it is only the sign of $\dfrac{d^2y}{dx^2}$ that matters and not its value.

$\quad x = \dfrac{3\pi}{4} \quad\quad \dfrac{d^2y}{dx^2} = e^{\frac{3\pi}{4}} \cdot \cos \dfrac{3\pi}{4} = (+\text{ve})(-\text{ve}) \quad = -\text{ve} \quad \therefore \text{ maximum}$

DIFFERENTIAL CALCULUS — PURE MATHEMATICS

$$x = \frac{7\pi}{4} \qquad \frac{d^2y}{dx^2} = e^{\frac{7\pi}{4}} \cdot \cos\frac{7\pi}{4} = (+\text{ve}).(+\text{ve}) = +\text{ve} \therefore \text{minimum}$$

d) Distance, velocity, acceleration

Velocity (v) is the rate of change of distance (s) with respect to time (t). Therefore, if distance is a function of time, ie $s = f(t)$, it follows that by differentiating $\frac{ds}{dt} = v$

Acceleration (a) is the rate of change of velocity with respect of time (t). Therefore, assuming velocity is a function of time again, ie $v = f(t)$, it follows that by differentiating $\frac{dv}{dt} = a$

$$\boxed{\text{So } v = \frac{ds}{dt} \text{ and } a = \frac{dv}{dt}}$$

Example 19

A point moves along a straight line so that, at the end of t seconds, its distance from a fixed point on the line is $t^3 - 2t^2 + t$ metres. Find the velocity and acceleration at the end of 3 seconds.

Solution

So $\qquad s = t^3 - 2t^2 + t$

$\therefore \qquad$ velocity $= \frac{ds}{dt} = 3t^2 - 4t + 1$

when $t = 3$, $\quad v = 3(3)^2 - 4(3) + 1 = 27 - 12 + 1 = 16 \text{ ms}^{-1}$

and acceleration $= \frac{dv}{dt} = 6t - 4$

when $t = 3$, $\quad a = 6(3) - 4 = 18 - 4 = 14 \text{ ms}^{-2}$

The next example is similar but contains one or two extra points.

Example 20

A point moves along a straight line OX so that its distance OX cm from the point 0 at time t seconds is given by the formula $x = t^3 - 6t^2 + 9t$. Find:

(1) at what time and in what positions the point will have zero velocity;

(2) its acceleration at those instants;

(3) the velocity when its acceleration is zero.

Solution

$\qquad x = t^3 - 6t^2 + 9t$

$\therefore \qquad v = \frac{dx}{dt} = 3t^2 - 12t + 9$

and $\qquad a = \frac{dv}{dt} = 6t - 12$

PURE MATHEMATICS — DIFFERENTIAL CALCULUS

(1) Zero velocity means $v = 3t^2 - 12t + 9 = 0$

$\therefore \quad 3(t^2 - 4t + 3) = 0$

$\therefore \quad t^2 - 4t + 3 = 0$

$(t - 1)(t - 3) = 0$

either $t - 1 = 0$ or $t - 3 = 0$

$t = 1$ $\qquad\qquad t = 3$

and $x = (1)^3 - 6(1)^2 + 9(1)$ $\qquad x = (3)^3 - 6(3)^2 + 9(3)$

$x = 1 - 6 + 9 = 4$ $\qquad\qquad x = 27 - 54 + 27 = 0$

The point has zero velocity after 1 second at a distance of 4 cm from 0, and after 3 seconds at 0 itself.

(2) $a = 6t - 12$

when $t = 1 \quad a = 6(1) - 12 = 6 - 12 = -6 \text{ cms}^{-2}$

when $t = 3 \quad a = 6(3) - 12 = 18 - 12 = 6 \text{ cms}^{-2}$

(3) Zero acceleration means $a = 6t - 12 = 0$

$6t = 12$

$t = 2$ seconds

and the velocity at this instance $v = 3(2)^2 - 12(2) + 9$

$= 12 - 24 + 9$

$\therefore \quad v = -3$ cm/s (- sign shows velocity is directed towards 0)

---

e) *Rates of change*

Just as distance and velocity can change with respect to time so can other factors such as the volume of a sphere, the surface area of a cylinder, the length of a piece of string.

The following examples, the second of which is an 'A' level question, will show how differentiation can be applied to this type of problem.

Example 21

The area of a circle is increasing at the rate of 3 cm^2 per second. Find the rate of change of the circumference when the radius is 2 cm.

Solution

$\qquad$ Area $A = \pi r^2$

Differentiating both sides with respect to time gives

$\qquad \dfrac{dA}{dt} = \dfrac{dA}{dr} \cdot \dfrac{dr}{dt}$

$\qquad \dfrac{dA}{dt} = 2\pi r \cdot \dfrac{dr}{dt}$

DIFFERENTIAL CALCULUS   PURE MATHEMATICS

but $\frac{dA}{dt}$ is the rate of increase of area with respect to time, ie 3 cm²s⁻¹

$$\therefore \quad 3 = 2\pi r \cdot \frac{dr}{dt}$$

$$\therefore \quad \frac{3}{2\pi r} = \frac{dr}{dt}$$

So when r = 2cm $\quad \frac{dr}{dt} = \frac{3}{2\pi(2)} = \frac{3}{4\pi} \text{ cm s}^{-1}$

Circumference C = 2πr

Differentiating both sides with respect to time gives

$$\frac{dC}{dt} = \frac{dC}{dr} \cdot \frac{dr}{dt}$$

$$\frac{dC}{dt} = 2\pi \cdot \left(\frac{3}{4\pi}\right) = \frac{3}{2}$$

Rate of increase of circumference = $\frac{3}{2}$ or 1.5 cms⁻¹

Example 22

The volume of a sphere is increasing at the constant rate 7 cm³s⁻¹

(1) Find in terms of π the rate at which the radius of the sphere is increasing at the instant when the radius is 5 cm.

(2) Show that, at any instant, the rate of increase of the surface area of the sphere is inversely proportional to the radius of the sphere.

Solution

(1)   V = $\frac{4}{3}\pi r^3$ for a sphere

Differentiating both sides with respect to time.

$$\frac{dV}{dt} = \frac{dV}{dr} \cdot \frac{dr}{dt}$$

$$= \frac{4}{3}\pi \cdot 3r^2 \cdot \frac{dr}{dt}$$

$$\therefore \quad \frac{dV}{dt} = 4\pi r^2 \cdot \frac{dr}{dt}$$

but the volume is increasing at a constant 7 cm³s⁻¹

$$\therefore \quad \frac{dV}{dt} = 4\pi r^2 \frac{dr}{dt} = 7$$

$$\therefore \quad \frac{dr}{dt} = \frac{7}{4\pi r^2}$$

so when r = 5, $\quad \dfrac{dr}{dt} = \dfrac{7}{4\pi 5^2} = \dfrac{7}{100\pi}$

$\dfrac{dr}{dt}$ is the rate of increase of the radius of the sphere, ie $\dfrac{7}{100\pi}$ cm s^{-1}

(2) Surface area $A = 4\pi r^2$

Differentiating both sides with respect to time

$$\dfrac{dA}{dt} = \dfrac{dA}{dr} \cdot \dfrac{dr}{dt}$$

$$\dfrac{dA}{dt} = 4\pi . 2r \left(\dfrac{7}{4\pi r^2}\right) = \dfrac{14}{r}$$

so the $\dfrac{dA}{dt}$ depends on $\dfrac{1}{r}$ at any instant

$\therefore \quad \dfrac{dA}{dt} \propto \dfrac{1}{r}$ and $\dfrac{dA}{dt} = k.\dfrac{1}{r}$ $\quad (k = 14)$

---

f) *Example* ('A' level question)

A closed container takes the form of a cylinder of radius r and height h surmounted by a hemi-spherical cap.

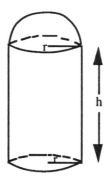

Given that the volume of this container is 45π, express h in terms of r and show that the total surface area of the container is

$$A = \dfrac{5}{3}\pi r^2 + \dfrac{90\pi}{r}$$

Find the radius if the total surface area is to be a minimum.

Solution

Volume = volume (cylinder) + volume (hemisphere)

$\qquad\qquad\qquad = \pi r^2 h + \dfrac{2}{3}\pi r^3$

but Volume = 45π

$\therefore \quad \pi r^2 h + \dfrac{2}{3}\pi r^3 = 45\pi \quad$ (dividing by π)

DIFFERENTIAL CALCULUS — PURE MATHEMATICS

$$r^2h + \frac{2}{3}r^3 = 45 \quad \text{(multiplying by 3)}$$

$$3r^2h + 2r^3 = 135$$

$$\therefore \quad 3r^2h = 135 - 2r^3$$

$$\therefore \quad h = \frac{135 - 2r^3}{3r^2}$$

this is the expression of h in terms of r.

Surface area A = area (cylinder) + area (hemisphere) + area (base)

$$= 2\pi rh + 2\pi r^2 + \pi r^2$$

$$= 2\pi rh + 3\pi r^2$$

$$= 2\pi r \left(\frac{135 - 2r^3}{3r^2} \right) + 3\pi r^2$$

$$= \frac{270\pi r}{3r^2} - \frac{4\pi r^4}{3r^2} + 3\pi r^2$$

$$= \frac{90\pi}{r} - \frac{4\pi r^2}{3} + 3\pi r^2$$

$$\therefore \quad A = \frac{90\pi}{r} + \frac{5}{3}\pi r^2$$

Area is a function of radius. Therefore, for minimum area $\frac{dA}{dr} = 0$

$$A = 90\pi r^{-1} + \frac{5}{3}\pi r^2$$

$$\frac{dA}{dr} = 90\pi(-1)r^{-2} + \frac{5}{3}\pi 2r$$

$$= \frac{-90\pi}{r^2} + \frac{10}{3}\pi r$$

So $\frac{dA}{dr} = 0$ when $\frac{-90\pi}{r^2} + \frac{10\pi r}{3} = 0$

$$\therefore \quad \frac{10\pi r}{3} = \frac{90\pi}{r^2} \quad \text{(cross multiplying)}$$

$$10\pi r^3 = 270\pi$$

$$\therefore \quad r^3 = \frac{270\pi}{10\pi} = 27$$

$$r = \sqrt[3]{27} = 3 \text{ cm}$$

This question does not ask for proof that this value of r gives minimum surface area. However, it will be included for completeness of solution.

$$\frac{dA}{dr} = -90\pi r^{-2} + \frac{10}{3}\pi r$$

$$\frac{d^2A}{dr^2} = -90\pi(-2)r^{-3} + \frac{10}{3}\pi = \frac{180\pi}{r^3} + \frac{10}{3}\pi$$

If r = 3 cm $\qquad \frac{d^2A}{dr^2} = \frac{180\pi}{3^3} + \frac{10\pi}{3} > 0$ ∴ minimum value for A when r = 3

―――――――――――――――――

5 INTEGRAL CALCULUS

   5.1   Introduction to integration
   5.2   Further integration
   5.3   Area under a curve
   5.4   Further applications of integration

5.1   Introduction to integration

a)   *The inverse of differentiation*

There are many occasions when the derived function $\left(\frac{dy}{dx}\right)$ is given and it is necessary to work in reverse to obtain the original function of x ie f(x). This process is called integration and the function of f(x) is called the integral with respect to x.

Some simple examples will now be considered in order to establish the general rules for integrating algebraic functions:

If $\quad \frac{dy}{dx} = 2x \quad$ then $y = x^2 + c$

If $\quad \frac{dy}{dx} = 6x^2 \quad$ then $y = 2x^3 + c \quad$ check by differentiation

If $\quad \frac{dy}{dx} = 5 \quad$ then $y = 5x + c$

where c is an arbitrary constant of integration, whose value cannot be determined unless further information is available concerning x and y. This constant arises because when a function is differentiated any constant terms go to zero and so disappear from the derived function. Therefore, when integrating it is necessary to allow for the fact that the original function of x may have contained a constant.

Consider the derivatives of the following functions of x:

(a) $y = x^2 + 6$ (b) $y = x^2 - 5$ (c) $y = x^2$

$\therefore \quad \frac{dy}{dx} = 2x \qquad \therefore \quad \frac{dy}{dx} = 2x \qquad \therefore \quad \frac{dy}{dx} = 2x$

For each function the first derivative is the same, namely 2x. So integrating in each case would give the term in x^2 but it would not give the constant term, ie + 6, - 5, 0. Therefore, it is assumed that a constant, c, exists but, unless further information is also given, its value cannot be calculated.

So, if $\frac{dy}{dx} = 2x$, then $y = x^2 + c$

This can be written as $\int 2x\,dx = x^2 + c$ which means the integral of 2x with respect of x is $x^2 + c$.

The rule for integrating a term in x is to add one on to the power of x and divide by the new power:

$$\int kx^n dx = k\int x^n dx = \frac{kx^{n+1}}{n+1} + c$$

This is valid for all values of n, except -1. n = -1 will be considered as a separate topic.

So $\quad \int 2x\,dx = \frac{2x^{1+1}}{1+1} + c \quad = \frac{2x^2}{2} + c \quad = x^2 + c$

PURE MATHEMATICS — INTEGRAL CALCULUS

and $\quad \int 6x^2\, dx = \dfrac{6x^{2+1}}{2+1} + c \quad = \dfrac{6x^3}{3} + c \quad = 2x^3 + c$

and $\quad \int 5\, dx = \dfrac{5x^{0+1}}{0+1} + c \quad = \dfrac{5x^1}{1} + c \quad = 5x + c$

Example 1

Integrate the following functions:

(1) $\quad x^2 + 3x$

(2) $\quad (2x + 3)^2$

(3) $\quad x^{-5}$

(4) $\quad \dfrac{1}{t^2} + 3 + 2t$

Solution

(1) $\quad \int x^2 + 3x\, dx \quad = \dfrac{x^{2+1}}{2+1} + \dfrac{3x^{1+1}}{1+1} + c$

$\quad\quad\quad\quad\quad\quad\quad\quad = \dfrac{x^3}{3} + \dfrac{3x^2}{2} + c$

(2) $\quad \int (2x+3)^2 dx$ - the terms must first be separated out by multiplying the brackets.

$\quad \int (2x+3)(2x+3)\, dx \quad = \int (4x^2 + 12x + 9)\, dx$

$\quad\quad\quad\quad\quad\quad\quad\quad\quad = \dfrac{4x^{2+1}}{2+1} + \dfrac{12x^{1+1}}{1+1} + \dfrac{9x^{0+1}}{0+1} + c$

$\quad\quad\quad\quad\quad\quad\quad\quad\quad = \dfrac{4x^3}{3} + \dfrac{12x^2}{2} + \dfrac{9x}{1} + c$

$\therefore \quad \int (2x+3)^2 dx \quad = \dfrac{4x^3}{3} + 6x^2 + 9x + c$

---

(3) $\quad \int x^{-5} dx = \dfrac{x^{-5+1}}{-5+1} + c$

$\quad\quad\quad\quad = \dfrac{x^{-4}}{-4} + c$

$\quad\quad\quad\quad = -\dfrac{1}{4x^4} + c$

INTEGRAL CALCULUS   PURE MATHEMATICS

(4) $\int \left(\dfrac{1}{t^2} + 3 + 2t\right) dt = \int (t^{-2} + 3 + 2t)\, dt$

$$= \dfrac{t^{-2+1}}{-2+1} + 3t + \dfrac{2t^{1+1}}{1+1} + c$$

$$= \dfrac{t^{-1}}{-1} + 3t + \dfrac{2t^2}{2} + c$$

$\therefore \quad \int \dfrac{1}{t^2} + 3 + 2t\, dt = -\dfrac{1}{t} + 3t + t^2 + c$

Example 2

Integrate the following functions:

(1) $2x^{1/5}$ (2) $4\sqrt{x}$

(3) $3\sqrt{x^2}$ (4) $\dfrac{2}{5\sqrt{x}}$

(5) $\sqrt{x} + \dfrac{2}{\sqrt{x}}$ (6) $(\sqrt{x} + 2)(\sqrt{x} - 3)$

Solution

(1) $\int 2x^{\frac{1}{5}}\, dx = \dfrac{2x^{(\frac{1}{5} + 1)}}{\frac{1}{5} + 1} + c$

$$= \dfrac{2x^{\frac{6}{5}}}{\frac{6}{5}} + c$$

$$= \dfrac{5(2x^{\frac{6}{5}})}{6} + c$$

$$= \dfrac{10x^{\frac{6}{5}}}{6} + c$$

$\therefore \quad \int 2x^{\frac{1}{5}}\, dx = \dfrac{5}{3} x^{\frac{6}{5}} + c$

---

(2) $\int 4\sqrt{x}\, dx = \int x^{\frac{1}{4}}\, dx$

$$= \dfrac{x^{\frac{1}{4} + 1}}{\frac{1}{4} + 1} + c$$

PURE MATHEMATICS — INTEGRAL CALCULUS

$$= \frac{x^{\frac{5}{4}}}{\frac{5}{4}} + c$$

$$\therefore \int \sqrt[4]{x}\, dx = \frac{4x^{\frac{5}{4}}}{5} + c$$

---

(3) $\int \sqrt[3]{x^2}\, dx = \int x^{\frac{2}{3}}\, dx$

$$= \frac{x^{(\frac{2}{3}+1)}}{\frac{2}{3}+1} + c$$

$$= \frac{x^{\frac{5}{3}}}{\frac{5}{3}} + c$$

$$\therefore \int \sqrt[3]{x^2}\, dx = \frac{3}{5} x^{\frac{5}{3}} + c$$

---

(4) $\int \dfrac{2}{\sqrt[5]{x}}\, dx = \int 2x^{-\frac{1}{5}}\, dx$

$$= \frac{2x^{(-\frac{1}{5}+1)}}{-\frac{1}{5}+1} + c$$

$$= \frac{2x^{\frac{4}{5}}}{\frac{4}{5}} + c$$

$$= \frac{10x^{\frac{4}{5}}}{4} + c$$

$$\therefore \int \frac{2}{\sqrt[5]{x}}\, dx = \frac{5x^{\frac{4}{5}}}{2} + c$$

INTEGRAL CALCULUS   PURE MATHEMATICS

(5) $\quad \int \left(\sqrt{x} + \dfrac{2}{\sqrt{x}}\right) dx \quad = \int x^{\frac{1}{2}} + 2x^{-\frac{1}{2}} dx$

$$= \dfrac{x^{(\frac{1}{2}+1)}}{\frac{1}{2}+1} + \dfrac{2x^{(-\frac{1}{2}+1)}}{-\frac{1}{2}+1} + c$$

$$= \dfrac{x^{\frac{3}{2}}}{\frac{3}{2}} + \dfrac{2x^{\frac{1}{2}}}{\frac{1}{2}} + c$$

$$= \dfrac{2x^{\frac{3}{2}}}{3} + 4x^{\frac{1}{2}} + c$$

$\therefore \quad \int \sqrt{x} + \dfrac{2}{\sqrt{x}} \, dx \quad = \dfrac{2}{3}\sqrt{x^3} + 4\sqrt{x} + c$

---

(6) $\quad \int (\sqrt{x}+2)(\sqrt{x}-3) \, dx \quad = \int ((\sqrt{x})^2 + 2\sqrt{x} - 3\sqrt{x} - 6) \, dx$

$$= \int (x - \sqrt{x} - 6) \, dx$$

$$= \int (x - x^{\frac{1}{2}} - 6) \, dx$$

$$= \dfrac{x^{1+1}}{2} - \dfrac{x^{\frac{1}{2}+1}}{\frac{1}{2}+1} - \dfrac{6x^{0+1}}{0+1} + c$$

$$= \dfrac{x^2}{2} - \dfrac{x^{\frac{3}{2}}}{\frac{3}{2}} - \dfrac{6x}{1} + c$$

$$= \dfrac{x^2}{2} - \dfrac{2x^{\frac{3}{2}}}{3} - 6x + c$$

$\therefore \quad \int(\sqrt{x}+2)(\sqrt{x}-3) \, dx = \dfrac{x^2}{2} - \dfrac{2}{3}\sqrt{x^3} - 6x + c$

The next examples will illustrate how the constant of integration can be calculated when more information is available.

Example 3

If $\dfrac{dy}{dx} = 2x + 5$, and $x = 3$ when $y = -1$ find y as a function of x.

Solution

$$\frac{dy}{dx} = 2x + 5 \qquad \therefore y = \int (2x + 5)\, dx$$

$$= \frac{2x^2}{2} + 5x + c$$

$$\therefore y = x^2 + 5x + c$$

However, knowing that x = 3 when y = -1 means that c can be calculated as follows:

$$y = x^2 + 5x + c \qquad \text{becomes}$$

$$(-1) = (3)^2 + 5(3) + c$$

$$-1 = 9 + 15 + c$$

$$-1 - 24 = c$$

$$\therefore -25 = c$$

So $y = x^2 + 5x - 25$ is the required function

Example 4

Find A in terms of x if $\frac{dA}{dx} = \frac{(3x + 1)(x^2 - 1)}{x^5}$

What is the value of A when x = 2, given that A = 0 when x = 1?

Solution

$$\frac{dA}{dx} = \frac{(3x + 1)(x^2 - 1)}{x^5}$$

$$= \frac{3x^3 + x^2 - 3x - 1}{x^5}$$

$$= \frac{3x^3}{x^5} + \frac{x^2}{x^5} - \frac{3x}{x^5} - \frac{1}{x^5}$$

$$\therefore \frac{dA}{dx} = \frac{3}{x^2} + \frac{1}{x^3} - \frac{3}{x^4} - \frac{1}{x^5}$$

$$\therefore A = \int (3x^{-2} + x^{-3} - 3x^{-4} - x^{-5})\, dx$$

$$= \frac{3x^{-1}}{-1} + \frac{x^{-2}}{-2} - \frac{3x^{-3}}{-3} - \frac{x^{-4}}{-4} + c$$

$$\therefore A = -\frac{3}{x} - \frac{1}{2x^2} + \frac{1}{x^3} + \frac{1}{4x^4} + c$$

Knowing that A = 0 when x = 1 means that c can be calculated as follows:

$$A = -\frac{3}{x} - \frac{1}{2x^2} + \frac{1}{x^3} + \frac{1}{4x^4} + c \qquad \text{becomes}$$

$$0 = -\frac{3}{1} - \frac{1}{2(1)^2} + \frac{1}{(1)^3} + \frac{1}{4(1)^4} + c$$

$$0 = -3 - \frac{1}{2} + 1 + \frac{1}{4} + c$$

$$0 = \frac{-12 - 2 + 4 + 1}{4} + c$$

$$\therefore c = \frac{9}{4}$$

So $\quad A = -\frac{3}{x} - \frac{1}{2x^2} + \frac{1}{x^3} + \frac{1}{4x^4} + \frac{9}{4}$

∴ when x = 2

$$A = -\frac{3}{2} - \frac{1}{2(2)^2} + \frac{1}{(2)^3} + \frac{1}{4(2)^4} + \frac{9}{4}$$

$$= -\frac{3}{2} - \frac{1}{8} + \frac{1}{8} + \frac{1}{64} + \frac{9}{4}$$

$$= \frac{-96 - 8 + 8 + 1 + 144}{64}$$

$$\therefore A = \frac{49}{64} \text{ when } x = 2$$

---

b) *Integration of simple trigonometric functions*

The derivatives of sine and cosine have already been given. They are:

$$\frac{d}{dx}(\sin x) = \cos x$$

$$\frac{d}{dx}(\cos x) = -\sin x$$

These lead to:

$$\int (-\sin x)\, dx = \cos x + c$$

$$\boxed{\int \cos x \, dx = \sin x + c \quad \text{and} \quad \int \sin x \, dx = -\cos x + c}$$

If $y = \sin(4x + 3)$ then $\frac{dy}{dx} = 4\cos(4x + 3)$.

It therefore follows that $\quad \int 4\cos(4x + 3)dx = \sin(4x + 3) + c$

and so $\quad \int \cos(4x + 3)dx = \frac{\sin(4x + 3)}{4} + c$

PURE MATHEMATICS — INTEGRAL CALCULUS

Therefore, in general terms:

$$\int \cos(ax+b)\,dx = \frac{\sin(ax+b)}{a} + c$$

$$\int \sin(ax+b)\,dx = \frac{-\cos(ax+b)}{a} + c$$

where a, b and c are constants

Example 5

Integrate the following functions:

(a) $\sin 3x$

(b) $6 \cos 4x$

(c) $\sin(2x + 1)$

(d) $\frac{2}{3} \sin \frac{x}{2}$

Solution

(a) $\int \sin 3x\, dx = \frac{-\cos 3x}{3} + c$

(b) $\int 6 \cos 4x\, dx = 6\left(\frac{\sin 4x}{4}\right) + c$

$$= \frac{3 \sin 4x}{2} + c$$

(c) $\int \sin(2x + 1)\,dx = \frac{-\cos(2x+1)}{2} + c$

(d) $\int \frac{2}{3} \sin \frac{x}{2}\, dx = \frac{2}{3}\left(\frac{-\cos \frac{x}{2}}{\frac{1}{2}}\right) + c$

$$= \frac{2}{3}\left(-2 \cos \frac{x}{2}\right) + c$$

$$= -\frac{4}{3} \cos \frac{x}{2} + c$$

In some problems the trigonometric function has to be altered using one of the many formulae already given in chapter 3.

The following example will show the approach needed.

INTEGRAL CALCULUS — PURE MATHEMATICS

Example 6

Integrate these functions:

(a) $\cos^2 x$

(b) $\sin^2 x$

(c) $\sin x \cos 3x$

(d) $\sin 4x \sin x$

Solution

(a) $\int (\cos^2 x)\, dx$ This cannot be integrated directly.

However, using $\cos 2x = 2\cos^2 x - 1$

gives $\cos^2 x = \frac{1}{2}(\cos 2x + 1)$

$\therefore \int (\cos^2 x)\, dx = \int \frac{1}{2}(\cos 2x + 1)\, dx$

$= \frac{1}{2} \int (\cos 2x + 1)\, dx$

$= \frac{1}{2}\left(\frac{\sin 2x}{2} + x\right) + c$

$\therefore \int (\cos^2 x)\, dx = \frac{\sin 2x}{4} + \frac{x}{2} + c$

(b) $\int (\sin^2 x)\, dx$ as before this cannot be integrated.

However, using $\cos 2x = 1 - 2\sin^2 x$

gives $\sin^2 x = \frac{1}{2}(1 - \cos 2x)$

$\therefore \int (\sin^2 x)\, dx = \int \frac{1}{2}(1 - \cos 2x)\, dx$

$= \frac{1}{2} \int (1 - \cos 2x)\, dx$

$= \frac{1}{2}\left(x - \frac{\sin 2x}{2}\right) + c$

$\therefore \int (\sin^2 x)\, dx = \frac{x}{2} - \frac{\sin 2x}{4} + c$

(c) $\int \sin x \cos 3x\, dx$ using the factor formula for the difference of two sines:

$\sin P - \sin Q = 2\cos\left(\frac{P+Q}{2}\right)\sin\left(\frac{P-Q}{2}\right)$

$\therefore \quad \frac{1}{2}(\sin P - \sin Q) = \cos 3x \sin x$ so $\frac{P+Q}{2} = 3x$ $\therefore P + Q = 6x$

PURE MATHEMATICS — INTEGRAL CALCULUS

$$\text{and} \quad \frac{P - Q}{2} = x \qquad \therefore P - Q = 2x$$

$$\therefore \quad P = 4x \text{ and } Q = 2x$$

$$\cos 3x \sin x = \frac{1}{2}(\sin 4x - \sin 2x)$$

$$\therefore \int (\cos 3x \sin x)\, dx = \int \frac{1}{2}(\sin 4x - \sin 2x)\, dx$$

$$= \frac{1}{2}\int (\sin 4x - \sin 2x)\, dx$$

$$= \frac{1}{2}\left[\frac{-\cos 4x}{4} - \left(\frac{-\cos 2x}{2}\right)\right] + c$$

$$\therefore \int \cos 3x \sin x\, dx = \frac{\cos 2x}{4} - \frac{\cos 4x}{8} + c$$

(d) $\int \sin 4x \sin x\, dx$ using the factor formula for the difference of two cosines:

$$\cos P - \cos Q = -2 \sin\left(\frac{P + Q}{2}\right).\sin\left(\frac{P - Q}{2}\right)$$

$$\therefore \quad -\frac{1}{2}(\cos P - \cos Q) = \sin 4x.\sin x \quad \text{so} \quad \frac{P + Q}{2} = 4x \qquad \therefore P + Q = 8x$$

$$\text{and} \quad \frac{P - Q}{2} = x \qquad \therefore P - Q = 2x$$

$$\therefore \quad P = 5x \text{ and } Q = 3x$$

$$\sin 4x.\sin x = -\frac{1}{2}(\cos 5x - \cos 3x)$$

$$\therefore \int (\sin 4x \sin x)\, dx = -\frac{1}{2}\int (\cos 5x - \cos 3x)\, dx$$

$$= -\frac{1}{2}\left[\frac{\sin 5x}{5} - \frac{\sin 3x}{3}\right] + c$$

$$\therefore \int \sin 4x.\sin x\, dx = \frac{\sin 3x}{6} - \frac{\sin 5x}{10} + c$$

c) Integration of simple exponential functions

The derivative of e^x is e^x, ie $\frac{d}{dx}(e^x) = e^x$

$$\boxed{\int e^x\, dx = e^x + c}$$

If $y = e^{2x+1}$ then $\frac{dy}{dx} = 2.e^{2x+1}$

It therefore follows that $\int (2e^{2x+1})\, dx = e^{2x+1} + c$

and so $\int (e^{2x+1})\, dx = \frac{e^{2x+1}}{2} + c$

INTEGRAL CALCULUS PURE MATHEMATICS

So in general terms:

$$\int e^{ax+b} dx = \frac{e^{ax+b}}{a} + c$$

Example 7

Integrate the following functions:

(a) $3e^{\frac{x}{2}}$

(b) e^{-x}

(c) $e^{\frac{x}{3}}$

(d) $2e^{3x-1}$

Solution

(a) $\int \left(3e^{\frac{x}{2}}\right) dx = 3\int e^{\frac{x}{2}} dx$

$$= 3\left(\frac{e^{\frac{x}{2}}}{\frac{1}{2}}\right) + c$$

$$= 6e^{\frac{x}{2}} + c$$

(b) $\int (e^{-x}) dx = \frac{e^{-x}}{-1} + c$

$$= -e^{-x} + c$$

(c) $\int e^{\frac{x}{3}} dx = \frac{e^{\frac{x}{3}}}{\frac{1}{3}} + c$

$$= 3e^{\frac{x}{3}} + c$$

(d) $\int 2e^{3x-1} dx = 2\int e^{3x-1} dx$

$$= 2\left(\frac{e^{3x-1}}{3}\right) + c$$

$$= \frac{2e^{3x-1}}{3} + c$$

PURE MATHEMATICS — INTEGRAL CALCULUS

The derivative of a^x is $a^x \log_e a$. This can be shown as follows:

Let $\quad y = a^x$

then $\quad \log_a y = x \quad$ and $\quad \log_a y = \dfrac{\log_e y}{\log_e a}$

$\therefore \quad x = \dfrac{1}{\log_e a} \cdot \log_e y$

$\therefore \quad \dfrac{dx}{dy} = \dfrac{1}{\log_e a} \cdot \dfrac{1}{y}$

$\therefore \quad \dfrac{dy}{dx} = y \cdot \log_e a = a^x \log_e a$

Hence, on integrating,

$$\int a^x \log_e a \, dx = a^x + c'$$

and so $\quad \boxed{\int a^x \, dx = \dfrac{a^x}{\log_e a} + c} \quad$ where $c = \dfrac{c'}{\log_e a}$

This means $\quad \int 5^x dx = \dfrac{5^x}{\log_e 5} + c$

d) Integration leading to simple logarithmic functions

The derivative of $\log_e x$ is $\dfrac{1}{x}$, $\quad$ ie $\quad \dfrac{d}{dx}(\log_e x) = \dfrac{1}{x}$

$\therefore \quad \boxed{\int \dfrac{1}{x} \, dx = \log_e x + c}$

By considering the derivatives of several other logarithmic functions, it will be possible to find a more general rule of integration.

eg $\quad y = \log_e(2x + 1)$

$\dfrac{dy}{dx} = \dfrac{2}{2x+1} \quad \therefore \quad \int \left(\dfrac{2}{2x+1}\right) dx = \log_e(2x+1) + c$

and $\quad y = \log_e(x^2 + 4)$

$\dfrac{dy}{dx} = \dfrac{2x}{x^2+4} \quad \therefore \quad \int \left(\dfrac{2x}{x^2+4}\right) dx = \log_e(x^2+4) + c$

and $\quad y = \log_e (\cos x)$

$\dfrac{dy}{dx} = \dfrac{-\sin x}{\cos x} = -\tan x \quad \therefore \quad \int \dfrac{-\sin x}{\cos x} dx = \log_e(\cos x) + c$

By looking at the integrals it can be seen that the top line is always the derivative of the bottom line, ie

$\dfrac{d}{dx}(2x+1) = 2, \quad \dfrac{d}{dx}(x^2+4) = 2x, \quad \dfrac{d}{dx}(\cos x) = -\sin x$

INTEGRAL CALCULUS   PURE MATHEMATICS

So the general rule of integration is:

$$\boxed{\int \frac{f'(x)}{f(x)} dx = \log_e f(x) + c}$$ where f'(x) is an alternative notation for $\frac{dy}{dx}$ ie $\frac{dy}{dx} = f'(x)$

Example 8

Integrate the following functions:

(a) $\dfrac{1}{4x}$

(b) $\dfrac{1}{2x - 3}$

(c) $\dfrac{1}{3 - 2x}$

(d) $\dfrac{2x - 3}{3x^2 - 9x + 4}$

Solution

(a) $\int \dfrac{1}{4x} dx = \dfrac{1}{4} \int \dfrac{1}{x} dx$

$= \dfrac{1}{4} (\log_e x) + c$

$= \log_e x^{\frac{1}{4}} + c$

$= \log_e \sqrt[4]{x} + c$

(b) $\int \dfrac{1}{2x - 3} dx$ however $\dfrac{d}{dx}(2x - 3) = 2$ so it is necessary to adjust the integral slightly.

$\int \left(\dfrac{1}{2x - 3}\right) dx = \dfrac{1}{2} \int \dfrac{2}{2x - 3} dx$ numerator is derivative of denominator

$= \dfrac{1}{2} (\log_e (2x - 3)) + c$

$\therefore \quad \int \dfrac{1}{2x - 3} dx = \log_e (2x - 3)^{\frac{1}{2}} + c$

$= \log_e \sqrt{2x - 3} + c$

(c) $\int \dfrac{1}{3 - 2x} dx$ again a slight adjustment is needed

$\int \dfrac{1}{3 - 2x} dx = \int \dfrac{-2}{-2(3 - 2x)} dx$

$= -\dfrac{1}{2} \int \dfrac{-2}{(3 - 2x)} dx$ numerator is derivative of denominator

178

PURE MATHEMATICS — INTEGRAL CALCULUS

$$= -\frac{1}{2}(\log_e(3 - 2x)) + c$$

$$\therefore \int \frac{1}{3 - 2x} dx = \log_e(3 - 2x)^{-\frac{1}{2}} + c$$

$$= \log_e\left(\frac{1}{\sqrt{3 - 2x}}\right) + c$$

(d) $\int \frac{2x - 3}{3x^2 - 9x + 4} dx$ but $\frac{d}{dx}(3x^2 - 9x + 4) = 6x - 9 = 3(2x - 3)$

$$\therefore \int \frac{2x - 3}{3x^2 - 9x + 4} dx = \int \frac{3(2x - 3)}{3(3x^2 - 9x + 4)} dx$$

$$= \frac{1}{3} \int \left(\frac{3(2x - 3)}{3x^2 - 9x + 4}\right) dx$$

$$= \frac{1}{3} \log_e(3x^2 - 9x + 4) + c$$

$$\therefore \int \frac{2x - 3}{3x^2 - 9x + 4} dx = \log_e\left(\sqrt[3]{3x^2 - 9x + 4}\right) + c$$

Some expressions require more alteration before integration is possible. This will be seen in the next example.

Example 9

Integrate the following functions:

(a) $\dfrac{x}{x + 2}$

(b) $\dfrac{2x}{3 - x}$

(c) $\dfrac{3 - 2x}{x - 4}$

Solution

(a) $\dfrac{x}{x + 2}$ by dividing out this becomes

$$\frac{x}{x + 2} = 1 - \frac{2}{x + 2}$$

$$\therefore \int \left(\frac{x}{x + 2}\right) dx = \int \left(1 - \frac{2}{x + 2}\right) dx$$

$$= x - 2\log_e(x + 2) + c$$

$$\therefore \int \frac{x}{x + 2} dx = x - \log_e(x + 2)^2 + c$$

(b) $\dfrac{2x}{3 - x} = \dfrac{2}{-1}\left(\dfrac{x}{x - 3}\right) = -2\left(1 + \dfrac{3}{x - 3}\right)$

$$\therefore \int \left(\frac{2x}{3-x}\right) dx = -2\int \left(1 + \frac{3}{x-3}\right) dx$$

$$= -2(x + 3\log_e(x-3)) + c$$

$$= -2x - 6\log_e(x-3) + c$$

$$\therefore \int \frac{2x}{3-x} dx = -2x - \log_e(x-3)^6 + c$$

(c) $\quad \dfrac{3-2x}{x-4} = -\dfrac{(2x-3)}{x-4} = -\left(2 + \dfrac{5}{x-4}\right)$

$$\therefore \int \left(\frac{3-2x}{x-4}\right) dx = -\int \left(2 + \frac{5}{x-4}\right) dx$$

$$= -(2x + 5\log_e(x-4)) + c$$

$$= -2x - 5\log_e(x-4) + c$$

$$\therefore \int \frac{3-2x}{x-4} dx = -2x - \log_e(x-4)^5 + c$$

e) *Example* ('A' level question)

Find:

(a) $\quad \int \sin 5x \cos 4x \, dx$

(b) $\quad \int \dfrac{a}{(1+x)^2} dx$

Solution

(a) $\sin 5x \cos 4x$ can be written as the sum of two sines:

$$\sin P + \sin Q = 2 \sin \left(\frac{P+Q}{2}\right) \cos \left(\frac{P-Q}{2}\right)$$

Let $\quad \dfrac{P+Q}{2} = 5x \qquad \therefore P + Q = 10x$

$\dfrac{P-Q}{2} = 4x \qquad \therefore P - Q = 8x$

ie $\quad P = 9x$ and $Q = x$

$\therefore \sin 9x + \sin x = 2 \sin 5x \cos 4x$

$\therefore \int \sin 5x \cos 4x \, dx = \dfrac{1}{2} \int \sin 9x + \sin x \, dx$

$$= \frac{1}{2}\left[-\frac{\cos 9x}{9} - \cos x\right] + c$$

$\therefore \int \sin 5x \cos 4x \, dx = -\dfrac{\cos 9x}{18} - \dfrac{\cos x}{2} + c$

PURE MATHEMATICS INTEGRAL CALCULUS

(b) $\dfrac{x}{(1+x)^2}$

This appears to be a logarithmic integral but some adjustment is needed because $\dfrac{d}{dx}(1+x)^2 = 2(1+x)$ and the numerator as given only contains x and not $2(1+x)$.

So the following stages of algebra are necessary to obtain a form that can be integrated.

$$\frac{x}{(1+x)^2} = \frac{x+1-1}{(1+x)^2}$$

$$\therefore \quad \frac{x}{(1+x)^2} = \frac{(x+1)}{(x+1)^2} - \frac{1}{(1+x)^2}$$

$$= \frac{1}{1+x} - (1+x)^{-2}$$

$$\therefore \quad \int \frac{x}{(1+x)^2}\, dx = \int \left(\frac{1}{1+x} - (1+x)^{-2}\right) dx$$

$$= \log_e(1+x) - \frac{(1+x)^{-1}}{-1} + c$$

$$= \log_e(1+x) + \frac{1}{1+x} + c$$

5.2 Further integration

a) *Introduction*

All the functions that have been integrated so far have been relatively simple. However, just as it was necessary to have set methods for differentiating more complicated functions, so it is necessary to have special methods for integrating other than the simplest of functions.

In this section two particular methods will be introduced. The first is integration using 'partial fractions', used when the function is in the form of a quotient. The rules for deriving partial fractions have already been detailed in chapter 1, d). The second method of integration is using a 'change of variable'. Often a function cannot be integrated as it stands but, by changing from one variable to another, it is possible to derive a function that can be integrated. This method in particular requires a considerable amount of practice because only certain changes of variable will work in particular problems.

As yet more methods of integration will be introduced at a later stage, it is essential therefore, to gain expertise in each method as it arises.

b) Integration using partial fractions

Integrating a function that has been split into partial fractions often results in logarithmic terms together with algebraic terms.

Considering a complicated function such as:

$\dfrac{3x^3 + x+1}{(x-2)(x+1)^3}$ which would be split into four partial fractions of the form

$$\frac{A}{x-2} + \frac{B}{x+1} + \frac{C}{(x+1)^2} + \frac{D}{(x+1)^3}$$

Working this through in the usual manner gives A = 1, B = 2, C = -3 and D = 1 (check for yourself).

INTEGRAL CALCULUS   PURE MATHEMATICS

$$\therefore \quad \frac{3x^3 + x + 1}{(x-2)(x+1)^3} \equiv \frac{1}{x-2} + \frac{2}{x+1} - \frac{3}{(x+1)^2} + \frac{1}{(x+1)^3}$$

Integrating both sides of this with respect to x:

$$\int \left(\frac{3x^3 + x + 1}{(x-2)(x+1)^3}\right) dx = \int \left(\frac{1}{x-2} + \frac{2}{x+1} - 3(x+1)^{-2} + (x+1)^{-3}\right) dx$$

$$= \log_e(x-2) + 2\log_e(x+1) - \frac{3(x+1)^{-1}}{-1} + \frac{(x+1)^{-2}}{-2} + c$$

$$= \log_e(x-2) + \log_e(x+1)^2 + \frac{3}{x+1} - \frac{1}{2(x+1)^2} + c$$

$$= \log_e((x-2)(x+1)^2) + \left(\frac{3(2(x+1)) - 1}{2(x+1)^2}\right) + c$$

$$\therefore \quad \int \frac{3x^3 + x + 1}{(x-2)(x+1)^3} = \log_e((x-2)(x+1)^2) + \frac{6x+5}{2(x+1)^2} + c$$

Example 10

Integrate $\dfrac{x^3 - 18x - 21}{(x+2)(x-5)}$ with respect to x.

Solution

This is an improper fraction with two linear factors in the denominator. Therefore, first it must be divided out:

$$\therefore \quad \frac{x^3 - 18x - 21}{(x+2)(x-5)} = \frac{x^3 - 18x - 21}{x^2 - 3x - 10} = x + 3 + \frac{x+9}{(x+2)(x-5)}$$

but $\dfrac{x+9}{(x+2)(x-5)} \equiv \dfrac{A}{x+2} + \dfrac{B}{x-5}$ in partial fractions

$$x + 9 \equiv A(x-5) + B(x+2)$$

let x = 5 $5 + 9 = A(0) + B(5+2)$ let x = -2 $-2 + 9 = A(-2-5) + 2(0)$

$14 = B(7)$ $7 = A(-7)$

$\therefore \quad 2 = B$ $-1 = A$

So the function to be integrated is:

$$\int \left(x + 3 - \frac{1}{x+2} + \frac{2}{x-5}\right) dx = \int \left(x + 3 - \frac{1}{(x+2)} + \frac{2}{(x-5)}\right) dx$$

$$= \frac{x^2}{2} + 3x - \log_e(x+2) + 2\log_e(x-5) + c$$

$$\therefore \quad \int \left(\frac{x^3 - 18x - 21}{(x+2)(x-5)}\right) dx = \frac{x^2}{2} + 3x + \log_e\left[\frac{(x-5)^2}{x+2}\right] + c$$

PURE MATHEMATICS INTEGRAL CALCULUS

c) *Integration using change of variable*

There are many functions that still cannot be integrated, but this next method of changing the variable has many applications in algebraic, trigonometric and inverse trigonometric integration.

Example 11

By using a suitable change of variable integrate these functions:

(a) $3x\sqrt{4x-1}$

(b) $x(2x-1)^6$

(c) $\dfrac{x-1}{\sqrt{2x+3}}$

Solution

(a) $\int 3x\sqrt{4x-1}\,dx$ so far there is no way of integrating this function.

Let $\sqrt{4x-1} = u$

squaring $4x - 1 = u^2$

$\therefore \quad x = \dfrac{u^2 + 1}{4}$

Differentiating $\dfrac{d}{dx}(4x-1) = \dfrac{d}{dx}(u^2)$

$\therefore \quad 4 = 2u\dfrac{du}{dx} \quad dx = \dfrac{2u\,du}{4}$

So the integral can be changed from a function of x to a function of u by using:

$$x = \dfrac{u^2+1}{4}, \quad \sqrt{4x-1} = u \text{ and } dx = \dfrac{2u}{4}du$$

$\int (3x\sqrt{4x-1})\,dx$ becomes $\int \left(3\left(\dfrac{u^2+1}{4}\right)(u)\left(\dfrac{2u}{4}\right)\right)du$

$= \dfrac{3}{8}\int (u^2+1)u^2\,du$

$= \dfrac{3}{8}\int (u^4+u^2)\,du$

$= \dfrac{3}{8}\left(\dfrac{u^5}{5} + \dfrac{u^3}{3}\right) + c$

However, the original function was given in terms of x and that is how the answer must also be given:

$$\int (3x\sqrt{4x-1})\,dx = \dfrac{3}{8}\left(\dfrac{(\sqrt{4x-1})^5}{5} + \dfrac{(\sqrt{4x-1})^3}{3}\right) + c$$

This can be simplified algebraically as follows:

$$\frac{3}{8}\left(\frac{(4x-1)^{\frac{5}{2}}}{5} + \frac{(4x-1)^{\frac{3}{2}}}{3}\right) = \frac{3}{8}(4x-1)^{\frac{3}{2}}\left(\frac{4x-1}{5} + \frac{1}{3}\right)$$

$$= \frac{3}{8}(4x-1)^{\frac{3}{2}}\left(\frac{3(4x-1)+5}{15}\right)$$

$$= \frac{3}{8}(4x-1)^{\frac{3}{2}}\left(\frac{12x+2}{15}\right)$$

$$= \frac{3}{8}(4x-1)^{\frac{3}{2}}\left(\frac{2(6x+1)}{15}\right)$$

$$\therefore \int 3x\sqrt{4x-1}\,dx = \frac{1}{20}(4x-1)^{\frac{3}{2}}(6x+1) + c$$

(b) $\int x(2x-1)^6\,dx$ unless a change of variable is used, this can only be integrated by multiplying out the brackets - very tedious.

Let $2x - 1 = u$ then $2x = u + 1$ and $x = \dfrac{u+1}{2}$

Also $\dfrac{d}{dx}(2x-1) = \dfrac{d}{dx}(u)$

$$\therefore \quad 2 = \frac{du}{dx} \quad \text{and} \quad dx = \frac{du}{2}$$

$\int (x(2x-1)^6)\,dx$ becomes $\int \dfrac{(u+1)}{2}(u)^6 \left(\dfrac{du}{2}\right)$

$$= \frac{1}{4}\int (u^7 + u^6)\,du$$

$$= \frac{1}{4}\left(\frac{u^8}{8} + \frac{u^7}{7}\right) + c$$

$\therefore \int (x(2x-1)^6)\,dx = \dfrac{1}{4}\left(\dfrac{(2x-1)^8}{8} + \dfrac{(2x-1)^7}{7}\right) + c$

$$= \frac{1}{4}(2x-1)^7\left(\frac{2x-1}{8} + \frac{1}{7}\right) + c$$

$$= \frac{1}{4}(2x-1)^7\left(\frac{7(2x-1)+8}{56}\right) + c$$

$$= \frac{1}{4}(2x-1)^7\left(\frac{14x+1}{56}\right) + c$$

$\therefore \int (x(2x-1)^6)\,dx = \dfrac{1}{224}(2x-1)^7(14x+1) + c$

PURE MATHEMATICS — INTEGRAL CALCULUS

(c) $\int \dfrac{x-1}{\sqrt{2x+3}}\, dx$ this does not integrate using any of the previous methods so a change of variable will be tried.

Let $\sqrt{2x+3} = u$ $\therefore$ $2x + 3 = u^2$

$\therefore \quad 2x = u^2 - 3 \quad \therefore \quad x = \dfrac{u^2 - 3}{2}$

Also $\quad \dfrac{d}{dx}(2x+3) = \dfrac{d}{dx}(u^2)$

$2 = 2u \dfrac{du}{dx} \quad \therefore \quad dx = \dfrac{2u}{2} du = u\, du$

$\int \dfrac{x-1}{\sqrt{2x+3}}\, dx$ becomes $\int \left[\dfrac{\left(\dfrac{u^2-3}{2}\right) - 1}{u} \right] (u\, du) = \int \left[\left(\dfrac{u^2-3}{2}\right) - 1\right] du$

$= \int \left[\dfrac{u^2 - 5}{2}\right] du \quad = \dfrac{1}{2}\int [u^2 - 5]\, du$

$= \dfrac{1}{2}\left(\dfrac{u^3}{3} - 5u\right) + c$

$\therefore \quad \int \left[\dfrac{x-1}{\sqrt{2x+3}}\right] dx = \dfrac{1}{2}\left(\dfrac{(2x+3)^{\frac{3}{2}}}{3} - 5(2x+3)^{\frac{1}{2}}\right) + c$

$= \dfrac{1}{2}(2x+3)^{\frac{1}{2}}\left(\dfrac{2x+3}{3} - 5\right) + c$

$= \dfrac{1}{2}(2x+3)^{\frac{1}{2}}\left(\dfrac{2x-12}{3}\right) + c$

$\therefore \quad = \dfrac{1}{2}(2x+3)^{\frac{1}{2}} \cdot 2\dfrac{(x-6)}{3} + c$

$\int \left[\dfrac{x-1}{\sqrt{2x+3}}\right] dx = \dfrac{1}{3}(x-6)\sqrt{2x+3} + c$

So far all the changes of variable have been connected with algebraic functions. However, they can also be used to simplify trigonometric functions so that they may be integrated. The next example will show the types of substitution that may be used in these cases.

Example 12

By using a suitable change of variable integrate these functions:

(a) $\sin^2 4x \cdot \cos 4x$

(b) $\sin^5 x$

(c) $\cos^3 2x$

INTEGRAL CALCULUS   PURE MATHEMATICS

Solution

(a) $\int \sin^2 4x \cos 4x \, dx$ this is different from any of the trigonometric integrals that have been met so far.

Let $\sin 4x = u$

$$\frac{d}{dx}(\sin 4x) = \frac{d}{dx}(u) \quad \therefore \quad 4\cos 4x = \frac{du}{dx} \quad \therefore \quad dx = \frac{du}{4\cos 4x}$$

$\therefore \int \sin^2 4x \cos 4x \, dx$ becomes $\int u^2 \cdot \cos 4x \cdot \dfrac{du}{4\cos 4x}$

$$= \frac{1}{4}\int u^2 \, du$$

$$= \frac{1}{4}\left(\frac{u^3}{3}\right) + c$$

$\therefore \int \sin^2 4x \cos 4x \, dx = \dfrac{1}{12} \cdot \sin^3 4x + c$

(b) $\int \sin^5 x \, dx$ this cannot be integrated as though it were x^5

Let $\cos x = u$ $\therefore \dfrac{d}{dx}(\cos x) = \dfrac{d}{dx}(u)$ $\therefore -\sin x = \dfrac{du}{dx}$ $\therefore dx = \dfrac{du}{-\sin x}$

$\therefore \int \sin^5 x \, dx$ becomes $\int \sin^5 x \cdot \dfrac{du}{(-\sin x)} = -\int \sin^4 x \, du$

If $\cos x = u$ then, using $\cos^2 x + \sin^2 x = 1$ gives $\sin^2 x = 1 - u^2$

So $\quad -\int \sin^4 x \, du = -\int (1 - u^2)^2 \, du$

$$= -\int (1 - 2u^2 + u^4) \, du$$

$$= -\left[u - \frac{2u^3}{3} + \frac{u^5}{5}\right] + c$$

$\therefore \int \sin^5 x \, dx = -\left(\cos x - \dfrac{2\cos^3 x}{3} + \dfrac{\cos^5 x}{x}\right) + c$

(c) $\int \cos^3 2x \, dx$ - this is similar to the last example. Therefore:

let $\sin 2x = u$

$\therefore \dfrac{d}{dx}(\sin 2x) = \dfrac{d}{dx}(u)$

$\therefore 2\cos 2x = \dfrac{du}{dx} \quad \therefore dx = \dfrac{du}{2\cos 2x}$

$\int \cos^3 2x \, dx$ becomes $\int \cos^3 2x \cdot \dfrac{du}{2\cos 2x} = \int \dfrac{\cos^2 2x}{2} \, du$

If $\sin 2x = u$ then, using $\cos^2 2x + \sin^2 2x = 1$ $\therefore \cos^2 2x = 1 - u^2$

So $\frac{1}{2}\int \cos^2 2x \, du = \frac{1}{2}\int [1 - u^2] \, du$

$\qquad\qquad\qquad\qquad = \frac{1}{2}\left(u - \frac{u^3}{3}\right) + c$

$\therefore \quad \int \cos^3 2x \, dx = \frac{1}{2}\left(\sin 2x - \frac{\sin^3 2x}{3}\right) + c$

Substitutions can also be used in certain integrals to give inverse trigonometrical functions. These are integrals such as

$\int \dfrac{1}{\sqrt{a^2 - b^2 x^2}} \, dx$ leading to functions of $\sin^{-1}$ and

$\int \dfrac{1}{a^2 + b^2 x^2} \, dx$ leading to functions of $\tan^{-1}$.

The next example will show the change of variable that is necessary to achieve these results.

Example 13

By using a suitable change of variable integrate the following functions:

(a) $\dfrac{1}{\sqrt{1 - x^2}}$

(b) $\dfrac{1}{\sqrt{9 - 4x^2}}$

(c) $\dfrac{1}{1 + x^2}$

(d) $\dfrac{1}{3 + 4x^2}$

Solution

(a) $\int \dfrac{1}{\sqrt{1 - x^2}}$ it might appear that a suitable change of variable would be $u = \sqrt{1 - x^2}$.

However, using this gives $\int \dfrac{-1}{\sqrt{1 - u^2}} \, du$ which is no better. Therefore:

let $\qquad x = \sin u$

then $\qquad \dfrac{d(x)}{dx} = \dfrac{d}{dx}(\sin u) \qquad \therefore \quad 1 = \cos u \cdot \dfrac{du}{dx}$ and $dx = \cos u \, du$

$\qquad\qquad \sqrt{1 - x^2} = \sqrt{1 - \sin^2 u} = \sqrt{\cos^2 u} = \cos u$

$\therefore \quad \int \dfrac{1}{\sqrt{1 - x^2}} \, dx$ becomes $\int \dfrac{1}{\cos u}(\cos u \, du)$

$\qquad\qquad\qquad\qquad = \int du$

$\qquad\qquad\qquad\qquad = u + c$

INTEGRAL CALCULUS   PURE MATHEMATICS

But what does u equal?

Well, if $x = \sin u$ then u is the angle whose sine is x.

$\therefore \quad \sin^{-1} x = u$

$$\int \frac{1}{\sqrt{1-x^2}} dx = \sin^{-1} x + c$$

(b) $\int \frac{1}{\sqrt{9-4x^2}} dx$ this is similar to the previous example, with the added complication of 9 and 4!

Let $\quad x = \frac{3}{2} \sin u$

then $\quad \frac{d}{dx}(x) = \frac{d}{dx}\left(\frac{3}{2} \sin u\right) \quad \therefore \quad 1 = \frac{3}{2} \cos u \cdot \frac{du}{dx}$ and $dx = \frac{3}{2} \cos u \, du$

$\sqrt{9-4x^2} = \sqrt{9 - 4\left(\frac{3}{2}\sin u\right)^2} = \sqrt{9 - 4 \cdot \frac{9}{4} \sin^2 u}$

$\qquad\qquad\quad = \sqrt{9 - 9\sin^2 u} = \sqrt{9(1-\sin^2 u)}$

$\qquad\qquad\quad = \sqrt{9 \cos^2 u} = 3 \cos u$

$\therefore \quad \int \frac{1}{\sqrt{9-4x^2}} dx \quad$ becomes $\quad \int \frac{1}{3 \cos u} \left(\frac{3}{2} \cos u \, du\right)$

$\qquad\qquad\qquad\qquad = \int \frac{1}{2} du \quad = \quad \frac{1}{2} u + c$

Well, if $\quad x = \frac{3}{2} \sin u$

then $\quad \frac{2x}{3} = \sin u$ and so $u = \sin^{-1} \frac{2x}{3}$

$\therefore \quad \int \frac{1}{\sqrt{9-4x^2}} dx = \frac{1}{2} \cdot \sin^{-1} \frac{2x}{3} + c$

(c) $\int \frac{1}{1+x^2} du$ - this is similar to the previous two examples but without the square root.

Let $\quad x = \tan u$

then $\quad \frac{d}{dx}(x) = \frac{d}{dx}(\tan u)$

$\qquad 1 = \sec^2 u \cdot \frac{du}{dx} \quad \therefore \quad dx = \sec^2 u \cdot du$

$\qquad 1 + x^2 = 1 + \tan^2 u = \sec^2 u$

$$\therefore \quad \int \frac{1}{1+x^2} dx \text{ becomes } \int \frac{1}{\sec^2 u} (\sec^2 u \, du)$$

$$= \int du$$

$$= u + c$$

$$\therefore \quad \int \frac{1}{1+x^2} dx = \tan^{-1} x + c$$

(d) $\int \frac{1}{3+4x^2} dx$ as before, but with 3 and 4 to take into account.

Let $x = \frac{\sqrt{3}}{2} \tan u$

$$\frac{d}{dx}(x) = \frac{d}{dx}\left(\frac{\sqrt{3}}{2} \tan u\right)$$

$$1 = \frac{\sqrt{3}}{2} \sec^2 u \cdot \frac{du}{dx} \qquad \therefore \quad dx = \frac{\sqrt{3}}{2} \cdot \sec^2 u \, du$$

$$3 + 4x^2 = 3 + 4\left(\frac{\sqrt{3}}{3} \tan u\right)^2 = 3 + 4 \cdot \frac{3}{4} \tan^2 u$$

$$= 3 + 3 \tan^2 u \qquad = 3(1 + \tan^2 u) \qquad = 3 \sec^2 u$$

$$\int \frac{1}{3+4x^2} dx \text{ becomes } \int \frac{1}{3 \sec^2 u} \left(\frac{\sqrt{3}}{2} \sec^2 u \, du\right)$$

$$= \int \frac{\sqrt{3}}{6} du$$

$$= \frac{\sqrt{3}}{6} u + c$$

$$\therefore \quad \int \frac{1}{3+4x^2} dx = \frac{\sqrt{3}}{6} \tan^{-1} \frac{2x}{\sqrt{3}} + c$$

Many different changes of variable have been introduced here: it is only with practice that the correct one will be recognised easily. There are also occasions when a certain change of variable is tried and found to be of no use, in which case an alternative substitution must be tried.

d) *Example* ('A' level question)

Integrate these functions:

(a) $(\sin x + \cos x)^2$

(b) $\frac{1}{e^x + e^{-x}}$

INTEGRAL CALCULUS   PURE MATHEMATICS

Solution

(a) $\int (\sin x + \cos x)^2 \, dx = \int (\sin^2 x + 2\sin x \cos x + \cos^2 x) \, dx$

$= \int (1 + \sin 2x) \, dx$

$\therefore \int (\sin x + \cos x)^2 \, dx = x - \dfrac{\cos 2x}{2} + c$

(b) $\int \dfrac{1}{e^x + e^{-x}} \, dx = \int \dfrac{1}{\left(e^x + \dfrac{1}{e^x}\right)} \, dx$

$= \int \left(\dfrac{e^x}{(e^x)^2 + 1}\right) dx$

Although it may not be obvious at first sight, this integral is similar to

$\int \dfrac{1}{1 + x^2} \, dx$

Therefore, try the change of variable

$e^x = \tan u \quad \text{thus} \quad \dfrac{d}{dx}(e^x) = \dfrac{d}{dx}(\tan u) \quad \text{and} \quad e^x = \sec^2 u \cdot \dfrac{du}{dx}$

$\therefore \quad dx = \dfrac{\sec^2 u}{e^x} \, du$

$1 + (e^x)^2 = 1 + (\tan u)^2 = \sec^2 u$

$\therefore \quad \int \dfrac{1}{e^x + e^{-x}} \, dx = \int \left(\dfrac{e^x}{(e^x)^2 + 1}\right) dx \quad \text{becomes} \quad \int \dfrac{e^x}{\sec^2 u} \left(\dfrac{\sec^2 u}{e^x} \, du\right)$

$= \int du = u + c$

$\therefore \quad \int \dfrac{1}{e^x + e^{-x}} \, dx = \tan^{-1} e^x + c$

5.3   Area under a curve

a)   *Curve sketching*

As has already been mentioned, it is often useful to sketch the graph of an equation so that the curve can be visualised and the problem it relates to understood more easily.

On certain occasions it is *essential* to draw a sketch graph so that mistakes are avoided. The calculation of 'areas under a curve' is one such occasion as it is very important to ascertain on which side of the axis the area is and whether the curve crosses the axis in the area under consideration.

The important points to find when sketching a curve are the points of intersection of the curve and the axes.

On the x axis y = 0, so replacing y with 0 in y = f(x) means that corresponding values of x can be calculated.

Similarly on the y axis x = 0, so replacing x with 0 in y = f(x) means that the corresponding y values can be calculated.

PURE MATHEMATICS — INTEGRAL CALCULUS

If the turning points have already been found they also can be marked in on the graph - but do not find them specially.

If the shape of the curve is still not obvious, then it is worth calculating one or two points on the graph and marking them in fairly accurately. This should then give a reliable picture of the curve from which to work.

It is not intended that a sketch graph should be used to find intermediate values of the variables - an accurate graph drawn on graph paper would be needed for this. A sketch graph is purely intended to convey the shape of the curve and where, in particular, it crosses the axes.

As several sketch graphs will be drawn in the rest of this section, no example is included here.

b) *Definite integrals*

Nearly all the integrals that have been found so far have been indefinite integrals because they contained the constant c. However, when an integral is given with limits on, then it is called a definite integral, eg

$\int_{-1}^{2} (x^2 + 3)\, dx$ is called the integral of the function $x^2 + 3$ between the limits of -1 and 2.

Therefore, first the function is integrated $\int_{-1}^{2}(x^2 + 3)\, dx = \left[\frac{x^3}{3} + 3x + c\right]_{-1}^{2}$

and then the values $x = 2$ and $x = -1$ are substituted in turn and the answers subtracted, ie

$$\left[\frac{x^3}{3} + 3x + c\right]_{-1}^{2} = \left(\frac{2^3}{3} + 3(2) + c\right) - \left(\frac{(-1)^3}{3} + 3(-1) + c\right)$$

$$= \frac{8}{3} + 6 + c - \left(-\frac{1}{3} - 3 + c\right)$$

$$= \frac{8}{3} + 6 + \cancel{c} + \frac{1}{3} + 3 - \cancel{c}$$

$\therefore \quad \int_{-1}^{2} (x^2 + 3)\, dx = 12$

After integrating, the top limit (2 in this case) is always substituted first and then the bottom limit (-1 in this case). In the course of subtracting the two values of the integral found in this way, the constant c always disappears.

Example 14

Evaluate the following definite integrals:

(a) $\quad \int_{1}^{4} (y^2 + \sqrt{y})\, dy$

(b) $\quad \int_{\frac{1}{2}}^{1} \left(\frac{1}{\sqrt{1 - x^2}}\right) dx$

INTEGRAL CALCULUS — PURE MATHEMATICS

(c) $\displaystyle\int_{2}^{3}\left(\frac{x-9}{x(x-1)(x+3)}\right)dx$

Solution

(a) $\displaystyle\int_{1}^{4}(y^2+\sqrt{y})\,dy$ the first step is always to integrate:

$$\int_{1}^{4}\left(y^2+y^{\frac{1}{2}}\right)dy = \left[\frac{y^3}{3}+\frac{y^{\frac{3}{2}}}{\frac{3}{2}}+c\right]_{1}^{4}$$

$$= \left[\frac{y^3}{3}+\frac{2y^{\frac{3}{2}}}{3}+c\right]_{1}^{4}$$

When definite integrals are to be evaluated, there is no need to simplify the algebraic expression. It is a waste of time as the next step is to substitute the numerical values for y (4 first, then 1) and subtract:

$$\int_{1}^{4}\left(y^2+y^{\frac{1}{2}}\right)dy = \left[\frac{(4)^3}{3}+\frac{2(4)^{\frac{3}{2}}}{3}+c\right]-\left[\frac{(1)^3}{3}+\frac{2(1)^{\frac{3}{2}}}{3}+c\right]$$

$$= \left[\frac{64}{3}+\frac{2(8)}{3}+c\right]-\left[\frac{1}{3}+\frac{2(1)}{3}+c\right]$$

$$= \frac{64}{3}+\frac{16}{3}+\cancel{c}-\frac{1}{3}-\frac{2}{3}-\cancel{c}$$

$$\therefore \int_{1}^{4}(y^2+\sqrt{y})\,dy = \frac{77}{3} = 25\frac{2}{3}$$

(b) $\displaystyle\int_{\frac{1}{2}}^{1}\left(\frac{3}{\sqrt{1-x^2}}\right)dx$

this is nearly the same as an earlier example under inverse trigonometric functions, viz:

$$\int_{\frac{1}{2}}^{1}\left(\frac{1}{\sqrt{1-x^2}}\right)dx = \sin^{-1}x + c$$

$$\therefore \int_{\frac{1}{2}}^{1}\left(\frac{3}{\sqrt{1-x^2}}\right)dx = 3\int_{\frac{1}{2}}^{1}\frac{1}{\sqrt{1-x^2}}\,dx$$

$$= 3\left[\sin^{-1}x + c\right]_{\frac{1}{2}}^{1}$$

$$= 3\left[\sin^{-1}(1) + c\right] - 3\left[\sin^{-1}\left(\frac{1}{2}\right) + c\right]$$

$\sin^{-1}(1)$ means the angle whose sine is 1, ie $\frac{\pi}{2}$ radians (90°) (all differentiation and integration is performed with angles in radians not degrees). $\sin^{-1}\left(\frac{1}{2}\right)$ means the angle whose sine is $\frac{1}{2}$, ie $\frac{\pi}{6}$ radians (30°).

$$\therefore \int_{\frac{1}{2}}^{1}\left(\frac{3}{\sqrt{1-x^2}}\right)dx = 3\left[\frac{\pi}{2} + c\right] - 3\left[\frac{\pi}{6} + c\right]$$

$$= \frac{3\pi}{2} + 3\cancel{c} - \frac{3\pi}{6} - 3\cancel{c}$$

$$\therefore \int_{\frac{1}{2}}^{1}\left(\frac{3}{\sqrt{1-x^2}}\right)dx = \pi$$

It is quite usual for answers to be left in terms of π and not evaluated numerically.

(c) $\int_{2}^{3} \frac{x-9}{x(x-1)(x+3)} dx$ this function needs to be put into partial fractions before it can be integrated.

It is not an improper fraction and the denominator contains three linear factors. Therefore, the method needed is:

$$\frac{x-9}{x(x-1)(x+3)} \equiv \frac{A}{x} + \frac{B}{x-1} + \frac{C}{x+3}$$

$$x - 9 \equiv A(x-1)(x+3) + Bx(x+3) + Cx(x-1)$$

Let $x = 0$ $0 - 9 = A(0-1)(0+3) + B(0)(0+3) + C(0)(0-1)$

         $-9 \;\; = A(-3) \;\; \therefore \;\;\; 3 = A$

Let $x = 1$ $1 - 9 = 3(0)(1+3) + B(1)(1+3) + C(1)(0)$

         $-8 \;\; = 4B \;\;\;\;\;\;\;\;\; -2 = B$

Let $x = -3$ $-3 - 9 = 3(-3-1)(0) - 2(-3)(0) + C(-3)(-3-1)$

         $-12 \;\; = 12C \;\; \therefore \;\; -1 = C$

$$\therefore \int_{2}^{3}\left[\frac{x-9}{x(x-1)(x+3)}\right]dx = \int_{2}^{3}\left[\frac{3}{x} - \frac{2}{x-1} - \frac{1}{x+3}\right]dx$$

$$= [3\log_e x - 2\log_e(x-1) - \log_e(x+3) + c]_2^3$$

$$= [\log_e x^3 - \log_e(x-1)^2 - \log_e(x+3) + c]_2^3$$

$$= \left[\log_e\left(\frac{x^3}{(x-1)^2(x+3)}\right) + c\right]_2^3$$

$$= \left[\log_e\left(\frac{3^3}{(3-1)^2(3+3)}\right) + c\right] - \left[\log_e\frac{2^3}{(2-1)^2(2+3)} + c\right]$$

$$= \log_e\left(\frac{27}{4(6)}\right) + c - \log_e\left(\frac{8}{1(5)}\right) - c$$

$$= \log_e\left(\frac{9}{8}\right) - \log_e\left(\frac{8}{5}\right) = \log_e\left(\frac{9/8}{8/5}\right) = \log_e\left(\frac{9}{8} \times \frac{5}{8}\right)$$

$$\therefore \int_2^3 \left(\frac{x-9}{x(x-1)(x+3)}\right) dx = \log_e\left(\frac{45}{64}\right)$$

$$= \log_e 0.703125$$

Answers are often left in this form, though the exact value can be calculated from tables or by calculator, $\log_e 0.703125 = -0.352$.

The main application of definite integrals with limits, such as in this last example, is in determining the area between a curve and an axis. This is usually called 'finding the area under the curve'.

c) *Area by integration*

(a)

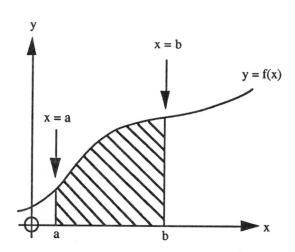

The area between the curve, the x axis and the lines x = a and x = b can be found by evaluating the definite integral:

$$A = \int_a^b f(x)\, dx$$

(b)

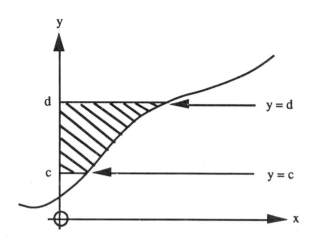

The area between the curve, the y axis and the lines y = c and y = d can be found by evaluating the definite integral:

$$A = \int_c^d g(y)\, dy$$

As equations are usually given in the form y = f(x), it is necessary to re-arrange them in this second case so that x is a function of y, ie x = g(y) - say.

It is necessary to sketch a curve before calculating the required area because areas above the x axis are positive - as shown in (a) but areas below the x axis are negative.

Similarly for (b), areas to the right of the y axis are positive and those to the left are negative.

(a)

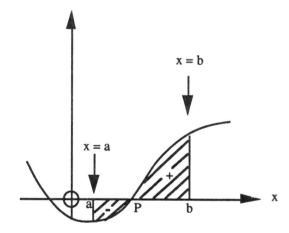

To find the shaded area between x = a and x = b, it would be necessary to find P - the point where the curve crosses the x axis, and evaluate

$$\int_a^P f(x)\, dx \;+\; \int_P^b f(x)\, dx$$

These are added ignoring the negative sign.

(b)

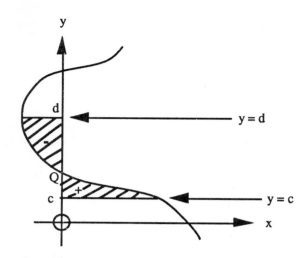

Having found Q - the point of intersection of the curve and the y axis,

$$\int_c^Q g(y)\,dy \;+\; \int_Q^d g(y)\,dy$$

is evaluated, ignoring the negative sign.

Example 15

The curve $y = 6 - x - x^2$ cuts the x axis in two points, A and B. Find the area enclosed by the x axis and the portion of the curve between A and B.

Solution

Sketch graph: on y axis $x = 0$ $\therefore y = 6 - 0 - 0 = 6$ $\therefore$ (0, 6) is a point on the curve

on x axis $y = 0$ $\therefore 0 = 6 - x - x^2$ or $x^2 + x - 6 = 0$

thus $(x + 3)(x - 2) = 0$

$\therefore$ either $x + 3 = 0$ or $x - 2 = 0$

$x = -3$ $x = 2$

$\therefore$ (-3, 0) and (2, 0) are the points A and B on the curve.

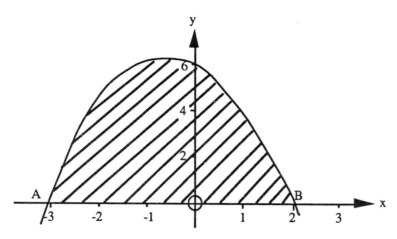

Since the required area is completely above the x axis and the curve does not cross the axis between A and B, it can be found as follows:

$$\text{Area} = \int_{-3}^{2} (6 - x - x^2)\, dx$$

$$= \left[6x - \frac{x^2}{2} - \frac{x^3}{3} + c \right]_{-3}^{2}$$

$$= \left[6(2) - \frac{(2)^2}{2} - \frac{(2)^3}{3} + c \right] - \left[6(-3) - \frac{(-3)^2}{2} - \frac{(-3)^3}{3} + c \right]$$

$$= 12 - 2 - \frac{8}{3} + \cancel{c} + 18 + \frac{9}{2} - 9 - \cancel{c}$$

$$\therefore \quad \text{Area} = 20\frac{5}{6}$$

Example 16

Find the area enclosed by the curve $x = (y - 1)(y - 4)$ and the y axis.

Solution

Sketch graph: On y axis $x = 0$ $\therefore$ $(y - 1)(y - 4) = 0$

$\therefore$ either $y - 1 = 0$ or $y - 4 = 0$
            $y = 1$ $y = 4$ } giving points (0, 1) and (0, 4).

On x axis $y = 0$ $\therefore (0 - 1)(0 - 4) = x$
                          $4 = x$ } giving the point (4, 0)

INTEGRAL CALCULUS   PURE MATHEMATICS

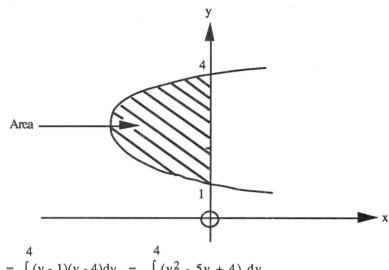

$$\text{Area} = \int_1^4 (y-1)(y-4)\,dy = \int_1^4 (y^2 - 5y + 4)\,dy$$

$$= \left[\frac{y^3}{3} - \frac{5y^2}{2} + 4y + c\right]_1^4$$

$$= \left[\frac{4^3}{3} - \frac{5(4)^2}{2} + 4(4) + c\right] - \left[\frac{1^3}{3} - \frac{5(1)^2}{2} + 4(1) + c\right]$$

$$= \frac{64}{3} - 40 + 16 + \not c - \frac{1}{3} + \frac{5}{2} - 4 - \not c$$

∴   Area = (-)4.5

(The negative sign showing that the area is to the left of the y axis as can be seen from the sketch graph.)

d)   *Area using trapezium rule*

Integration is the method to use when an area is required accurately. However, there are times when an estimate of the area is all that is needed, in which case one of two rules can be used: either the trapezium rule, which will be considered first, or Simpson's rule, which will be explained in the next part of this chapter. These rules are also used when the function cannot be integrated but, even so, an estimate of the area is required.

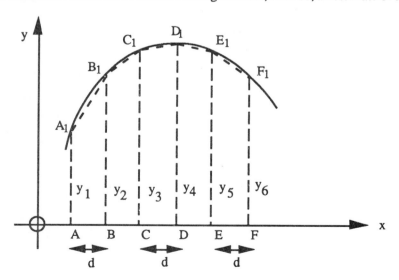

A, B, C, D, E and F are six equally spaced points on the x axis. They are such that the distance from A to B = BC = CD = DE = EF = d(say).

PURE MATHEMATICS — INTEGRAL CALCULUS

A^1, B^1, C^1, D^1, E^1 and F^1 are the corresponding points on the curve and they have been joined with straight lines.

AA^1 = value of the y coordinate at A^1, y_1 say

BB^1 = value of the y coordinate at B^1, y_2 say

Then the area of the trapezium $AA^1B^1B = \frac{1}{2}(y_1 + y_2)d$

So the total area of the five trapezia will be:

$$\text{Area} = \frac{1}{2}(y_1 + y_2)d + \frac{1}{2}(y_2 + y_3)d + \frac{1}{2}(y_3 + y_4)d + \frac{1}{2}(y_4 + y_5)d + \frac{1}{2}(y_5 + y_6)d$$

$$= \frac{1}{2}(y_1 + 2y_2 + 2y_3 + 2y_4 + 2y_5 + y_6)d$$

and this is known as the trapezium rule for six ordinates (the ordinates being y_1, y_2, y_3, y_4, y_5 and y_6).

In general terms the trapezium rule for n ordinates (making n - 1 trapezia) is:

$$\boxed{\text{Area} = \frac{1}{2}(y_1 + 2y_2 + 2y_3 + \ldots + 2y_{n-1} + y_n)d}$$

This is not a difficult rule to apply as the next example will show.

Example 17

Evaluate A where $A = \int_{0.1}^{0.5} e^{-x} \, dx$

Estimate the percentage error correct to 1 decimal place in evaluating A using the trapezium rule with 5 equally spaced ordinates.

Solution

To start with a sketch graph of $y = e^{-x}$

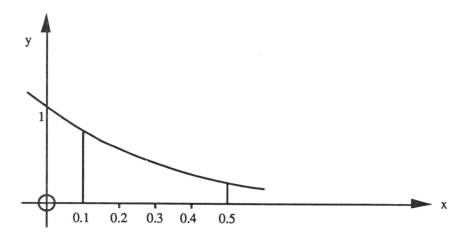

$A = \int_{0.1}^{0.5} e^{-x} \, dx = \left[\frac{e^{-x}}{-1} + c \right]_{0.1}^{0.5}$

$$= [-e^{-x} + c]_{0.1}^{0.5}$$

$$= -e^{-0.5} + c - [-e^{-0.1} + c]$$

$$= -e^{-0.5} + e^{-0.1}$$

$$= -0.6065 + 0.9048$$

$$\therefore A = 0.2983$$

'Five equally spaced ordinates' means ordinates at $x = 0.1, 0.2, 0.3, 0.4, 0.5$. Therefore, $d = 0.1$.

To find the values of the ordinates, ie y_1, y_2, y_3, y_4 and y_5, it is best to draw up a table of values:

x	0.1	0.2	0.3	0.4	0.5
$y = e^{-x}$	0.9048	0.8187	0.7408	0.6703	0.6065
	y_1	y_2 (x 2)	y_3 (x 2)	y_4 (x 2)	y_5

$\therefore$ Area by trapezium rule $= \frac{1}{2}(0.9048 + 2 \times 0.8187 + 2 \times 0.7408 + 2 \times 0.6703 + 0.6065)0.1$

$$= \frac{1}{2}(5.9709)(0.1)$$

$$= 0.2985$$

$\therefore$ Actual error $= 0.2985 - 0.2983$

$$= 0.0002$$

$\therefore$ Percentage error $= \frac{0.0002}{0.2983} \times 100$

$$= 0.067\%$$

$$= 0.1\% \text{ to 1 decimal place}$$

Therefore, in this particular case, the error involved in using the approximate method is very slight.

e) *Area using Simpson's rule*

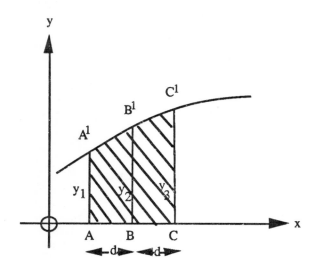

PURE MATHEMATICS — INTEGRAL CALCULUS

A, B and C are three equally-spaced points on the x axis such that $AB = BC = d$, and A^1, B^1 and C^1 are the corresponding points on the curve such that AA^1 = ordinate y_1, $BB^1 = y_2$, $CC^1 = y_3$.

Simpson's rule for 3 ordinates gives the area under the curve as:

$$\text{Area} = \tfrac{1}{3} d (y_1 + 4y_2 + y_3)$$

This formula is deduced by assuming that the points A^1, B^1 and C^1 can be joined by a smooth curve rather than straight lines as in the trapezium rule. It will not actually be derived here as it is more complicated to derive than the trapezium rule.

Simpson's rule can be applied to more than 3 ordinates as long as there is an odd number of ordinates, eg

with 5 ordinates (apply Simpson's rule twice)

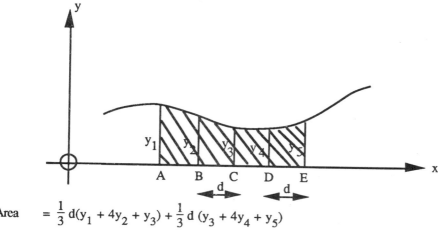

$$\begin{aligned}\text{Area} &= \tfrac{1}{3} d(y_1 + 4y_2 + y_3) + \tfrac{1}{3} d(y_3 + 4y_4 + y_5) \\ &= \tfrac{1}{3} d(y_1 + 4y_2 + 2y_3 + 4y_4 + y_5)\end{aligned}$$

with 7 ordinates (apply Simpson's rule three times)

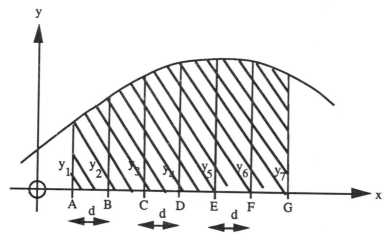

$$\begin{aligned}\text{Area} &= \tfrac{1}{3} d(y_1 + 4y_2 + y_3) + \tfrac{1}{3} d(y_3 + 4y_4 + y_5) + \tfrac{1}{3} d(y_5 + 4y_6 + y_7) \\ &= \tfrac{1}{3} d(y_1 + 4y_2 + 2y_3 + 4y_4 + 2y_5 + 4y_6 + y_7)\end{aligned}$$

INTEGRAL CALCULUS   PURE MATHEMATICS

The next example will show how to calculate the area under a curve using this rule.

Example 18

Using Simpson's rule with 6 spaces to find an approximate value for

$$\int_0^\pi \left(\sqrt{1 + 2 \sin x}\right) dx$$

giving the answer to 3 significant figures.

Solution

When drawing a sketch graph of a trigonometric function it is simplest to work out a few points.

$y = \sqrt{1 + 2 \sin x}$ when $x = 0$ $y = \sqrt{1 + 2 \sin 0} = \sqrt{1} = 1$

when $x = \frac{\pi}{2}$ $y = \sqrt{1 + 2 \sin \frac{\pi}{2}} = \sqrt{3} = 1.732$

when $x = \pi$ $y = \sqrt{1 + 2 \sin \pi} = \sqrt{1} = 1$

Giving the points $(0, 1)$ $\left(\frac{\pi}{2}, 1.732\right)$ $(\pi, 1)$

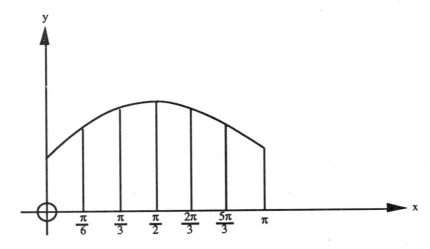

Six spaces mean seven ordinates at $x = 0, \frac{\pi}{6}, \frac{\pi}{3}, \frac{\pi}{2}, \frac{2\pi}{3}, \frac{5\pi}{6}, \pi$ and $d = \frac{\pi}{6}$

x	0	$\frac{\pi}{6}$	$\frac{\pi}{3}$	$\frac{\pi}{2}$	$\frac{2\pi}{3}$	$\frac{5\pi}{6}$	π
sin x	0	0.5	0.866	1.0	0.866	0.5	0
2 sin x	0	1.0	1.732	2.0	1.732	1.0	0
2 sin x + 1	1	2.0	2.732	3.0	2.732	2.0	1
$\sqrt{1 + 2 \sin x}$	1	1.414	1.653	1.732	1.653	1.414	1
	y_1 (1)	$4y_2$ (5.656)	$2y_3$ (3.306)	$4y_4$ (6.928)	$2y_5$ (3.306)	$4y_6$ (5.656)	y_7 (1)

$\therefore$ Area $= \frac{1}{3}\left(\frac{\pi}{6}\right)(1 + 5.656 + 3.306 + 6.928 + 3.306 + 5.656 + 1)$

∴ Area $= \dfrac{1}{3}\left(\dfrac{\pi}{6}\right)(26.852)$

$= 4.69$ (to 3 significant figures)

Example 19

Sketch the graphs of $y^2 = 16x$ and $y = x - 5$. Find:

(a) the coordinates of the points of intersection;

(b) the area of the region enclosed between the graphs.

Solution

$y^2 = 16x$ when $x = 0$, $y^2 = 16(0) = 0$ giving $(0, 0)$

when $x = 1$, $y^2 = 16$ ∴ $y = \pm 4$ giving $(1, 4)$ $(1, -4)$

$y = x - 5$ when $x = 0$, $y = 0 - 5 = -5$ giving $(0, -5)$

when $y = 0$, $0 = x - 5$ ∴ $x = 5$ giving $(5, 0)$

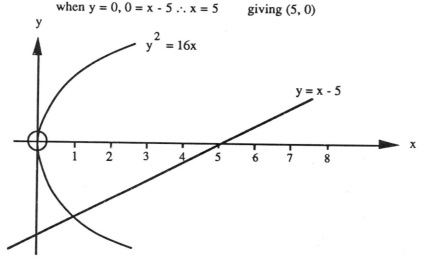

(a) The coordinates of the points of intersection are found by solving the equations:

$y^2 = 16x$ -------- (1)

$y = x - 5$ -------- (2)

Replacing for y from (2) into (1):

$(x - 5)^2 = 16x$ $x^2 - 10x + 25 = 16x$

$x^2 - 26x + 25 = 0$ $(x - 1)(x - 25) = 0$

Either $x - 1 = 0$ or $x - 25 = 0$

∴ $x = 1$ $x = 25$

$y = 1 - 5$ $y = 25 - 5$

$= -4$ $= 20$

So the points of intersection are $(1, -4)$ and $(25, 20)$.

(b) A second sketch graph must be drawn to show the area that is to be calculated, ie the shaded area below.

Also:

$$y^2 = 16x$$

$$\therefore y = \pm\sqrt{16x}$$

$$= \pm 4\sqrt{x}$$

$y = +4\sqrt{x}$ is the equation of the curve above x axis

$y = -4\sqrt{x}$ is the equation of the curve below x axis

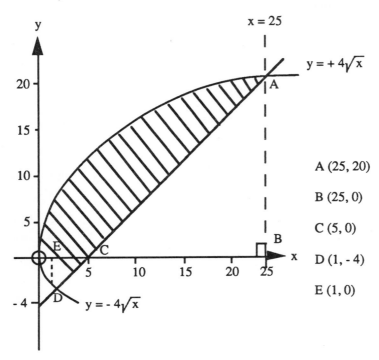

A (25, 20)
B (25, 0)
C (5, 0)
D (1, -4)
E (1, 0)

To find the area above the x axis, ie the area from OACO:

Integrating the functions between the limits $x = 0$ and $x = 25$ will give all the area between the curve, the x axis and the line $x = 25$. However, this is too much area and it is necessary to subtract the area of the right angled triangle ABC.

$$\text{Area OABO} = \int_0^{25} 4\sqrt{x}\, dx = \int_0^{25} 4.x^{\frac{1}{2}}\, dx = \left[\frac{4x^{\frac{3}{2}}}{\frac{3}{2}} + c\right]_0^{25}$$

$$= \left[\frac{8x^{\frac{3}{2}}}{3} + c\right]_0^{25} = \frac{8}{3}(25)^{\frac{3}{2}} + c - \frac{8}{3}(0) - c = \frac{8}{3}.(125)$$

$$= 333\frac{1}{3} \text{ square units}$$

Area of $\triangle ABC = \frac{1}{2}(20)(20)$

$= 200$ square units

$\therefore$ Area of OACO $= 333\frac{1}{3} - 200$

$= 133\frac{1}{3}$ square units

To find the area below the x axis, ie the area from ODCO:

Integrating the function from $x = 0$ to $x = 1$ will give the area ODEO and in addition the area of the right-angled triangle DCE is required.

$$\text{Area ODEO} = \int_0^1 -4\sqrt{x}\, dx$$

$$= \int_0^1 -4x^{\frac{1}{2}} dx = \left[\frac{-4x^{\frac{3}{2}}}{\frac{3}{2}} + c\right]_0^1$$

$$= \left[\frac{-8x^{\frac{3}{2}}}{3} + c\right]_0^1$$

$$= \frac{-8(1)^{\frac{3}{2}}}{3} + c - \frac{-8(0)}{3} - c = -\frac{8}{3}$$

$$= -2\frac{2}{3} \text{ square units (negative sign denoting area below x axis)}$$

Area of $\triangle DCE = \frac{1}{2}(4)(4) = 8$ square units

$\therefore$ Area of ODCO $= 2\frac{2}{3} + 8 = 10\frac{2}{3}$ square units

$\therefore$ Shaded area $= 133\frac{1}{3} + 10\frac{2}{3}$ square units

$= 144$ square units

INTEGRAL CALCULUS — PURE MATHEMATICS

5.4 Further applications of integration

a) Volumes of revolution

When the area under a curve is rotated about the x axis a solid of revolution is formed, and the volume of this solid can be determined by integration.

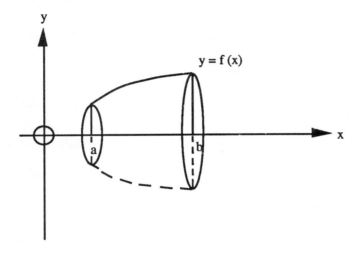

When this curve $y = f(x)$ is rotated through 360° about the x axis the volume of the solid generated is given by:

$$V = \int_a^b \pi y^2 \, dx \qquad \text{where } y = f(x)$$

A curve can equally well be rotated about the y axis and the volume so generated is given by:

$$V = \int_c^d \pi x^2 \, dy \qquad \text{where } x = g(y)$$

Example 20

The area bounded by the arc of the curve $y = x(3 - x)$ between the points where $x = 0$ and $x = 2$, the x axis, and the line $x = 2$ is rotated about the x axis. Find the volume of the solid so generated.

PURE MATHEMATICS — INTEGRAL CALCULUS

Solution

First to sketch the curve over the range of interest.

On the y axis, $x = 0$ $\therefore$ $y = 0(3 - 0) = 0$ giving $(0, 0)$

On the x axis, $y = 0$ $\therefore$ $0 = x(3 - x)$

    either $x = 0$ giving $(0, 0)$

    or $x = 3$ giving $(3, 0)$

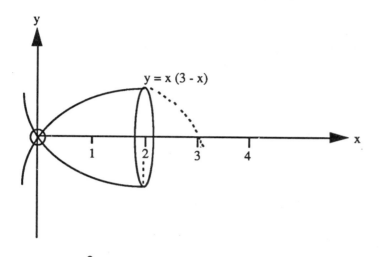

Volume $= \int_0^2 \pi y^2 \, dx$

$= \pi \int_0^2 x^2(3 - x)^2 \, dx$

$= \pi \int_0^2 x^2(9 - 6x + x^2) \, dx$

$= \pi \int_0^2 (9x^2) - 6x^3 + x^4 \, dx$

$= \pi \left[\dfrac{9x^3}{3} - \dfrac{6x^4}{4} + \dfrac{x^5}{5} + c \right]_0^2 = \pi \left[3x^2 - \dfrac{3x^4}{2} + \dfrac{x^5}{5} + c \right]_0^2$

$= \pi \left[3(2)^3 - \dfrac{3(2)^4}{2} + \dfrac{(2)^5}{5} + c \right] - \pi \left[3(0) - \dfrac{3(0)}{2} + \dfrac{0}{5} + c \right]$

$= \pi \left(24 - 24 + \dfrac{32}{5} + c \right) - \pi(c)$

Volume $= \pi(6{\cdot}4)$ cubic units

INTEGRAL CALCULUS — PURE MATHEMATICS

Example 21

The area under $y = \frac{1}{9}x^2 + 1$ from $x = 0$ to $x = 3$, and the area enclosed by $y = 0$, $y = 2$, $x = 3$ and $x = 4$ are rotated about the y axis. Calculate the volume of the solid so generated.

Solution

A sketch graph:

On y axis $\quad x = 0 \quad\quad \therefore\ y = \frac{1}{9}(0) + 1 = 1 \quad\quad$ giving $(0, 1)$

On x axis $\quad y = 0 \quad\quad \therefore\ 0 = \frac{1}{9}x^2 + 1 \quad\quad \therefore\ 0 = x^2 + 9$

$\quad\quad\quad\quad\quad\quad\quad\quad\quad\quad \therefore\ x^2 = -9$

This is impossible. Therefore, curve does not cross x axis.

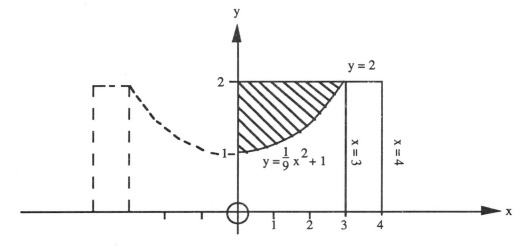

This graph represents a section through the centre of the solid of revolution. When the line $x = 4$ is rotated about the y axis between $y = 0$ and $y = 2$ a cylinder would be formed with volume (V).

$$V = \pi r^2 h = \pi(4)^2(2) = 32\pi$$

But this is too much volume and the volume (V_1) generated by rotating the shaded area about the y axis must be subtracted from V.

Volume of shaded area (V_1) is given by

$$V_1 = \int_1^2 \pi x^2\ dy$$

But $\quad y = \frac{x^2}{9} + 1 \quad\quad \therefore\quad x^2 = 9y - 9$

Hence $\quad V_1 = \int_1^2 \pi (9y - 9)\ dy = 9\pi \int_1^2 (y - 1)\ dy$

$$= 9\pi \left[\frac{y^2}{2} - y\right]_1^2$$

$$= 9\pi \left[\left(\frac{(2)^2}{2} - 2\right) - \left(\frac{(1)^2}{2} - 1\right)\right]$$

$$= 9\pi \left(0 + \frac{1}{2}\right) = \frac{9}{2}\pi$$

$\therefore$ Required volume $= V - V_1 = 32\pi - \frac{9}{2}\pi$

$$= 27\frac{1}{2}\pi \text{ cubic units}$$

---

b) *Mean values*

The mean value of a function $y = f(x)$ over the range $a \leq x \leq b$ is defined as:

$$\boxed{\text{Mean value} = \frac{1}{b-a} \int_a^b f(x)\, dx}$$

Example 22

Calculate the mean value of $\sin^2 x$ over the interval $0 \leq x \leq \frac{\pi}{3}$

Solution

$$\text{Mean value} = \frac{1}{\frac{\pi}{3} - 0} \int_0^{\frac{\pi}{3}} \sin^2 x \, dx \qquad \text{but} \qquad \cos 2x = 1 - 2\sin^2 x$$

$$\therefore \quad 2\sin^2 x = 1 - \cos 2x$$

$$\therefore \quad \sin^2 x = \frac{1 - \cos 2x}{2}$$

$$= \frac{1}{\frac{\pi}{3}} \int_0^{\frac{\pi}{3}} \frac{1 - \cos 2x}{2} \, dx$$

$$= \frac{3}{2\pi} \int_0^{\frac{\pi}{3}} (1 - \cos 2x) \, dx$$

$$= \frac{3}{2\pi} \left[x - \frac{\sin 2x}{2} + c\right]_0^{\frac{\pi}{3}}$$

$$= \frac{3}{2\pi} \left[\left(\frac{\pi}{3} - \frac{\sin 2\left(\frac{\pi}{3}\right)}{2} + c\right) - \left(0 - \frac{\sin 2(0)}{2} + c\right)\right]$$

INTEGRAL CALCULUS   PURE MATHEMATICS

$$= \frac{3}{2\pi}\left[\frac{\pi}{3} - \frac{\sqrt{3}}{2} + c - 0 + 0 - c\right]$$

$$= \frac{3}{2\pi}\left(\frac{\pi}{3} - \frac{\sqrt{3}}{4}\right)$$

$$= \frac{3}{2\pi}\left(\frac{\pi}{3}\right) - \frac{3}{2\pi}\left(\frac{\sqrt{3}}{4}\right)$$

∴ Mean value $= \frac{1}{2} - \frac{3\sqrt{3}}{8\pi}$

Example 23

Use Simpson's rule with 7 equally spaced ordinates to find an estimate of the mean value of the function $\log_{10}(1 + x^3)$ between the values of $x = 1$ and $x = 19$. Show the working in the form of a table and answer to 3 significant figures.

Solution

$$\text{Mean value} = \frac{1}{19 - 1} \int_1^{19} \log_{10}(1 + x^3)\, dx$$

Simpson's rule can be used to estimate a value of the integral, ie of

$$\int_1^{19} \log_{10}(1 + x^3)\, dx$$

For 3 ordinates (y_1, y_2, y_3) Simpson's rule states integral $= \frac{1}{3} d(y_1 + 4y_2 + y_3)$, so for 7 ordinates the rule must be applied 3 times.

The ordinates are at $x = 1, 4, 7, 10, 13, 16$ and 19. $d = 3$.

x	1	4	7	10	13	16	19
x^3	1	64	343	1,000	2,197	4,096	6,859
$1 + x^3$	2	65	344	1,001	2,198	4,097	6,860
$\log_{10}(1 + x^3)$	0.3010	1.8129	2.5366	3.0000	3.3420	3.6125	3.8363
	y_1	y_2	y_3	y_4	y_5	y_6	y_7
	y_1	$4y_2$	$2y_3$	$4y_4$	$2y_5$	$4y_6$	y_7
	0.3010	7.2516	5.0732	12.0000	6.6840	14.4500	3.8363

Applying Simpson's rule 3 times gives:

Integral $= \frac{1}{3} d(y_1 + 4y_2 + y_3) + \frac{1}{3} d(y_3 + 4y_4 + y_5) + \frac{1}{3} d(y_5 + 4y_6 + y_7)$

$= \frac{1}{3} d(y_1 + 4y_2 + 2y_3 + 4y_4 + 2y_5 + 4y_6 + y_7)$

$= \frac{1}{3}.3(0.3010 + 7.2516 + 5.0732 + 12 + 6.6840 + 14.4500 + 3.8363)$

$= 1 . (49.5961)$

PURE MATHEMATICS — INTEGRAL CALCULUS

Mean value $= \dfrac{1}{19-1}(49.5961)$

$= 2.76$ (to 3 significant figures)

c) Distance, velocity, acceleration

If distance (s) is a function of time, ie $s = f(t)$

then velocity: $v = \dfrac{ds}{dt}$ and so $s = \int v\, dt$

and acceleration: $a = \dfrac{dv}{dt}$ and so $v = \int a\, dt$

$\therefore \quad \boxed{v = \int a\, dt \text{ and } s = \int v\, dt}$

Example 24

A racing car starts from rest and its acceleration after t seconds is $\left(k - \dfrac{1}{6}t\right)$ m/sec², until it reaches a velocity of 60 m/sec at the end of one minute. Find the value of k and the distance travelled in the first minute.

Solution

$a = k - \dfrac{1}{6}t \quad \therefore v = \int a\, dt = \int k - \dfrac{1}{6}t\, dt$

$\therefore v = kt - \dfrac{1}{6}\dfrac{t^2}{2} + c$

There are apparently two constants (k and c) to be calculated but, since the car starts from rest,

$t = 0, v = 0 \quad \therefore 0 = k(0) - \dfrac{1}{12}(0) + c \quad \therefore c = 0$

so $v = kt - \dfrac{1}{12}t^2$

Also velocity = 60 m/sec when t = 1 min, ie 60 sec.

so $60 = k(60) - \dfrac{1}{12}(60)^2$

$\therefore k = 6$

So $a = 6 - \dfrac{1}{6}t$

and $v = 6t - \dfrac{t^2}{12} \quad \therefore s = \int v\, dt = \int \left(6t - \dfrac{t^2}{12}\right) dt = \dfrac{6t^2}{2} - \dfrac{t^3}{12.3} + c$

$\therefore s = 3t^2 - \dfrac{t^3}{36} + c$

but when the car starts $s = 0$ and $t = 0 \therefore 0 = 3(0) - \dfrac{0}{36} + c \quad \therefore c = 0$

INTEGRAL CALCULUS — PURE MATHEMATICS

$$\text{so } s = 3t^2 - \frac{t^3}{36}$$

At the end of the first minute $t = 60$ secs. $\therefore s = 3(60)^2 - \frac{60^3}{36} = 10{,}800 - 6{,}000$

$$\therefore s = 4{,}800\text{m}$$

d) *Example* ('A' level question)

The region enclosed by the curve $y = \tan x$, the x axis and the ordinate $x = \frac{\pi}{4}$, is rotated completely about the x axis. Find the volume of the solid formed.

Also evaluate approximately this volume, using Simpson's rule with two intervals. Calculate the percentage error in the result.

Solution

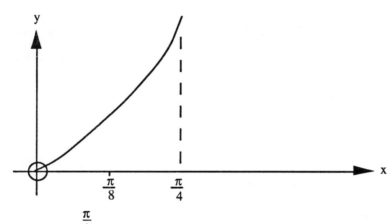

Volume $\displaystyle = \pi \int_0^{\pi/4} y^2 \, dx$

$\displaystyle = \pi \int_0^{\pi/4} \tan^2 x \, dx$ but $1 + \tan^2 x = \sec^2 x$ $\therefore$ $\tan^2 x = \sec^2 x - 1$

$\displaystyle = \pi \int_0^{\pi/4} (\sec^2 x - 1) \, dx$

$\displaystyle = \pi \left[\tan x - x + c\right]_0^{\pi/4}$

$\displaystyle = \pi \left[\tan \frac{\pi}{4} - \frac{\pi}{4} + c - (\tan 0 - 0 + c)\right]$

$\displaystyle = \pi \left(1 - \frac{\pi}{4}\right)$

$\therefore$ Volume = 0.6742

Simpson's rule is more normally used to evaluate areas under a curve but it can be used to evaluate approximately any integral.

$$\text{Volume} = \pi \int_0^{\frac{\pi}{4}} \tan^2 x \, dx$$

x	0	$\frac{\pi}{8}$	$\frac{\pi}{4}$
tan x	0	0.4142	1
$\tan^2 x$	0	0.1716	1

Using Simpson's rule:

$$\text{Integral} = \frac{1}{3} d (y_1 + 4y_2 + y_3)$$

$$\therefore \int_0^{\frac{\pi}{4}} \tan^2 x \, dx = \frac{1}{3}\left(\frac{\pi}{8}\right)(0 + 4(0.1716) + 1)$$

$$= \frac{\pi}{24}(1.6864)$$

$$= 0.2207$$

$$\text{Volume} = \pi \int_0^{\frac{\pi}{4}} \tan^2 x \, dx$$

$$= \pi(0.2207)$$

$$= 0.6933$$

$$\text{Percentage error} = \frac{\text{Actual error}}{\text{True value}} \times 100\%$$

$$= \frac{0.6933 - 0.6742}{0.6742} \times 100\%$$

$$= \frac{0.0191}{0.6742} \times 100\%$$

$$= 2.83\%$$

6 LINE, CIRCLE, PARABOLA

    6.1   Graphical work
    6.2   Introduction to the straight line
    6.3   Loci
    6.4   Circle and parabola

6.1 Graphical work

a) *Equation of a straight line*

The equation of any straight line is determined by two facts about the line. They are:

(a)   the slope of the line ie the 'gradient' (m);

(b)   the point (0, C) at which the line crosses the y axis - ie the 'intercept' on the y axis.

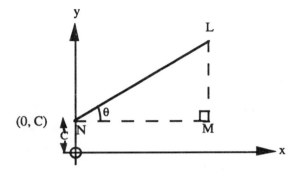

The gradient of the line is found by drawing a large right angled triangle LMN under the line.

Then $m = \dfrac{LM}{MN} = \dfrac{\text{change in y's}}{\text{change in x's}}$

Also $\tan L\hat{N}M = \dfrac{\text{Opposite}}{\text{Adjacent}} = \dfrac{LM}{MN} = m$

So the gradient of a line is the same as the tangent of the angle that the line makes with the horizontal. From this it follows that if $0° < \theta < 90°$ the gradient will be positive because $\tan \theta$ is positive but if $90° < \theta < 180°$ the gradient will be negative because $\tan \theta$ is negative.

In general the equation of any straight line can be written as:

$$\boxed{y = mx + c}$$

Example 1

Draw the graph of $2y - 4x = 6$ and find the gradient and the intercept from the graph.

Solution

Before drawing the graph of $2y - 4x = 6$, the equation will be rearranged to make y the subject.

$$2y - 4x = 6$$
$$2y = 4x + 6$$
$$y = 2x + 3$$

PURE MATHEMATICS — LINE, CIRCLE, PARABOLA

Comparing this with the general equation

$$y = mx + c$$

it follows that the gradient should be 2 and the intercept on the y axis (0, 3).

x	0	1	2	3	4	5	6
2x	2	2	4	6	8	10	12
3	3	3	3	3	3	3	3
y = 2x + 3	3	5	7	9	11	13	15

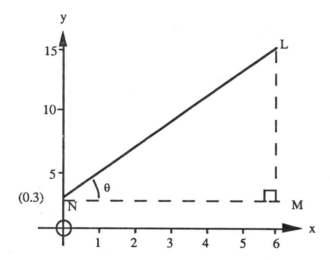

The point at which the line crosses the y axis is (0, 3) ∴ C = 3

Gradient = $\dfrac{LM}{MN} = \dfrac{12}{6}$ ∴ m = 2

Answers as above.

Example 2

Draw the graph of $2y + 3x - 4 = 0$ and find the intercept and the gradient.

Solution

$$2y + 3x - 4 = 0$$

$$2y = -3x + 4$$

$$y = -\dfrac{3}{2}x + 2$$

x	0	1	2	3	4	5	6
$-\dfrac{3}{2}x$	0	$-\dfrac{3}{2}$	-3	$-\dfrac{9}{2}$	-6	$-\dfrac{15}{2}$	-9
+2	2	2	2	2	2	2	2
$y = -\dfrac{3}{2}x + 2$	2	$\dfrac{1}{2}$	-1	$-\dfrac{5}{2}$	-4	$-\dfrac{11}{2}$	-7

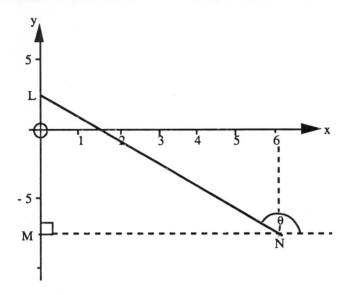

The point at which the line crosses the y axis is (0, 2). ∴ C = 2.

Gradient = $\dfrac{LM}{MN}$ = $\dfrac{9}{6}$ = $\dfrac{3}{2}$ ... m = -$\dfrac{3}{2}$

Negative sign because θ is more than 90°.

These two examples have shown practically that y = mx + c is, in general terms, the equation of a straight line. The real advantage of knowing this is that the equation of a straight line can be determined directly from its graph.

Example 3

Determine the equation of this straight line from the graph:

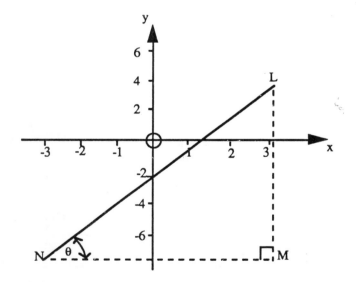

Solution

Since this is a straight line with equation y = mx + c it is necessary to find m and c.

This crosses the y axis at (0, -2) ∴ c = -2.

Drawing the right angled triangle LMN

$$\text{Gradient} = \frac{LM}{MN} = \frac{3-(-7)}{3-(-3)} = \frac{10}{6}$$

$$\therefore \quad m = \frac{5}{3} \quad (\text{positive} \because \theta < 90°)$$

$$\therefore \quad \text{Equation is } y = \frac{5}{3}x - 2 \quad \text{Multiplying by 3}$$

$$3y = 5x - 6$$

b) *Graphical determination of a law*

Sometimes the graph given relates to experimental data, and the straight line equation is used to determine whether or not the variables are related in a mathematical way.

Example 4

A marble was dropped from height h_1 cm and observed to rise to a height h_2 cm, giving the results:

h_1	4	9	16	22
h_2	$\frac{3}{2}$	3	$\frac{11}{2}$	$\frac{15}{2}$

Draw a graph of these results and determine the law connecting h_1 and h_2 - if any law exists.

Solution

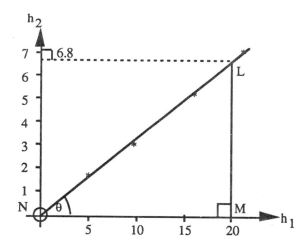

Because these are experimental results the points do not lie perfectly on a straight line. However, by drawing in the line of best fit, it can be assumed that h_1 and h_2 are related by a law of the form $h_2 = mh_1 + c$ (h_1 and h_2 replacing x and y).

From the graph $c = 0$ and $m = \frac{LM}{MN} = \frac{6.8}{20} = 0.34$

$$\therefore \quad h_2 = 0.34 h_1 \text{ is the required law}$$

On many occasions two variables are related in a non-linear way. However, by plotting an appropriate straight line graph it is possible to determine the law connecting the variables. The next example will help to make this clear.

Example 5

A marble was allowed to run down a sloping sheet of glass and the time - t secs - taken to roll a distance - s metres - from rest was measured. The following results were found:

s	1	2	3	4	5
t	1.4	2	2.5	2.8	3.2
t^2	1.96	4	6.25	7.84	10.24

Draw graphs of:

(1)   t against s;

(2) t^2 against s.

Determine, if possible, the law relating s and t.

Solution

(1)

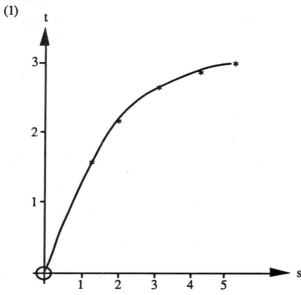

(2)

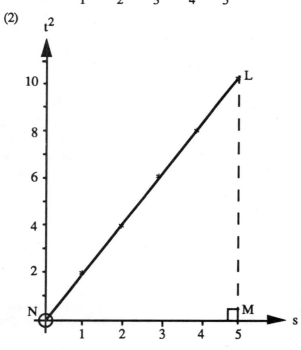

(1) Since the graph of t against s is a smooth curve it implies that there is a law connecting these variables but *not* a linear law.

(2) The graph of t^2 against s is a straight line. Therefore, a law exists of the form

$$t^2 = ms + c \text{ (s and } t^2 \text{ replacing x and y)}$$

From the graph $c = 0$

and $m = \dfrac{LM}{MN} = \dfrac{10.25}{5} = 2.05$

$\therefore t^2 = 2.05s$ is the required law.

Now for a harder example.

Example 6

The following pairs of values for x and y have been found by experiment and some values of $yx^{\frac{3}{2}}$ have been calculated and tabulated:

x	1.0	2.6	3.2	4.0	6.2
y	5.5	2.0	1.7	1.4	1.0
$yx^{\frac{3}{2}}$	5.50		9.73		15.44

Calculate the missing values. It is believed that x and y are connected by a law of the form $yx^{\frac{3}{2}} = ax + b$, where a, b are constants. Show graphically that for these data this law is approximately valid and use the graph to estimate:

(1) values for a and b;

(2) y when x = 5.1.

Solution

(1) When x = 2.6 and y = 2.0 $yx^{\frac{3}{2}} = 2.0(2.6)^{\frac{3}{2}} = 2.0(4.19) = 8.38$

When x = 4.0 and y = 1.4 $yx^{\frac{3}{2}} = 1.4(4.0)^{\frac{3}{2}} = 1.4(8.0) = 11.20$

The graph to be plotted is $yx^{\frac{3}{2}}$ against x.

Therefore, using:

x	1.0	2.6	3.2	4.0	6.2
$yx^{\frac{3}{2}}$	5.50	8.38	9.73	11.20	15.44

LINE, CIRCLE, PARABOLA   PURE MATHEMATICS

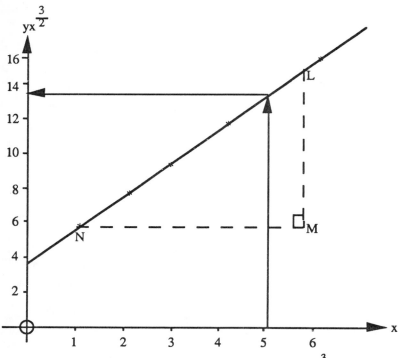

(1)   Since the points lie very nearly on a straight line the law $yx^{\frac{3}{2}} = ax + b$ is valid.

Comparing this with 'y' $= mx + c$'

b: intercept on the y axis $= 3.6$

a: gradient $= \dfrac{LM}{MN} = \dfrac{9.5}{5} = 1.9$

∴   equation is $yx^{\frac{3}{2}} = 1.9x + 3.6$

(2) $x = 5.1$ gives a value of 13.2 on the vertical axis but this is not the value of y. It is the value of $yx^{\frac{3}{2}}$ where $x = 5.1$.

∴ $y(5.1)^{\frac{3}{2}} = 13.2$

$$y = \dfrac{13.2}{(5.1)^{\frac{3}{2}}}$$

$$y = 1.15$$

---

c)   *Graphical determination involving logarithms*

The method just used for non-linear laws is very limited because the power of x and/or y may not in practice be known, in which case the equation may be of the type:

$y = ax^n$ where $n = ?$

However, by taking logarithms of each side it follows that:

$\log y = \log(ax^n)$

$\log y = \log a + \log(x^n)$

$$\log y = \log a + n \log x$$

So the graph to plot is log y against log x.

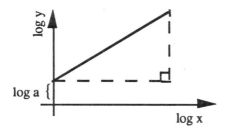

and, assuming a straight line results, by comparing log y = n log x + log a with 'y = mx + c' it follows that the intercept on the y axis will give log a and the gradient will give n, and hence the law connecting x and y will be established completely.

Logarithms are usually taken to base 10 of each side as their numerical values can easily be ascertained.

Example 7

Corresponding values of s and t are as shown:

s	1.5	3	5	8	10
t	275	800	2,000	3,550	5,000

By drawing a suitable graph show that 's' and 't' are related approximately by the equation $t = As^p$ where A and p are constants.

Find suitable values for A and p.

Solution

$t = As^p$: since p is an unknown *and* is an index, the best graph to draw will be a $\log_{10}$ graph.

∴ taking logarithms to base 10 gives:

$$\log_{10} t = \log_{10}(As^p)$$
$$= \log_{10} A + \log_{10} s^p$$
$$= \log_{10} A + p \log_{10} s$$

∴ $\log_{10} t = p \log_{10} s + \log_{10} A$

$\log_{10} s$	0.176	0.477	0.699	0.903	1.000
$\log_{10} t$	2.439	2.903	3.301	3.550	3.699

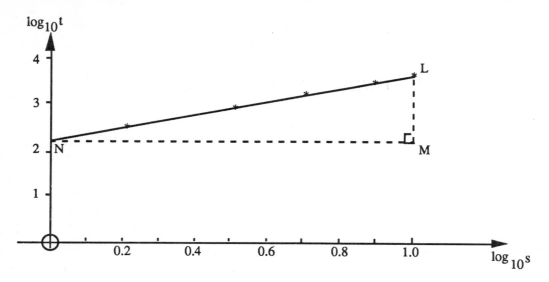

As a straight line gives the best fit the relationship holds.

Comparing $\log_{10}t = p \log_{10}s + \log_{10}A$ with

$$y = mx + c$$

it follows that p is the gradient of the line and $\log_{10}A$ is the intercept on the vertical axis.

At N $\therefore \log_{10}A = 2.2$ (approximately)

$\therefore A = 10^{2.2} = 158.5$

When calculating the gradient of the line it is best to draw the largest possible right-angled triangle.

$$\therefore P = \frac{LM}{MN} = \frac{3.699 - 2.2}{1 - 0} = \frac{1.5}{1} = 1.5$$

So the relationship is $t = 158.5s^{1.5}$ (approximately)

Example 8

The table below gives corresponding values of variables x and y:

x	1	2	3	4	5
y	0.71	0.87	1.06	1.30	1.59

Verify graphically that these values of x and y satisfy approximately a relationship of the form $ay^2 = b^x$ where a and b are constants. From the graph obtain approximate values for a and b.

Solution

$ay^2 = b^x$: again a $\log_{10}$ graph is required because one of the variables - x - is an index.

$\log_{10}(ay^2) = \log_{10}(b^x)$

$\log_{10}a + \log_{10}(y^2) = x \log_{10}b$

$\therefore \log_{10}a + 2\log_{10}y = x \log_{10}b$

$$\therefore \quad \log_{10}y = \left(\frac{\log_{10}b}{2}\right)x - \left(\frac{\log_{10}a}{2}\right)$$

x	1	2	3	4	5
$\log_{10}y$	$\bar{1}.8513$ / -0.149	$\bar{1}.9395$ / -0.060	0.025	0.114	0.201

$\bar{1}.8513 = -1 + 0.8513 = -0.1487$

$\bar{1}.9395 = -1 + 0.9395 = -0.0605$

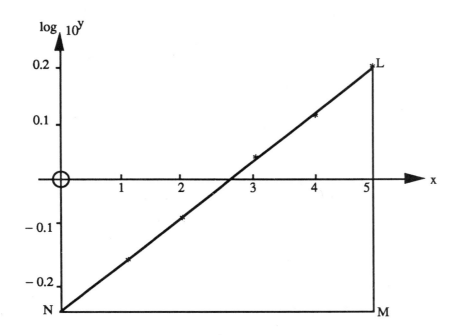

Since all the points lie on a straight line the relationship holds.

Comparing $\log_{10}y = \left(\frac{\log_{10}b}{2}\right)x - \left(\frac{\log_{10}a}{2}\right)$ with $y = mx + c$

It follows that $\frac{\log_{10}b}{2}$ is the gradient of the line and $\frac{-\log_{10}a}{2}$ is the intercept on the y axis.

At N $\quad \frac{-\log_{10}a}{2} = -0.235$

$\therefore \quad \log_{10}a = 0.470 \qquad \therefore \quad a = 10^{0.47} = 2.95$

Drawing the largest possible triangle gives

$$\frac{\log_{10}b}{2} = \frac{LM}{MN} = \frac{.201 - (-0.235)}{5 - 0} = \frac{0.436}{5} = 0.087$$

$\therefore \quad \log_{10}b = 2(0.087) = 0.174$

$\therefore \quad b = 10^{0.174} = 1.49$

LINE, CIRCLE, PARABOLA — PURE MATHEMATICS

So the relationship is $2.95y^2 = 1.49^x$

or rounding this becomes $3y^2 = 1.5^x$

(*Note*: In this example it was only necessary to take logarithms of one of the variables (y) because the other variable (x), being an index in the original equation $ay^2 = b^x$, came down outside the $\log_{10}b$ when logarithms of both sides were taken.)

---

d) *Example* ('A' level question)

The following corresponding values of x and y are believed to be related by the equation $y = (x + ax^b)$

x	2	3	4	5	6	10
y	5.54	7.33	9.00	10.59	12.12	17.90

Draw a suitable graph to show that this may be so and hence find probable values for a and b.

Solution

Just taking logarithms of both sides does not help as $y = x + ax^b$

gives $\log_{10}y = \log_{10}(x + ax^b)$

thus a and b are both 'trapped' inside the logarithmic function and a graph still cannot be drawn. Therefore:

$$y = x + ax^b \quad \text{has to be rearranged to give}$$

$$y - x = ax^b \quad \text{Now taking logarithms}$$

$$\log_{10}(y - x) = \log_{10}ax^b$$

$$\therefore \quad \log_{10}(y - x) = \log_{10}a + \log_{10}x^b$$

$$\therefore \quad \log_{10}(y - x) = b\log_{10}x + \log_{10}a$$

The graph that needs to be drawn is of $\log_{10}(y - x)$ against $\log_{10}x$.

$\log_{10}x$	0.30	0.48	0.60	0.70	0.78	1.00
$(y - x)$	3.54	4.33	5.00	5.59	6.12	7.90
$\log_{10}(y - x)$	0.55	0.64	0.70	0.75	0.79	0.90

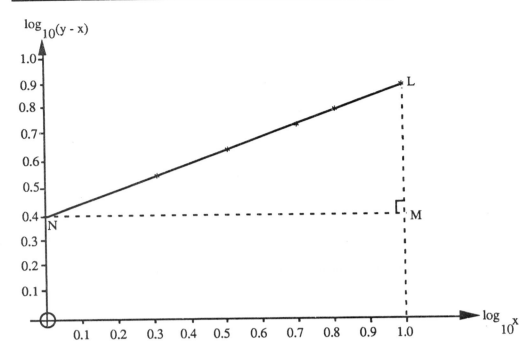

So, comparing $\log_{10}(y - x) = b \log_{10}x + \log_{10}a$ with 'y = mx + c'

it follows that intercept $\log_{10}a = 0.4000$ $\therefore$ $a = 10^{0.04} = 2.512 = 2.5$ say

and gradient $b = \dfrac{LM}{MN} = \dfrac{0.9 - 0.4}{1 - 0} = \dfrac{0.5}{1} = 0.5$

So the relationship that exists is $y = x + 2.5x^{0.5}$.

6.2 Introduction to the straight line

a) *Coordinates*

The coordinates of a point $P = (a, b)$ say - fix the position of that point on a plane by reference to axes Ox and Oy at right angles to each other.

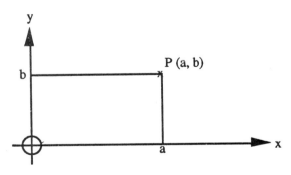

If the distance between two points $A(-2, 4)$ and $B(3, -2)$ is required, then the theorem of Pythagoras can be applied as follows:

$AB^2 = AC^2 + CB^2$ where $CB = -2 - 3 = -5$

$ = 6^2 + (-5)^2$

= 36 + 25   AC = 4 - (- 2) = 6

= 61

∴   AB = $\sqrt{61}$

= 7.8

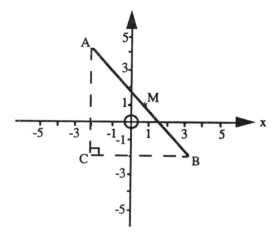

Also the mid-point M of AB will be $\left(\dfrac{(-2)+3}{2}, \dfrac{4+(-2)}{2}\right) = \left(\dfrac{1}{2}, 1\right)$

In general terms this can be given as:

$$AB = \sqrt{(x_1 - x_2)^2 + (y_1 - y_2)^2} \qquad M \text{ is } \left(\dfrac{x_1 + x_2}{2}, \dfrac{y_1 + y_2}{2}\right)$$

where A is the point (x_1, y_1) and B is the point (x_2, y_2)

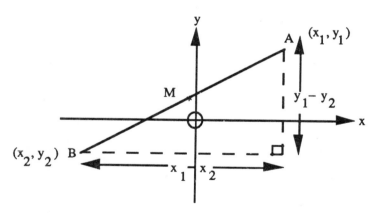

Example 9

PQR are the points (5, - 3) (- 6, 1) (1, 8) respectively. Show that △PQR is isosceles, and find the coordinates of the mid-pont of the base.

Solution

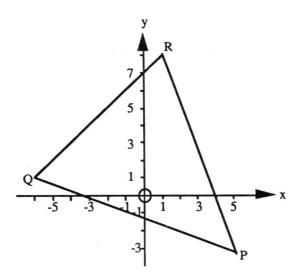

To show that the triangle is isosceles it is necessary to find the lengths of the sides:

$$PQ^2 = [5 - (-6)]^2 + (-3 - 1)^2 = 11^2 + (-4)^2$$
$$= 121 + 16 = 137$$

∴ $PQ = \sqrt{137}$

$$QR^2 = (-6 - 1)^2 + (1 - 8)^2 = (-7)^2 + (-7)^2$$
$$= 49 + 49 = 98$$

∴ $QR = \sqrt{98}$

$$RP^2 = (1 - 5)^2 + [8 - (-3)]^2 = (-4)^2 + 11^2$$
$$= 16 + 121 = 137$$

∴ $RP = \sqrt{137}$

So PQ = RP and ΔPQR is isosceles.

This means that QR is the base of the triangle and the mid-pont is $\left(\dfrac{-6 + 1}{2}, \dfrac{1 + 8}{2}\right) = \left(-2\dfrac{1}{2}, 4\dfrac{1}{2}\right)$

---

b)   *Gradients*

The gradient of a straight line has already been defined in earlier work. It is included at this stage so that the work on coordinates and the straight line will be complete.

The gradient of the line joining A(1, 2) and B(7, 6) is calculated as follows:

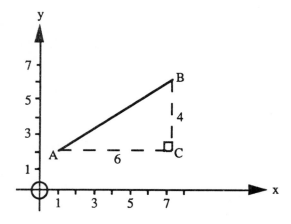

gradient of AB $= \dfrac{(6-2)}{(7-1)} = \dfrac{4}{6} = \dfrac{2}{3}$

In general terms:

$$\boxed{\text{Gradient of AB} = \dfrac{y_1 - y_2}{x_1 - x_2}}$$ where A is (x_1, y_1) and B is (x_2, y_2)

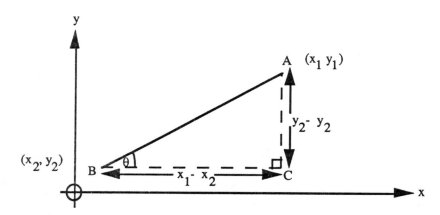

The gradient of a line is the same as the tangent of the angle that the line makes with the horizontal.

In the above diagram $\tan \theta = \dfrac{AC}{BC} = \dfrac{y_1 - y_2}{x_1 - x_2} = $ gradient

Example 10

Calculate the gradients of the lines joining:

(a)   (-2, -3) and (4, 6)

(b)   (5, 6) and (10, 2)

Solution

(a)

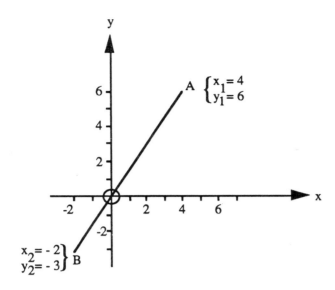

gradient of AB $= \dfrac{6-(-3)}{4-(-2)} = \dfrac{9}{6}$ ∴ gradient of AB $= \dfrac{3}{2}$

(b)

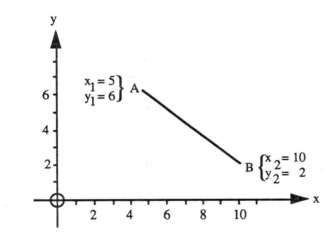

gradient of AB $= \dfrac{6-2}{5-10} = \dfrac{4}{-5}$ ∴ gradient of AB $= \dfrac{-4}{5}$

Line (a) has a positive gradient because θ is acute and tan θ > 0°. However, line (b) has a negative gradient because θ is obtuse and tan θ < 0°.

Parallel lines have the same gradient, so if the gradients are m_1 and m_2 then

$$\boxed{m_1 = m_2 \text{ for parallel lines}}$$

However, perpendicular lines are such that the product of their gradients is always -1.

So, if the gradients of the two lines are m_1 and m_2 then

$$\boxed{m_1 m_2 = -1 \text{ for perpendicular lines}}$$

This can easily be derived as follows:

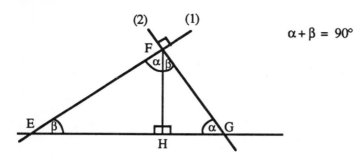

$\alpha + \beta = 90°$

Gradient of line (1) $m_1 = \dfrac{FH}{EH} = \tan \beta$ (from $\triangle EFH$)

Gradient of line (2) $m_2 = \dfrac{-FH}{GH} = -\tan \alpha$ (from $\triangle GFH$)

but $\tan \alpha = \dfrac{EH}{FH}$ (from $\triangle EFH$)

$\therefore \quad m_2 = \dfrac{-EH}{FH} = -\dfrac{1}{m_1}$

$\therefore \quad m_1 m_2 = -1$

Example 11

Show that P(1, 7), Q(7, 5), R(6, 2), S(0, 4) are the vertices of a rectangle. Find the point of intersection of the diagonals.

Solution

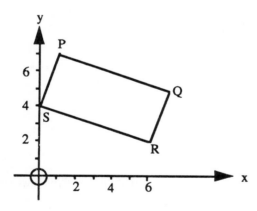

It looks like a rectangle but it needs to be shown more mathematically.

Calculating the gradient of PQ $= \dfrac{7-5}{1-7} = \dfrac{2}{-6} = -\dfrac{1}{3}$

gradient of QR $= \dfrac{5-2}{7-6} = \dfrac{3}{6} = 3$

gradient of RS $= \dfrac{2-4}{6-0} = \dfrac{-2}{6} = -\dfrac{1}{3}$

gradient of SP $= \dfrac{4-7}{0-1} = \dfrac{-3}{-1} = 3$

Therefore, lines PQ and SR are parallel (gradients $-\frac{1}{3}$)

and lines QR and PS are also parallel (gradients 3).

Also adjacent lines are perpendicular to one another because $3 \times \left(-\frac{1}{3}\right) = -1$.

Therefore, PQRS is a rectangle.

The point of intersection is the mid-point of PR or QS. Calculating both to show that they are the same point:

$$\text{Mid-point PR} = \left(\frac{1+6}{2}, \frac{7+2}{2}\right) = \left(3\frac{1}{2}, 4\frac{1}{2}\right)$$

$$\text{Mid-point QS} = \left(\frac{7+0}{2}, \frac{5+4}{2}\right) = \left(3\frac{1}{2}, 4\frac{1}{2}\right)$$

c) *Equation of a straight line*

This has already been met in the form $y = mx + c$ where m is the gradient of the line of c, the intercept on the y axis. Points worth noting are:

(a) $y = mx$ must be a line of gradient m passing through the origin because $c = 0$

(b) $y = c$ must be a line of zero gradient and so is parallel to the x axis

(c) $x = $ constant is a line parallel to the y axis.

Example 12

Find the equation of the straight line joining the origin to the mid-point of the line A(3,2) B(5, -1).

Solution

Since the line goes through the origin $c = 0$, the mid-point of AB is

$$\left(\frac{3+5}{2}, \frac{2+(-1)}{2}\right) \left(4, \frac{1}{2}\right)$$

Therefore, to find the equation, which will be of the form $y = mx$, it is necessary to find the gradient m of the line joining $(0, 0)$ and $(4, \frac{1}{2})$.

$$m = \frac{\left(\frac{1}{2} - 0\right)}{(4 - 0)} = \frac{\frac{1}{2}}{4} = \frac{1}{8}$$

$\therefore \quad y = \frac{1}{8}x \qquad \therefore \ 8y = x$ is the required equation

The next two examples will illustrate how the equation of a straight line can be found even when m and c are not given as such.

Example 13

Find the equation of a straight line of gradient $\frac{1}{3}$, passing through (2, -5).

Solution

This is the method to use when the gradient and one point on the line are known, and that point is *not* the intercept on the y axis.

Using $y = mx + c$ gives $y = \frac{1}{3}x + c$. Now the problem is how to find c.

Knowing the point passes through $x = 2$, $y = -5$ means c can be found as follows:

$$y = \frac{1}{3}x + c \quad \text{becomes}$$

$$-5 = \frac{1}{3}(2) + c$$

$$-\frac{17}{3} = c$$

So the equation is $y = \frac{1}{3}x - \frac{17}{3}$ or multiplying through by 3

$$3y = x - 17$$

Example 14

Find the equation of the straight line passing through $(-3, 4)$ and $(8, 1)$.

Solution

This is the method to use when only two points are known and nothing else.

Knowing the two points means that the gradient of the line can be found using

$$m = \frac{y_1 - y_2}{x_1 - x_2} = \frac{4 - 1}{-3 - 8} = \frac{3}{-11}$$

Therefore, the equation is $y = -\frac{3}{11}x + c$. Now c can be found using either of the points on the line as in the last example. So, taking $(8, 1)$:

$$y = 1 \text{ when } x = 8 \quad \therefore \quad y = -\frac{3}{11}x + c \text{ becomes } 1 = -\frac{3}{11}(8) + c$$

$$\therefore \quad \frac{35}{11} = c$$

So the equation is $\quad y = -\frac{3}{11}x + \frac{35}{11}$ or multiplying through by 11

$$11y = -3x + 35$$

And now for a slightly harder example using any of the methods outlined so far.

Example 15

Find the equations of the lines which pass through the point of intersection of the lines $x - 3y = 4$, $3x + y = 2$ and are respectively parallel and perpendicular to the line $3x + 4y = 0$.

PURE MATHEMATICS — LINE, CIRCLE, PARABOLA

Solution

First, to find the intersection of the two lines:

$x - 3y = 4$ ---------- (1)

$3x + y = 2$ ---------- (2) multiply by 3

$$\begin{aligned} x - 3y &= 4 \\ 9x + 3y &= 6 \end{aligned} \} \text{ ADD}$$

$$10x = 10$$
$$x = 1$$

Replacing in (2) to get y

$$3(1) + y = 2$$
$$\therefore y = 2 - 3 = -1$$

Therefore, the lines cross at $(1, -1)$ and so this must be a point on each of the two new lines.

Considering $3x + 4y = 0$ $\therefore 4y = -3x$

$$\therefore y = -\frac{3}{4}x$$

So the line parallel to $3x + 4y = 0$ has gradient $-\frac{3}{4}$ and equation $y = -\frac{3}{4}x + c$. Using $(1, -1)$ to find c gives:

$$-1 = -\frac{3}{4}(1) + c \qquad \therefore c = -\frac{1}{4}$$

$\therefore$ Equation of parallel line is $y = -\frac{3}{4}x - \frac{1}{4}$

or multiplying through by 4 $\qquad 4y = -3x - 1$

The line perpendicular to $3x + 4y = 0$ has gradient $\frac{4}{3}$ because $\left(-\frac{3}{4}\right)\left(\frac{4}{3}\right) = -1$

The equation is $y = \frac{4}{3}x + c$.

Using $(1, -1)$ to find c gives: $\qquad -1 = \frac{4}{3}(1) + c \qquad \therefore c = -\frac{7}{3}$

$\therefore$ Equation of perpendicular line is $y = \frac{4}{3}x - \frac{7}{3}$

or multiplying through by 3 $\qquad 3y = 4x - 7$

---

d) *Topics associated with the straight line*

There are several other important methods to be learnt when studying the straight line. These methods will be detailed now.

(a) The angle between two lines: problems often require the calculation of the angle between two straight lines. A formula can be derived for this using

$$\tan(A - B) = \frac{\tan A - \tan B}{1 + \tan A \tan B}$$

LINE, CIRCLE, PARABOLA   PURE MATHEMATICS

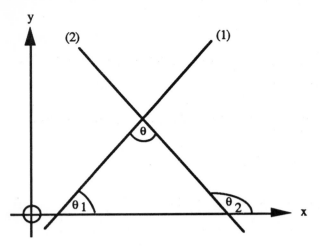

Line (1) makes an angle of θ_1 with the positive direction of the x axis, and line (2) makes an angle of θ_2.

θ is the angle between the lines.

$\therefore \quad \theta_1 + \theta = \theta_2 \qquad \therefore$ exterior angle of a triangle

$\therefore \quad \theta = \theta_2 - \theta_1$

$\therefore \quad \tan \theta = \tan(\theta_2 - \theta_1)$

$$= \frac{\tan \theta_2 - \tan \theta_1}{1 + \tan \theta_1 \tan \theta_2}$$

but $\tan \theta_1 = m_1$ gradient of line (1)

and $\tan \theta_2 = m_2$ gradient of line (2)

$$\boxed{\therefore \quad \tan \theta = \frac{m_2 - m_1}{1 + m_1 m_2} \text{ or } \theta = \tan^{-1}\left(\frac{m_2 - m_1}{1 + m_1 m_2}\right)}$$

Example 16

Find the acute angle between this pair of straight lines:

$$4x - 2y + 3 = 0 \qquad 6x - 2y + 1 = 0$$

Solution

Rearranging both equations gives:

$4x - 2y + 3 = 0 \quad \therefore \quad y = \frac{4x}{2} - \frac{3}{2} \quad \therefore \quad$ gradient = 2

$6x - 2y + 1 = 0 \quad \therefore \quad y = \frac{6x}{2} + 1 \quad \therefore \quad$ gradient = 3

So, using $m_1 = 2$ and $m_2 = 3$

$$\tan \theta = \frac{3 - 2}{1 + 2(3)} = \frac{1}{7}$$

PURE MATHEMATICS — LINE, CIRCLE, PARABOLA

$$\theta = \tan^{-1}\left(\frac{1}{7}\right) = 8°08'$$

If the obtuse angle had been required, then the answer would have been $180 - 8°08' = 171°52'$.

(b) The distance of a point from a line: this is always taken to mean the perpendicular distance of the point from the line, ie PQ in this example:

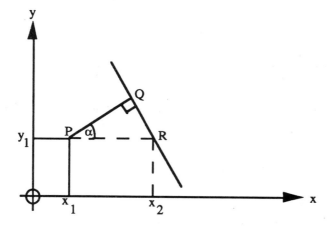

Let the point be $P(x_1, y_1)$

and the equation of the line $ax + by + c = 0$

and angle $QPR = \alpha$

Rearranging to find the gradient: $\quad by = -ax - c$

$$\therefore y = -\frac{ax}{b} - \frac{c}{b}$$

As gradient of line $= -\frac{a}{b}$ and $\left(-\frac{a}{b}\right)\left(\frac{b}{a}\right) = -1$

so gradient of perpendicular PQ is $\frac{b}{a}$

Gradient of $PQ = \frac{b}{a} = \tan \alpha$

But $\sec^2\alpha = 1 + \tan^2\alpha \quad \therefore \frac{1}{\cos^2\alpha} = 1 + \left(\frac{b}{a}\right)^2 = \frac{a^2 + b^2}{a^2}$

$$\therefore \cos^2\alpha = \frac{a^2}{a^2 + b^2}$$

and $\cos \alpha = \frac{PQ}{PR}$

$$\therefore \left(\frac{PQ}{PR}\right)^2 = \frac{a^2}{a^2 + b^2} \quad \therefore PQ^2 = \frac{a^2(PR)^2}{a^2 + b^2}$$

LINE, CIRCLE, PARABOLA — PURE MATHEMATICS

Now to find an expression for the distance PR. Being a horizontal distance, this is the difference between the x coordinates of P and R, ie $PR = x_2 - x_1$.

As R is on the line $ax + by + c = 0$ and has y coordinate y_1.

$$\therefore \quad x_2 = \frac{-by_1 - c}{a}$$

So $\quad PR = x_2 - x_1 = \left(\frac{-by_1 - c}{a}\right) - x_1$

$$= \frac{-by_1 - c - ax_1}{a}$$

Rearranging $\quad PR = \frac{-(ax_1 + by_1 + c)}{a}$

$$PQ^2 = \frac{a^2}{a^2 + b^2}\left(\frac{-(ax_1 + by_1 + c)}{a}\right)^2$$

$$PQ^2 = \frac{a^2}{a^2 + b^2} \cdot \frac{(ax_1 + by_1 + c)^2}{a^2}$$

Square rooting

$$\boxed{PQ = \pm \frac{ax_1 + by_1 + c}{\sqrt{a^2 + b^2}}}$$

Example 17

Find the distance of the point $(-1, 7)$ from the line $2x = 5y + 1$

Solution

$2x = 5y + 1 \quad$ ie $\quad 2x - 5y - 1 = 0 \quad$ which gives $a = 2, b = -5, c = -1$

Point $(-1, 7)$ gives $x_1 = -1$ and $y_1 = 7$

$$\therefore \quad \text{Distance} = \pm\left[\frac{2(-1) + (-5)7 + (-1)}{\sqrt{(2)^2 + (-5)^2}}\right] = \pm \frac{(-38)}{\sqrt{29}}$$

$\therefore \quad$ Ignoring signs, distance is $\dfrac{38}{\sqrt{29}}$

PURE MATHEMATICS — LINE, CIRCLE, PARABOLA

The easiest way to remember the equation is in the form:

$$(x - a)^2 + (y - b)^2 = r^2$$

The following example will show several ways in which the equation can be used.

Example 31

(a) Find the equation of a circle centre (0, -5) radius 5.

(b) Find the centre and radius of the circle $x^2 + y^2 + 3x - 4y - 6 = 0$.

(c) Find the radii of the two circles, with centres at the origin, which touch the circle $x^2 + y^2 - 8x - 6y + 24 = 0$.

Solution

(a) Centre (0, -5) radius 5. $\therefore a = 0, b = -5, r = 5$

$$(x - 0)^2 + (y - (-5))^2 = 5^2$$
$$x^2 + y^2 + 10y + 25 = 25$$
$\therefore \quad x^2 + y^2 + 10y + 25 - 25 = 0$

$\therefore \quad x^2 + y^2 + 10y = 0$ is the required equation

(b) To find the centre and radius of this circle the equation must be rearranged into the form $(x - a)^2 + (y - b)^2 = r^2$

$$x^2 + y^2 + 3x - 4y - 6 = 0$$
$\therefore \quad x^2 + 3x + y^2 - 4y = 6$

Now $x^2 + 3x = \left(x + \frac{3}{2}\right)^2 - \frac{9}{4}$ } Remember 'completing
                                                                                                } the square' in
and $y^2 - 4y = (y - 2)^2 - 4$ } chapter 1.

$$\left(x + \frac{3}{2}\right)^2 - \frac{9}{4} + (y - 2)^2 - 4 = 6$$

$\therefore \quad \left(x + \frac{3}{2}\right)^2 + (y - 2)^2 = 6 + \frac{9}{4} + 4$

$\therefore \quad \left(x - \left(-\frac{3}{2}\right)\right)^2 + (y - 2)^2 = \frac{49}{4}$

So $a = -\frac{3}{2}$, $b = 2$ and $r^2 = \frac{49}{4}$. Therefore, $r = \frac{7}{2}$.

Centre is $(-\frac{3}{2}, 2)$ and radius is $\frac{7}{2}$.

LINE, CIRCLE, PARABOLA — PURE MATHEMATICS

(c) In order to find the radii of the two circles, which touch the given circle, it is necessary to find the centre and radius of the given circle and then draw a sketch graph to help understand which circles are required.

$x^2 + y^2 - 8x - 6y + 24 = 0$ is the given circle

∴ $x^2 - 8x + y^2 - 6y = -24$

but $x^2 - 8x = (x - 4)^2 - 16$

and $y^2 - 6y = (y - 3)^2 - 9$ } Replacing

$(x - 4)^2 - 16 + (y - 3)^2 - 9 = -24$

∴ $(x - 4^2) + (y - 3)^2 = -24 + 16 + 9$

∴ $(x - 4)^2 + (y - 3)^2 = 1$

∴ $a = 4, b = 3$ and $r^2 = 1$

∴ Centre $(4, 3)$ and $r = 1$

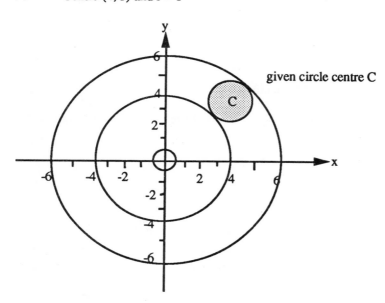

The distance between the origin 0 and the centre C of the given circle is:

$$OC = \sqrt{(4 - 0)^2 + (3 - 0)^2} = \sqrt{16 + 9} = \sqrt{25} = 5$$

so the required radii are $5 - 1 = 4$ units and $5 + 1 = 6$ units

b)   *Problems relating to the circle*

Many of the problems given on the circle concern tangents to the circle. Two important geometrical facts are:

(a)   from any point outside the circle it is possible to draw two equal tangents to the circle; and

(b)   a tangent and a radius meet at right angles on the circumference of a circle.

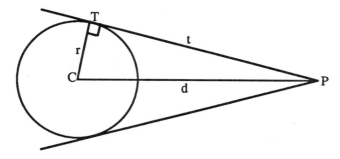

If P is a point outside a circle and T the point of contact of the tangent from P to the circle centre C then PTC forms a right-angled triangle and Pythagoras theorem can be used:

$$PC^2 = PT^2 + TC^2 \quad \text{if } PC = d, PT = t, TC = r$$

$$\boxed{d^2 = t^2 + r^2}$$

Example 32

State the equation of the straight line which has gradient m and which passes through P(0, 18).

Show that this line is a tangent to the circle centre (4, 6) and radius 10 provided that m satisfies the equation $21m^2 - 24m - 11 = 0$.

Find the product of the gradients of the tangents from P to this circle.

Solution

Equation of the straight line is $y = mx + 18$.

Sketch graph:

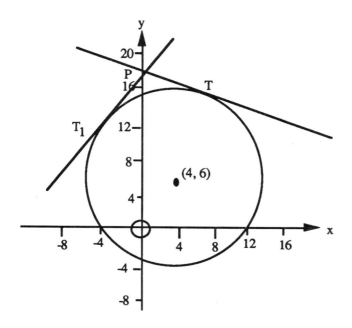

LINE, CIRCLE, PARABOLA — PURE MATHEMATICS

The equation of the circle is:

$$(x - 4)^2 + (y - 6)^2 = 10^2$$

$$x^2 - 8x + 16 + y^2 - 12y + 36 = 100$$

$$x^2 - 8x + y^2 - 12y = 100 - 16 - 36$$

$$x^2 - 8x + y^2 - 12y = 48$$

Now to study the intersection of this curve and the line $y = mx + 18$; replacing for y gives:

$$x^2 - 8x + (mx + 18)^2 - 12(mx + 18) = 48$$

$$x^2 - 8x + m^2x^2 + 36mx + 324 - 12mx - 216 = 48$$

$$x^2(1 + m^2) + x(-8 + 36m - 12m) + 60 = 0$$

$$x^2(1 + m^2) + 8(x)(3m - 1) + 60 = 0$$

ie quadratic equation where $a = 1 + m^2$, $b = 8(3m - 1)$, $c = 60$

For the line to be a tangent to the circle this quadratic equation must have equal roots and '$b^2 = 4ac$'.

$\therefore$ $[8(3m - 1)]^2 = 4(1 + m^2)(60)$

$64(9m^2 - 6m + 1) = 240(1 + m^2)$ Dividing both sides by 16

$4(9m^2 - 6m + 1) = 15(1 + m^2)$

$36m^2 - 15m^2 - 24m + 4 - 15 = 0$

$21m^2 - 24m - 11 = 0$

The roots of this quadratic equation would be the values of the gradients of the lines PT and PT_1.

Using the formulae for the sum and product of the roots of a quadratic equation, ie

$$`\alpha + \beta = -\frac{b}{a}` \quad `\alpha\beta = \frac{c}{a}`$$

assuming the gradients are m_1 and m_2 then

$$m_1 + m_2 = -\left(-\frac{24}{21}\right) = \frac{8}{7} \text{ and } m_1 m_2 = -\frac{11}{21}$$

Other problems on the circle concern the intersection of two circles at points A and B, giving rise to a common chord AB.

The equation of AB can be found by subtracting the equations of the circles, for example,

If two circles have equations:

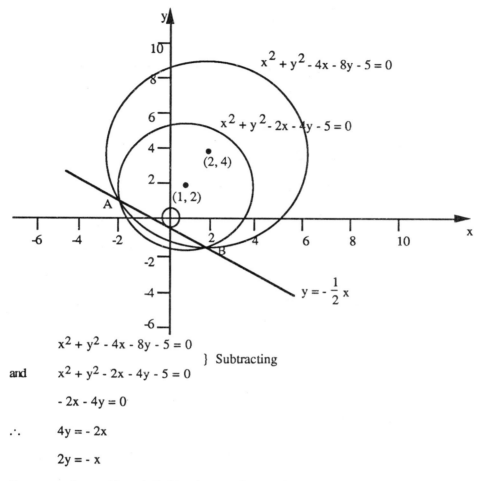

$$x^2 + y^2 - 4x - 8y - 5 = 0$$
and $$x^2 + y^2 - 2x - 4y - 5 = 0$$ } Subtracting

$$-2x - 4y = 0$$
$$\therefore \quad 4y = -2x$$
$$2y = -x$$

This is a linear equation and is satisfied by the coordinates of A and B since they lie on both circles. Therefore, it is the equation of the line joining A and B, ie the common chord.

If the tangents to the two circles at A (and B) are perpendicular, the circles are said to be orthogonal and it means that the radius of one circle is a tangent to the other circle.

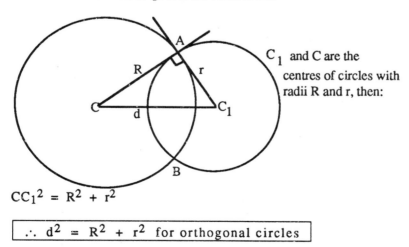

C_1 and C are the centres of circles with radii R and r, then:

$$CC_1^2 = R^2 + r^2$$

$$\boxed{\therefore \ d^2 = R^2 + r^2 \text{ for orthogonal circles}}$$

LINE, CIRCLE, PARABOLA — PURE MATHEMATICS

Example 33

Find the centres and radii of the circles $x^2 + y^2 + 8x + 10y - 4 = 0$ and $x^2 + y^2 - 2x - 4 = 0$.

Find also the distance between their centres and hence:

(a) show that the circles intersect at right angles;

(b) find the length of the common chord.

Solution

$$x^2 + y^2 + 8x + 10y - 4 = 0$$

$$x^2 + 8x + y^2 + 10y = 4$$

$$\left. \begin{array}{l} x^2 + 8x = (x + 4)^2 - 16 \\ y^2 + 10y = (y + 5)^2 - 25 \end{array} \right\} \text{Replacing}$$

$\therefore \quad (x + 4)^2 - 16 + (y + 5)^2 - 25 = 4$

$\therefore \quad (x + 4)^2 + (y + 5)^2 = 4 + 16 + 25$

$$(x + 4)^2 + (y + 5)^2 = 45$$

$\therefore$ Centre $(-4, -5)$ and radius is $\sqrt{45} = R$

$$x^2 + y^2 - 2x - 4 = 0$$

$$x^2 - 2x + y^2 = 4$$

$$\left. \begin{array}{l} x^2 - 2x = (x - 1)^2 - 1 \\ y^2 = (y - 0)^2 \end{array} \right\} \text{Replacing}$$

$$(x - 1)^2 - 1 + (y - 0)^2 = 4$$

$$(x - 1)^2 + (y - 0)^2 = 5$$

$\therefore$ Centre $(1, 0)$ and radius is $\sqrt{5} = r$

Distance between centres $d = \sqrt{(-4 - 1)^2 + (-5 - 0)^2}$

$$= \sqrt{(-5)^2 + (-5)^2}$$

$$d = \sqrt{50}$$

(a) If the circles intersect at right angles, then:

$$R^2 + r^2 = d^2$$

$\therefore \quad 45 + 5 = 50 \quad$ correct $\therefore$ orthogonal circles

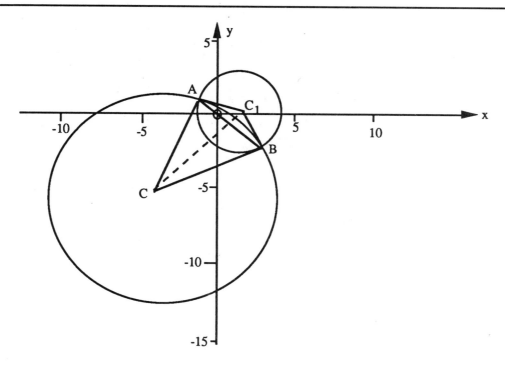

AC_1BC is a kite shape whose area is

$\frac{1}{2}$ (product of diagonals) $= \frac{1}{2} \cdot AB \cdot CC_1$

$\qquad\qquad\qquad\qquad\qquad = \frac{1}{2} \cdot AB \cdot \sqrt{50}$

But the area of $\triangle CAC_1 = \frac{1}{2} \cdot AC \cdot AC_1$

$\qquad\qquad\qquad\qquad\quad = \frac{1}{2} \cdot \sqrt{45} \sqrt{5}$

$\therefore \quad$ Area of $AC_1BC = 2\left(\frac{1}{2}\sqrt{45}\sqrt{5}\right)$

$\qquad\qquad\qquad\qquad = \sqrt{45}\sqrt{5}$

$\therefore \quad \frac{1}{2} AB \sqrt{50} = \sqrt{45}\sqrt{5}$

$\qquad AB = 2\sqrt{\frac{45.5}{50}}$

$\qquad AB = 2\sqrt{\frac{9}{2}}$

$\qquad AB = 2\left(\frac{3}{\sqrt{2}}\right)$

$\qquad\qquad = \frac{6}{\sqrt{2}}$

LINE, CIRCLE, PARABOLA — PURE MATHEMATICS

c) *Equation of the parabola*

The parabola is the locus of a point, P, which moves so that it is equidistant from a fixed point, S, and a fixed line, L.

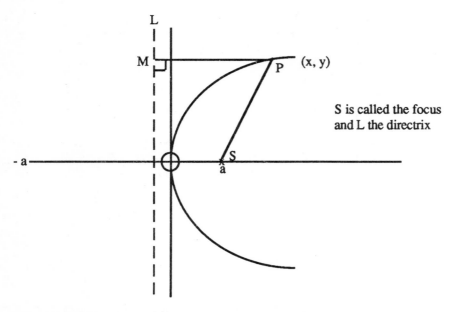

S is called the focus and L the directrix

The fixed point S has coordinates (a, 0) and the fixed line has equation $x = -a$, the variable point is $P(x, y)$.

$$PM = x + a \qquad PS = \sqrt{(x-a)^2 + (y-0)^2}$$

$$PS = \sqrt{(x-a)^2 + y^2}$$

Since $PM = PS$ it follows that

$$PM^2 = PS^2$$

$\therefore \quad (x+a)^2 = (x-a)^2 + y^2$

$\quad x^2 + 2ax + a^2 = x^2 - 2ax + a^2 + y^2$

$\therefore \quad 4ax = y^2$

So the equation of a parabola is, in general terms:

$$\boxed{y^2 = 4ax}$$

The coordinates of any point P on the parabola are often given in terms of a parameter t where:

$$\boxed{x = at^2 \quad \text{and} \quad y = 2at}$$

A standard piece of book work that is often required is finding the equations of the tangents and normals to the parabola $y^2 = 4ax$ at the point $(at^2, 2at)$.

Differentiate with respect to x: $\quad 2y\dfrac{dy}{dx} = 4a$

$\therefore \quad$ gradient of tangent $\quad \dfrac{dy}{dx} = \dfrac{4a}{2y} = \dfrac{4a}{2(2at)} = \dfrac{1}{t}$

Equation of tangent is $y = \dfrac{1}{t}x + c$

and finding c using $(at^2, 2at)$ gives $2at = \dfrac{1}{t}(at^2) + c$

$\therefore$ $2at = at + c$

$\therefore$ $at = c$

So the equation of the tangent is $y = \dfrac{1}{t}x + at$

$\therefore$ $\boxed{ty = x + at^2}$

Gradient of normal $= -t$ because (gradient of normal).(gradient of tangent) $= -1$

Equation of normal is $y = -tx + c$

and finding c using $(at^2, 2at)$ gives $2at = -t(at^2) + c$

$\therefore$ $2at = -at^3 + c$

$\therefore$ $2at + at^3 = c$

So the equation of the normal is:

$\boxed{y = -xt + 2at + at^3}$

d) *Problems relating to the parabola*

Most of the problems on the parabola relate to tangents, normals or chords, as will be seen in the following examples.

Example 34

Show that the equation of the normal to the parabola $y^2 = 4ax$ at the point $P(at^2, 2at)$ is $y + tx = 2at + at^3$.

If this normal meets the x axis at Q show that the mid-point M of PQ has coordinates $(a + at^2, at)$.

If P is a variable point on the parabola, find the cartesian equation of the locus of M.

LINE, CIRCLE, PARABOLA — PURE MATHEMATICS

Solution

The equation of the normal, $y + tx = 2at + at^3$ has just been derived and so will not be proved again.

Sketch graph to show P, M and Q:

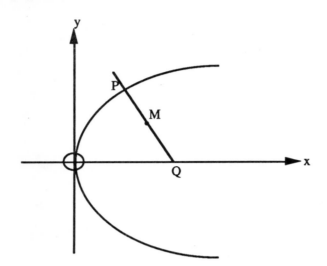

At Q, y = 0   ∴ $y + tx = 2at + at^3$ becomes

∴ $0 + tx = 2at + at^3$

∴ $x = \dfrac{2at + at^3}{t}$

∴ $x = 2a + at^2$

Q is the point $(2a + at^2, 0)$ and P $(at^2, 2at)$

∴   M is $\left(\dfrac{2a + at^2 + at^2}{2}, \dfrac{0 + 2at}{2}\right) = \left(\dfrac{2a + 2at^2}{2}, \dfrac{2at}{2}\right)$

∴   coordinates of M are $(a + at^2, at)$ as required

Since P is a variable point, the locus of M can be found by eliminating t from the parametric equations $x = a + at^2$, $y = at$

$y = at$ gives $\dfrac{y}{a} = t$

Replacing in the other parametric equation:

$x = a + at^2 \quad = a + a\left(\dfrac{y}{a}\right)^2$

∴ $x = a + \dfrac{ay^2}{a^2} \quad = a + \dfrac{y^2}{a}$

Multiplying $ax = a^2 + y^2$

∴ $y^2 = ax - a^2$

∴ $y^2 = a(x - a)$ is the required equation

PURE MATHEMATICS — LINE, CIRCLE, PARABOLA

Example 36

Show that the equation of the chord joining $P(at_1^2, 2at_1)$ and $Q(at_2^2, 2at_2)$ which lie on the parabola $y^2 = 4ax$ is:

$$(t_1 + t_2)y = 2x + 2at_1t_2$$

The tangents to the parabola at P and Q meet at T. When P and Q move on the parabola such that PQ passes through the fixed point (4a, a) show that the locus of T is a straight line.

Solution

Sketch graph to show P, Q, T, point (4a, a)

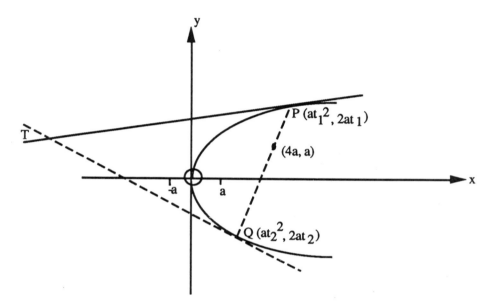

Equation of the chord $\quad y = mx + c$

but $\quad m = \dfrac{2at_1 - 2at_2}{at_1^2 - at_2^2} = \dfrac{2a(t_1 - t_2)}{a(t_1^2 - t_2^2)} = \dfrac{2a(t_1 - t_2)}{a(t_1 - t_2)(t_1 + t_2)}$

$\therefore \quad m = \dfrac{2}{t_1 + t_2}$

$\therefore \quad y = \dfrac{2}{(t_1 + t_2)} \cdot x + c$

Replacing $(at_1^2, 2at_1)$ to find c $\quad 2at_1 = \dfrac{2}{(t_1 + t_2)}(at_1^2) + c$

$\quad 2at_1 = \dfrac{2at_1^2}{(t_1 + t_2)} + c$

$\quad c = 2at_1 - \dfrac{2at_1^2}{t_1 + t_2}$

$\quad = \dfrac{2at_1(t_1 + t_2) - 2at_1^2}{t_1 + t_2} = \dfrac{2at_1^2 + 2at_1t_2 - 2at_1^2}{t_1 + t_2}$

$\therefore \quad c = \dfrac{2at_1t_2}{t_1 + t_2}$

LINE, CIRCLE, PARABOLA — PURE MATHEMATICS

Equation of PQ is $\qquad y = \dfrac{2}{(t_1 + t_2)}x + \dfrac{2at_1t_2}{(t_1 + t_2)}$

Multiplying through by $t_1 + t_2$: $\quad (t_1 + t_2)y = 2x + 2at_1t_2 \qquad$ - as required

Equation of tangent at P is $\qquad t_1 y = x + at_1^2$

Equation of tangent at Q is $\qquad t_2 y = x + at_2^2$

To find the coordinates of T, these must be solved simultaneously for x and y.

Subtracting $\qquad t_1 y - t_2 y = at_1^2 - at_2^2$

$\qquad\qquad\qquad (t_1 - t_2)y = a(t_1 - t_2)(t_1 + t_2)$

$\qquad\qquad\qquad y = a(t_1 + t_2)$

Replacing y to find x gives $\qquad x = t_1 y - at_1^2$

$\qquad\qquad\qquad x = t_1 a(t_1 + t_2) - at_1^2$

$\qquad\qquad\qquad = at_1^2 + at_1t_2 - at_1^2$

$\therefore \qquad\qquad x = at_1t_2$

$\therefore \qquad$ T is the point $(at_1t_2, a(t_1 + t_2))$

One fact that has not been used is that the chord PQ passes through $(4a, a)$.

$\therefore \qquad$ Taking equation of PQ and replacing gives:

$\qquad\qquad (t_1 + t_2)(a) = 2(4a) + 2at_1t_2 \quad$ dividing by a

$\qquad\qquad t_1 + t_2 = 8 + 2t_1t_2 \qquad$ --------- (1)

Returning to T and $\qquad x = at_1t_2 \qquad$ --------- (2)

and $\qquad\qquad y = a(t_1 + t_2) \qquad$ --------- (3)

$\therefore$ using (1) and (3) $\qquad y = a(8 + 2t_1t_2)$

$\qquad\qquad y = 8a + 2at_1t_2$

Now using (2) $\qquad y = 8a + 2(x)$

$\therefore \qquad$ Equation of locus of T is $\quad y = 2x + 8a \qquad$ - a linear equation

---

e) *Example ('A' Level question)*

Find the points (P and Q) of intersection of the curves whose equations are $x^2 + y^2 = 4$ and $y^2 = 3x$.

Prove that angle $P\hat{O}Q = \dfrac{2\pi}{3}$ where O is the origin. Hence, or otherwise, find the area of the smaller finite region enclosed by the two curves leaving the answer in terms of π and $\sqrt{3}$.

Solution

Solving $\quad x^2 + y^2 = 4$

and $\quad y^2 = 3x$

gives $\quad x^2 + 3x = 4$

$\quad\quad x^2 + 3x - 4 = 0$

$\quad\quad (x + 4)(x - 1) = 0$

∴ either $\quad x + 4 = 0 \quad$ or $\quad x - 1 = 0$

$\quad\quad\quad x = -4 \quad\quad\quad x = 1$

Replacing in $y^2 = 3x$ to find the corresponding y coordinates gives:

$\quad\quad y^2 = 3(-4)$

$\quad\quad\quad = -12$

∴ $\quad y = \pm \sqrt{-12} \quad$ impossible

or $\quad y^2 = 3(1)$

$\quad\quad\quad = 3$

∴ $\quad y = \pm\sqrt{3}$

So the points are $P(1, \sqrt{3})\ \ Q(1, -\sqrt{3})$

Sketch graph: $\quad x^2 + y^2 = 4$ is a circle centre (0, 0), radius = 2 and $y^2 = 3x$ is a parabola.

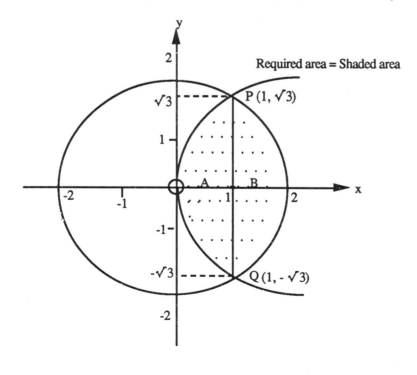

LINE, CIRCLE, PARABOLA   PURE MATHEMATICS

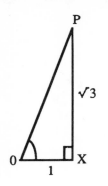

$$P\hat{O}Q = 2P\hat{O}X \quad \text{- by symmetry}$$

and $\tan P\hat{O}X = \dfrac{\sqrt{3}}{1}$

$\therefore \quad P\hat{O}X = 60° \text{ or } \dfrac{\pi}{3} \text{ radians}$

$\therefore \quad P\hat{O}Q = 2 \times 60° \text{ or } 2\dfrac{\pi}{3} \text{ radians}$

The enclosed area divides neatly into two parts:

A (area under parabola) $= 2\displaystyle\int_0^1 \sqrt{3x}\, dx = 2\displaystyle\int_0^1 \sqrt{3} \cdot x^{\frac{1}{2}} dx$

$= 2\sqrt{3}\left[\dfrac{x^{\frac{3}{2}}}{\frac{3}{2}} + c\right]_0^1$

$= 2\sqrt{3}\left[\dfrac{2}{3} x^{\frac{3}{2}} + c\right]_0^1$

$= 2\sqrt{3}\left[\dfrac{2(1)}{3} + c - \dfrac{2(0)}{3} - c\right]$

$\therefore \quad \text{Area A} = \dfrac{4\sqrt{3}}{3}$

B (area under circle): This can be found by integration of $2\displaystyle\int_1^2 \sqrt{4 - x^2}\, dx$ but a simpler method is to find the area of a segment of a circle, ie

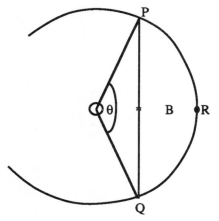

area of sector OPRQ - area of △OPQ

Area of sector OPRQ $= \dfrac{1}{2} r^2 \theta$

$= \dfrac{1}{2}(2)^2 \left(\dfrac{2\pi}{3}\right)$

$= \dfrac{4\pi}{3}$

Area of △OPQ $= \dfrac{1}{2}(PQ)(OX) = 2(2\sqrt{3}\,(1) = \sqrt{3}$

$\therefore \quad \text{Area of B} = \dfrac{4\pi}{3} - \sqrt{3}$

∴ Total area $= \dfrac{4\sqrt{3}}{3} + \dfrac{4\pi}{3} - \sqrt{3}$

$= \dfrac{4\sqrt{3} - 3\sqrt{3}}{3} + \dfrac{4\pi}{3}$

$= \dfrac{\sqrt{3}}{3} + \dfrac{4\pi}{3}$

7 APPROXIMATIONS AND SERIES

7.1 Solutions of equations
7.2 Hyperbolic functions
7.3 Further work on series

7.1 Solutions of equations

a) *General solution of trigonometric equations*

Trigonometric equations have been solved in chapter. The solutions have always been given within a certain range of values of θ such as $0° \leq \theta \leq 360°$ or $0 \leq \theta \leq 2\pi$ radians, $-180° \leq \theta \leq 180°$ or $-\pi \leq \theta \leq \pi$ radians. It is now necessary to take the working one stage further and find an expression which represents all the angles which satisfy the given equation. This expression is known as the general solution of the equation and it is, in fact, an infinite set of angles.

In order to explain the method, a simple equation will be taken and solved, ie $\cos \theta = \frac{1}{2}$. Cosine is positive in the first and fourth quadrants so the solutions of this equation are $\theta = 60°$ and $-60°$.

for $-180° \leq \theta < 180°$

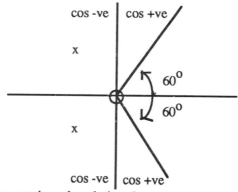

However, these are not the only solutions because:

$$\cos(360° + 60°) = \cos 420° = \frac{1}{2}$$

$$\cos(360° - 60°) = \cos 300° = \frac{1}{2}$$

and $$\cos(720° + 60°) = \cos 780° = \frac{1}{2}$$

$$\cos(720° - 60°) = \cos 660° = \frac{1}{2}$$

and ... etc.

So the general solution is $\theta = 360°n \pm 60°$ where n is an integer

    n = 0   gives $\theta = \pm 60°$ ie 60° and -60°

    n = 1   gives $\theta = 360° \pm 60°$ ie 420° and 300°

    n = 2   gives $\theta = 360° \times 2 \pm 60° = 720° \pm 60°$ ie 780° and 660°

                            = 720° ± 60°   ie 780° and 660°

    n = 3   ... etc.

PURE MATHEMATICS — APPROXIMATIONS AND SERIES

More complicated equations are solved using any of the methods that have already been detailed in earlier working. The general solution is always determined by studying specific solutions in the range 0° to 360° (ie 0 to 2π radians) and then extending the range of values, as in the above example.

Example 1

Find the general solution in radians of the equation:

$$2\sqrt{3}\sin^2\theta - \sin 2\theta = 0$$

Solution

$2\sqrt{3}\sin^2\theta - \sin 2\theta = 0$ but $\sin 2\theta = 2\sin\theta\cos\theta$

$2\sqrt{3}\sin^2\theta - 2\sin\theta\cos\theta = 0$

$2\sin\theta(\sqrt{3}\sin\theta - \cos\theta) = 0$

∴ either $2\sin\theta = 0$

∴ $\sin\theta = 0$

∴ $\theta = 0, \pi, 2\pi, 3\pi, \dots$ etc.

ie $\theta = n\pi$ where n is an integer

or $\sqrt{3}\sin\theta - \cos\theta = 0$

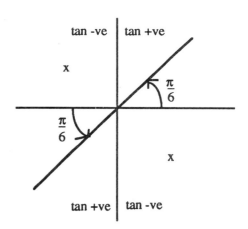

∴ $\sqrt{3}\sin\theta = \cos\theta$

∴ $\dfrac{\sqrt{3}\sin\theta}{\cos\theta} = 1$

∴ $\sqrt{3}\tan\theta = 1$

∴ $\tan\theta = \dfrac{1}{\sqrt{3}}$

∴ $\theta = \dfrac{\pi}{6},\ \pi + \dfrac{\pi}{6},\ 2\pi + \dfrac{\pi}{6},\ 3\pi + \dfrac{\pi}{6} \dots$

∴ $\theta = n\pi + \dfrac{\pi}{6}$ where n is an integer

∴   General solutions are $n\pi,\ n\pi + \dfrac{\pi}{6}$

Example 2

Find the general solution of the equation:

$\sin 5x - \sin 3x = \sin 4x - \sin 2x$ (x in radians)

APPROXIMATIONS AND SERIES — PURE MATHEMATICS

Solution

Using the factor formula:

$$\sin 5x - \sin 3x = 2 \cos\left(\frac{5x + 3x}{2}\right) \sin\left(\frac{5x - 3x}{2}\right)$$

$$\sin 4x - \sin 2x = 2 \cos\left(\frac{4x + 2x}{2}\right) \sin\left(\frac{4x - 2x}{2}\right)$$

∴ $\quad \sin 5x - \sin 3x = \sin 4x - \sin 2x$

becomes $\quad 2 \cos 4x \sin x = 2 \cos 3x \sin x$

∴ $\quad 2 \cos 4x \sin x - 2 \cos 3x \sin x = 0$

∴ $\quad 2 \sin x (\cos 4x - \cos 3x) = 0$

But $\quad \cos 4x - \cos 3x = -2 \sin\left(\frac{4x + 3x}{2}\right) \sin\left(\frac{4x - 3x}{2}\right)$

∴ $\quad 2 \sin x (\cos 4x - \cos 3x) = 0$

becomes $\quad 2 \sin x \left(-2 \sin \frac{7x}{2} \sin \frac{x}{2}\right) = 0$

∴ $\quad -4 \sin x \sin \frac{7x}{2} \sin \frac{x}{2} = 0$

either $\quad \sin x = 0$

∴ $\quad x = 0, \pi, 2\pi, 3\pi, 4\pi \ldots$ etc.

∴ $\quad x = n\pi$

or $\quad \sin \frac{7x}{2} = 0$

∴ $\quad \frac{7x}{2} = 0, \pi, 2\pi, 3\pi, 4\pi \ldots$ etc.

∴ $\quad x = 0, \frac{2\pi}{7}, \frac{4\pi}{7}, \frac{6\pi}{7}, \frac{8\pi}{7} \ldots$ etc.

∴ $\quad x = \frac{2n\pi}{7}$

or $\quad \sin \frac{x}{2} = 0$

∴ $\quad \frac{x}{2} = 0, \pi, 2\pi, 3\pi, 4\pi \ldots$ etc.

∴ $\quad x = 0, 2\pi, 4\pi, 6\pi, 8\pi \ldots$ etc.

but these solutions are already contained in $x = n\pi$ and do not need repeating.

Therefore, general solutions are $n\pi, \frac{2n\pi}{7}$

PURE MATHEMATICS — APPROXIMATIONS AND SERIES

b) *Linear approximations*

When the rules of mathematics are applied to practical or experimental problems, the functions, f(x), established may be so complicated that only approximate values can be calculated.

The 'linear approximation' method is a formula which enables approximate values of a function to be calculated easily. Since the formula is not too complicated it will be derived from the following diagram:

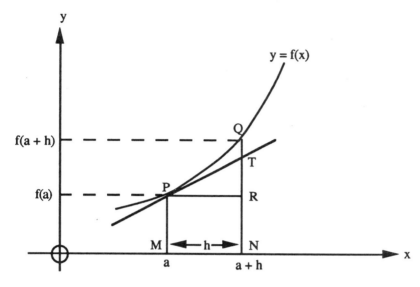

This diagram represents part of the graph of y = f(x), ie any function of x. P and Q are two points on the curve with coordinates.

$$P: x = a, \quad y = f(a) \quad \therefore P(a, f(a))$$

$$Q: x = a + h, \quad y = f(a + h) \quad \therefore Q(a + h, f(a + h))$$

and h is very small, therefore, P and Q are very close together on the curve.

PT is a tangent to the curve at P.

The aim is to establish an approximate relationship between f(a) and f(a + h).

$$f(a + h) = NQ \simeq NT \quad \text{approximately} \quad \text{assuming h is very small}$$

but NT = NR + RT and NR = MP = f(a)

∴ NT = f(a) + RT

Now it is necessary to find an expression for RT and this can be done by considering triangle PRT.

$$\text{Tan } T\hat{P}R = \frac{RT}{PR} \quad \text{but PR = MN = h}$$

∴ $$\text{Tan } T\hat{P}R = \frac{RT}{h}$$

∴ $$h \cdot \text{Tan } T\hat{P}R = RT$$

but Tan T$\hat{P}$R is the same as the gradient of the tangent to the curve at P where:

$$\text{gradient} = \frac{dy}{dx} = f'(x)$$

APPROXIMATIONS AND SERIES — PURE MATHEMATICS

∴ gradient at P = f'(a)

∴ RT = h f'(a)

∴ $\boxed{\text{if h is small } f(a + h) \simeq f(a) + h f'(a)}$

This is called the linear approximation because the straight line PT has been considered in place of the curve PQ.

Another way of expressing this same relationship is given by $x \simeq a$. The function f(x) may be expressed in the approximate linear form of:

$\boxed{f(x) \simeq f(a) + f'(a)(x - a)}$

Example 3

Find approximate values of:

(a) sin 31°

(b) $e^{1.08}$ (given e = 2.7183)

(c) $\log_e 2.001$ (given $\log_e 2 = 0.6931$)

Solution

(a) It is necessary to work in radians because of using the derivative of sin x in the working.

$31° = 30° + 1° = \frac{\pi}{6} + 0.0175$ radians   as $1° = \frac{\pi}{180}$ radians = 0.0175

= a + h

∴ $a = \frac{\pi}{6}$ and h = 0.0175

Using $f(a + h) \simeq f(a) + h f'(a)$

where   f(x) = sin x   ∴ $f\left(\frac{\pi}{6}\right) = \sin\frac{\pi}{6} = 0.5$

f'(x) = cos x   ∴ $f'\left(\frac{\pi}{6}\right) = \cos\frac{\pi}{6} = 0.866$

∴ $\sin\left(\frac{\pi}{6} + 0.0175\right) \simeq \sin\frac{\pi}{6} + 0.0175 \cos\frac{\pi}{6}$

= 0.5 + 0.0175 × 0.866

= 0.515155

∴ sin 31° $\simeq$ 0.5152 (to 4 decimal places)

(*Note*: Calculator value for sin 31° = 0.5150 (to 4 decimal places).)

(b)   1.08 = 1.00 + 0.08

= a + h

∴   a = 1.00 and h = 0.08

PURE MATHEMATICS — APPROXIMATIONS AND SERIES

Using $f(a + h) \simeq f(a) + h f'(a)$

where $f(x) = e^x$ $\therefore f(1) = e^1 = 2.7183$

and $f'(x) = e^x$ $\therefore f'(1) = e^1 = 2.7183$

$\therefore e^{1.08} \simeq e^1 + 0.08\, e^1$

$\qquad = 2.7183 + 0.08 \times 2.7183$

$\qquad = 2.7183 + 0.2175$

$\therefore e^{1.08} \simeq 2.9358$ (to 4 decimal places)

(*Note*: Calculator value for $e^{1.08} = 2.9447$ (to 4 decimal places).)

(c) $\quad 2.001 = 2.000 + 0.001$

$\qquad\qquad = a + h$

$\therefore a = 2.000$ and $h = 0.001$

Using $f(a + h) \simeq f(a) + h f'(a)$

where $f(x) = \log_e x$ $\therefore f(2) = \log_e 2.000 = 0.6931$

and $f'(x) = \dfrac{1}{x}$ $\therefore f'(2) = \dfrac{1}{2} = 0.5$

$\therefore \log_e 2.001 \simeq \log_e 2.000 + 0.001 \times \dfrac{1}{2}$

$\qquad\qquad = 0.6931 + 0.001 \times 0.5$

$\qquad\qquad = 0.6931 + 0.0005$

$\therefore \log_e 2.001 \simeq 0.6936$

(*Note*: Calculator value for $\log_e 2.001 = 0.6936$ (to 4 decimal places).)

Example 4

Prove:

(a) if $x \simeq 0$, $\quad e^x = 1 + x$

(b) if $x \simeq 2$, $\quad \dfrac{1}{(1 + x)^2} \simeq \dfrac{1}{27}(7 - 2x)$

Solution

Using $x \simeq a$ then $f(x) \simeq f(a) + f'(a)(x - a)$.

(a) If $x \simeq 0$ and $f(x) = e^x$ then $f(0) = e^0 = 1$

and $f'(x) = e^x$ then $f'(0) = e^0 = 1$

APPROXIMATIONS AND SERIES — PURE MATHEMATICS

Using $f(x) \simeq f(0) + f'(0).(x - 0)$

$= 1 + 1.x$

$\therefore \quad f(x) \simeq 1 + x$

(b) If $x \simeq 2$ and $f(x) = \dfrac{1}{(1 + x)^2}$ then $f(2) = \dfrac{1}{(1 + 2)^2} = \dfrac{1}{9}$

and $f'(x) = \dfrac{-2}{(1 + x)^3}$ then $f'(2) = \dfrac{-2}{(1 + 2)^3} = \dfrac{-2}{27}$

Using $f(x) \simeq f(2) + f'(2).(x - 2)$

$= \dfrac{1}{9} - \dfrac{2}{27}.(x - 2)$

$= \dfrac{1}{9} - \dfrac{2x}{27} + \dfrac{4}{27}$

$= \dfrac{3 + 4}{27} - \dfrac{2x}{27}$

$\therefore \quad f(x) \simeq \dfrac{7}{27} - \dfrac{2x}{27}$

$= \dfrac{1}{27}(7 - 2x)$

c) *Approximate solutions to an equation - Newton's method*

When an equation cannot be solved directly, it is necessary to find an approximate solution by some means. One possibility is to draw a graph of the function. However, this will only give a very rough value and a better approximation will be found using a numerical rather than a graphical method. One such approach is Newton's method which is used to obtain successive approximations to the root of an equation once it has been roughly located.

If the equation of $f(x) = 0$ is to be solved and it has been discovered that $f(a) < 0$ and $f(a + 1) > 0$, then $f(x)$ is equal to zero somewhere between a and a + 1. Therefore, there is at least one root of the equation $f(x) = 0$ between a and a + 1.

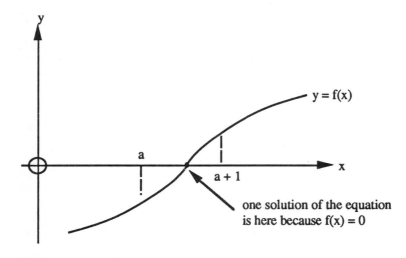

Having roughly located the root, then Newton's method is used as follows:

PURE MATHEMATICS — APPROXIMATIONS AND SERIES

Let the root be $a + h$, then $f(a + h) = 0$

but, from the previous work, if h is small $f(a + h) \simeq f(a) + h f'(a)$

$\therefore \quad f(a) + h f'(a) \simeq 0$

$\therefore \quad h f'(a) \simeq - f(a)$

$\therefore \quad h \simeq \dfrac{-f(a)}{f'(a)}$

This gives a second approximation to the root of $f(x)$, ie $\quad a + h = a - \dfrac{f(a)}{f'(a)}$

This process can be repeated several times to obtain closer and closer approximations.

It must be realised that an equation such as $f(x) = 0$ may have more than one root and so Newton's method would have to be used more than once to find all the roots. However, questions usually only ask for an approximate value of one root as will be seen from the following examples.

Example 5

Check that one root of the equation $x^3 - 4x^2 - x - 12 = 0$ lies between 4 and 5. Use Newton's method twice to find a better approximation for this root of the equation.

Solution

$$f(x) = x^3 - 4x^2 - x - 12$$

when $x = 4 \quad f(4) = 4^3 - 4(4)^2 - 4 - 12 = -16$

$\therefore \quad f(4) < 0$

when $x = 5 \quad f(5) = 5^3 - 4(5)^2 - 5 - 12 = 8$

$\therefore \quad f(5) > 0$

$\therefore \quad f(x) = 0$ between 4 and 5, and closer to 5 than 4.

<u>Newton's method (1)</u>

Let $x = 5 + h$ be the root. $\quad \therefore \quad f(5 + h) = 0$

$\therefore \quad h = \dfrac{-f(5)}{f'(5)}$

$f(x) = x^3 - 4x^2 - x - 12 \quad \therefore \quad f(5) = 8$

$f'(x) = 3x^2 - 8x - 1 \quad \therefore \quad f'(5) = 34$

$\therefore \quad h = \dfrac{-8}{34}$

$= -0.24$

$\therefore \quad x = 5 - 0.24$

$= 4.76$

APPROXIMATIONS AND SERIES — PURE MATHEMATICS

Newton's method (2)

Let $x = 4.76 + h$ be the root. $\therefore$ $f(4.76 + h) = 0$

$$\therefore h = \frac{-f(4.76)}{f'(4.76)}$$

$f(4.76) = 4.76^3 - 4(4.76)^2 - 4.76 - 12$

$ = 107.85 - 90.63 - 4.76 - 12$

$ = 0.46$

$f'(4.76) = 3(4.76)^2 - 8(4.76) - 1$

$ = 67.97 - 38.08 - 1$

$ = 28.89$ $\therefore$ $h = \frac{-0.46}{28.89} = -0.016$

$\therefore x = 4.76 - 0.016$

$ = 4.744$

The first approximation was $x = 5$, the second approximation was $x = 4.76$ and the third approximation $x = 4.744 = 4.74$ (to 3 significant figures).

Example 6

An arc AB of a circle subtends an angle of θ radians at the centre. The length of the arc is s and the length of the chord AB is d. If $s : d = 4 : 3$ show that $3\theta = 8 \sin \frac{\theta}{2}$.

Show that θ is approximately 2.5. Taking this as a first approximation use Newton's method once to find a second approximation.

Solution

Arc length $s = r\theta$ where r = radius

Using $\triangle OMB$ $\sin \frac{\theta}{2} = \frac{\frac{d}{2}}{r} = \frac{d}{2r}$

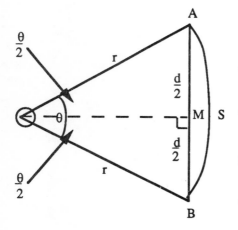

$\therefore d = 2r \sin \frac{\theta}{2}$

$\therefore s : d = r\theta : 2r \sin \frac{\theta}{2}$

$\therefore \frac{s}{d} = \frac{r\theta}{2r \sin \frac{\theta}{2}} = \frac{\theta}{2 \sin \frac{\theta}{2}}$

But $s : d = 4 : 3$

$\therefore \frac{s}{d} = \frac{4}{3}$

So it follows that

$$\frac{4}{3} = \frac{\theta}{2 \sin \frac{\theta}{2}}$$

$$\therefore \quad 8 \sin \frac{\theta}{2} = 3\theta$$

To show that θ is approximately 2.5

Let $\quad f(\theta) = 8 \sin \frac{\theta}{2} - 3\theta$

when $\theta = 2.5 \quad f(2.5) = 8 \sin \frac{2.5}{2} - 3(2.5)$

$$= 8(0.9490) - 7.5$$

$$= 7.592 - 7.5$$

$$\therefore \quad f(2.5) = 0.092$$

Since $f(2.5)$ is very close to zero,

$\theta = 2.5$ is an approximate value for the root of the equation. $\quad 8 \sin \frac{\theta}{2} - 3\theta = 0$

Using Newton's method let the root of the equation be

$$\theta = 2.5 + h \qquad \therefore \quad f(2.5 + h) = 0$$

$$\therefore \quad h = \frac{-f(1.5)}{f'(2.5)}$$

$f(\theta) = 8 \sin \frac{\theta}{2} - 3\theta \qquad \therefore \quad f(2.5) = 0.092$

$f'(\theta) = 8 \left(\cos \frac{\theta}{2} \right) \frac{1}{2} - 3$

$\therefore \quad f'(\theta) = 4 \cos \frac{\theta}{2} - 3 \qquad$ and $\qquad f'(2.5) = 4 \cos \frac{2.5}{2} - 3 = -1.739$

$$\therefore \quad h = \frac{-0.092}{-1.739}$$

$$= 0.053$$

Therefore, a better approximation is $\theta = 2.5 + 0.053$

$$= \quad 2.55 \text{ (to 3 significant figures)}$$

Example 7

Show that the equation $5x - e^{-x^2} = 0$ has a root near $x = 0.2$.

Taking $x = 0.2$ as a first approximation, apply Newton's method once to find an improved value of this approximation.

APPROXIMATIONS AND SERIES   PURE MATHEMATICS

Solution

$$f(x) = 5x - e^{-x^2}$$

$$\therefore f(0.2) = 5(0.2) - e^{(-0.2)^2}$$

$$= 1 - 0.961$$

$$\therefore f(0.2) = 0.039$$

Since $f(0.2)$ is very close to zero, $x = 0.2$ is an approximate value for the root of the equation:

$$5x - e^{-x^2} = 0$$

Using Newton's method let the root of the equation be

$$x = 0.2 + h \qquad \therefore f(0.2 + h) = 0$$

$$\therefore h = \frac{-f(0.2)}{f'(0.2)}$$

$$f(x) = 5x - e^{-x^2} \qquad \therefore f(0.2) = 0.039$$

$$f'(x) = 5 - e^{-x^2}(-2x)$$

$$\therefore f'(x) = 5 + 2x\, e^{-x^2} \qquad \therefore f'(0.2) = 5 + 2(0.2)e^{(-0.2)^2}$$

$$\therefore f'(0.2) = 5 + 0.384$$

$$= 5.384$$

$$\therefore h = \frac{-0.039}{5.384}$$

$$= -0.007$$

A better approximation is $x = 0.2 - 0.007$

$$= 0.193 \text{ (to 3 significant figures)}$$

d) *Example ('A' level question)*

(a) Show that the equation $x^2 + 8\cos 3x = 0$ has a root between 0.5 and 0.6. Using Newton's method with $x = 0.5$ as a first approximation find a second approximation to the root.

(b) Find in radians the general solution of the equation

$$4\sin x + 4\cos x = \sec x + \csc x$$

PURE MATHEMATICS   APPROXIMATIONS AND SERIES

Solution

(a) $f(x) = x^2 + 8 \cos 3x$

when $x = 0.5$
then $f(0.5) = (0.5)^2 + 8 \cos [3(0.5)]$

$\qquad = 0.25 + 8 \cos (.15)$

$\qquad = 0.25 + 8 (0.0707)$

$\qquad = 0.8156$

when $x = 0.6$
then $f(0.6) = (0.6)^2 + 8 \cos [3(0.6)]$

$\qquad = 0.36 + 8 \cos (1.8)$

$\qquad = 0.36 + 8(-0.2272)$

$\therefore \quad f(0.5) > 0$ $\qquad\qquad \therefore \quad f(0.6) < 0$

The root of the equation will occur where $f(x) = 0$ and this must be somewhere between $x = 0.5$ and $x = 0.6$ since $f(x)$ changes from positive to negative over this range.

Applying Newton's method

Let $x = 0.5 + h$ be the root $\therefore \quad f(0.5 + h) = 0$

$$\therefore \quad h = \frac{-f(0.5)}{f'(0.5)}$$

$f(x) = x^2 + 8 \cos 3x \qquad \therefore \quad f(0.5) = 0.8156$

$f'(x) = 2x + 8 (3) (-\sin 3x)$

$\qquad = 2x - 24 \sin 3x \qquad \therefore \quad f'(0.5) = 2 (0.5) - 24 \sin [3 (0.5)]$

$\qquad\qquad\qquad = 1 - 24 \sin 1.5$

$\qquad\qquad\qquad = -22.94$

$$\therefore \quad h = \frac{-0.8156}{-22.94}$$

$\qquad = 0.036$

$\therefore \quad x = 0.5 + h = 0.5 + 0.036$

$\therefore \quad x = 0.536$ is a better approximation to the root of the equation $x^2 + 8 \cos 3x = 0$

(b) $4 \sin x + 4 \cos x = \sec x + \csc x$

$4(\sin x + \cos x) = \dfrac{1}{\cos x} + \dfrac{1}{\sin x}$

$4(\sin x + \cos x) = \dfrac{\sin x + \cos x}{\cos x \sin x}$

$\therefore \quad 4(\sin x + \cos x) \cos x \sin x = \sin x + \cos x$

Resist the temptation to divide both sides by $(\sin x + \cos x)$ as a solution will be lost.

$\therefore \quad 4(\sin x + \cos x) \cos x \sin x - (\sin x + \cos x) = 0$

$\therefore \quad (\sin x + \cos x) (4 \cos x \sin x - 1) = 0$

APPROXIMATIONS AND SERIES — PURE MATHEMATICS

either $\sin x + \cos x = 0$

$\therefore \quad \sin x = -\cos x$

$\therefore \quad \tan x = -1$

$\therefore \quad x = \dfrac{3\pi}{4},\ \dfrac{7\pi}{4},\ 2\pi + \dfrac{3\pi}{4},\ 2\pi + \dfrac{7\pi}{4},\ \ldots$ etc.

but $\dfrac{3\pi}{4} = \pi - \dfrac{\pi}{4}$ and $\dfrac{7\pi}{4} = 2\pi - \dfrac{\pi}{4}$ so the solutions can be written as:

$x = \pi - \dfrac{\pi}{4},\ 2\pi - \dfrac{\pi}{4},\ 2\pi + \left(\pi - \dfrac{\pi}{4}\right),\ 2\pi + \left(2\pi - \dfrac{\pi}{4}\right),\ \ldots$ etc.

$\quad = \pi - \dfrac{\pi}{4},\ 2\pi - \dfrac{\pi}{4},\ 3\pi - \dfrac{\pi}{4},\ 4\pi - \dfrac{\pi}{4},\ \ldots$ etc.

In general $x = n\pi - \dfrac{\pi}{4}$ where $n = 1, 2, 3 \ldots$ etc.

or $\quad 4\cos x \sin x - 1 = 0$

$\therefore \quad 4\cos x \sin x = 1 \quad$ but $2\cos x \sin x = \sin 2x$

$\therefore \quad 2 \sin 2x = 1$

$\therefore \quad \sin 2x = \dfrac{1}{2}$

$\therefore \quad 2x = \dfrac{\pi}{6},\ \dfrac{5\pi}{6},\ 2\pi + \dfrac{\pi}{6},\ 2\pi + \dfrac{5\pi}{6},\ \ldots$ etc.

$\therefore \quad x = \dfrac{\pi}{12},\ \dfrac{5\pi}{12},\ \pi + \dfrac{\pi}{12},\ \pi + \dfrac{5\pi}{12},\ \ldots$ etc.

In general $x = n\pi + \dfrac{\pi}{12},\ n\pi + \dfrac{5\pi}{12}$ where $n = 0, 1, 2 \ldots$ etc.

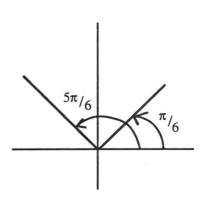

7.2 Hyperbolic functions

a) *Introduction to sinh and cosh*

The exponential functions e^x and e^{-x} were introduced much earlier in the work. They are now going to be used in the definition of two entirely new functions called sinh (pronounced shine) and cosh.

The hyperbolic sine of x, $\sinh x = \dfrac{1}{2}(e^x - e^{-x})$

The hyperbolic cosine of x, $\cosh x = \dfrac{1}{2}(e^x + e^{-x})$

Whilst these are not trigonometrical functions, there are many similaries between these new functions and their trigonometrical equivalents. These similarities will become more obvious as the work progresses and certain identities are developed and proved.

PURE MATHEMATICS — APPROXIMATIONS AND SERIES

Their main application is in integration. They will be used in later work to integrate functions such as $\dfrac{1}{\sqrt{x^2-1}}$ and $\dfrac{1}{\sqrt{x^2+1}}$ which can not be integrated by any of the methods studied so far.

The graphs of e^x, e^{-x}, $\cosh x$ and $\sinh x$ will now be drawn and should be learnt as they may be useful when answering questions.

Graph (1) contains the graphs of $y = e^x$ and $y = e^{-x}$.

Graph (2) contains the graphs of $y = \sinh x$ and $y = \cosh x$.

(1)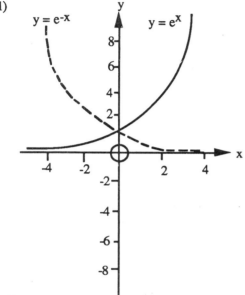

(2)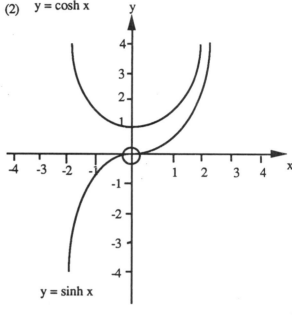

Points to note:

(1) e^x and $e^{-x} > 0$ for all x

        $e^x \to 0$ as $x \to -\infty$

        $e^{-x} \to 0$ as $x \to +\infty$

(2) $\cosh x \geq 1$ for all x

        $\sinh x \to \cosh x$ as $x \to \infty$

        $\sinh x \to -\cosh x$ as $x \to -\infty$

        $\cosh x$ contains one minimum turning point

If $\sinh x = \dfrac{1}{2}(e^x - e^{-x})$ and $\cosh x = \dfrac{1}{2}(e^x + e^{-x})$

then $\tanh x = \dfrac{\sinh x}{\cosh x}$ $\therefore \tanh x = \dfrac{\frac{1}{2}(e^x - e^{-x})}{\frac{1}{2}(e^x + e^{-x})} = \dfrac{e^x - e^{-x}}{e^x + e^{-x}}$

and $\operatorname{cosech} x = \dfrac{1}{\sinh x}$ $\therefore \operatorname{cosech} x = \dfrac{2}{e^x - e^{-x}}$

and $\operatorname{sech} x = \dfrac{1}{\cosh x}$ $\therefore \operatorname{sech} x = \dfrac{2}{e^x + e^{-x}}$

and $\coth x = \dfrac{1}{\tanh x}$ $\therefore \coth x = \dfrac{e^x + e^{-x}}{e^x - e^{-x}}$

APPROXIMATIONS AND SERIES — PURE MATHEMATICS

One of the most important identities to be proved is:

$$\cosh^2 x - \sinh^2 x = 1$$

The proof is as follows:

$$\cosh^2 x - \sinh^2 x = \left[\tfrac{1}{2}(e^x + e^{-x})\right]^2 - \left[\tfrac{1}{2}(e^x - e^{-x})\right]^2$$

$$= \tfrac{1}{4}(e^{2x} + 2 + e^{-2x}) - \tfrac{1}{4}(e^{2x} - 2 + e^{-2x})$$

$$= \tfrac{1}{4}(e^{2x} + 2 + e^{-2x} - e^{2x} + 2 - e^{-2x})$$

$$= \tfrac{1}{4}(4)$$

$\therefore$ $\boxed{\cosh^2 x - \sinh^2 x = 1}$

Two further identities that are worth proving are:

$$2 \sinh x . \cosh x = \sinh 2x$$

and $$\cosh^2 x + \sinh^2 x = \cosh 2x$$

ie $$2 \sinh x . \cosh x = 2 . \tfrac{1}{2}(e^x - e^{-x}) \tfrac{1}{2}(e^x + e^{-x})$$

$$= \tfrac{1}{2}(e^{2x} + 1 - 1 - e^{-x2})$$

$$= \tfrac{1}{2}(e^{2x} - e^{-2x})$$

$\therefore$ $\boxed{2 \sinh x . \cosh x = \sinh 2x}$

and $$\cosh^2 x + \sinh^2 x = \left[\tfrac{1}{2}(e^x + e^{-x})\right]^2 + \left[\tfrac{1}{2}(e^x - e^{-x})\right]^2$$

$$= \tfrac{1}{4}(e^{2x} + 2 + e^{-2x}) + \tfrac{1}{4}(e^{2x} - 2 + e^{-2x})$$

$$= \tfrac{1}{4}(e^{2x} + 2 + e^{-2x} + e^{2x} - 2 + e^{-2x})$$

$$= \tfrac{1}{4}(2e^{2x} + 2e^{-2x})$$

$$= \tfrac{1}{2}(e^{2x} + e^{-2x})$$

$\therefore$ $\boxed{\cosh^2 x + \sinh^2 x = \cosh 2x}$

Although it must be stressed again that cosh x and sinh x are in no way trigonometric functions, the similarities between the two types of functions - hyperbolic and trigonometric - is considerable.

The next example will show how hyperbolic equations are solved.

PURE MATHEMATICS — APPROXIMATIONS AND SERIES

Example 8

Solve the equations:

(a) $\quad 2\cosh x + \sinh x = 2$

(b) $\quad 12\cosh^2 x + 7\sinh x = 24$

Solution

(a) $\quad 2\cosh x + \sinh x = 2$

$\therefore \quad 2 \cdot \dfrac{1}{2}(e^x + e^{-x}) + \dfrac{1}{2}(e^x - e^{-x}) = 2$

$\therefore \quad e^x + e^{-x} + \dfrac{e^x}{2} - \dfrac{e^{-x}}{2} = 2$

$\therefore \quad \dfrac{3e^x}{2} + \dfrac{e^{-x}}{2} = 2$

$\therefore \quad 3e^x + e^{-x} = 4$

$\therefore \quad 3e^x + \dfrac{1}{e^x} = 4 \qquad$ multiplying by e^x

$\therefore \quad 3e^{2x} + 1 = 4e^x$

$\therefore \quad 3e^{2x} - 4e^x + 1 = 0 \qquad$ quadratic equation in e^x

$\therefore \quad (3e^x - 1)(e^x - 1) = 0$

either $\quad 3e^x - 1 = 0 \qquad\qquad$ or $\quad e^x - 1 = 0$

$\therefore \quad 3e^x = 1 \qquad\qquad\qquad \therefore \quad e^x = 1$

$\therefore \quad e^x = \dfrac{1}{3} \qquad\qquad\qquad \therefore \quad x = 0$ because $e^0 = 1$

$\therefore \quad x = \log_e\left(\dfrac{1}{3}\right)$

Therefore, the solutions are $x = 0$ or $\log_e\left(\dfrac{1}{3}\right)$

(b) $\quad 12\cosh^2 x + 7\sinh x = 24 \qquad$ but $\cosh^2 x = 1 + \sinh^2 x$

$\therefore \quad 12(1 + \sinh^2 x) + 7\sinh x = 24$

$\therefore \quad 12 + 12\sinh^2 x + 7\sinh x - 24 = 0$

$\therefore \quad 12\sinh^2 x + 7\sinh x - 12 = 0 \qquad$ a quadratic in $\sinh x$

$\therefore \quad (3\sinh x + 4)(4\sinh x - 3) = 0$

either $\quad 3\sinh x + 4 = 0 \qquad\qquad$ or $\quad 4\sinh x - 3 = 0$

$\therefore \quad 3\sinh x = -4 \qquad\qquad\qquad \therefore \quad 4\sinh x = 3$

$$\therefore \quad \sinh x = \frac{-4}{3} \qquad \therefore \quad \sinh x = \frac{3}{4}$$

However, these are not the values of x, so more working is needed to find x itself.

Using $\sinh x = \frac{-4}{3}$ gives $\frac{1}{2}(e^x - e^{-x}) = \frac{-4}{3}$

$$\therefore \quad 3(e^x - e^{-x}) = -8$$

$$\therefore \quad 3e^x - 3e^{-x} = -8$$

$$\therefore \quad 3e^x - \frac{3}{e^x} = -8$$

$$\therefore \quad 3e^{2x} - 3 = -8e^x$$

$$\therefore \quad 3e^{2x} + 8e^x - 3 = 0$$

$$\therefore \quad (3e^x - 1)(e^x + 3) = 0$$

either $3e^x - 1 = 0$ \qquad or $e^x = -3$

$\therefore \quad 3e^x = 1$ \qquad - no real solution

$\therefore \quad e^x = \frac{1}{3}$

$\therefore \quad x = \log_e\left(\frac{1}{3}\right)$

Using $\sinh x = \frac{3}{4}$ \qquad $\therefore \quad \frac{1}{2}(e^x - e^{-x}) = \frac{3}{4}$

$$\therefore \quad 2(e^x - e^{-x}) = 3$$

$$\therefore \quad 2e^{2x} - 2 = 3e^x$$

$$\therefore \quad 2e^{2x} - 3e^x - 2 = 0$$

$$\therefore \quad (2e^x + 1)(e^x - 2) = 0$$

Either $2e^x + 1 = 0$ \qquad or $e^x - 2 = 0$

$\therefore \quad 2e^x = -1$ \qquad $\therefore \quad e^x = 2$

$\therefore \quad e^x = -\frac{1}{2}$ \qquad $\therefore \quad x = \log_e 2$

$\qquad = -\frac{1}{2}$ - no real solution

Therefore, the solutions are $x = \log_e\left(\frac{1}{3}\right)$ and $\log_e 2$.

b) *Inverse hyperbolic functions*

If $x = \cosh y$ then $\cosh^{-1} x = y$

and if $x = \sinh y$ then $\sinh^{-1} x = y$

and if $x = \tanh y$ then $\tanh^{-1} x = y$

$\cosh^{-1} x$, $\sinh^{-1} x$ and $\tanh^{-1} x$ are the inverse hyperbolic functions. The graphs of these are as follows:

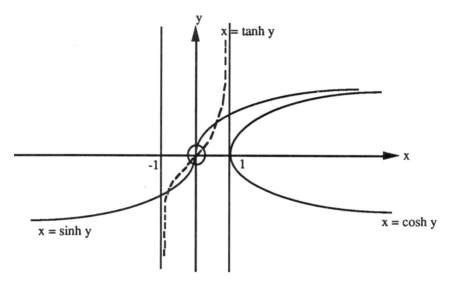

The graphs of $\sinh^{-1} x$ and $\cosh^{-1} x$ are identical to those drawn earlier with x and y interchanged.

$\cosh^{-1} x$ and $\sinh^{-1} x$ can also be expressed in logarithmic form as follows:

$\cosh^{-1} x = y$ means $x = \cosh y$

$\therefore \quad x = \frac{1}{2}(e^y + e^{-y})$

$\therefore \quad 2x = e^y + e^{-y}$

$\therefore \quad e^y + \frac{1}{e^y} - 2x = 0$

$\therefore \quad e^{2y} - 2xe^y + 1 = 0$

This is a quadratic in e^y which will not factorise. Therefore, it is necessary to use the formula

$\frac{-b \pm \sqrt{b^2 - 4ac}}{2a}$ with $a = 1$, $b = -2x$, $c = 1$.

$\therefore \quad e^y = \frac{-(-2x) \pm \sqrt{(-2x)^2 - 4 \cdot (1) \cdot (1)}}{2 \cdot 1}$

$= \frac{2x \pm \sqrt{4x^2 - 4}}{2}$

$= \frac{2x \pm 2\sqrt{x^2 - 1}}{2}$

APPROXIMATIONS AND SERIES — PURE MATHEMATICS

$$\therefore \quad e^y = x \pm \sqrt{x^2 - 1}$$

$$\therefore \quad y = \log_e(x \pm \sqrt{x^2 - 1})$$

$$\therefore \quad \cosh^{-1}x = \log_e(x \pm \sqrt{x^2 - 1})$$

The $\pm$ sign arises because for every value of $x > 1$ there are two equal and opposite values of y, ie of $\cosh^{-1}x$. This can be seen from the graph which is symmetrical about the x axis for values of $x > 1$. The expression for $\cosh^{-1}x$ can also be written as:

$$\boxed{\cosh^{-1}x = \pm \log_e(x + \sqrt{x^2 - 1})}$$

$\sinh^{-1}x = y$ means $x = \sinh y$

$$\therefore \quad x = \frac{1}{2}(e^y - e^{-y}) \ldots \text{etc.}$$

The method is identical to that just used for $\cosh^{-1}x$ and it leads to:

$$\sinh^{-1}x = \log_e(x \pm \sqrt{x^2 + 1})$$

but $x - \sqrt{x^2 + 1} < 0$ and logarithms do not exist for negative numbers. Therefore, there is only one value of $\sinh^{-1}x$ for each value of x. This is confirmed by studying the graph.

$$\therefore \quad \boxed{\sinh^{-1}x = \log_e(x + \sqrt{x^2 + 1})}$$

The following examples show the type of problems that arise concerning inverse hyperbolic functions.

Example 9

Find x if $\sinh^{-1}x = \log_e 3$

Solution

$$\sinh^{-1}x = \log_e 3$$

$$\therefore \quad x = \sinh(\log_e 3)$$

$$\therefore \quad x = \frac{1}{2}(e^{\log_e 3} - e^{-\log_e 3}) \quad \text{and} \quad e^{-\log_e 3} = \frac{1}{e^{\log_e 3}}$$

$$= \frac{1}{2}\left(3 - \frac{1}{3}\right)$$

$$\therefore \quad x = \frac{1}{2}\left(\frac{8}{3}\right)$$

$$= \frac{4}{3}$$

Example 10

Solve the equation $\cosh^{-1}5x = \sinh^{-1}4x$

Solution

Let $\cosh^{-1}5x = y$ and $\sinh^{-1}4x = y$

$\therefore \quad 5x = \cosh y \qquad 4x = \sinh y$

Using $\cosh^2 y - \sinh^2 y = 1$

gives $(5x)^2 - (4x)^2 = 1$

$25x^2 - 16x^2 = 1$

$\therefore \quad 9x^2 = 1$

$\therefore \quad x^2 = \dfrac{1}{9}$

$\therefore \quad x = \pm\dfrac{1}{3}$

but $x = -\dfrac{1}{3}$ gives $y = \cosh^{-1}5\left(-\dfrac{1}{3}\right) = \cosh^{-1}\left(-\dfrac{5}{3}\right)$ which is impossible as can be seen from the graph of $y = \cosh^{-1}x$. Therefore, the only solution is $x = \dfrac{1}{3}$.

Example 11

Draw the graph of $\sinh y = x$ for $0 \le y \le 3$.

Show graphically that the equation $x = 2\sinh^{-1}x$ has only one positive root.

Solution

To draw the graph accurately it is necessary to work out corresponding values of x and y.

y	0	0.5	1.0	1.5	2.0	2.5	3.0
sinh y = x	0	0.52	1.18	2.13	3.63	6.05	10.02

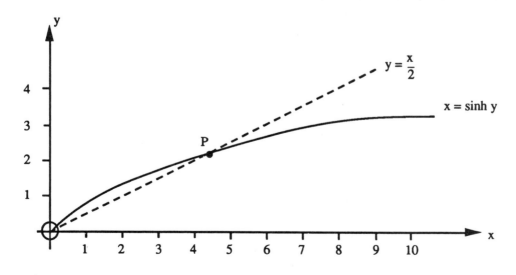

APPROXIMATIONS AND SERIES   PURE MATHEMATICS

Now $x = 2 \sinh^{-1} x$ gives $\dfrac{x}{2} = \sinh^{-1} x$

and the root of this equation is the point of intersection of the line $y = \dfrac{x}{2}$ and the curve $y = \sinh^{-1} x$. The dotted line on the graph is $y = \dfrac{x}{2}$ and P is, therefore, the root of the equation $x = 2 \sinh^{-1} x$ as requested ie $P \simeq 4.3$.

c) *Differentiation of hyperbolic functions*

(a) The hyperbolic functions have just been defined as:

$$\sinh x = \tfrac{1}{2}(e^x - e^{-x}), \qquad \cosh x = \tfrac{1}{2}(e^x + e^{-x}),$$

$$\therefore \tfrac{d}{dx}(\sinh x) = \tfrac{1}{2}(e^x - (-1)e^{-x}) \qquad \tfrac{d}{dx}(\cosh x) = \tfrac{1}{2}(e^x + (-1)e^{-x})$$

$$= \tfrac{1}{2}(e^x + e^{-x}) \qquad\qquad\qquad = \tfrac{1}{2}(e^x - e^{-x})$$

$$= \cosh x \qquad\qquad\qquad\qquad = \sinh x$$

> The derivative of $\sinh x$ ie $\tfrac{d}{dx}(\sinh x) = \cosh x$ and the derivative of $\cosh x$ ie $\tfrac{d}{dx}(\cosh x) = \sinh x$

Example 12

Differentiate:

(a) $\tanh x$

(b) $\coth x$

(c) $\operatorname{sech} x$

(d) $\operatorname{cosech} x$

Solution

(a) $y = \tanh x = \dfrac{\sinh x}{\cosh x}$

$$\therefore \dfrac{dy}{dx} = \dfrac{\cosh x (\cosh x) - \sinh x (\sinh x)}{(\cosh x)^2}$$

$$= \dfrac{\cosh^2 x - \sinh^2 x}{\cosh^2 x} \qquad \text{but } \cosh^2 x - \sinh^2 x = 1$$

$$= \dfrac{1}{\cosh^2 x}$$

$$\therefore \dfrac{dy}{dx} = \operatorname{sech}^2 x$$

(b) $y = \coth x = \dfrac{\cosh x}{\sinh x}$

$$\therefore \dfrac{dy}{dx} = \dfrac{\sinh x (\sinh x) - \cosh x (\cosh x)}{(\sinh x)^2}$$

PURE MATHEMATICS   APPROXIMATIONS AND SERIES

$$= \frac{\sinh^2 x - \cosh^2 x}{\sinh^2 x}$$

$$= \frac{-(\cosh^2 x - \sinh^2 x)}{\sinh^2 x}$$

$$= \frac{-1}{\sinh^2 x}$$

$$\therefore \quad \frac{dy}{dx} = -\operatorname{cosech}^2 x$$

(c) $y = \operatorname{sech} x = \dfrac{1}{\cosh x}$

$\therefore \quad y = (\cosh x)^{-1}$

$\therefore \quad \dfrac{dy}{dx} = -1(\cosh x)^{-2}(\sinh x)$

$$= -\frac{\sinh x}{\cosh^2 x}$$

$$= -\frac{\sinh x}{\cosh x} \cdot \frac{1}{\cosh x}$$

$$= -\tanh x \cdot \operatorname{sech} x$$

(d) $y = \operatorname{cosech} x = \dfrac{1}{\sinh x}$

$\therefore \quad y = (\sinh x)^{-1}$

$\therefore \quad \dfrac{dy}{dx} = -1(\sinh x)^{-2}(\cosh x)$

$$= -\frac{\cosh x}{\sinh^2 x}$$

$$= -\frac{\cosh x}{\sinh x} \cdot \frac{1}{\sinh x}$$

$\therefore \quad \dfrac{dy}{dx} = -\coth x \cdot \operatorname{cosech} x$

Example 13

If $y = A \cosh kx + B \sinh kx$, prove that $\dfrac{d^2y}{dx^2} = k^2 y$. Hence find y as a function of x given that $\dfrac{d^2y}{dx^2} = 4y$ and when $x = 0$, $y = 2$ and $\dfrac{dy}{dx} = 2$.

Solution

$$y = A\cosh kx + B\sinh kx$$

$$\therefore \frac{dy}{dx} = Ak\sinh kx + Bk\cosh kx$$

$$\therefore \frac{d^2y}{dx^2} = Ak \cdot k\cosh kx + Bk \cdot k\sinh kx$$

$$= Ak^2\cosh kx + Bk^2\sinh kx$$

$$= k^2(A\cosh kx + B\sinh kx)$$

$$\therefore \frac{d^2y}{dx^2} = k^2 y$$

but $\frac{d^2y}{dx^2} = 4y$ $\quad\therefore\quad k^2 = 4$ and $k = 2$

$k = 2$ gives $\quad y = A\cosh 2x + B\sinh 2x$

$x = 0, y = 2.$ $\quad\therefore\quad 2 = A\cosh 2(0) + B\sinh 2(0)$

But $\cosh 2x = \frac{1}{2}(e^{2x} + e^{-2x})$ and $\sinh 2x = \frac{1}{2}(e^{2x} - e^{-2x})$

$$\sinh 0 = \frac{1}{2}(1 - 1) = 0$$

$$\therefore\quad \cosh 0 = \frac{1}{2}(1 + 1) = 1$$

$$\therefore\quad 2 = A(1) + B(0)$$

$$\therefore\quad 2 = A$$

$x = 0, \frac{dy}{dx} = 2$ $\quad 2 = A(2)\sinh 2(0) + B(2)\cosh 2(0)$

$$\therefore\quad 2 = 4\sinh 0 + 2B\cosh 0$$

$$\therefore\quad 2 = 4(0) + 2B(1)$$

$$\therefore\quad 1 = B$$

So $y = 2\cosh 2x + \sinh 2x$

(b) Now to study the method of differentiating the inverse functions,

ie $\sinh^{-1}x$, $\cosh^{-1}x$, $\tanh^{-1}x$ etc.

(1) Let $y = \sinh^{-1}x$

then $\sinh y = x$

$$\therefore\quad \frac{d}{dy}(\sinh y) = 1$$

$$(\cosh y)\frac{dy}{dx} = 1$$

and $\quad \dfrac{dy}{dx} = \dfrac{1}{\cosh y} \quad$ but $\cosh^2 y - \sinh^2 y = 1$

$\therefore \quad \cosh^2 y = 1 + \sinh^2 y$

$\qquad\qquad\quad = 1 + x^2$

$\therefore \quad \cosh y = \pm\sqrt{1 + x^2}$

$\therefore \quad \dfrac{dy}{dx} = \dfrac{1}{\sqrt{1 + x^2}}$

Note: The negative sign has been dropped because $y = \sinh^{-1} x$ has a positive gradient for all values of x. $\therefore \dfrac{dy}{dx} > 0$ for all x.

(2)   Let $\quad y = \cosh^{-1} x$

then $\quad \cosh y = x$

$\therefore \quad \dfrac{d}{dx}(\cosh y) = 1$

$\therefore \quad (\sinh y)\dfrac{dy}{dx} = 1$

$\therefore \quad \dfrac{dy}{dx} = \dfrac{1}{\sinh y} \quad$ but $\sinh y = \pm\sqrt{x^2 - 1}$

$\therefore \quad \dfrac{dy}{dx} = \dfrac{\pm 1}{\sqrt{x^2 - 1}}$

Note: From the graph of $y = \cosh^{-1} x$ it can be seen that the gradient is positive for $y > 0$, ie $\cosh^{-1} x > 0$ and the gradient is negative for $y < 0$, ie $\cosh^{-1} x < 0$.

(3)   Let $\quad y = \tanh^{-1} x$

$\therefore \quad \tanh y = x$

$\therefore \quad \dfrac{d}{dx}(\tanh y) = 1$

$\therefore \quad (\text{sech}^2 y)\dfrac{dy}{dx} = 1$

$\therefore \quad \dfrac{dy}{dx} = \dfrac{1}{\text{sech}^2 y} \qquad \cosh^2 y - \sinh^2 y = 1$

$$\dfrac{\cosh^2 y}{\cosh^2 y} - \dfrac{\sinh^2 y}{\cosh^2 y} = \dfrac{1}{\cosh^2 y}$$

$\therefore \qquad 1 - \tanh^2 y = \text{sech}^2 y$

$\therefore \qquad 1 - x^2 = \text{sech}^2 y$

APPROXIMATIONS AND SERIES — PURE MATHEMATICS

$$\therefore \quad \frac{dy}{dx} = \frac{1}{1-x^2}$$

To summarise these results:

$$\frac{d}{dx}(\sinh^{-1} x) = \frac{1}{\sqrt{1+x^2}}$$

$$\frac{d}{dx}(\cosh^{-1} x) = \frac{1}{\sqrt{x^2-1}} \text{ when } \cosh^{-1} x > 0$$

$$\frac{d}{dx}(\cosh^{-1} x) = \frac{-1}{\sqrt{x^2-1}} \text{ when } \cosh^{-1} x < 0$$

$$\frac{d}{dx}(\tanh^{-1} x) = \frac{1}{1-x^2}$$

Example 14

Differentiate:

$\cosh^{-1}(\sinh 2x)$

Solution

Let $\quad y = \cosh^{-1}(\sinh 2x)$

$\therefore \quad \cosh y = \sinh 2x \quad$ differentiating

$$\frac{d}{dx}(\cosh y) = 2 \cdot \cosh 2x$$

$\therefore \quad (\sinh y)\dfrac{dy}{dx} = 2 \cosh 2x$

$\therefore \quad \dfrac{dy}{dx} = \dfrac{2 \cosh 2x}{\sinh y} \quad$ using $\cosh^2 y - \sinh^2 y = 1$

$\therefore \quad (\sinh 2x)^2 - 1 = \sinh^2 y$

$\quad \pm \sqrt{\sinh^2 2x - 1} = \sinh y$

$\therefore \quad \dfrac{dy}{dx} = \dfrac{\pm 2 \cosh 2x}{\sqrt{\sinh^2 2x - 1}}$

d) *Integration of hyperbolic functions*

$$\boxed{\int \cosh x \, dx = \sinh x + c \quad \int \sinh x \, dx = \cosh x + c}$$ — since integration is the reverse of differentiation

Some other results that might be useful when integrating are:

$\quad \cosh^2 x - \sinh^2 x = 1 \quad \text{---------} \quad (1)$

$\quad \cosh^2 x + \sinh^2 x = \cosh 2x \quad \text{---------} \quad (2)$

$\quad (1) + (2)$ gives $\quad 2\cosh^2 x = 1 + \cosh 2x$

$\quad (2) - (1)$ gives $\quad 2\sinh^2 x = \cosh 2x - 1$

$\quad$ also $\quad \sinh 2x = 2 \sinh x \cosh x$

PURE MATHEMATICS   APPROXIMATIONS AND SERIES

Example 15

Integrate the following functions:

(a) $\sinh^2 x$,   (b) $\cosh^3 x$,   (c) $e^x \cosh x$.

Solution

(a) $\quad \int \sinh^2 x \, dx = \int \frac{1}{2}(\cosh 2x - 1) dx$

$\qquad\qquad\qquad = \frac{1}{2}\left(\frac{\sinh 2x}{2} - x\right) + c$

$\qquad\qquad\qquad = \frac{1}{2}\left(\frac{\sinh 2x - 2x}{2}\right) + c$

$\therefore \int \sinh^2 x \, dx = \frac{1}{4}(\sinh 2x - 2x) + c$

(b) $\quad \int \cosh^3 x \, dx \qquad$ Let $\sinh x = u$

$\qquad\qquad \therefore \quad \cosh x = \frac{du}{dx}$

$\qquad\qquad \therefore \quad dx = \frac{du}{\cosh x}$

$\therefore \int \cosh^3 x \, dx = \int \cosh^3 x \left(\frac{du}{\cos dx}\right)$

$\qquad\qquad = \int \cosh^2 x \, du \quad$ but $\cosh^2 x = \sinh^2 x + 1 = u^2 + 1$

$\qquad\qquad = \int (u^2 + 1) \, du$

$\qquad\qquad = \frac{u^3}{3} + u + c$

$\therefore \int \cosh^3 x \, dx = \frac{\sinh^3 x}{3} + \sinh x + c$

(c) $\quad \int e^x \cosh x \, dx = \int e^x \left[\frac{1}{2}(e^x + e^{-x})\right] dx$

$\qquad\qquad\qquad = \frac{1}{2} \int (e^x \cdot e^x + e^x \cdot e^{-x}) \, dx$

$\qquad\qquad\qquad = \frac{1}{2} \int (e^{2x} + e^0) \, dx \qquad e^0 = 1$

$\qquad\qquad\qquad = \frac{1}{2}\left[\frac{e^{2x}}{2} + x\right] + c$

$\qquad\qquad\qquad = \frac{1}{2}\left[\frac{e^{2x} + 2x}{2}\right] + c$

$\therefore \int e^x \cosh x \, dx = \frac{1}{4}(e^{2x} + 2x) + c$

APPROXIMATIONS AND SERIES — PURE MATHEMATICS

It also follows that:

$$\int \frac{1}{\sqrt{1+x^2}}\,dx = \sinh^{-1}x + c$$

$$\int \frac{\pm 1}{\sqrt{x^2-1}}\,dx = \cosh^{-1}x + c$$

$$\int \frac{1}{1-x^2}\,dx = \tanh^{-1}x + c$$

Functions of the type $\sqrt{a^2 - b^2x^2}$ were integrated using the change of variable $x = \frac{a}{b}\sin u$ and led to answers involving $\sin^{-1}\left(\frac{bx}{a}\right)$. Functions of the type $a^2 + b^2x^2$ were integrated using the change of variable $x = \frac{a}{b}\tan u$ and led to answers involving $\tan^{-1}\left(\frac{bx}{a}\right)$. [cf chapter 5, page 187 et seq.]

Now integrals involving $\sqrt{b^2x^2 - a^2}$ and $\sqrt{a^2 + b^2x^2}$ can be integrated using substitutions involving hyperbolic functions, eg

$$\int \frac{1}{\sqrt{1+9x^2}}\,dx \qquad \text{let } 3x = \sinh u$$

$$\therefore \quad 3\frac{dx}{du} = \cosh u$$

$$\therefore \quad dx = \frac{\cosh u}{3}\,du$$

$$\therefore \int \frac{1}{\sqrt{1+9x^2}}\,dx = \int \frac{1}{\sqrt{1+\sinh^2 u}} \cdot \frac{\cosh u}{3}\,du \qquad \text{using } \cosh^2 u = 1 + \sinh^2 u$$

$$= \int \frac{1}{\cosh u} \cdot \frac{\cosh u}{3}\,du$$

$$= \int \frac{1}{3}\,du$$

$$= \frac{u}{3} + c \qquad \text{but if } 3x = \sinh u \qquad \text{then } u = \sinh^{-1}3x$$

$$\therefore \int \frac{1}{\sqrt{1+9x^2}}\,dx = \frac{1}{3}\sinh^{-1}3x + c$$

Considering the integral $\int \frac{1}{\sqrt{x^2-9}}\,dx$, the substitution needed this time is $\frac{x}{3} = \cosh u$

$$\therefore \quad x = 3\cosh u$$

$$\therefore \quad \frac{dx}{du} = 3\sinh u$$

$$\therefore \quad dx = 3\sinh u\,du$$

$$\therefore \int \frac{1}{\sqrt{x^2 - 9}} dx = \int \frac{1}{\sqrt{9 \cosh^2 u - 9}} \cdot 3 \sinh u \, du$$

$$= \int \frac{1}{3\sqrt{\cosh^2 u - 1}} \cdot 3 \sinh u \, du \text{ using } \cosh^2 u - 1 = \sinh^2 u$$

$$= \int \frac{1}{\sinh u} \cdot \sinh u \, du$$

$$= \int du$$

$$= u + c \qquad \text{but if } \frac{x}{3} = \cosh u \quad \text{then } u = \cosh^{-1} \frac{x}{3}$$

$$\therefore \int \frac{1}{\sqrt{x^2 - 9}} dx = \cosh^{-1} \frac{x}{3} + c$$

Therefore, the general results are:

$$\boxed{\begin{aligned} \int \frac{1}{\sqrt{a^2 + b^2 x^2}} dx &= \frac{1}{b} \sinh^{-1} \frac{bx}{a} + c \\ \int \frac{1}{\sqrt{b^2 x^2 - a^2}} dx &= \frac{1}{b} \cosh^{-1} \frac{bx}{a} + c \end{aligned}}$$

These two functions can now be integrated using the results quoted above, or they can be integrated from first principles as shown above.

For practice the following examples will be worked from first principles to give extra experience of these types of integral.

Example 16

Integrate the following functions using a hyperbolic substitution:

(a) $\int \sqrt{9x^2 + 4} \, dx$

(b) $\int (4x^2 - 1)^{\frac{3}{2}} dx$

Solution

(a) $\int \sqrt{9x^2 + 4} \, dx \qquad \text{let} \quad \frac{3x}{2} = \sinh u$

$$\therefore x = \frac{2}{3} \sinh u$$

$$\therefore \frac{dx}{du} = \frac{2}{3} \cosh u$$

$$\therefore dx = \frac{2}{3} \cosh u \, du$$

$$\therefore \int \sqrt{9x^2 + 4} \, dx = \int \left[\sqrt{9 \left(\frac{2}{3} \sinh u\right)^2 + 4} \right] \left(\frac{2}{3} \cosh u \, du\right)$$

APPROXIMATIONS AND SERIES — PURE MATHEMATICS

$$= \int \left[\sqrt{9\left(\tfrac{4}{9}\sinh^2 u\right) + 4}\,\right]\left(\tfrac{2\cosh u}{3}\,du\right)$$

$$= \int (2\sqrt{\sinh^2 u + 1})\,\tfrac{2\cosh u}{3}\,du; \quad \text{but as } \sinh^2 u + 1 = \cosh^2 u$$

$$= \int 2\cosh u \cdot \tfrac{2\cosh u}{3}\,du$$

$$= \int \tfrac{4}{3}\cosh^2 u\,du \quad \text{but } \cosh^2 u = \tfrac{1}{2}(1 + \cosh 2u)$$

Hence $\int \sqrt{9x^2 + 4}\,dx = \int \tfrac{4}{3}\cdot\tfrac{1}{2}(1 + \cosh 2u)\,du$

$$= \tfrac{2}{3}\left(u + \tfrac{1}{2}\sinh 2u\right) + c$$

but $\sinh 2u = 2\sin u \cosh u$

$$= 2\left(\tfrac{3x}{2}\right)\sqrt{1 + \left(\tfrac{3x}{2}\right)^2}$$

$$= 3x\sqrt{1 + \tfrac{9x^2}{4}}$$

$$\int \sqrt{9x^2 + 4}\,dx = \tfrac{2}{3}\left(\sinh^{-1}\tfrac{3x}{2} + \tfrac{1}{2}\left(3x\sqrt{1 + \tfrac{9x^2}{4}}\,\right)\right) + c$$

$$= \tfrac{2}{3}\sinh^{-1}\tfrac{3x}{2} + x\sqrt{1 + \tfrac{9x^2}{4}} + c$$

(b) $\int (4x^2 - 1)^{\tfrac{3}{2}}\,dx$ let $2x = \cosh u$

$$\therefore\quad x = \tfrac{1}{2}\cosh u$$

$$\therefore\quad \tfrac{dx}{du} = \tfrac{1}{2}\sinh u$$

$$\therefore\quad dx = \tfrac{1}{2}\sinh u\,du$$

$$(4x^2 - 1)^{\tfrac{3}{2}}\,dx = \int \left(4\left(\tfrac{1}{2}\cosh u\right)^2 - 1\right)^{\tfrac{3}{2}} \cdot \tfrac{1}{2}\sinh u\,du$$

$$= \int (\cosh^2 u - 1)^{\tfrac{3}{2}}\,\tfrac{1}{2}\sinh u\,du; \quad \text{but } \sinh^2 u = \cosh^2 u - 1$$

$$= \int (\sinh^2 u)^{\tfrac{3}{2}}\cdot\tfrac{1}{2}\sinh u\,du$$

$$= \int \sinh^3 u \cdot \tfrac{1}{2}\sinh u\,du$$

PURE MATHEMATICS — APPROXIMATIONS AND SERIES

$$= \int \tfrac{1}{2} \sinh^4 u \, du \quad \text{using } \sinh^2 u = \tfrac{1}{2}(\cosh 2u - 1)$$

$$\therefore \quad \sinh^4 u = \left(\tfrac{1}{2}(\cosh 2u - 1)\right)^2$$

$$= \int \tfrac{1}{2} \left(\tfrac{1}{2}(\cosh 2u - 1)\right)^2 du$$

$$= \int \tfrac{1}{2} \left(\tfrac{1}{4}(\cosh^2 2u - 2\cosh 2u + 1)\right) du$$

$$= \tfrac{1}{8} \int (\cosh^2 2u - 2\cosh 2u + 1) \, du$$

If $\cosh^2 u = \tfrac{1}{2}(1 + \cosh 2u)$ then $\cosh^2 2u = \tfrac{1}{2}(1 + \cosh 4u)$

$$\therefore \int (4x^2 - 1)^{\tfrac{3}{2}} dx = \tfrac{1}{8} \int \left(\tfrac{1}{2}(1 + \cosh 4u) - 2\cosh 2u + 1\right) du$$

$$= \tfrac{1}{8} \int \left(\tfrac{1}{2} + \tfrac{1}{2}\cosh 4u - 2\cosh 2u + 1\right) du$$

$$= \tfrac{1}{8} \int \left(\tfrac{3}{2} + \tfrac{1}{2}\cosh 4u - 2\cosh 2u\right) du$$

$$= \tfrac{1}{8} \left[\tfrac{3u}{2} + \tfrac{1}{2}\cdot\tfrac{\sinh 4u}{4} - \tfrac{2\sinh 2u}{2}\right] + c$$

$$= \tfrac{1}{8} \left[\tfrac{3u}{2} + \tfrac{\sinh 4u}{8} - \sinh 2u\right] + c$$

$$= \tfrac{1}{8} \left(\tfrac{3}{2}\cosh^{-1} 2x + \tfrac{2\sinh 2u \cosh 2u}{8} - \sinh 2u\right) + c$$

$$= \tfrac{3}{16} \cosh^{-1} 2x + \tfrac{1}{8}\sinh 2u \left(\tfrac{\cosh 2u}{4} - 1\right) + c$$

but $\sinh 2u = 2\sinh u \cosh u = 2\sqrt{4x^2 - 1} \cdot 2x = 4x\sqrt{4x^2 - 1}$

$\cosh 2u = \cosh^2 u + \sinh^2 u = 4x^2 + (\sqrt{4x^2-1})^2 = 4x^2 + 4x^2 - 1 = 8x^2 - 1$

$$\therefore \int (4x^2 - 1)^{\tfrac{3}{2}} dx = \tfrac{3}{16}\cosh^{-1} 2x + \tfrac{1}{8} \cdot 4x\sqrt{4x^2 - 1} \left(\tfrac{8x^2 - 1}{4} - 1\right) + c$$

$$= \tfrac{3}{16}\cosh^{-1} 2x + \tfrac{x}{8}\sqrt{4x^2 - 1}\,(8x^2 - 5) + c$$

Functions such as $\sqrt{x^2 + 4x + 5}$ have to be rewritten as $\sqrt{(x+2)^2 + 1}$ before integration, then $x + 2 = \sinh u$ is the necessary substitution and the method proceeds as in the other examples.

$\sqrt{x^2 + 4x + 3} = \sqrt{(x+2)^2 - 1}$ and then $x + 2 = \cosh u$

$\sqrt{x^2 + 6x + 5} = \sqrt{(x+3)^2 - 4}$ and then $\tfrac{x+3}{2} = \cosh u$

etc.

APPROXIMATIONS AND SERIES — PURE MATHEMATICS

e) *Example ('A' level question - part)*

Integrate $\dfrac{1}{\sqrt{9x^2 - 12x - 1}}$ with respect to x.

Solution

$9x^2 - 12x - 1$ needs re-arranging before any attempt at integration can be made.

Using the method of completing the square

$9x^2 - 12x - 1 = 9x^2 - 12x (+4 - 4) - 1 = (9x^2 - 12x + 4) - 4 - 1 = (3x - 2)^2 - 5$.

Hence $\displaystyle\int \dfrac{dx}{\sqrt{9x^2 - 12x - 1}} = \int \dfrac{dx}{\sqrt{(3x - 2)^2 - 5}}$

$\sqrt{(3x - 2)^2 - 5}$ implies a hyperbolic substitution.

Let $\dfrac{3x - 2}{\sqrt{5}} = \cosh u \qquad \therefore \quad u = \cosh^{-1}\left(\dfrac{3x - 2}{\sqrt{5}}\right)$

$\therefore \quad 3x - 2 = \sqrt{5} \cdot \cosh u$

$\therefore \quad x = \dfrac{\sqrt{5} \cdot \cosh u}{3} + \dfrac{2}{3}$

and $\dfrac{dx}{du} = \dfrac{\sqrt{5} \sinh u}{3}$

$\therefore \quad dx = \dfrac{\sqrt{5}}{3} \sinh u \, du$

Thus $\displaystyle\int \dfrac{dx}{\sqrt{9x^2 - 12x - 1}} = \int \dfrac{1}{\sqrt{(\sqrt{5} \cosh u)^2 - 5}} \left(\dfrac{\sqrt{5}}{3} \sinh u \, du\right)$

$\qquad = \displaystyle\int \dfrac{1}{\sqrt{5 \cosh^2 u - 5}} \cdot \dfrac{\sqrt{5}}{3} \sinh u \, du$

$\qquad = \displaystyle\int \dfrac{1}{\sqrt{5(\cosh^2 u - 1)}} \cdot \dfrac{\sqrt{5}}{3} \sinh u \, du \quad$ and as $\cosh^2 u - 1 = \sinh^2 u$

$\qquad = \displaystyle\int \dfrac{1}{\sqrt{5} \cdot (\sinh u)} \cdot \dfrac{\sqrt{5} (\sinh u)}{3} du$

$\qquad = \displaystyle\int \dfrac{du}{3}$

$\qquad = \dfrac{u}{3} + c$

$\therefore \quad \displaystyle\int \dfrac{dx}{\sqrt{9x^2 - 12x - 1}} = \dfrac{1}{3} \cosh^{-1}\left(\dfrac{3x - 2}{\sqrt{5}}\right) + c$

PURE MATHEMATICS — APPROXIMATIONS AND SERIES

7.3 Further work on series

a) *Σ notation*

$1^2 + 2^2 + 3^2 + \ldots + n^2$ can be written more briefly as $\sum_{1}^{n} r^2$ which means the sum of all the terms - r^2 - from $r = 1$ to $r = n$. So $1^3 + 2^3 + 3^3 + \ldots + n^3 = \sum_{1}^{n} r^3$, ie the sum of all the terms - r^3 - from $r = 1$ to $r = n$.

Example 17

Write the following out in full:

(a) $\sum_{1}^{4} r^2$

(b) $\sum_{1}^{3} \frac{1}{r(r+1)}$

(c) $\sum_{2}^{5} 2^r$

(d) $\sum_{1}^{n} r^r$

Solution

(a) $\sum_{1}^{4} r^2 = 1^2 + 2^2 + 3^2 + 4^2$

(b) $\sum_{1}^{3} \frac{1}{r(r+1)} = \frac{1}{1(1+1)} + \frac{1}{2(2+1)} + \frac{1}{3(3+1)} = \frac{1}{1.2} + \frac{1}{2.3} + \frac{1}{3.4}$

(c) $\sum_{2}^{5} 2^r = 2^2 + 2^3 + 2^4 + 2^5$

(d) $\sum_{1}^{n} r^r = 1^1 + 2^2 + 3^3 + \ldots + n^n = 1 + 2^2 + 3^3 + \ldots + n^n$

Example 18

Write the following in Σ notation:

(a) $1 + 2 + 3 + \ldots + n$

(b) $3^2 + 3^3 + 3^4 + 3^5$

(c) $2(7) + 3(8) + 4(9) + 5(10) + 6(11)$

(d) $1^4 + 2^4 + 3^4 + \ldots + n^4 + (n + 1)^4$

APPROXIMATIONS AND SERIES PURE MATHEMATICS

Solution

(a) $\quad 1 + 2 + 3 + \ldots + n = \sum_{1}^{n} r$

(b) $\quad 3^2 + 3^3 + 3^4 + 3^5 = \sum_{2}^{5} 3^r$

(c) $\quad 2(7) + 3(8) + 4(9) + 5(10) + 6(11) = \sum_{2}^{6} r(r + 5)$

(d) $\quad 1^4 + 2^4 + 3^4 + \ldots + n^4 + (n + 1)^4 = \sum_{1}^{n+1} r^4$

Two other important points about the Σ notation are:

(a) $\quad \sum_{1}^{n} ar \text{ (where a is a constant)} = a \sum_{1}^{n} r$

(b) $\quad \sum_{1}^{n} (r + r^2) = \sum_{1}^{n} r + \sum_{1}^{n} r^2$ - ie the terms can be summed separately.

b) *Proof by induction*

Sometimes a result can be found by a method which is not, in itself, a proof.

In 'proof by induction' it is shown that if the result holds for some particular value of n - say, k - then it also holds for n = k + 1. It is then verified that the result does hold for some particular value of n - usually 1 or 2.

This may sound complicated but the general method to follow will be detailed in the following examples.

Example 19

Prove by induction that: $\quad \dfrac{1}{1.2} + \dfrac{1}{2.3} + \dfrac{1}{3.4} + \ldots + \dfrac{1}{n(n + 1)} = \dfrac{n}{n+1}$

Solution

Start by assuming this result is true for some value of n - say, k, ie

$$\dfrac{1}{1.2} + \dfrac{1}{2.3} + \ldots + \dfrac{1}{k(k + 1)} = \dfrac{k}{k + 1}$$

Then add the next term to each side of the sum giving:

$$\dfrac{1}{1.2} + \dfrac{1}{2.3} + \ldots + \dfrac{1}{k(k + 1)} + \dfrac{1}{(k + 1)(k + 2)} = \dfrac{k}{k + 1} + \dfrac{1}{(k + 1)(k + 2)}$$

$$= \dfrac{k(k + 2) + 1}{(k + 1)(k + 2)}$$

$$= \dfrac{k^2 + 2k + 1}{(k + 1)(k + 2)}$$

$$= \dfrac{(k + 1)^2}{(k + 1)(k + 2)} \text{ cancelling by } (k + 1)$$

$\therefore \quad \dfrac{1}{1.2} + \dfrac{1}{2.3} + \ldots + \dfrac{1}{(k+1)(k+2)}$

This is the original expression $\left(\dfrac{n}{n+1}\right)$ with n = k + 1. Therefore, if the result is true for n = k it is also true for n = k + 1. So trying n = 1

$$\text{Left hand side (LHS)} = \dfrac{1}{1.2} = \dfrac{1}{2}$$
$$\text{Right hand side (RHS)} = \dfrac{1}{1+1} = \dfrac{1}{2}$$

$\} \therefore$ LHS = RHS

So the result is true for n = 1 and by induction it is also true for n = 2. Since it is true for n = 2 by induction it is also true for n = 3 and so on for all integral values of n.

This then is the 'proof of induction' method.

The following examples are all worked out in exactly the same way.

Example 20

Prove by induction $\sum\limits_{1}^{n} r(r!) = (n+1)! - 1$

Solution

Assuming this result is true for some value of n - k, then:

$$\sum_{1}^{k} r(r!) = (k+1)! - 1$$

ie $\quad 1(1!) + 2(2!) + 3(3!) + \ldots + k(k!) = (k+1)! - 1$

Adding the next term (k + 1)(k + 1)! to both sides gives:

$1(1!) + 2(2!) + 3(3!) + \ldots + k(k!) + (k+1)(k+1)! = (k+1)! - 1 + (k+1)(k+1)!$

$\qquad\qquad\qquad\qquad\qquad\qquad\qquad = (k+1)! + (k+1)(k+1)! - 1$

$\qquad\qquad\qquad\qquad\qquad\qquad\qquad = (k+1)!(1 + k + 1) - 1$

$\qquad\qquad\qquad\qquad\qquad\qquad\qquad = (k+1)!(k+2) - 1$

$\therefore \quad \sum\limits_{1}^{k+1} r(r!) = (k+2)! - 1$

This is the original expression with n = k + 1. Therefore, if the formula is true for n = k it is also true for n = k + 1. So trying n = 1

$\qquad\qquad$ LHS = 1(1!) = 1 $\qquad$ RHS = (1 + 1)! - 1 = 2! - 1 = 1

$\qquad \therefore \quad$ LHS = RHS

So the result is true for n = 1 and by induction it is true for n = 2. Since it is true for n = 2, by induction it is true for n = 3 and so on for all integral values of n.

APPROXIMATIONS AND SERIES — PURE MATHEMATICS

Example 21

Prove by induction $1^2 + 2^2 + 3^3 + \ldots + n^2 = \dfrac{n^3}{3} + \dfrac{n^2}{2} + \dfrac{n}{6}$

Solution

Assuming this result is true for some value of n - k, say:

$$1^2 + 2^2 + 3^2 + \ldots + k^2 = \dfrac{k^3}{3} + \dfrac{k^2}{2} + \dfrac{k}{6}$$

$$= \dfrac{2k^3 + 3k^2 + k}{6}$$

Adding the next term $(k + 1)^2$ to both sides gives:

$$1^2 + 2^2 + 3^2 + \ldots + k^2 + (k+1)^2 = \dfrac{2k^3 + 3k^2 + k}{6} + (k+1)^2$$

$$= \dfrac{2k^3 + 3k^2 + k}{6} + k^2 + 2k + 1$$

$$= \dfrac{2k^3 + 3k^2 + k + 6k^2 + 12k + 6}{6}$$

$$= \dfrac{2k^3 + 9k^2 + 13k + 6}{6}$$

It is now necessary to rearrange this expression to obtain the form required for the answer.

$\therefore \quad 1^2 + 2^2 + 3^2 + \ldots + k^2 + (k+1)^2 = \dfrac{2k^3 + (6k^2 + 3k^2) + (6k + 6k + k) + (2, + 3 + 1)}{6}$

$$= \dfrac{2k^3 + (6k^2 + 6k + 2) + (3k^2 + 6k + 3) + (k + 1)}{6}$$

$$= \dfrac{2(k^3 + 3k^2 + 3k + 1) + 3(k^2 + 2k + 1) + (k + 1)}{6}$$

$$= \dfrac{2(k+1)^3}{6} + \dfrac{3(k+1)^2}{6} + \dfrac{k+1}{6}$$

$\therefore \quad 1^2 + 2^2 + 3^2 + \ldots + k^2 + (k+1)^2 = \dfrac{(k+1)^3}{6} + \dfrac{(k+1)^2}{2} + \dfrac{k+1}{6}$

This is the original expression with n = k + 1. Therefore, if the formula is true for n = k, it is also true for n = k + 1. So, trying n = 1

$\quad$ LHS $= 1^2 = 1 \quad$ RHS $= \dfrac{1^3}{3} + \dfrac{1^2}{2} + \dfrac{1}{6} = \dfrac{2 + 3 + 1}{6} = \dfrac{6}{6} = 1$

$\quad \therefore \quad$ LHS $=$ RHS

So the result is true for n = 1 and, by induction, it is also true for n = 2. Since it is true for n = 2, by induction it is true for n = 3 and so on for all integral values of n.

PURE MATHEMATICS — APPROXIMATIONS AND SERIES

c) *Summation of series using set expansions*

Before continuing further it is worth listing again the series expansions that were given in earlier work:

$$e^x = 1 + x + \frac{x^2}{2!} + \frac{x^3}{3!} + \frac{x^4}{4!} + \ldots + \frac{x^n}{n!} + \ldots$$

This is valid for all values of x

$$\log_e(1+x) = x - \frac{x^2}{2} + \frac{x^3}{3} - \ldots$$

This is valid for values of x in the range $-1 < x \leq +1$

$$\log_e(1-x) = -x - \frac{x^2}{2} - \frac{x^3}{3} - \ldots$$

This is valid for values of x in the range $-1 < x < +1$

$$\sin x = x - \frac{x^3}{3!} + \frac{x^5}{5!} - \frac{x^7}{7!} + \ldots$$

$$\cos x = 1 - \frac{x^2}{2!} + \frac{x^4}{4!} - \frac{x^6}{6!} + \ldots$$

Both of these are valid for all values of x. Three other useful results are:

The sum of the first n integers	$\sum_{1}^{n} r = \frac{n(n+1)}{2}$	.... (i)
The sum of the squares of the first n integers	$\sum_{1}^{n} r^2 = \frac{n(n+1)(2n+1)}{6}$	.... (ii)
The sum of the cubes in the first n integers	$\sum_{1}^{n} r^3 = \frac{n^2(n+1)^2}{4}$	.... (iii)

(i) $\quad \sum_{1}^{n} r = 1 + 2 + 3 + 4 + \ldots + n \qquad$ - this is an AP

with $a = 1$, $d = 1$ using $\quad S_n = \frac{n}{2}(2a + (n-1)d)$

$$= \frac{n}{2}(2(1) + (n-1)1)$$

$$= \frac{n}{2}(2 + n - 1)$$

$\therefore \quad \sum_{1}^{n} r = \frac{n(n+1)}{2} \qquad$ as above

(ii) $\quad \sum_{1}^{n} r^2 = 1^2 + 2^2 + 3^2 + \ldots + n^2 = \frac{n(n+1)(2n+1)}{6}$

as was proved by induction in the previous example.

Note:

$$\frac{n(n+1)(2n+1)}{6} = \frac{n(2n^2 + 3n + 1)}{6}$$

$$= \frac{2n^3 + 3n^2 + n}{6} \quad \text{was used in the previous example.}$$

(iii) $\sum_{1}^{n} r^3 = 1^3 + 2^3 + 3^3 + \ldots + n^3 = \frac{n^2(n+1)^2}{4}$

will be proved by induction in the next example and then all three results will be used.

Example 22

Use the method of induction to prove $\sum_{1}^{n} r^3 = \frac{n^2(n+1)^2}{4}$

Evaluate $3^3 + 5^3 + \ldots + (2n+1)^3$

Solution

Assuming the result is true for some value of n - k say, then:

$$\sum_{1}^{k} r^3 = \frac{k^2(k+1)^2}{4}$$

$$1^3 + 2^3 + 3^3 + \ldots + k^3 = \frac{k^2(k+1)^2}{4}$$

Adding the next term $(k+1)^3$ to both sides gives:

$$1^3 + 2^3 + 3^3 + \ldots + k^3 + (k+1)^3 = \frac{k^2(k+1)^2}{4} + (k+1)^3$$

$$= (k+1)^2 \left[\frac{k^2}{4} + (k+1) \right]$$

$$= (k+1)^2 \left[\frac{k^2 + 4k + 4}{4} \right]$$

$$\therefore \sum_{1}^{k+1} r^3 = \frac{(k+1)^2(k+2)^2}{4}$$

This is the original expression with $n = k + 1$. Therefore, if the formula is true for $n = k$ it is also true for $n = k + 1$. So trying $n = 1$.

$$\text{LHS} = 1^3 = 1 \qquad \text{RHS} = \frac{1^2(1+1)^2}{4} = \frac{1(4)}{4} = 1$$

$$\therefore \quad \text{LHS} = \text{RHS}$$

So the result is true for $n = 1$ and by induction it is true for $n = 2$. Since it is true for $n = 2$, by induction it is true for $n = 3$ and so for all integral values of n.

PURE MATHEMATICS — APPROXIMATIONS AND SERIES

$3^3 + 5^3 + ... + (2n + 1)^3$ can be written in Σ notation as

$$\sum_1^n (2r + 1)^3 = \sum_1^n [(2r)^3 + 3(2r)^2 + 3(2r) + 1]$$

$$= \sum_1^n [8r^3 + 12r^2 + 6r + 1]$$

$$\sum_1^n (2r + 1)^3 = \sum_1^n 8r^3 + \sum_1^n 12r^2 + \sum_1^n 6r + \sum_1^n 1$$

but

$$\sum_1^n 8r^3 = 8 \sum_1^n r^3 = \frac{8n^2 (n + 1)^2}{4}$$

$$\sum_1^n 12r^2 = 12 \sum_1^n r^2 = \frac{12n(n + 1)(2n + 1)}{6}$$

$$\sum_1^n 6r = 6 \sum_1^n r = \frac{6n(n + 1)}{2}$$

$$\sum_1^n 1 = 1 + 1 + 1 + ... + 1 = n$$

$$\therefore \sum_1^n (2r + 1)^3 = \frac{8n^2(n + 1)^2}{4} + \frac{12n(n + 1)(2n + 1)}{6} + \frac{6n(n + 1)}{2} + n$$

$$= 2n^2(n + 1)^2 + 2n(n + 1)(2n + 1) + 3n(n + 1) + n$$

$$= 2n^2(n^2 + 2n + 1) + 2n(2n^2 + 3n + 1) + 3n^2 + 3n + n$$

$$= 2n^4 + 4n^3 + 2n^2 + 4n^3 + 6n^2 + 2n + 3n^2 + 3n + n$$

$$\therefore \sum_1^n (2r + 1)^3 = 2n^4 + 8n^3 + 11n^2 + 6n$$

Example 23

Evaluate $\sum_1^n [3^r + (r - 1)(r + 1)]$

Solution

$$\sum_1^n [3^r + (r - 1)(r + 1)] = \sum_1^n [3^r + r^2 - 1]$$

$$= \sum_1^n 3^r + \sum_1^n r^2 - \sum_1^n 1$$

APPROXIMATIONS AND SERIES — PURE MATHEMATICS

$$\sum_{1}^{n} 3^r = 3 + 3^2 + 3^3 + \ldots + 3^n \quad \text{- a GP with } a = 3, r = 3$$

Using $S_n = \dfrac{a(r^n - 1)}{(r - 1)}$ then $\sum_{1}^{n} 3^r = \dfrac{3(3^n - 1)}{(3 - 1)} = \dfrac{3(3^n - 1)}{2}$

$$\sum_{1}^{n} r^2 = \dfrac{n(n + 1)(2n + 1)}{6}$$

$$\sum_{1}^{n} 1 = 1 + 1 + 1 + \ldots + 1 = n$$

$$\sum_{1}^{n} [3^r + (r - 1)(r + 1)] = \dfrac{3(3^n - 1)}{2} + \dfrac{n(n + 1)(2n + 1)}{6} - n$$

This cannot easily be simplified and so it will be left in this form.

d) *Summation of series using partial fractions*

Partial fractions can be used to help sum series such as $\sum_{1}^{n} \dfrac{1}{r^2 + 3r}$. The same rules apply that were outlined earlier.

Example 24

The rth term of a series is $\dfrac{1}{r^2 + 3r}$. Find the sum of the first n terms of the series and hence find its sum to infinity.

Solution

The sum of the first n terms $= \sum_{1}^{n} \dfrac{1}{r^2 + 3r}$

This must be expressed in partial fractions before any attempt at summation can be made.

$$\therefore \quad \dfrac{1}{r(r + 3)} \equiv \dfrac{A}{r} + \dfrac{B}{r + 3}$$

Multiplying by $r(r + 3)$ $\quad 1 \equiv A(r + 3) + Br$

Let $\quad r = 0 \quad\quad 1 = A(0 + 3) + B(0) \quad \therefore \quad A = \dfrac{1}{3}$

Let $\quad r = -3 \quad\quad 1 = \dfrac{1}{3}(0) + B(-3) \quad \therefore \quad B = -\dfrac{1}{3}$

So $\quad \dfrac{1}{r^2 + 3r} = \dfrac{1}{3r} - \dfrac{1}{3(r + 3)} = \dfrac{1}{3}\left(\dfrac{1}{r} - \dfrac{1}{r + 3}\right)$

PURE MATHEMATICS — APPROXIMATIONS AND SERIES

Summing this series for values of r from 1 to n gives the sum of the first n terms:

$$\sum_{1}^{n} \frac{1}{r^2 + 3r} = \sum_{1}^{n} \frac{1}{3}\left(\frac{1}{r} - \frac{1}{r+3}\right)$$

$$= \frac{1}{3} \sum_{1}^{n} \left(\frac{1}{r} - \frac{1}{r+3}\right)$$

$$= \frac{1}{3}\left[\left(\frac{1}{1} - \frac{1}{1+3}\right) + \left(\frac{1}{2} - \frac{1}{2+3}\right) + \left(\frac{1}{3} - \frac{1}{3+3}\right) + \ldots + \left(\frac{1}{n} - \frac{1}{n+3}\right)\right]$$

$$= \frac{1}{3}\left[\left(1 - \frac{1}{4}\right) + \left(\frac{1}{2} - \frac{1}{5}\right) + \left(\frac{1}{3} - \frac{1}{6}\right) + \left(\frac{1}{4} - \frac{1}{7}\right) + \left(\frac{1}{5} - \frac{1}{8}\right) \ldots + \left(\frac{1}{n} - \frac{1}{n+3}\right)\right]$$

Most of the terms cancel out leaving the first three and last three terms:

$$\sum_{1}^{n} \frac{1}{r^2 + 3r} = \frac{1}{3}\left[1 + \frac{1}{2} + \frac{1}{3} - \frac{1}{(n-2)+3} - \frac{1}{(n-1)+3} - \frac{1}{n+3}\right]$$

$$= \frac{1}{3}\left[1 + \frac{1}{2} + \frac{1}{3} - \frac{1}{n+1} - \frac{1}{n+2} - \frac{1}{n+3}\right]$$

$$= \frac{1}{3}\left[\frac{6+3+2}{6} - \frac{1}{n+1} - \frac{1}{n+2} - \frac{1}{n+3}\right]$$

$$\therefore \quad \sum_{1}^{n} \frac{1}{r^2 + 3r} = \frac{1}{3}\left[\frac{11}{6} - \left(\frac{1}{n+1} + \frac{1}{n+2} + \frac{1}{n+3}\right)\right]$$

As $n \to \infty$, $\frac{1}{n+1} \to 0$, $\frac{1}{n+2} \to 0$, $\frac{1}{n+3} \to 0$

$$\therefore \quad \sum_{1}^{\infty} \frac{1}{r^2 + 3r} = \frac{1}{3}\left(\frac{11}{6} - 0\right) = \frac{11}{18}$$

Example 25

Given that $S_n = \sum_{1}^{n} \frac{1}{(3r-2)(3r+1)}$ find:

(a) S_n in terms of n

(b) $\lim_{n \to \infty} S_n$

APPROXIMATIONS AND SERIES — PURE MATHEMATICS

Solution

(a) $$S_n = \sum_{1}^{n} \frac{1}{(3r-2)(3r+1)}$$

Using partial fractions to simplify this expression:

$$\frac{1}{(3r-2)(3r+1)} \equiv \frac{A}{3r-2} + \frac{B}{3r+1}$$

Multiplying $\quad 1 \equiv A(3r+1) + B(3r-2)$

Let $r = -\frac{1}{3}$ $\quad 1 = A(0) + B(-1-2) \quad \therefore B = -\frac{1}{3}$

Let $r = \frac{2}{3}$ $\quad 1 = A(2+1) + B(0) \quad \therefore A = \frac{1}{3}$

$$\therefore \sum_{1}^{n} \frac{1}{(3r-2)(3r+1)} = \sum_{1}^{n} \left(\frac{1}{3(3r-2)} - \frac{1}{3(3r+1)} \right)$$

$$= \sum_{1}^{n} \frac{1}{3} \left[\frac{1}{3r-2} - \frac{1}{3r+1} \right]$$

$$= \frac{1}{3} \sum_{1}^{n} \left[\frac{1}{3r-2} - \frac{1}{3r+1} \right]$$

$$\therefore \sum_{1}^{n} \frac{1}{(3r-2)(3r+1)} =$$

$$\frac{1}{3}\left[\left(\frac{1}{3.1-2} - \frac{1}{3.1+1}\right) + \left(\frac{1}{3.2-2} - \frac{1}{3.2+1}\right) + \left(\frac{1}{3.3-2} - \frac{1}{3.3+1} + \ldots + \frac{1}{3n-2} - \frac{1}{3n+1}\right) \right]$$

$$= \frac{1}{3}\left[\left(\frac{1}{1} - \frac{1}{4}\right) + \left(\frac{1}{4} - \frac{1}{7}\right) + \left(\frac{1}{7} - \frac{1}{9}\right) + \ldots + \left(\frac{1}{3n-2} - \frac{1}{3n+1}\right) \right]$$

Most of the terms cancel out leaving

$$S_n = \frac{1}{3}\left[1 - \frac{1}{3n+1} \right]$$

(b) As $n \to \infty$, $\frac{1}{3n+1} \to 0$ and $S_n \to \frac{1}{3}[1-0]$

$$\therefore \lim_{n \to \infty} S_n = \frac{1}{3}$$

PURE MATHEMATICS — APPROXIMATIONS AND SERIES

e) *MacLaurin's and Taylor's theorems*

As has already been seen many functions can be expressed as a series expansion. These include $\log_e(1 + x)$, e^x, $\sin x$, $\cos x$ ... etc. However, the expansions have just been stated each time without any attempt being made to prove them.

The theorem about to be derived - MacLaurin's theorem - is one method that can be used to obtain a series expansion of a function of x - f(x).

First it is necessary to assume that f(x) can be expanded as a power series of x, ie

$$f(x) = a_0 + a_1 x + a_2 x^2 + a_3 x^3 + a_4 x^4 + a_5 x^5 + \ldots$$

Therefore, the problem becomes one of finding values for the coefficients $a_0, a_1, a_2, a_3 \ldots$ etc.

If $x = 0$ \quad $f(0) = a_0 + a_1(0) + a_2(0) + \ldots \quad = a_0$

Differentiating \quad $f'(x) = a_1 + 2a_2 x + 3a_3 x^2 + 4a_4 x^3 + 5a_5 x^4 + \ldots$

If $x = 0$ \quad $f'(0) = a_1 + 2a_2(0) + 3a_3(0) + \ldots \quad = a_1$

Differentiating again \quad $f''(x) = 2a_2 + 6a_3 x + 12a_4 x^2 + 20a_5 x^3 + \ldots$

If $x = 0$ \quad $f''(0) = 2a_2 + 6a_3(0) + 12a_4(0) + \ldots \quad = 2a_2$

Differentiating again \quad $f'''(x) = 6a_3 + 24a_4 x + 60a_5 x^2 + \ldots$

If $x = 0$ \quad $f'''(0) = 6a_3 + 24a_4(0) + 60a_5(0) \quad = 6a_3$

Differentiating again \quad $f''''(x) = 24a_4 + 120a_5 x + \ldots$

If $x = 0$ \quad $f''''(0) = 24a_4 + 120a_5(0) + \ldots \quad = 24a_4$

etc.

$\therefore \quad a_0 = f(0)$

$\therefore \quad a_1 = f'(0)$

$\therefore \quad 2a_2 = f''(0) \quad \therefore \quad a_2 = \dfrac{f''(0)}{2!}$

$\therefore \quad 6a_3 = f'''(0) \quad \therefore \quad a_3 = \dfrac{f'''(0)}{3!}$

$\therefore \quad 24a_4 = f''''(0) \quad \therefore \quad a_4 = \dfrac{f''''(0)}{4!}$

$\therefore \quad \boxed{f(x) = f(0) + f'(0)x + \dfrac{f''(0)}{2!}x^2 + \dfrac{f'''(0)}{3!}x^3 + \dfrac{f''''(0)}{4!}x^4 + \ldots}$

APPROXIMATIONS AND SERIES   PURE MATHEMATICS

This is known as a MacLaurin expansion and it will now be used to derive series expansions for:

(a) e^x;

(b) $\log_e(1+x)$;

(c) $\cos x$.

(a) $f(x) = e^x$ $\therefore$ $f(0) = e^0 = 1$

$f'(x) = e^x$ $\therefore$ $f'(0) = e^0 = 1$

$f''(x) = e^x$ $\therefore$ $f''(0) = e^0 = 1$

etc.

$\therefore \quad e^x = 1 + 1x + \frac{1}{2!}x^2 + \frac{1}{3!}x^3 + \frac{1}{4!}x^4 + \ldots$

$\qquad\qquad = 1 + x + \frac{x^2}{2!} + \frac{x^3}{3!} + \frac{x^4}{4!} + \ldots$ etc.

(b) $f(x) = \log_e(1+x)$ $\therefore$ $f(0) = \log_e(1+0) = 0$

$f'(x) = \frac{1}{1+x} = (1+x)^{-1}$ $\therefore$ $f'(0) = \frac{1}{1+0} = 1$

$f''(x) = \frac{-1}{(1+x)^2} = -(1+x)^{-2}$ $\therefore$ $f''(0) = \frac{-1}{(1+0)^2} = -1$

$f'''(x) = \frac{2}{(1+x)^3} = 2(1+x)^{-3}$ $\therefore$ $f'''(0) = \frac{2}{(1+0)^3} = 2$

$f''''(x) = \frac{-6}{(1+x)^{-4}} = -6(1+x)^{-4}$ $\therefore$ $f''''(0) = \frac{-6}{(1+0)^4} = -6 = -3!$

etc.

$\therefore \quad \log_e(1+x) = 0 + 1x + \frac{-1}{2!}x^2 + \frac{2}{3!}x^3 + \frac{-3!}{4!}x^4 + \ldots$

$\qquad\qquad = x - \frac{x^2}{2} + \frac{x^3}{3} - \frac{x^4}{4} + \ldots$ etc.

(c) $f(x) = \cos x$ $\therefore$ $f(0) = \cos 0 = 1$

$f'(x) = -\sin x$ $\therefore$ $f'(0) = -\sin 0 = 0$

$f''(x) = -\cos x$ $\therefore$ $f''(0) = -\cos 0 = -1$

$f'''(x) = \sin x$ $\therefore$ $f'''(0) = \sin 0 = 0$

$f''''(x) = \cos x$ $\therefore$ $f''''(0) = \cos 0 = 1$

etc.

$\therefore \quad \cos x = 1 + 0x + \frac{-1}{2!}x^2 + \frac{0}{3!}x^3 + \frac{1}{4!}x^4 + \ldots$

$\qquad\qquad = 1 - \frac{x^2}{2!} + \frac{x^4}{4!} - \ldots$ etc.

PURE MATHEMATICS — APPROXIMATIONS AND SERIES

MacLaurin's theorem cannot be applied to all problems - for example $\log_e x$ cannot be expanded as a series because $f(0) = \log_e 0$ cannot be evaluated. Some problems can be overcome by expanding $f(x)$ as a series of ascending powers of $(x - a)$.

$$\therefore \quad f(x) = a_0 + a_1(x-a) + a_2(x-a)^2 + a_3(x-a)^3 + a_4(x-a)^4 + \ldots$$

Using a method similar to that for obtaining MacLaurin's expansion gives the following expansion known as Taylor's Expansion.

$$\boxed{f(x) = f(a) + f'(a)(x-a) + \frac{f''(a)}{2!}(x-a)^2 + \frac{f'''(a)}{3!}(x-a)^3 + \frac{f''''(a)}{4!}(x-a)^4 + \ldots}$$

Example 26

Expand $\log_e x$ as a series of powers of $(x - a)$ up to the term containing $(x - a)^3$.

Use this expansion to evaluate $\log_e 1.01$ correct to 5 decimal places.

Solution

$$f(x) = \log_e x \qquad \therefore \quad f(a) = \log_e a$$

$$f'(x) = \frac{1}{x} = x^{-1} \qquad \therefore \quad f'(a) = \frac{1}{a}$$

$$f''(x) = \frac{-1}{x^2} = x^{-2} \qquad \therefore \quad f''(a) = \frac{-1}{a^2}$$

$$f'''(x) = \frac{2}{x^3} = 2x^{-3} \qquad \therefore \quad f'''(a) = \frac{2}{a^3}$$

$$\therefore \quad \log_e x = \log_e a + \frac{1}{a}(x-a) + \frac{-1}{a^2}\frac{(x-a)^2}{2!} + \frac{2}{a^3}\frac{(x-a)^3}{3!} \ldots$$

$$= \log_e a + \frac{x-a}{a} - \frac{(x-a)^2}{a^2 2!} + \frac{2(x-a)^3}{a^3 3!} \ldots$$

If $x = 1.01$ and $a = 1$ then $x - a = 1.01 - 1 = 0.01$

$$\therefore \quad \log_e 1.01 = \log_e 1 + \frac{0.01}{1} - \frac{(0.01)^2}{1^2 .2!} + \frac{2(0.01)^3}{1^3 .3!} - \ldots$$

$$= 0 + 0.01 - 0.00005 + 0.0000003 - \ldots$$

$$\therefore \quad \log_e 1.01 = 0.00995 \quad \text{(to 5 decimal places)}$$

APPROXIMATIONS AND SERIES — PURE MATHEMATICS

f) *Expansion of hyperbolic functions*

Since sinh x and cosh x are functions of e^x they can be expanded in a series of powers of x.

$$\sinh x = \tfrac{1}{2}(e^x - e^{-x})$$

$$= \tfrac{1}{2}\left[1 + x + \tfrac{x^2}{2!} + \tfrac{x^3}{3!} + \tfrac{x^4}{4!} + \tfrac{x^5}{5!} + \ldots - \left(1 + (-x) + \tfrac{(-x)^2}{2!} + \tfrac{(-x)^3}{3!} + \tfrac{(-x)^4}{4!} + \tfrac{(-x)^5}{5!}\right)\right]$$

$$= \tfrac{1}{2}\left[1 + x + \tfrac{x^2}{2!} + \tfrac{x^3}{3!} + \tfrac{x^4}{4!} + \tfrac{x^5}{5!} + \ldots - 1 + x - \tfrac{x^2}{2!} + \tfrac{x^3}{3!} - \tfrac{x^4}{4!} + \tfrac{x^5}{5!} \ldots\right]$$

$$= \tfrac{1}{2}\left[2x + \tfrac{2x^3}{3!} + \tfrac{2x^5}{5!} + \ldots\right]$$

$$\therefore \boxed{\sinh x = x + \tfrac{x^3}{3!} + \tfrac{x^5}{5!} + \ldots}$$

$$\cosh x = \tfrac{1}{2}(e^x + e^{-x})$$

$$= \tfrac{1}{2}\left[1 + x + \tfrac{x^2}{2!} + \tfrac{x^3}{3!} + \tfrac{x^4}{4!} + \tfrac{x^5}{5!} + \ldots + \left(1 + (-x) + \tfrac{(-x)^2}{2!} + \tfrac{(-x)^3}{3!} + \tfrac{(-x)^4}{4!} + \tfrac{(-x)^5}{5!}\right)\right]$$

$$= \tfrac{1}{2}\left[1 + x + \tfrac{x^2}{2!} + \tfrac{x^3}{3!} + \tfrac{x^4}{4!} + \tfrac{x^5}{5!} + \ldots + 1 - x + \tfrac{x^2}{2!} - \tfrac{x^3}{3!} + \tfrac{x^4}{4!} - \tfrac{x^5}{5!} + \ldots\right]$$

$$= \tfrac{1}{2}\left[2 + \tfrac{2x^2}{2!} + \tfrac{2x^4}{4!} + \ldots\right]$$

$$\therefore \boxed{\cosh x = 1 + \tfrac{x^2}{2!} + \tfrac{x^4}{4!} + \ldots}$$

g) *Example ('A' level question)*

The last part of this question has been included because, although it is not a summation, it uses the series expansions revised here and the idea of a limit.

Example

(a) Evaluate $\displaystyle\sum_{1}^{n} \frac{1}{r(r+1)}$ and deduce the sum to infinity.

(b) Find the sum of the series:

(i) $\displaystyle\sum_{1}^{\infty} \frac{a^r}{r}$

(ii) $\displaystyle\sum_{1}^{\infty} \frac{a^r}{r(r+1)}$ where $|a| < 1$

PURE MATHEMATICS — APPROXIMATIONS AND SERIES

(c) Find the sum to infinity of the series:

$$\frac{1}{2} + \frac{1}{2.4} + \frac{1}{2.4.6} + \ldots + \frac{1}{2.4.6\ldots 2r} + \ldots$$

and show that the sum to infinity of the series:

$$\frac{1}{2} + \frac{3}{2.4} + \frac{5}{2.4.6} + \ldots + \frac{2r-1}{2.4.6\ldots 2r} + \ldots \quad \text{is } 1.$$

(d) Evaluate $\displaystyle\lim_{x \to 0} \left[\frac{e^x - \cos x}{\sin \frac{1}{2} x} \right]$

Solution

(a) $\displaystyle\sum_{1}^{n} \frac{1}{r(r+1)}$ must be expressed in partial fractions before it can be summed.

$$\frac{1}{r(r+1)} \equiv \frac{A}{r} + \frac{B}{r+1}$$

Multiplying by $r(r+1)$ $\quad 1 \equiv A(r+1) + B(r)$

Let $r = 0$ $\qquad 1 = A(0+1) + B(0) \quad \therefore \quad 1 = A$

Let $r = -1$ $\qquad 1 = 1(0) + B(-1) \quad \therefore \quad -1 = B$

So $\quad \dfrac{1}{r(r+1)} = \dfrac{1}{r} - \dfrac{1}{r+1}$

$$\sum_{1}^{n} \frac{1}{r(r-1)} = \sum_{1}^{n} \left(\frac{1}{r} - \frac{1}{r+1} \right)$$

$$= \left(1 - \frac{1}{1+1}\right) + \left(\frac{1}{2} - \frac{1}{2+1}\right) + \left(\frac{1}{3} - \frac{1}{3+1}\right) + \ldots + \left(\frac{1}{n} - \frac{1}{n+1}\right)$$

$$= 1 - \frac{1}{2} + \frac{1}{2} - \frac{1}{3} + \frac{1}{3} - \frac{1}{4} + \ldots + \frac{1}{n} - \frac{1}{n+1}$$

$$\therefore \quad \sum_{1}^{n} \frac{1}{r(r+1)} = 1 - \frac{1}{n+1}$$

As $n \to \infty$, $\dfrac{1}{n+1} \to 0$, $\displaystyle\sum_{1}^{\infty} \frac{1}{r(r+1)} \to 1 - 0 = 1$

(b) (i) $\displaystyle\sum_{1}^{\infty} \frac{a^r}{r} = \frac{a}{1} + \frac{a^2}{2} + \frac{a^3}{3} + \ldots + \frac{a^n}{n} + \ldots$

APPROXIMATIONS AND SERIES — PURE MATHEMATICS

$$= -\left(-a - \frac{a^2}{2} - \frac{a^3}{3} - \ldots - \frac{a^n}{n} - \ldots\right)$$

$$= -\log_e(1-a)$$

(ii) $$\sum_1^\infty \frac{a^r}{r(r+1)} = \sum_1^\infty \left(\frac{a^r}{r} - \frac{a^r}{r+1}\right)$$

$$= \sum_1^\infty \frac{a^r}{r} - \sum_1^\infty \frac{a^r}{(r+1)}$$

$$= -\log_e(1-a) - \left(\frac{a}{1+1} + \frac{a^2}{2+1} + \frac{a^3}{3+1} + \ldots + \frac{a^n}{n+1} + \ldots\right)$$

$$= -\log_e(1-a) + \left(-\frac{a}{2} - \frac{a^2}{3} - \frac{a^3}{4} - \ldots - \frac{a^n}{n+1} - \ldots\right)$$

$$= -\log_e(1-a) + \frac{1}{a}\left(-\frac{a^2}{2} - \frac{a^3}{3} - \frac{a^4}{4} - \ldots - \frac{a^{n+1}}{n+1} - \ldots\right)$$

$$= -\log_e(1-a) + \frac{1}{a}\left(+a - a - \frac{a^2}{2} - \frac{a^3}{3} - \frac{a^4}{4} - \ldots - \frac{a^{n+1}}{n+1} - \ldots\right)$$

$$= -\log_e(1-a) + \frac{1}{a}(a) + \frac{1}{a}\left(-a - \frac{a^2}{2} - \frac{a^3}{3} - \frac{a^4}{4} - \ldots - \frac{a^{n+1}}{n+1} + \ldots\right)$$

$$\therefore \sum_1^\infty \frac{a^r}{r(r+1)} = -\log_e(1-a) + 1 + \frac{1}{a}\log_e(1-a)$$

(c) $$\frac{1}{2} + \frac{1}{2.4} + \frac{1}{2.4.6} + \ldots + \frac{1}{2.4.6 \ldots 2r} + \ldots$$

$$= \frac{1}{2} + \frac{1}{2^2(1.2)} + \frac{1}{2^3(1.2.3)} + \ldots + \frac{1}{2^r(1.2.3 \ldots r)}$$

$$= \frac{1}{2} + \frac{1}{2^2 2!} + \frac{1}{2^3 3!} + \ldots + \frac{1}{2^r r!} + \ldots$$

Compare this with $e^x = 1 + x + \frac{x^2}{2!} + \frac{x^3}{3!} + \ldots + \frac{x^r}{r!} + \ldots$

Replacing $x = \frac{1}{2}$ gives $e^{\frac{1}{2}} = 1 + \frac{1}{2} + \frac{\left(\frac{1}{2}\right)^2}{2!} + \frac{\left(\frac{1}{2}\right)^3}{3!} + \ldots + \frac{\left(\frac{1}{2}\right)^r}{r!} + \ldots$

$$\therefore e^{\frac{1}{2}} - 1 = \frac{1}{2} + \frac{1}{2^2 2!} + \frac{1}{2^3 3!} + \ldots + \frac{1}{2^r r!} + \ldots$$

$$\therefore \frac{1}{2} + \frac{1}{2.4} + \frac{1}{2.4.6} + \ldots + \frac{1}{2.4.6 \ldots 2r} + \ldots = e^{\frac{1}{2}} - 1$$

Also $\quad \dfrac{1}{2} + \dfrac{3}{2.4} + \dfrac{5}{2.4.6} + \ldots + \dfrac{2r-1}{2.4.6\ldots 2r} + \ldots$

$= \dfrac{2-1}{2} + \dfrac{4-1}{2.4} + \dfrac{6-1}{2.4.6} + \ldots + \dfrac{2r-1}{2.4.6\ldots 2r} + \ldots$

$= \dfrac{2}{2} + \dfrac{4}{2.4} + \dfrac{6}{2.4.6} + \ldots + \dfrac{2r}{2.4.6\ldots 2r} + \ldots \left(-\dfrac{1}{2} - \dfrac{1}{2.4} - \dfrac{1}{2.4.6}\ldots - \dfrac{1}{2.4.6\ldots 2r} - \ldots\right)$

$= 1 + \dfrac{1}{2} + \dfrac{1}{2.4} + \ldots + \dfrac{1}{2.4.6\ldots (2r-2)} + \ldots - (e^{\frac{1}{2}} - 1)$

$= 1 + \dfrac{1}{2} + \dfrac{1}{2^2 2!} + \ldots + \dfrac{1}{2^{r-1}(r-1)!} + \ldots - e^{\frac{1}{2}} + 1$

$= e^{\frac{1}{2}} - e^{\frac{1}{2}} + 1$

$= 1$

$\therefore \quad \dfrac{1}{2} + \dfrac{3}{2.4} + \dfrac{5}{2.4.6} + \ldots + \dfrac{2r-1}{2.4.6\ldots 2r} + \ldots = 1$

(d) $\quad \lim\limits_{x \to 0} \left[\dfrac{e^x - \cos x}{\sin \frac{1}{2}x}\right]$

Using $\quad e^x = 1 + x + \dfrac{x^2}{2!} + \dfrac{x^3}{3!} + \ldots + \dfrac{x^n}{n!} + \ldots$

and $\quad \cos x = 1 - \dfrac{x^2}{2!} + \dfrac{x^4}{4!} + \ldots$

and $\quad \sin \dfrac{1}{2}x = \left(\dfrac{1}{2}x\right) - \dfrac{\left(\frac{1}{2}x\right)^3}{3!} + \dfrac{\left(\frac{1}{2}x\right)^5}{5!} - \ldots$

$\therefore \quad \dfrac{e^x - \cos x}{\sin \frac{1}{2}x} = \dfrac{1 + x + \frac{x^2}{2!} + \frac{x^3}{3!} + \ldots - \left(1 - \frac{x^2}{2!} + \frac{x^4}{4!} - \ldots\right)}{\frac{1}{2}x - \frac{x^3}{(8)3!} + \frac{x^5}{(32)5!} - \ldots}$

$= \dfrac{1 + x + \frac{x^2}{2!} + \frac{x^3}{3!} + \ldots - 1 + \frac{x^2}{2!} - \frac{x^4}{4!} + \ldots}{\frac{1}{2}x - \frac{x^3}{(8)3!} + \frac{x^5}{(32)5!} - \ldots}$

$= \dfrac{x + \frac{2x^2}{2!} + \frac{x^3}{3!} + \ldots}{\frac{1}{2}x - \frac{x^3}{(8)3!} + \frac{x^5}{(32)5!} \ldots}$

$$\therefore \quad \frac{e^x - \cos x}{\sin \frac{1}{2} x} = \frac{x\left(1 + \frac{2x}{2!} + \frac{x^2}{3!} + \ldots\right)}{x\left(\frac{1}{2} - \frac{x^2}{(8)3!} + \frac{x^4}{(32)5!} - \ldots\right)}$$

In the limit as $x \to 0$ $\dfrac{e^x - \cos x}{\sin \frac{1}{2} x} \to \dfrac{1 + 0}{\frac{1}{2} - 0} = 2$

PURE MATHEMATICS — CONICS

8 CONICS

 8.1 Curve sketching
 8.2 Polar coordinates
 8.3 Ellipse
 8.4 Hyperbola

8.1 Curve sketching

a) *Inequalities*

When manipulating inequalities there are certain rules that must be observed. These have already been outlined but are included again as revision.

(a) Any number (positive or negative) may be added to each side of an inequality.

 ex. $6 + 2 > 3 + 2$ or $6 - 2 > 3 - 2$

 ie $8 > 5$ $\qquad 4 > 1$

(b) Each side of an inequality may be multiplied or divided by a positive number.

 ex. $5 \times 4 < 9 \times 4$

 ie $20 < 36$

(c) Each side of an inequality may be multiplied or divided by a negative number if, and only if, the inequality sign is reversed.

 ex. $7 > 2$ but on multiplying by -3 this becomes

 $7 \times -3 < 2x - 3$

 $-21 < -6$

The different ways in which inequalities arise in questions are illustrated by the following examples.

Example 1

Find the range of values for which the following inequalities are valid:

(a) $2x^2 + x - 15 < 0$

(b) $\dfrac{x+1}{2-x} < 1$

Solution

(a) $2x^2 + x - 15 < 0$

 $\therefore \quad (2x - 5)(x + 3) < 0$

 The function $f(x) = 2x^2 + x - 15$ is equal to zero when

 either $2x - 5 = 0$ or $x + 3 = 0$

 $\therefore \quad 2x = 5 \qquad \therefore \quad x = -3$

 $\therefore \quad x = \dfrac{5}{2}$

This gives three ranges of values to consider:

$x < -3$, $-3 < x < \frac{5}{2}$ and $x > \frac{5}{2}$

When $x < -3$ $(2x - 5)$ will be -ve, $(x + 3)$ will be -ve
(eg try $x = -4$)
$\therefore$ $(2x - 5)(x + 3)$ will be +ve $\because$ -ve times -ve = +ve

When $-3 < x < \frac{5}{2}$ $(2x - 5)$ will be -ve, $(x + 3)$ will be +ve
(eg try $x = 0$)
$\therefore$ $(2x - 5)(x + 3)$ will be -ve $\because$ -ve times +ve = -ve

When $x > \frac{5}{2}$ $(2x - 5)$ will be +ve, $(x + 3)$ will be +ve
(eg try $x = 3$)
$\therefore$ $(2x - 5)(x + 3)$ will be +ve $\because$ +ve times +ve = +ve

This information can be displayed more clearly in a table as follows:

	$x < -3$	$-3 < x < \frac{5}{2}$	$x > \frac{5}{2}$
$2x - 5$	-ve	-ve	+ve
$x + 3$	-ve	+ve	+ve
$f(x)$	+ve	-ve	+ve

$\therefore$ $f(x) = 2x^2 + x - 15 < 0$ when $-3 < x < \frac{5}{2}$

(b) $\frac{x + 1}{2 - x} < 1$:   the working has to be split into two parts for a function in a quotient form.

Assuming $2 - x > 0$ $\therefore x < 2$ then $x + 1 < 1.(2 - x)$

$\therefore$ $x + 1 < 2 - x$

$\therefore$ $2x < 1$

$\therefore$ $x < \frac{1}{2}$

which is in the given range of $x < 2$.

Assuming $2 - x < 0$ $\therefore x > 2$ then $x + 1 > 1.(2 - x)$

Note: The inequality sign has been reversed because of multiplying by a negative number.

$\therefore$ $x + 1 > 2 - x$

$\therefore$ $2x > 1$

$\therefore$ $x > \frac{1}{2}$

but the given range is $x > 2$.

Therefore, the function is less than 1 when $x < \frac{1}{2}$ or $x > 2$.

PURE MATHEMATICS — CONICS

Example 2

Find the range of values for y if:

$$(x^2 - x + 1)y = 2x$$

Solution

$$(x^2 - x + 1)y = 2x$$

$\therefore \quad x^2y - xy + y - 2x = 0$

$\therefore \quad x^2y - xy - 2x + y = 0$

$\therefore \quad yx^2 - (y + 2)x + y = 0$

This is a quadratic equation in x and, for such an equation to have real roots.

'$b^2 - 4ac \geq 0$' where $a = y$, $b = -(y + 2)$, $c = y$

$\therefore \quad [-(y + 2)]^2 - 4(y)(y) \geq 0$

$y^2 + 4y + 4 - 4y^2 \geq 0$

$4 + 4y - 3y^2 \geq 0$

$(2 + 3y)(2 - y) \geq 0$

This function $f(y) = (2 + 3y)(2 - y)$ is equal to zero when:

either $\quad 2 + 3y = 0 \quad$ or $\quad 2 - y = 0$

$\therefore \quad 3y = -2 \qquad \therefore \quad y = 2$

$\therefore \quad y = -\dfrac{2}{3}$

The ranges of values to be considered are:

	$y < -\dfrac{2}{3}$	$-\dfrac{2}{3} < y < 2$	$y > 2$
$(2 + 3y)$	-ve	+ve	+ve
$(2 - y)$	+ve	+ve	-ve
$f(y)$	-ve	+ve	-ve

$\therefore \quad$ if x is real $4 + 4y - 3y^2 \geq 0$ which occurs when $-\dfrac{2}{3} \leq y \leq 2$.

---

b) *Curve sketching - rules*

As has already been shown, curve sketching is a very important stage of working when dealing with complicated functions. There are certain features of a curve which can be determined before it is sketched. These are:

(a) The points of intersection of the curve and the axes. To find where the curve crosses the x axis, put y = 0 in the equation and, to find where the curve crosses the y axis, put x = 0.

(b) The maximum and minimum turning points. These are found by putting $\dfrac{dy}{dx}$ equal to zero; the nature of the turning points is determined from $\dfrac{d^2y}{dx^2}$. $\dfrac{d^2y}{dx^2} < 0$ for a maximum and $\dfrac{d^2y}{dx^2} > 0$ for a minimum.

(c) Points of inflexion. These are points where the nature of the curvature changes:

ex.

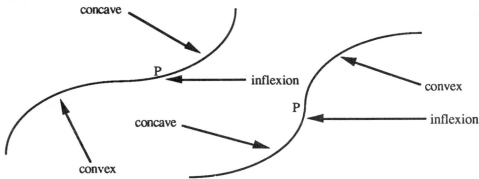

This curve changes from convex
to concave as it passes through P

This curve changes from concave
to convex as it passes through P.

In both cases this means that at a point of inflexion $\frac{d^2y}{dx^2} = 0$ and it changes sign before and after P.

Also $\frac{dy}{dx}$ does not change sign because, if the gradient is positive before the point of inflexion, it will still be positive after the point. Similarly, if it is negative before, it will be negative afterwards.

(d) The equations of the asymptotes. If, as either x or y becomes very large, the equation of the curve approximates to the equation of a straight line, that straight line is called an asymptote. The equations of the asymptotes are found by letting x and y tend to infinity in turn and considering what effect this has on the equation of the curve.

(e) Any regions where the curve cannot exist. Sometimes it is possible to find areas of the x - y plane where the graph cannot be plotted. These areas are usually found by considering the separate factors that make up the equation of the curve.

(f) Symmetry. If the equation of the curve is of the form $y^2 = f(x)$, therefore $y = \pm \sqrt{f(x)}$ or $x^2 = f(y)$, therefore $x = \pm \sqrt{f(y)}$, the curve may be symmetrical about one of the axes or about another straight line. This can be quite difficult to determine before sketching the curve and is only obvious once the curve has been sketched.

The next example will illustrate how each of these rules can be used to help sketch an actual curve.

Example 3

Sketch the curve whose equation is $y = \frac{x - 2}{x - 1}$

Solution

Following the procedure already given:

(a) Intersection with the axes:

when $x = 0$ $y = \frac{0 - 2}{0 - 1} = 2$ giving (0, 2)

when $y = 0$ $0 = \frac{x - 2}{x - 1}$

∴ $0(x - 1) = x - 2$

PURE MATHEMATICS   CONICS

$\therefore \quad 0 = x - 2$

$\therefore \quad 2 = x \quad\quad$ giving (2, 0)

(b) Turning points:

$$y = \frac{x-2}{x-1} \quad\quad \text{- a quotient form } \left(\frac{u}{v}\right)$$

$$\therefore \quad \frac{dy}{dx} = \frac{(x-1) \cdot 1 - (x-2) \cdot 1}{(x-1)^2}$$

$$= \frac{x - 1 - x + 2}{(x-1)^2}$$

$$= \frac{1}{(x-1)^2}$$

when $\frac{dy}{dx} = 0 \quad\quad \frac{1}{(x-1)^2} = 0$

$\therefore \quad 1 = 0(x-1)^2$

$\therefore \quad 1 = 0 \quad\quad$ - no real solutions

Therefore, there are no turning points.

(c) Points of inflexion:

$$y = \frac{x-2}{x-1} \quad\quad \frac{dy}{dx} = \frac{1}{(x-1)^2} = (x-1)^{-2}$$

$$\therefore \quad \frac{d^2y}{dx^2} = -2(x-1)^{-3}$$

$$= \frac{-2}{(x-1)^3}$$

when $\frac{d^2y}{dx^2} = 0 \quad\quad \frac{-2}{(x-1)^3} = 0$

$\therefore \quad -2 = 0(x-1)^3$

$\therefore \quad -2 = 0 \quad\quad$ - no real solutions.

Therefore, thre are no points of inflexion.

(d) Asymptotes:

$$y = \frac{x-2}{x-1}$$

Dividing out

$$\begin{array}{r} 1 \\ x-1\overline{\smash{)}\,x-2} \\ \underline{x-1} \\ -1 \end{array}$$

$\therefore \quad y = 1 - \frac{1}{x-1}$

As x becomes larger and larger $\frac{1}{x-1}$ becomes smaller and smaller, ie

as $x \to \pm\infty$, $\frac{1}{x-1} \to 0$ and $y \to 1$.

It follows that as $x \to 1$, $\frac{1}{x-1} \to \frac{1}{0} = \pm\infty$, ie $y \to \pm\infty$ when $x \to 1$.

So $y = 1$ and $x = 1$ are the equations of the asymptotes because the curve approximates to these straight lines as x and y respectively become large (ie tend to infinity).

(e) Nil regions:

$$y = \frac{x-2}{x-1}$$ gives two critical values for x, 1 and 2

because $x - 2 = 0$ when $x = 2$, and $x - 1 = 0$ when $x = 1$. There are, therefore, three areas to consider: $x < 1$, $1 < x < 2$ and $x > 2$.

	x < 1	1 < x < 2	x > 2
x - 2	-ve	-ve	+ve
x - 1	-ve	+ve	+ve
f(x)	+ve	-ve	+ve

where $f(x) = \frac{x-2}{x-1}$

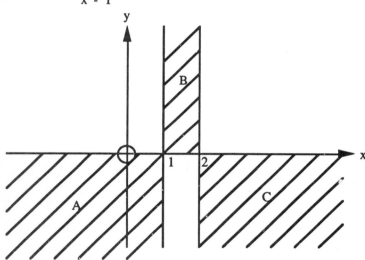

The shaded areas are those regions where the curve *cannot* exist.

A - for $x < 1$, f(x) is +ve. Therefore, the negative area has been shaded out.

B - for $1 < x < 2$, f(x) is -ve. Therefore, the positive area has been shaded out.

C - for $x > 2$, f(x) is +ve. Therefore, the negative area has been shaded out.

(f) Symmetry. The equation does not lead to any expectations of symmetry.

Now to sketch the curve:

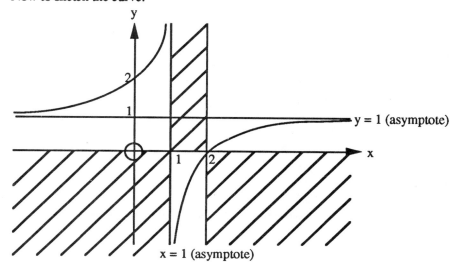

This method may seem rather long and tedious, especially in determining the points of inflexion - if any exist. This stage of the working is usually omitted unless the question specifically asks for such points to be determined as will be seen in later examples.

c) *Curve sketching - specific functions*

The types of functions that are likely to arise are:

$$y = \frac{ax^2 + bx + c}{mx^2 + nx + p}, \quad ay^2 = x^2(a^2 - x^2), \quad y = e^x \sin x$$

and parametric functions such as:

$$x = a\cos^3 t, \quad y = a\sin^3 t$$

The following examples will illustrate the types of questions that may be asked.

Example 4

Find the coordinates of the points of inflexion on the curve $y = e^{-x^2}$. Mark these points on a sketch of the curve.

Solution

$$y = e^{-x^2}$$

$$\therefore \quad \frac{dy}{dx} = e^{-x^2}(-2x)$$

$$= -2xe^{-x^2}$$

$$\therefore \quad \frac{d^2y}{dx^2} = -2x \cdot (e^{-x^2} \cdot (-2x)) + (-2)e^{-x^2}$$

$$= 4x^2 e^{-x^2} - 2e^{-x^2}$$

$$= e^{-x^2}(4x^2 - 2)$$

CONICS   PURE MATHEMATICS

For points of inflexion $\quad \dfrac{d^2y}{dx^2} = 0 \qquad \therefore e^{-x^2}(4x^2 - 2) = 0$

either $\quad e^{-x^2} = 0 \qquad$ or $\qquad 4x^2 - 2 = 0$

- no solutions $\qquad \therefore \qquad 4x^2 = 2$

$$\therefore \quad x^2 = \frac{2}{4} = \frac{1}{2}$$

$$\therefore \quad x = \pm \frac{1}{\sqrt{2}}$$

when $\quad x = \dfrac{1}{\sqrt{2}} \qquad y = e^{-\frac{1}{2}} = \dfrac{1}{\sqrt{e}}$

when $\quad x = -\dfrac{1}{\sqrt{2}} \qquad y = e^{-\frac{1}{2}} = \dfrac{1}{\sqrt{e}}$

So the points of inflexion are $\left(\dfrac{1}{\sqrt{2}}, \dfrac{1}{\sqrt{e}}\right)$ and $\left(-\dfrac{1}{\sqrt{2}}, \dfrac{1}{\sqrt{e}}\right)$

(To *prove* that these are the points of inflexion it is necessary to show that the sign of $\dfrac{d^2y}{dx^2}$ changes in going through the point and that $\dfrac{dy}{dx}$ does not change sign. This proof is included here although, strictly speaking, it is not required as part of the solution since the question only asks for the calculation of the points of inflexion.

For the point $\left(\dfrac{1}{\sqrt{2}}, \dfrac{1}{\sqrt{e}}\right) \quad x = \dfrac{1}{\sqrt{2}} = 0.707$, so taking a value each side of this:

$x = 0.5$ and $x = 1$, say:

$\dfrac{dy}{dx} = -2xe^{-x^2} \quad$ for $x = 0.5 \qquad \dfrac{dy}{dx} = -2(0.5)e^{-(0.5)^2} \qquad = -e^{-0.25} < 0$

$\qquad\qquad\qquad$ for $x = 1 \qquad \dfrac{dy}{dx} = -2(1)e^{-1^2} \qquad = -2e^{-1} < 0$

$\qquad\qquad \therefore \quad \dfrac{dy}{dx}$ does not change sign

$\dfrac{d^2y}{dx^2} = e^{-x^2}(4x^2 - 2) \quad$ for $x = 0.5 \quad \dfrac{d^2y}{dx^2} = e^{-(0.5)^2}(4(0.5)^2 - 2)$

$\qquad\qquad\qquad\qquad\qquad\qquad\qquad = e^{-0.25}(1 - 2) = -e^{-0.25} < 0$

$\qquad\qquad\qquad$ for $x = 1 \quad \dfrac{d^2y}{dx^2} = e^{-1^2}(4(1)^2 - 2) = e^{-1}(4 - 2) = e^{-1}(2) > 0$

$\qquad\qquad \therefore \quad \dfrac{d^2y}{dx^2}$ changes sign from -ve to +ve

The proof for the other points of inflexion $\left(-\dfrac{1}{\sqrt{2}}, \dfrac{1}{\sqrt{e}}\right)$ would be identical:

$x = -\frac{1}{\sqrt{2}} = -0.707$ so appropriate values to take would be $x = -1$ and $x = -0.5$, etc.

Before sketching the curve it is necessary to find some other points:

(a) Intersection with the axes:

when $x = 0$ $\quad y = e^{-0^2} = e^0 = 1$ $\qquad$ giving $(0, 1)$

when $y = 0$ $\quad 0 = e^{-x^2}$

This has no solution, therefore the curve does not cross the x axis.

(b) Turning points:

$\frac{dy}{dx} = -2xe^{-x^2}$ $\qquad \therefore \quad \frac{dy}{dx} = 0$ $\qquad$ when $2xe^{-x^2} = 0$

$\qquad\qquad\qquad\qquad\qquad \therefore \quad x = 0$ $\qquad$ because $e^{-x^2} \neq 0$

$\frac{d^2y}{dx^2} = e^{-x^2}(4x^2 - 2)$ when $x = 0$ $\quad \frac{d^2y}{dx^2} = e^0(4.0^2 - 2)$

$\qquad\qquad\qquad\qquad\qquad\qquad\qquad\qquad = -2e^0 < 0 \; \therefore \;$ maximum

So there is a maximum at $x = 0$, $y = e^{-x^2} = e^0 = 1$, ie at $(0, 1)$.

(c) Points of inflexion:

$\left(\frac{1}{\sqrt{2}}, \frac{1}{\sqrt{e}}\right) \left(-\frac{1}{\sqrt{2}}, \frac{1}{\sqrt{e}}\right)$

(d) Asymptotes:

$y = e^{-x^2}$ as x gets larger and larger, e^{-x^2} gets smaller and smaller, ie as $x \to \pm\infty$, $y \to 0$. Therefore, $y = 0$ is an asymptote.

Since the curve has a maximum turning point at $(0, 1)$ the curve does not exist for $y > 1$.

(e) Nil regions:

As has just been stated the curve does not exist for $y > 1$, also $e^{-x^2} = \frac{1}{e^{x^2}}$ cannot be negative.

Therefore, the curve does not exist for $y < 0$.

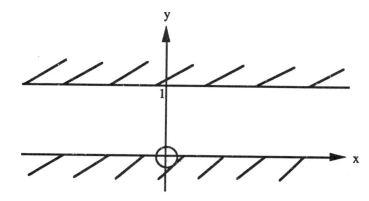

# CONICS	PURE MATHEMATICS

(f) Symmetry:

The curve is symmetrical about the y axis because x^2 is positive for all values of x, ie

$x = -3 \qquad x^2 = 9 \qquad \therefore \qquad e^{-x^2} = e^{-9}$

and $\quad x = 3 \qquad x^2 = 9 \qquad \therefore \qquad e^{-x^2} = e^{-9}$

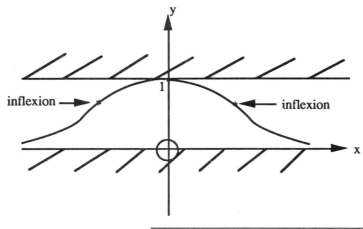

Example 5

Find the coordinates of the turning point on the graph of $y = \dfrac{4}{4 - x^2}$ and determine the nature of this turning point. Prove that the graph has no points of inflexion. Write down the equations of the asymptotes and sketch the graph.

Solution

$$y = \frac{4}{4 - x^2} \qquad \text{using quotient form } \frac{u}{v}$$

$$\therefore \quad \frac{dy}{dx} = \frac{(4 - x^2) \cdot 0 - 4 \cdot (-2x)}{(4 - x^2)^2}$$

$$= \frac{8x}{(4 - x^2)^2}$$

$$= \frac{8x}{16 - 8x^2 + x^4}$$

$$\frac{dy}{dx} = 0 \qquad \text{when } \frac{8x}{(4 - x^2)^2} = 0$$

$$\therefore \quad 8x = 0$$

$$\therefore \quad x = 0 \text{ and } y = \frac{4}{4 - 0^2} = 1$$

$$\therefore \quad \frac{dy^2}{dx^2} = \frac{(16 - 8x^2 + x^4) \cdot 8 - 8x(-16x + 4x^3)}{(16 - 8x^2 + x^4)^2}$$

$$= \frac{128 - 64x^2 + 8x^4 + 128x^2 - 32x^4}{(16 - 8x^2 + x^4)^2}$$

$$= \frac{128 + 64x^2 - 24x^4}{(16 - 8x^2 + x^4)^2}$$

For $x = 0$ $\quad \frac{d^2y}{dx^2} = \frac{128 + 64.0^2 - 24.0^4}{(16 - 8.0^2 + 0^4)^2} > 0 \quad \therefore$ minimum

Therefore, there is a minimum turning point at $(0, 1)$.

For points of inflexion $\frac{d^2y}{dx^2} = 0$

$\therefore \quad \frac{128 + 64x^2 - 24x^4}{(16 - 8x^2 + x^4)^2} = 0$

$\therefore \quad 128 + 64x^2 - 24x^4 = 0 \cdot (16 - 8x^2 + x^4)^2$

$\therefore \quad -8(3x^4 - 8x^2 - 16) = 0$

$\therefore \quad 3x^4 - 8x^2 - 16 = 0$

$\therefore \quad (3x^2 + 4)(x^2 - 4) = 0$

either $\quad 3x^2 + 4 = 0 \quad$ or $\quad x^2 - 4 = 0$

$\therefore \quad 3x^2 = -4 \quad\quad \therefore \quad x^2 = 4$

$\therefore \quad x^2 = -\frac{4}{3} \quad\quad \therefore \quad x = \pm 2$

- no solutions

It appears that there are points of inflexion where $x = 2$ and $x = -2$ BUT $y = \frac{4}{4 - x^2}$

$\therefore \quad$ for $x = 2 \quad y = \frac{4}{4 - 4} = \frac{4}{0} = \infty$

and $\quad$ for $x = -2 \quad y = \frac{4}{4 - 4} = \frac{4}{0} = \infty$

So no points of inflexion exist.

For the asymptotes $y = \frac{4}{4 - x^2} = \frac{4}{(2 - x)(2 + x)}$

As has been seen, as $x \to 2 \quad\quad y \to \infty$

and $\quad$ as $x \to -2 \quad\quad y \to \infty$

$\therefore \quad x = 2$ and $x = -2$ are asymptotes.

As x gets larger and larger $\frac{4}{4 - x^2}$ will get smaller and smaller.

$\therefore \quad$ as $x \to \pm \infty \to \frac{4}{4 - x^2} \to 0 \quad \therefore \quad y \to 0$

$\therefore \quad y = 0$ is also an asymptote.

CONICS — PURE MATHEMATICS

Going through the usual procedure for curve sketching gives:

(a) Points of intersection with axes:

$$x = 0 \quad y = \frac{4}{4 - 0} = 1 \quad \ldots \quad (0, 1)$$

$$y = 0 \quad 0 = \frac{4}{4 - x^2}$$

$$\therefore \quad 0(4 - x^2) = 4$$

$$\therefore \quad 0 = 4$$

- no solution. Therefore, curve does not cross x axis.

(b)
(c) } - as above.
(d)

(e) Nil regions:

$$y = \frac{4}{(2 - x)(2 + x)} \quad \ldots \quad x = 2 \text{ and } -2 \text{ are critical values}$$

There are three areas to consider $x < -2$, $-2 < x < 2$, $x > 2$.

	$x < -2$	$-2 < x < 2$	$x > 2$
$(2 - x)$	+ve	+ve	-ve
$(2 + x)$	-ve	+ve	+ve
$f(x)$	-ve	+ve	-ve

where $f(x) = \dfrac{4}{4 - x^2}$

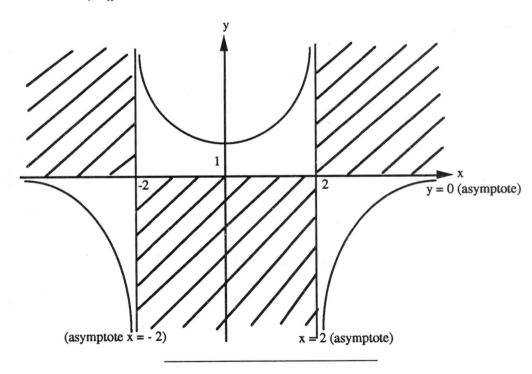

(asymptote $x = -2$) $x = 2$ (asymptote)

$y = 0$ (asymptote)

PURE MATHEMATICS — CONICS

d) *Example ('A' level question)*

If x is real, $x \neq 1$ and $y = \dfrac{x^2 + x - 1}{2(x - 1)}$ show that $y \leq \dfrac{1}{2}$ or $y \geq \dfrac{5}{2}$

Show that $y = \dfrac{x}{2} + 1$ is an asymptote to the curve $y = \dfrac{x^2 + x - 1}{2(x - 1)}$ and state the equation of the other asymptote to the curve.

Solution

$$y = \frac{x^2 + x - 1}{2(x - 1)}$$

$\therefore \quad 2(x - 1)y = x^2 + x - 1$

$\quad\quad 2xy - 2y = x^2 + x - 1$

$\therefore \quad 0 = x^2 + x - 2xy - 1 + 2y$

$\quad\quad 0 = x^2 + (1 - 2y)x - 1 + 2y$

This is a quadratic equation in x and for it to have real roots '$b^2 - 4ac \geq 0$' where

$a = 1 \quad b = (1 - 2y) \quad c = -1 + 2y$

$(1 - 2y)^2 - 4.1(-1 + 2y) \geq 0$

$1 - 4y + 4y^2 + 4 - 8y \geq 0$

$4y^2 - 12y + 5 > 0$

$(2y - 1)(2y - 5) \geq 0$

The function $f(y) = (2y - 1)(2y - 5)$ is equal to zero when

either $\quad 2y - 1 = 0 \quad$ or $\quad 2y - 5 = 0$

$\therefore \quad\quad 2y = 1 \quad\quad\quad\quad\quad 2y = 5$

$\therefore \quad\quad y = \dfrac{1}{2} \quad\quad\quad\quad\quad y = \dfrac{5}{2}$

The ranges of values to be considered are:

	$y < \dfrac{1}{2}$	$\dfrac{1}{2} < y < \dfrac{5}{2}$	$y > \dfrac{5}{2}$
(2y - 1)	-ve	+ve	+ve
(2y - 5)	-ve	-ve	+ve
f(y)	+ve	-ve	+ve

Therefore, if x is real, $4y^2 - 12y + 5 \geq 0$, which occurs when $y \leq \dfrac{1}{2}$ or $y \geq \dfrac{5}{2}$.

To find the equations of the asymptotes it is first necessary to divide out:

$$y = \frac{x^2 + x - 1}{2x - 2}$$

$$\begin{array}{r} \frac{x}{2} + 1 \\ 2x-2 \overline{\smash{)}\ x^2 + x - 1} \\ \underline{x^2 - x} \\ 2x - 1 \\ \underline{2x - 2} \\ 1 \end{array}$$

$$y = \frac{x}{2} + 1 + \frac{1}{2x - 2}$$

As x gets larger and larger $\frac{1}{2x-2}$ gets smaller and smaller, ie as $x \to \pm\infty$, $\frac{1}{2x-2} \to 0$ and $y = \frac{x}{2} + 1$ is the equation of the asymptote.

Also, as $x \to 1$, $2x - 2 \to 0$ and $\frac{1}{2x-2} \to \pm\infty$, ie $y \to \pm\infty$ as $x \to 1$ and so $x = 1$ is the equation of the other asymptote.

8.2 Polar coordinates

a) *Definition*

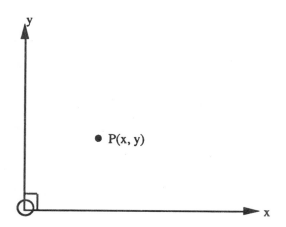

The cartesian coordinates of a point P are x and y, ie P is the point (x, y) when referred to the axes Ox and Oy.

There is, however, an alternative method of locating the position of P and that is using polar coordinates:

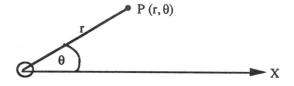

O is a fixed point called the pole and OX is a fixed line called the initial line. Then, if P is any point such that OP = r and $P\hat{O}X = \theta$, the polar coordinates of P are r and θ, ie P is the point (r, θ) when referred to the initial line OX.

θ is measured in the anti-clockwise direction from OX. Therefore, negative values of θ are measured in the clockwise direction from OX.

PURE MATHEMATICS — CONICS

Similarly r is positive when measured in the direction OP and negative when measured in the direction PO produced.

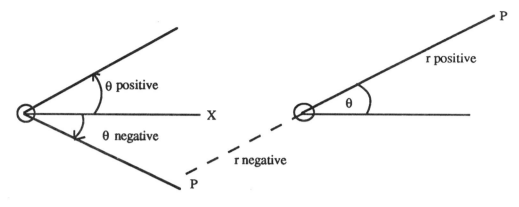

Example 6

Plot the following points on a diagram:

A $\left(3, \dfrac{\pi}{4}\right)$

B $\left(1, -\dfrac{\pi}{3}\right)$

C $\left(-2, \dfrac{\pi}{2}\right)$

D $\left(-1, -\dfrac{2\pi}{3}\right)$

Solution

A $\quad A\hat{O}X = \dfrac{\pi}{4} \quad OA = 3$

B $\quad B\hat{O}X = -\dfrac{\pi}{3} \quad OB = 1$

C $\quad C'\hat{O}X = \dfrac{\pi}{2} \quad OC = -2$

D $\quad D'\hat{O}X = \dfrac{2\pi}{3} \quad OD = -1$

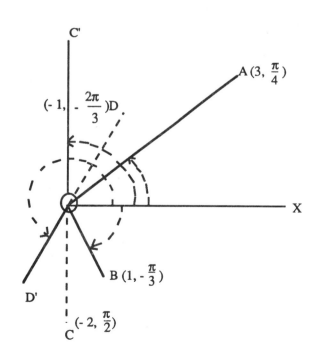

CONICS   PURE MATHEMATICS

b) *Relationship between polar and cartesian coordinates*

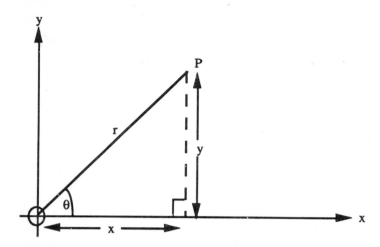

P has cartesian coordinates (x, y) and polar coordinates (r, θ) as shown in the diagram. By using Pythagoras theorem and trigonometry, it is possible to derive several useful relationships between the two sets of coordinates:

$$\left.\begin{array}{l} \sin\theta = \dfrac{y}{r} \qquad y = r\sin\theta \\ \\ \cos\theta = \dfrac{x}{r} \qquad x = r\cos\theta \end{array}\right\} \quad \therefore \ \dfrac{y}{x} = \tan\theta$$

Applying Pythagoras $\quad x^2 + y^2 = r^2 \quad \therefore \ r = \sqrt{x^2 + y^2}$

These relationships can be used to convert cartesian equations into polar form and vice versa.

Example 7

Find the polar equations of the curves whose cartesian equations are:

(a) $(x - a)^2 + (y - a)^2 = a^2$ where a is a constant

(b) $y^2 = 4x$

Solution

(a) $(x - a)^2 + (y - a)^2 = a^2$

$\therefore \quad x^2 - 2ax + a^2 + y^2 - 2ay + a^2 = a^2$

$\therefore \quad x^2 + y^2 - 2a(x + y) + 2a^2 = a^2$

$\therefore \quad r^2 - 2a(r\sin\theta + r\cos\theta) = a^2 - 2a^2$

$\therefore \quad r^2 - 2ar(\sin\theta + \cos\theta) = -a^2$

This cannot be simplified further.

PURE MATHEMATICS   CONICS

(b) $y^2 = 4x$

$\therefore \quad (r \sin \theta)^2 = 4r \cos \theta$

$\therefore \quad r^2 \sin^2 \theta = 4r \cos \theta$ - dividing by r gives

$r \sin^2 \theta = 4 \cos \theta$

$\therefore \quad r = \dfrac{4 \cos \theta}{\sin^2 \theta}$

Example 8

Find the cartesian equations of the curves whose polar equations are:

(a) $r = a \cos 2\theta$ where a is a constant

(b) $r = d \sec(\theta - \alpha)$ where d and α are constants

Solution

(a) $r = a \cos 2\theta$

$\therefore \quad r = a(2\cos^2 \theta - 1)$

$\therefore \quad (x^2 + y^2)^{\frac{1}{2}} = a\left(2\left(\dfrac{x}{r}\right)^2 - 1\right)$

$\therefore \quad (x^2 + y^2)^{\frac{1}{2}} = a\left(\dfrac{2x^2}{r^2} - 1\right)$

$\therefore \quad (x^2 + y^2)^{\frac{1}{2}} = a\left(\dfrac{2x^2 - r^2}{r^2}\right)$

$\therefore \quad (x^2 + y^2)^{\frac{1}{2}} r^2 = a(2x^2 - r^2)$ using $r^2 = x^2 + y^2$

$\therefore \quad (x^2 + y^2)^{\frac{1}{2}} \cdot (x^2 + y^2) = a(2x^2 - (x^2 + y^2))$

$\therefore \quad (x^2 + y^2)^{\frac{3}{2}} = a(2x^2 - x^2 - y^2)$

$\therefore \quad (x^2 + y^2)^{\frac{3}{2}} = a(x^2 - y^2)$

(b) $r = d \sec(\theta - \alpha)$

$\therefore \quad r = \dfrac{d}{\cos(\theta - \alpha)}$

$\therefore \quad r \cdot \cos(\theta - \alpha) = d$

$\therefore \quad r(\cos \theta \cdot \cos \alpha + \sin \theta \sin \alpha) = d$

$\therefore \quad r\left(\dfrac{x}{r} \cos \alpha + \dfrac{y}{r} \sin \alpha\right) = d$

$\therefore \quad x \cos \alpha + y \sin \alpha = d$

c) *Curve sketching*

The easiest way to sketch a curve with a polar equation is to take a range of values (from 0 to 2π) for θ and work out the corresponding values of r. The pairs of values (r, θ) can then be plotted on a graph.

However, it is possible to reduce the number of values that need to be calculated. The following points should be noted:

(a) If r is a function of $\cos \theta$ the curve is symmetrical about the line $\theta = 0$, ie about OX. This is because $\cos(-\theta) = \cos \theta$.

(b) If r is a function of $\sin \theta$ the curve is symmetrical about the line $\theta = \frac{\pi}{2}$, ie about a line perpendicular to OX at O. This is because $\sin(\pi - \theta) = \sin \theta$.

Example 9

Sketch the curves of:

(a) $r = a(1 + \cos \theta)$ where a is a constant

(b) $r = a(1 + \sin \theta)$

Solution

(a) $r = a(1 + \cos \theta)$ is a function of $\cos \theta$ and so is symmetrical about $\theta = 0$. Therefore, it is only necessary to work out values of r from $\theta = 0$ to $\theta = \pi$ and then reflect this curve in OX.

Taking values of θ at intervals of $\frac{\pi}{6}$ say (ie at intervals of 30° - but all working is usually done in radians rather than degrees):

θ	0	$\frac{\pi}{6}$	$\frac{\pi}{3}$	$\frac{\pi}{2}$	$\frac{2\pi}{3}$	$\frac{5\pi}{6}$	π
$\cos \theta$	1	0.9	0.5	0	-0.5	-0.9	-1
$1 + \cos \theta$	2	1.9	1.5	1	0.5	0.1	0
r	2a	1.9a	1.5a	a	0.5a	0.1a	0

where $r = a(1 + \cos \theta)$ and a is a constant.

Before sketching this curve it is necessary to draw in the angles that have been used to calculate the values of r.

PURE MATHEMATICS — CONICS

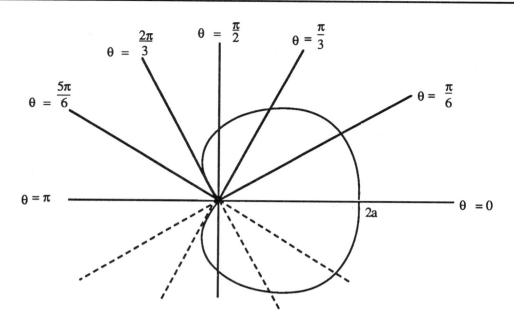

This curve is called a cardioid and the shape of the curve at 0 is called a cusp.

(b) $r = a(1 + \sin \theta)$ is a function of $\sin \theta$ and so is symmetrical about $\theta = \frac{\pi}{2}$. Therefore, it is only necessary to work out values of r from $\theta = -\frac{\pi}{2}$ to $\theta = \frac{\pi}{2}$ and then reflect this curve in $\theta = \frac{\pi}{2}$.

θ	$-\frac{\pi}{2}$	$-\frac{\pi}{3}$	$-\frac{\pi}{6}$	0	$\frac{\pi}{6}$	$\frac{\pi}{3}$	$\frac{\pi}{2}$
$\sin \theta$	-1	-0.9	-0.5	0	0.5	0.9	-1
$1 + \cos \theta$	0	0.1	0.5	1	1.5	1.9	2
r	0	0.1a	0.5a	a	1.5a	1.9a	2a

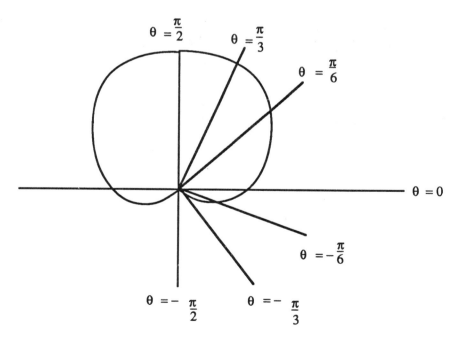

Another cardioid.

CONICS — PURE MATHEMATICS

Example 10

Sketch the curve of $r = 2 \cos 2\theta$

Solution

$$r = 2 \cos 2\theta$$

$\therefore \quad r = 2(2\cos^2\theta - 1)$

$\quad\quad = 4\cos^2\theta - 2$

or $\quad r = 2(1 - 2\sin^2\theta)$

$\quad\quad = 2 - 4\sin^2\theta$

Since r can depend on $\cos \theta$ this curve is symmetrical about the line $\theta = 0$. However, it can also depend on $\sin \theta$ and so it is symmetrical about $\theta = \frac{\pi}{2}$ as well. This means that it is only necessary to tabulate values of θ from $\theta = 0$ to $\theta = \frac{\pi}{2}$ and the complete curve will be obtained by reflecting in both axes of symmetry.

Since the range of values for θ is smaller, closer intervals will be taken.

θ	0	$\frac{\pi}{12}$	$\frac{\pi}{6}$	$\frac{\pi}{4}$	$\frac{\pi}{3}$	$\frac{5\pi}{12}$	$\frac{\pi}{2}$
$\cos 2\theta$	1	0.9	0.5	0	-0.5	-0.9	-1
r	2	1.8	1.0	0	-1.0	-1.8	-2

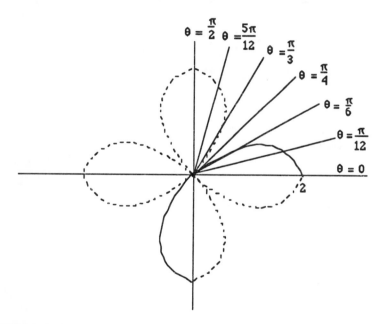

The solid line is the curve formed from plotting the points calculated in the table, ie

$(2, 0) \quad \left(1.8, \frac{\pi}{12}\right) \quad \left(1.0, \frac{\pi}{6}\right) \quad \left(0, \frac{\pi}{4}\right) \quad \left(-1.0, \frac{\pi}{3}\right) \quad \left(-1.8, \frac{5\pi}{12}\right) \quad \left(-2, \frac{\pi}{2}\right)$

The dotted lines complete the curve $r = 2 \cos 2\theta$ and are found by reflection in $\theta = 0$ and $\theta = \frac{\pi}{2}$.

PURE MATHEMATICS — CONICS

d)   *Area bounded by a polar curve*

The area of a polar curve can be found by integration using the formula:

$$A = \frac{1}{2} \int_{\alpha}^{\beta} r^2 \, d\theta$$

where $\theta = \alpha$ and $\theta = \beta$ are the boundary values.

This formula is not being derived but will now be used to find the area of several polar curves.

Example 11

Find the area of the cardioid $r = a(1 + \cos \theta)$

Solution

This has already been sketched, giving:

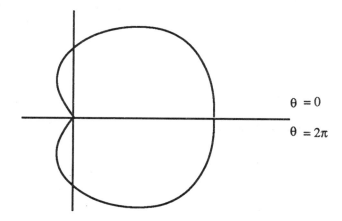

$\theta = 0$
$\theta = 2\pi$

The area can be calculated in two ways:

(a)   Area $= \frac{1}{2} \int_0^{2\pi} r^2 d\theta = \frac{1}{2} \int_0^{2\pi} a^2 (1 + \cos \theta)^2 \, d\theta$

$= \frac{a^2}{2} \int_0^{2\pi} (1 + 2 \cos \theta + \cos^2 \theta) \, d\theta$ but $\cos 2\theta = 2 \cos^2 \theta - 1$

$\therefore \quad \cos 2\theta + 1 = 2 \cos^2 \theta$

$\frac{1}{2}(\cos 2\theta + 1) = \cos^2 \theta$

$= \frac{a^2}{2} \int_0^{2\pi} \left(1 + 2 \cos \theta + \frac{1}{2}(\cos 2\theta + 1) \right) d\theta$

$= \frac{a^2}{2} \left[\theta + 2 \sin \theta + \frac{1}{2}\left(\frac{\sin 2\theta}{2} + \theta \right) \right]_0^{2\pi}$

$= \frac{a^2}{2} \left[2\pi + 2 \sin 2\pi + \frac{1}{2}\left(\frac{\sin 4\pi}{2} + 2\pi \right) \right] - \left[0 + 2 \sin 0 + \frac{1}{2}\left(\frac{\sin 0}{2} + 0 \right) \right]$

$$= \frac{a^2}{2} \left[2\pi + 0 + \frac{1}{2}\left(\frac{0}{2} + 2\pi\right) \right] - [0]$$

$$\therefore \quad \text{Area} = \frac{3\pi a^2}{2}$$

(b) The alternative method is based on the fact that the curve is symmetrical about $\theta = 0$. Therefore, the area of the top half can be found and, when doubled, will give the area of the

$$A = 2 \times \left[\frac{1}{2} \int_0^\pi r^2 \, d\theta \right] \text{ etc.}$$

Example 12

Find the area of one loop of the curve $r = 2\cos 2\theta$.

Solution

This has already been sketched - (one loop only).

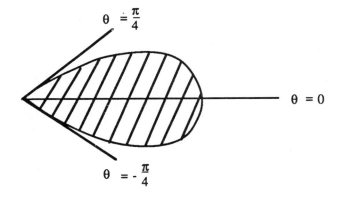

$$\text{Area} = \frac{1}{2} \int_{-\pi/4}^{\pi/4} r^2 \, d\theta = \frac{1}{2} \int_{-\pi/4}^{\pi/4} (2\cos 2\theta)^2 \, d\theta$$

$$= \frac{1}{2} \int_{-\pi/4}^{\pi/4} 4\cos^2 2\theta \, d\theta \qquad \text{but } \cos 4\theta = 2\cos^2 2\theta - 1$$

$$\therefore \cos 4\theta + 1 = 2\cos^2 2\theta$$

$$\therefore 2(\cos 4\theta + 1) = 4\cos^2 2\theta$$

$$\therefore A = \frac{1}{2} \int_{-\pi/4}^{\pi/4} 2(\cos 4\theta + 1) \, d\theta$$

$$= \int_{-\frac{\pi}{4}}^{\frac{\pi}{4}} (\cos 4\theta + 1) \, d\theta$$

$$= \left[\frac{\sin 4\theta}{4} + \theta \right]_{-\frac{\pi}{4}}^{\frac{\pi}{4}}$$

$$= \left[\frac{\sin 4\left(\frac{\pi}{4}\right)}{4} + \frac{\pi}{4} \right] - \left[\frac{\sin 4\left(-\frac{\pi}{4}\right)}{4} - \frac{\pi}{4} \right]$$

$$= \left[\frac{0}{4} + \frac{\pi}{4} \right] - \left[\frac{0}{4} - \frac{\pi}{4} \right]$$

$$\therefore \text{Area} = \frac{\pi}{4} + \frac{\pi}{4}$$

$$= \frac{\pi}{2}$$

e) *Example ('A' level question)*

Sketch the curve whose polar equation is $r^2 = 2 \sin 2\theta$. Find the area enclosed by a loop of the curve.

Find the cartesian equation of the curve and hence show that $x = \dfrac{2t}{1 + t^4}$ and $y = \dfrac{2t^3}{1 + t^4}$ may be taken as parametric equations of the curve.

Find the value of t for the points on the curve at which the tangents are parallel to the y axis.

Solution

(This question has been chosen as it involves several different topics - all in one question, ie polars, cartesians, parameters, gradients.)

$$r^2 = 2 \sin 2\theta \quad \therefore r = \pm \sqrt{2 \sin 2\theta}$$

For real values of r, $2 \sin 2\theta$ must be positive or zero, ie $\sin 2\theta \geq 0$. This will only be true for values of 2θ between 0 and π, and between 2π and 3π.

$$0 \leq 2\theta < \pi \quad \text{and} \quad 2\pi \leq 2\theta \leq 3\pi$$

$$0 \leq \theta \leq \frac{\pi}{2} \quad \text{and} \quad \pi \leq \theta \leq \frac{3\pi}{2}$$

So the curve does not exist between $\frac{\pi}{2}$ and π or between $\frac{3\pi}{2}$ and 2π.

Taking values of θ at intervals of $\frac{\pi}{12}$ gives the following results for $0 \leq \theta \leq \frac{\pi}{2}$

θ	0	$\frac{\pi}{12}$	$\frac{2\pi}{12}$	$\frac{3\pi}{12}$	$\frac{4\pi}{12}$	$\frac{5\pi}{12}$	$\frac{6\pi}{12}$
sin 2θ	0	0.5	0.9	1	0.9	0.5	0
$r^2 = 2 \sin 2\theta$	0	1.0	1.8	2	1.8	1.0	0
$r = \pm \sqrt{r^2}$	0	±1.0	±1.3	±1.4	±1.3	±1.0	0

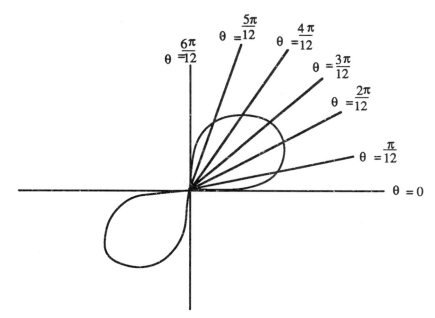

Area enclosed by loop $= \frac{1}{2} \int_0^{\frac{\pi}{2}} r^2 \, d\theta = \frac{1}{2} \int_0^{\frac{\pi}{2}} 2 \sin 2\theta \, d\theta$

$$= \frac{1}{2} \left[-\frac{2 \cos 2\theta}{2} \right]_0^{\frac{\pi}{2}}$$

$$= \frac{1}{2} \left[-\cos 2 \left(\frac{\pi}{2} \right) \right] - \frac{1}{2} [-\cos 0]$$

$$= \frac{1}{2} [-(-1)] - \frac{1}{2} [-1]$$

∴ Area = 1

To find the cartesian equation it is necessary to use $x^2 + y^2 = r^2$, $\sin \theta = \frac{y}{r}$ and $\cos \theta = \frac{x}{r}$

$r^2 = 2 \sin 2\theta$

$r^2 = 2 . 2 \sin \theta . \cos \theta$

$r^2 = 4 \sin \theta . \cos \theta$ replacing:

∴ $x^2 + y^2 = 4 . \frac{y}{r} . \frac{x}{r}$

PURE MATHEMATICS   CONICS

$\therefore \quad x^2 + y^2 = \dfrac{4xy}{r^2}$

$\therefore \quad x^2 + y^2 = \dfrac{4xy}{(x^2 + y^2)}$

$\therefore \quad (x^2 + y^2)^2 = 4xy$ is the required equation.

Replacing $x = \dfrac{2t}{1 + t^4}$ and $y = \dfrac{2t^3}{1 + t^4}$ into each side in turn gives:

$$\begin{aligned}
\text{LHS} &= (x^2 + y^2)^2 & \text{RHS} &= 4xy \\
&= \left[\left(\dfrac{2t}{1 + t^4}\right)^2 + \left(\dfrac{2t^3}{1 + t^4}\right)^2\right]^2 & &= 4 \cdot \dfrac{2t}{(1 + t^4)} \cdot \dfrac{2t^3}{(1 + t^4)} \\
&= \left[\dfrac{4t^2}{(1 + t^4)^2} + \dfrac{4t^6}{(1 + t^4)^2}\right]^2 & &= \dfrac{16t^4}{(1 + t^4)^2} \\
&= \left[\dfrac{4t^2 + 4t^6}{(1 + t^4)^2}\right]^2 & &= \left[\dfrac{4t^2}{1 + t^4}\right]^2 \\
&= \left[\dfrac{4t^2(1 + t^4)}{(1 + t^4)^2}\right]^2 \\
&= \left[\dfrac{4t^2}{1 + t^4}\right]^2
\end{aligned}$$

$\therefore \quad$ LHS = RHS

If the tangents are parallel to the y axis, then they will have an infinite gradient.

Using $\dfrac{dy}{dx} = \dfrac{dy}{dt} \cdot \dfrac{dt}{dx}$

with $\dfrac{dy}{dt} = \dfrac{(1 + t^4) \cdot 6t^2 - 2t^3(4t^3)}{(1 + t^4)^2}$ and $\dfrac{dx}{dt} = \dfrac{(1 + t^4)(2) - 2t(4t^3)}{(1 + t^4)^2}$

$\qquad\qquad\quad = \dfrac{6t^2 + 6t^6 - 8t^6}{(1 + t^4)^2} \qquad\qquad\qquad\quad = \dfrac{2 + 2t^4 - 8t^4}{(1 + t^4)^2}$

$\qquad\qquad\quad = \dfrac{6t^2 - 2t^6}{(1 + t^4)^2} \qquad\qquad\qquad\qquad\;\; = \dfrac{2 - 6t^4}{(1 + t^4)^2}$

$\therefore \quad \dfrac{dy}{dx} = \dfrac{6t^2 - 2t^6}{(1 + t^4)^2} \cdot \dfrac{(1 + t^4)^2}{2 - 6t^4}$

$\qquad\qquad = \dfrac{6t^2 - 2t^6}{2 - 6t^4}$

This will be infinite if either $t = \infty$ or $2 - 6t^4 = 0$, ie

$\qquad 6t^4 = 2$

$\therefore \quad t^4 = \dfrac{1}{3}$

CONICS — PURE MATHEMATICS

$$\therefore \quad t = \pm \sqrt[4]{\tfrac{1}{3}}$$

$$\therefore \quad t = \left(\tfrac{1}{3}\right)^{\tfrac{1}{4}} \quad \text{or} \quad -\left(\tfrac{1}{3}\right)^{\tfrac{1}{4}}$$

or this can be written as:

$$t = (3^{-1})^{\tfrac{1}{4}} \quad \text{or} \quad -(3^{-1})^{\tfrac{1}{4}}$$

$$\therefore \quad t = 3^{-\tfrac{1}{4}} \quad \text{or} \quad -3^{\tfrac{1}{4}}$$

8.3 Ellipse

a) *Equation of the ellipse*

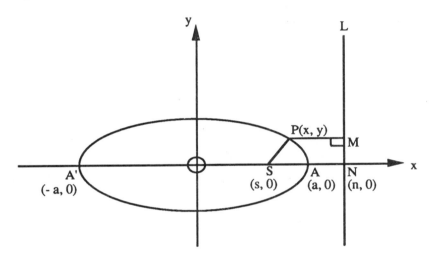

An ellipse is the locus of a point, P, which moves so that the ratio of its distance from a fixed point to its distance from a fixed line is constant and less than 1, ie

$$\frac{PS}{PM} = e < 1 \qquad \text{where S is the fixed point (s, 0) called the focus and L is the fixed line } (x = n) \text{ called the directrix.}$$

The ratio of the two distances (e) is called the eccentricity of the ellipse; it can take different values for different ellipses but it is always less than 1 and remains a constant for any particular ellipse. **It must not be confused with the mathematical constant e, used in logs, hyperbolic functions etc.**

If A and A' have coordinates (a, 0) and (- a, 0) respectively, then

$$\frac{AS}{AN} = e \qquad\qquad \text{and} \quad \frac{A'S}{A'N} = e$$

$$AS = a - s \qquad\qquad\qquad A'S = a + s$$

$$AN = n - a \qquad\qquad\qquad A'N = n + a$$

$$\therefore \quad \frac{a - s}{n - a} = e \qquad\qquad\qquad \frac{a + s}{n + a} = e$$

$$\therefore \quad a - s = (n - a)e \qquad\qquad a + s = (n + a)e$$

PURE MATHEMATICS — CONICS

∴ $a - s = ne - ae$ $\qquad\qquad$ $a + s = ne + ae$

Combining these two equations gives:

$\left.\begin{array}{l} a - s = ne - ae \\ a + s = ne + ae \end{array}\right\}$ add $\qquad$ $\left.\begin{array}{l} a - s = ne - ae \\ a + s = ne + ae \end{array}\right\}$ subtract

$\qquad 2a = 2ne \qquad\qquad\qquad\qquad -2s = -2ae$

∴ $\qquad \dfrac{2a}{2e} = n \qquad\qquad\qquad\qquad s = \dfrac{-2ae}{-2}$

∴ $\qquad \dfrac{a}{e} = n \qquad\qquad\qquad\qquad s = ae$

The equation of the directrix is $x = \dfrac{a}{e}$ and the focus is the point $(ae, 0)$.

To find the equation of the ellipse a general point $P(x, y)$ on the ellipse is taken and then:

$$\dfrac{PS}{PM} = e$$

∴ $\qquad PS = ePM$

Using Pythagoras theorem:

$$PS^2 = (x - ae)^2 + (y - 0)^2$$
$$= x^2 - 2aex + a^2e^2 + y^2$$

and $\qquad PM = \dfrac{a}{e} - x \qquad$ where $\dfrac{a}{e}$ is the x coordinate of M

∴ $\qquad PM^2 = \left(\dfrac{a}{e} - x\right)^2$

$$= \dfrac{a^2}{e^2} - \dfrac{2ax}{e} + x^2$$

Replacing in $\qquad (PS)^2 = (ePM)^2$

ie $\qquad PS^2 = e^2 PM^2$

gives $\quad x^2 - 2aex + a^2e^2 + y^2 = e^2\left(\dfrac{a^2}{e^2} - \dfrac{2ax}{e} + x^2\right)$

$\qquad x^2 - 2aex + a^2e^2 + y^2 = a^2 - 2aex + e^2x^2$

$\qquad x^2 - 2aex + y^2 + 2aex - e^2x^2 = a^2 - a^2e^2$

$\qquad x^2(1 - e^2) + y^2 = a^2(1 - e^2)$

∴ $\qquad \dfrac{x^2(1 - e^2)}{a^2(1 - e^2)} + \dfrac{y^2}{a^2(1 - e^2)} = \dfrac{a^2(1 - e^2)}{a^2(1 - e^2)}$

∴ $\qquad \dfrac{x^2}{a^2} + \dfrac{y^2}{a^2(1 - e^2)} = 1$

and this is usually written as:

$$\frac{x^2}{a^2} + \frac{y^2}{b^2} = 1 \qquad \text{where } b^2 = a^2(1 - e^2)$$

Since the equation contains x^2 and y^2 the curve is symmetrical about both axes. Therefore, a second focus exists at $(-ae, 0)$ and a second directrix at $x = -\frac{a}{e}$. Also, if $x = 0$, $\frac{y^2}{b^2} = 1$. Therefore $y = \pm b$, so the graph contains all the important points shown below:

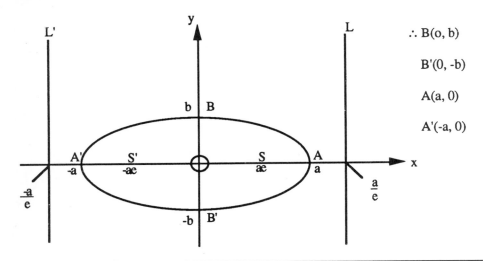

∴ B(0, b)

B'(0, -b)

A(a, 0)

A'(-a, 0)

Foci: S(ae, 0) and S'(-ae, 0)   Directrices: L $x = \frac{a}{e}$ and L' $x = -\frac{a}{e}$

The axes of symmetry cross at the centre of the ellipse, ie at 0, in the above diagram.

Any chord passing through the centre of the ellipse is called a diameter.

AA' is called the major axis and BB' the minor axis of the ellipse.

Example 13

For the ellipse $4x^2 + 9y^2 = 36$, find the foci and the directrices

Solution

Rewriting $4x^2 + 9y^2 = 36$ to make the RHS = 1

gives $\quad \dfrac{4x^2}{36} + \dfrac{9y^2}{36} = \dfrac{36}{36}$

∴ $\quad \dfrac{x^2}{9} + \dfrac{y^2}{4} = 1$

∴ $\quad a^2 = 9$ and $b^2 = 4$ but $b^2 = a^2(1 - e^2)$

∴ $\quad 4 = 9(1 - e^2)$

∴ $\quad 4 = 9 - 9e^2$

∴ $\quad 9e^2 = 9 - 4 = 5$

PURE MATHEMATICS — CONICS

$\therefore \quad e^2 = \dfrac{5}{9}$

$\therefore \quad e = \dfrac{\sqrt{5}}{3}$

The foci are given by (ae, 0) and (- ae, 0), so

$$ae = 3 \cdot \dfrac{\sqrt{5}}{3} = \sqrt{5}$$

$\therefore \quad$ foci $= (\sqrt{5}, 0)\ (-\sqrt{5}, 0)$

The directrices are given by $x = \dfrac{a}{e}$ and $x = -\dfrac{a}{e}$, so

$$\dfrac{a}{e} = \dfrac{3}{\frac{\sqrt{5}}{3}} = \dfrac{9}{\sqrt{5}}$$

$\therefore \quad$ directrices $x = \dfrac{9}{\sqrt{5}}$ and $x = -\dfrac{9}{\sqrt{5}}$

---

b) *Parametric coordinates*

When x and y are expressed in terms of a third variable - known as a parameter - then the equation of the curve is given in terms of two parametric equations.

Since the equation of the ellipse is of the form

$$(\ )^2 + (\ )^2 = 1$$

a trigonometric parameter θ (called the eccentric angle) is an obvious choice.

because $\cos^2\theta + \sin^2\theta = 1$

leads to $\dfrac{a^2\cos^2\theta}{a^2} + \dfrac{b^2\sin^2\theta}{b^2} = 1 \quad$ ie $x^2 = a^2\cos^2\theta$ and $y^2 = b^2\sin^2\theta$

Therefore, the parametric coordinates of any point P on an ellipse are (a cos θ, b sin θ) giving corresponding parametric equations of:

$$\boxed{x = a\cos\theta \qquad y = b\sin\theta}$$

Example 14

The eccentric angles of two points P and Q on an ellipse differ by a constant k. Find the locus of the mid-point of PQ.

CONICS PURE MATHEMATICS

Solution

Say   P is the point (a cos θ, b sin θ)

then   Q will be the point (a cos(θ + k), b sin(θ + k))

and the coordinates of M, the mid-point of PQ, will be given by:

$$x = \frac{(a\cos\theta + a\cos(\theta + k))}{2}$$

and $$y = \frac{(b\sin\theta + b\sin(\theta + k))}{2}$$

These can be simplified using the factor formulae for the sum of two sines and the sum of the two cosines.

∴ $x = \frac{a}{2}\left(2\cos\left(\frac{2\theta + k}{2}\right) \cdot \cos\frac{k}{2}\right)$ $y = \frac{b}{2}\left(2\sin\left(\frac{2\theta + k}{2}\right) \cdot \cos\frac{k}{2}\right)$

∴ $x = a\cos\left(\theta + \frac{k}{2}\right) \cdot \cos\frac{k}{2}$ $y = b\sin\left(\theta + \frac{k}{2}\right) \cdot \cos\frac{k}{2}$

∴ $\dfrac{x}{a\cos\frac{k}{2}} = \cos\left(\theta + \frac{k}{2}\right)$ $\dfrac{y}{b\cos\frac{k}{2}} = \sin\left(\theta + \frac{k}{2}\right)$

Squaring and adding gives:

$$\left(\frac{x}{a\cos\frac{k}{2}}\right)^2 + \left(\frac{y}{b\cos\frac{k}{2}}\right)^2 = \cos^2\left(\theta + \frac{k}{2}\right) + \sin^2\left(\theta + \frac{k}{2}\right)$$

∴ $$\frac{x^2}{a^2\cos^2\frac{k}{2}} + \frac{y^2}{b^2\cos^2\frac{k}{2}} = 1$$

∴ $\dfrac{x^2}{a^2} + \dfrac{y^2}{b^2} = \cos^2\frac{k}{2}$ is the required locus.

---

c)   *Equations of tangent and normal*

The equation of the tangent to an ellipse is found in the usual way, ie by finding the gradient (m) from $\frac{dy}{dx}$ and then using the equation of a straight line: y = mx + c

Example 15

Find the equations of the tangent and normal to the ellipse $9x^2 + 16y^2 = 25$ at the point (1, 1).

Solution

To find $\frac{dy}{dx}$ first

$$16y^2 = 25 - 9x^2$$

∴ $16 \cdot 2y \dfrac{dy}{dx} = -18x$

PURE MATHEMATICS　　　　　　　　　　　　CONICS

$\therefore \quad \dfrac{dy}{dx} = \dfrac{-18x}{32y}$

At the point (1, 1) $\quad \dfrac{dy}{dx} = \dfrac{-18(1)}{32(1)}$

$$= \dfrac{-9}{16}$$

Therefore, the equation of the tangent is:

$$y = mx + c$$

$\therefore \quad y = \dfrac{-9x}{16} + c$

Using (1, 1) to find c $\quad 1 = \dfrac{-9}{16}(1) + c$

$\therefore \quad 1 + \dfrac{9}{16} = c$

$\therefore \quad \dfrac{25}{16} = c$

Therefore, the equation of the tangent is:

$$y = \dfrac{-9}{16}x + \dfrac{25}{16}$$

or $\quad 16y = -9x + 25$

The gradient of the normal is:

$$m' = \dfrac{-1}{m}$$

$\therefore \quad m' = \dfrac{-1}{\frac{-9}{16}}$

$$= \dfrac{16}{9}$$

Therefore, the equation of the normal is:

$$y = \dfrac{16x}{9} + c$$

Using (1, 1) to find c $\quad 1 = \dfrac{16}{9}(1) + c$

$\therefore \quad 1 - \dfrac{16}{9} = c$

$\therefore \quad \dfrac{-7}{9} = c$

CONICS   PURE MATHEMATICS

Therefore, the equation of the normal is:

$$y = \frac{16x}{9} - \frac{7}{9}$$

or $$9y = 16x - 7$$

However, equations of tangents and normals are often required in terms of the parameter θ. The method is identical although it probably seems more complicated.

Parametric form of the equation of the tangent to the ellipse:

$$\frac{x^2}{a^2} + \frac{y^2}{b^2} = 1 \text{ at } (a\cos\theta, \, b\sin\theta)$$

$$\frac{dy}{dx} = \frac{dy}{d\theta} \cdot \frac{d\theta}{dx} \qquad\qquad x = a\cos\theta \quad \text{and} \quad y = b\sin\theta$$

$$\therefore \qquad\qquad\qquad\qquad \frac{dx}{d\theta} = -a\sin\theta \quad \therefore \quad \frac{dy}{d\theta} = b\cos\theta$$

$$= b\cos\theta \cdot \frac{1}{-a\sin\theta}$$

$$\therefore \quad \frac{dy}{dx} = \frac{-b\cos\theta}{a\sin\theta} = m$$

Using $y = mx + c$

$$y = \frac{-b\cos\theta}{a\sin\theta} x + c \qquad \text{and to find } c$$

$$b\sin\theta = \frac{-b\cos\theta}{a\sin\theta}(a\cos\theta) + c$$

$$\therefore \quad b\sin\theta = \frac{-b\cos^2\theta}{\sin\theta} + c$$

$$\therefore \quad b\sin\theta + \frac{b\cos^2\theta}{\sin\theta} = c$$

$$\therefore \quad \frac{b\sin^2\theta + b\cos^2\theta}{\sin\theta} = c$$

$$\therefore \quad \frac{b(\sin^2\theta + \cos^2\theta)}{\sin\theta} = c \qquad \text{but } \sin^2\theta + \cos^2\theta = 1$$

$$\therefore \quad \frac{b}{\sin\theta} = c$$

Therefore, the equation becomes:

$$y = \frac{-b\cos\theta \, x}{a\sin\theta} + \frac{b}{\sin\theta} \quad \text{multiplying by } a\sin\theta$$

or $$\boxed{(a\sin\theta)y = -(b\cos\theta)x + ab}$$

PURE MATHEMATICS — CONICS

Parametric form of the equation of the normal to:

$$\frac{x^2}{a^2} + \frac{y^2}{b^2} = 1 \qquad \text{at } (a\cos\theta,\ b\sin\theta)$$

gradient $m' = \dfrac{-1}{m}$

$$= \frac{-1}{\frac{-b\cos\theta}{a\sin\theta}}$$

$\therefore\ m' = \dfrac{a\sin\theta}{b\cos\theta}$

Using $y = m'x + c$

$$y = \frac{a\sin\theta}{b\cos\theta}\, x + c \qquad \text{and to find } c$$

$$b\sin\theta = \frac{a\sin\theta}{b\cos\theta}(a\cos\theta) + c$$

$\therefore\ b\sin\theta = \dfrac{a^2\sin\theta}{b} + c$

$\therefore\ b\sin\theta - \dfrac{a^2\sin\theta}{b} = c$

$\therefore\ \dfrac{b^2\sin\theta - a^2\sin\theta}{b} = c$

Therefore, the equation becomes:

$$y = \frac{a\sin\theta}{b\cos\theta}\, x + \frac{(b^2\sin\theta - a^2\sin\theta)}{b}$$

multiplying by $b\cos\theta$

$$(b\cos\theta)y = (a\sin\theta)x + (b^2\sin\theta - a^2\sin\theta)\cos\theta$$

$\therefore\ \boxed{(b\cos\theta)y = (a\sin\theta)x + (b^2 - a^2)\sin\theta\cos\theta}$

These equations are not easy to derive because of the amount of trigonometry involved so great care is needed at every stage of the work.

Example 16

The normal at the point $(a\cos\theta,\ b\sin\theta)$ on the ellipse $\dfrac{x^2}{a^2} + \dfrac{y^2}{b^2} = 1$ meets the axes at L and M. Find the locus of the mid-point at LM.

CONICS   PURE MATHEMATICS

Solution

Using the equation just derived, ie

$$(b\cos\theta)\cdot y = (a\sin\theta)\cdot x + (b^2 - a^2)\sin\theta\cos\theta$$

on x axis, $y = 0$:

$\therefore \quad b\cos\theta(0) = (a\sin\theta)\cdot x + (b^2 - a^2)\sin\theta\cos\theta$

$\therefore \quad 0 = (a\sin\theta)x + (b^2 - a^2)\sin\theta\cos\theta$

$\therefore \quad -(a\sin\theta)x = (b^2 - a^2)\sin\theta\cos\theta$

$\therefore \quad x = \dfrac{(b^2 - a^2)\sin\theta\cos\theta}{-a\sin\theta}$

Therefore, L is the point $\left(-\dfrac{(b^2 - a^2)\cos\theta}{a},\ 0\right)$

on y axis $x = 0$:

$\therefore \quad (b\cos\theta)y = a\sin\theta(0) + (b^2 - a^2)\sin\theta\cos\theta$

$\therefore \quad (b\cos\theta)\cdot y = 0 + (b^2 - a^2)\sin\theta\cos\theta$

$\therefore \quad y = \dfrac{(b^2 - a^2)\sin\theta\cos\theta}{b\cos\theta}$

Therefore, M is the point $\left(0,\ \dfrac{(b^2 - a^2)\sin\theta}{b}\right)$

The mid-point of LM, P say, will have coordinates

$$\left(\dfrac{-\dfrac{(b^2 - a^2)\cdot\cos\theta}{a} + 0}{2},\ \dfrac{0 + \dfrac{(b^2 - a^2)\cdot\sin\theta}{b}}{2}\right)$$

$\therefore \quad x = \dfrac{-(b^2 - a^2)\cos\theta}{2a} \quad \text{and} \quad y = \dfrac{(b^2 - a^2)\sin\theta}{2b}$

$\therefore \quad \dfrac{2ax}{(b^2 - a^2)} = -\cos\theta \quad \text{and} \quad \dfrac{2by}{(b^2 - a^2)} = \sin\theta$

Squaring and adding to eliminate θ gives:

$$\left(\dfrac{2ax}{b^2 - a^2}\right)^2 + \left(\dfrac{2by}{b^2 - a^2}\right)^2 = (-\cos\theta)^2 + \sin^2\theta$$

$\therefore \quad \dfrac{4a^2x^2}{(b^2 - a^2)^2} + \dfrac{4b^2y^2}{(b^2 - a^2)^2} = 1$

or $\quad 4a^2x^2 + 4b^2y^2 = (b^2 - a^2)^2 \quad$ is the required locus of the mid-point of LM

PURE MATHEMATICS — CONICS

d) *Examples*

'A' level questions on the ellipse can often be quite long and very complicated. Therefore, two examples have been included here - the second of which uses parameters.

Example A

The line $y = mx + c$ cuts the ellipse $b^2x^2 + a^2y^2 = a^2b^2$ at the points P and Q. Find the coordinates of M, the mid-point of P and Q. Show that the locus of M as c varies is a straight line $y = m'x$ where $mm' = \dfrac{-b^2}{a^2}$.

Hence find the equation of the chord of the ellipse $\dfrac{x^2}{3} + \dfrac{y^2}{2} = 1$ with mid-point $(1, 1)$.

Find also the equations of the tangents to this ellipse which are parallel to the diameter $y = x$.

Solution

First it is necessary to find the points of intersection of the line and the ellipse:

$$y = mx + c$$
$$\therefore \quad y^2 = (mx + c)^2$$
$$= m^2x^2 + 2mcx + c^2$$

Replacing for y^2 in the equation of the ellipse gives:

$$b^2x^2 + a^2(m^2x^2 + 2mcx + c^2) = a^2b^2$$
$$\therefore \quad b^2x^2 + a^2m^2x^2 + 2a^2mcx + a^2c^2 - a^2b^2 = 0$$
$$\therefore \quad (b^2 + a^2m^2)x^2 + 2a^2mcx + a^2(c^2 - b^2) = 0$$

The solution (or roots) of this quadratic equation would give the x coordinates (x_1 and x_2) of the points of intersection of the straight line and the ellipse. However, it is obviously a very complicated equation to solve so, before attempting to solve it, a sketch will be drawn:

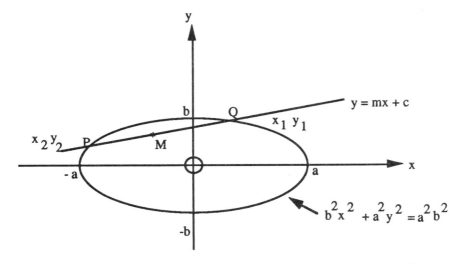

The aim of the question is to find the locus of M as c varies and the coordinates of M are given by

$$x = \dfrac{x_1 + x_2}{2} \quad \text{and} \quad y = \dfrac{y_1 + y_2}{2}$$

because it is the mid-point of PQ.

CONICS　　　　　　　　　　　　　　　　　　　　　　PURE MATHEMATICS

Since the roots of the quadratic equation are x_1 and x_2, it follows that the sum of the roots $x_1 + x_2$ can be found directly from the equation using 'sum of roots $= -\frac{b}{a}$'.

$\therefore \qquad x_1 + x_2 = \frac{-2a^2mc}{b^2 + a^2m^2}$

$\therefore \qquad x = \frac{x_1 + x_2}{2} = \frac{-a^2mc}{b^2 + a^2m^2}$

Since M is on the line $y = mx + c$, the y coordinate of M can be found most easily by replacing $x = \frac{-a^2mc}{b^2 + a^2m^2}$ in the straight line equation.

$$y = m\left(\frac{-a^2mc}{b^2 + a^2m^2}\right) + c$$

$$y = \frac{-a^2m^2c + cb^2 + ca^2m^2}{b^2 + a^2m^2}$$

$\therefore \qquad y = \frac{cb^2}{b^2 + a^2m^2}$

Therefore, M is the point $\left(\frac{-a^2mc}{b^2 + a^2m^2}, \frac{cb^2}{b^2 + a^2m^2}\right)$

To find the locus of M as c varies, it is necessary to eliminate c between the equations

$$x = \frac{-a^2mc}{b^2 + a^2m^2} \qquad \text{and} \qquad y = \frac{cb^2}{b^2 + a^2m^2}$$

$\therefore \quad (b^2 + a^2m^2)x = a^2mc \qquad\qquad (b^2 + a^2m^2)y = cb^2$

$\therefore \quad \dfrac{(b^2 + a^2m^2)x}{-a^2m} = c \qquad\qquad \dfrac{(b^2 + a^2m^2)y}{b^2} = c$

Therefore, it follows that:

$$\frac{(b^2 + a^2m^2)x}{-a^2m} = \frac{(b^2 + a^2m^2)y}{b^2}$$

$\therefore \quad \dfrac{x}{-a^2m} = \dfrac{y}{b^2}$ because $(b^2 + a^2m^2)$ cancels

$\therefore \quad -\dfrac{b^2}{a^2m} x = y$

$\qquad y = m'x \qquad \text{where } m' = \dfrac{-b^2}{a^2m} \qquad \therefore \quad mm' = \dfrac{-b^2}{a^2}$

The equation of the chord of the ellipse will be of the form $y = mx + c$ and its mid-point of (1, 1) will be just one point on the locus $y = m'x$. Therefore $1 = m'1$ giving $m' = 1$.

But $m' = \dfrac{-b^2}{a^2m}$ and $a^2 = 3$ and $b^2 = 2$ from the equation of the ellipse; so replacing gives:

$$1 = -\frac{2}{3m}$$

PURE MATHEMATICS   CONICS

$\therefore \quad m = -\dfrac{2}{3}$

$\therefore \quad y = mx + c \qquad$ becomes

$y = -\dfrac{2}{3}x + c \qquad$ and since $(1, 1)$ lies on this line

$1 = -\dfrac{2}{3}(1) + c$

$\therefore \quad 3 = -2 + 3c$

$\therefore \quad 5 = 3c$

$\therefore \quad \dfrac{5}{3} = c$

Therefore, the equation is:

$y = -\dfrac{2}{3}x + \dfrac{5}{3}$

or $\quad 3y = -2x + 5$

(It should be stressed that $y = m'x$ is the locus of the mid-points of all the chords that could be drawn in the ellipse; $3y = -2x + 5$ is the equation of one specific chord with mid-point $(1, 1)$.)

If a chord has become a tangent to the ellipse it means that the quadratic equation

$$(b^2 + a^2m^2)x^2 + 2a^2mcx + a^2(c^2 - b^2) = 0$$

must have equal roots because the two values of x_1 and x_2 coincide. For equal roots '$b^2 = 4ac$' in general.

$\therefore \quad (2a^2mc)^2 = 4(b^2 + a^2m^2)a^2(c^2 - b^2)$

But from the equation of the ellipse $a^2 = 3$, $b^2 = 2$ and $m = 1$ because the tangent is parallel to $y = x$ with gradient 1.

$\therefore \quad [2(3)(1)(c)]^2 = 4[2 + 3(1)](3)(c^2 - 2)$

$\therefore \quad (6c)^2 = 12(5)(c^2 - 2)$

$\therefore \quad 36c^2 = 60c^2 - 120$

$\therefore \quad 36c^2 - 60c^2 = -120$

$\therefore \quad -24c^2 = -120$

$\therefore \quad c^2 = \dfrac{-120}{-24}$

$\qquad\qquad\quad = 5$

$\therefore \quad c = \pm\sqrt{5}$

Therefore, the equations of the tangents are:

$y = mx + c$

ie $\quad y = x \pm \sqrt{5}$

CONICS　　　　　　　　　　　　　　　　　　PURE MATHEMATICS

Example B

Show that the equation of the normal at the point $P(a\cos\theta, b\sin\theta)$ to the ellipse $b^2x^2 + a^2y^2 = a^2b^2$ is

$$ax\sec\theta - by\csc\theta = a^2 - b^2$$

The normal cuts the x axis at A and the y axis at B. Find the coordinates of A and B and show that the maximum area of triangle AOB is $\dfrac{(a^2-b^2)^2}{4ab}$

Given that the foci of the ellipse are S_1 and S_2 show that $S_1P + S_2P = 2a$.

Solution

The equation of the normal has already been derived as $(b\cos\theta)y = (a\sin\theta)x + (b^2-a^2)\sin\theta\cos\theta$ (in an examination question the actual derivation would have to be given).

Dividing through by $\sin\theta\cos\theta$:

$$\frac{(b\cos\theta)y}{\sin\theta\cos\theta} = \frac{(a\sin\theta)x}{\sin\theta\cos\theta} + \frac{(b^2-a^2)\sin\theta\cos\theta}{\sin\theta\cos\theta}$$

$$\frac{by}{\sin\theta} = \frac{ax}{\cos\theta} + (b^2 - a^2)$$

$\therefore\quad by\csc\theta = ax\sec\theta + (b^2 - a^2)$

giving $\quad ax\sec\theta - by\csc\theta = a^2 - b^2$

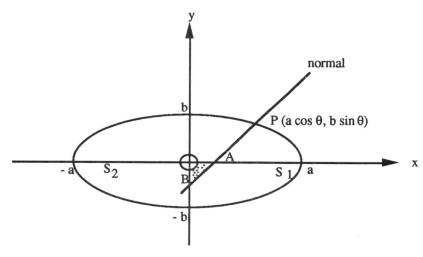

On the x axis $\quad y = 0$

$\therefore\quad ax\sec\theta = a^2 - b^2$

$\therefore\quad x = \dfrac{a^2 - b^2}{a\sec\theta}$ 　　　Therefore A is the position $\left(\dfrac{a^2-b^2}{a\sec\theta},\ 0\right)$

On the y axis $\quad x = 0$

$\therefore\quad -by\csc\theta = a^2 - b^2$

$\therefore\quad y = \dfrac{a^2 - b^2}{-b\csc\theta}$ 　　　Therefore B is the position $\left(0,\ \dfrac{a^2-b^2}{-b\csc\theta}\right)$

PURE MATHEMATICS — CONICS

The area of the triangle AOB

$$= \tfrac{1}{2}(OA)(OB)$$

$$= \tfrac{1}{2}\left(\frac{a^2 - b^2}{a \sec \theta}\right)\left(\frac{a^2 - b^2}{b \operatorname{cosec} \theta}\right)$$

$$= \tfrac{1}{2}\frac{(a^2 - b^2)^2}{ab \sec \theta \operatorname{cosec} \theta}$$

$$= \frac{(a^2 - b^2)^2}{2ab} \cos \theta \sin \theta$$

$$= \frac{(a^2 - b)^2}{2ab} \cdot \frac{\sin 2\theta}{2}$$

$$\therefore \text{Area } \triangle AOB = \frac{(a^2 - b^2)^2}{4ab} \sin 2\theta$$

This will be a maximum when $\sin 2\theta$ has its maximum value of 1, then

$$\text{Area} = \frac{(a^2 - b^2)^2}{4ab}$$

The easiest way to prove that $S_1P + S_2P = 2a$ is from the basic geometry of the ellipse, rather than finding S_1P and S_2P in terms of θ.

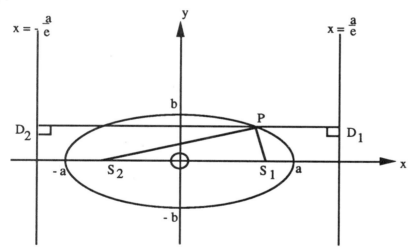

D_1 and D_2 are the directrices with equations $x = \frac{a}{e}$ and $x = -\frac{a}{e}$. From the basic definition of an ellipse:

$$\frac{PS_1}{PD_1} = e \qquad\qquad \therefore \quad \frac{PS_2}{PD_2} = e$$

$$PS_1 = ePD_1 \qquad\qquad \therefore \quad PS_2 = ePD_2$$

$\therefore \quad PS_1 + PS_2 = e(PD_1 + PD_2) \quad$ but $\quad PD_1 + PD_2 = \frac{a}{e} + \frac{a}{e} = \frac{2a}{e}$

$\therefore \qquad\qquad\qquad PS_1 + PS_2 = e \cdot \frac{2a}{e} = 2a$

8.4 Hyperbola

a) *Equation of the hyperbola*

A hyperbola is the locus of a point P which moves so that the ratio of its distance from a fixed point to its distance from a fixed line is a constant greater than 1.

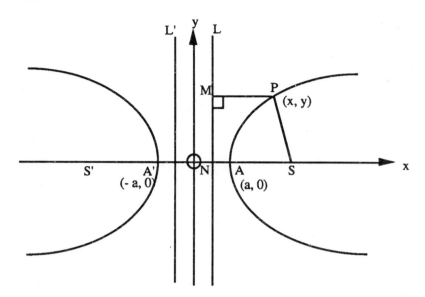

$$\therefore \quad \frac{PS}{PM} = e > 1 \quad \text{where S is the focus and M is a point on the directrix}$$

Since the curve is symmetrical about both axes, as was the ellipse, there are two foci at the points S and S', ie (ae, 0) and (- ae, 0), and two directrices with equations $x = \frac{a}{e}$ and $x = -\frac{a}{e}$

To derive the equation of the hyperbola a general point P(x, y) is taken and then:

$$PS = ePM$$

Therefore, by Pythagoras theorem:

$$PS^2 = (ae - x)^2 + (y - 0)^2$$
$$= a^2e^2 - 2aex + x^2 + y^2$$

and $\quad PM = x - \frac{a}{e}$

$$\therefore \quad PM^2 = \left(x - \frac{a}{e}\right)^2$$
$$= x^2 - \frac{2ax}{e} + \frac{a^2}{e^2}$$

Replacing in $\quad PS = ePM$

$$\therefore \quad PS^2 = e^2PM^2$$

gives $\quad a^2e^2 - 2eax + x^2 + y^2 = e^2\left(x^2 - \frac{2a}{e}x + \frac{a^2}{e^2}\right)$

$$a^2e^2 - 2aex + x^2 + y^2 = e^2x^2 - 2aex + a^2$$

PURE MATHEMATICS — CONICS

$\therefore \quad x^2 - e^2x^2 + y^2 - 2aex + 2aex = a^2 - a^2e^2$

$\therefore \quad x^2(1 - e^2) + y^2 = a^2(1 - e^2)$

However $(1 - e^2)$ will be negative because $e > 1$.

$\therefore \quad \dfrac{x^2(1 - e^2)}{a^2(1 - e^2)} + \dfrac{y^2}{a^2(1 - e^2)} = \dfrac{a^2(1 - e^2)}{a^2(1 - e^2)} = 1$

$\therefore \quad \dfrac{x^2}{a^2} - \dfrac{y^2}{a^2(e^2 - 1)} = 1$

and this is usually written as:

$$\boxed{\dfrac{x^2}{a^2} - \dfrac{y^2}{b^2} = 1 \quad \text{where } b^2 = a^2(e^2 - 1)}$$

AA' is called the major axis of the hyperbola and it should be noted that the curve does not exist for $-a < x < a$. 0 is the centre of the hyperbola.

Example 17

Find the equation of the hyperbola with centre at $(0, 0)$, one directrix $x = 2$ and the distance between the vertices 12.

Solution

The general equation is:

$$\dfrac{x^2}{a^2} - \dfrac{y^2}{b^2} = 1$$

The vertices are at A and A' so AA' = 2a.

$\therefore \quad 2a = 12$

$\therefore \quad a = 6$ so the vertices are $(6, 0)$ and $(-6, 0)$.

The equation of the directrix is:

$\therefore \quad \dfrac{a}{e} = 2$

$\therefore \quad \dfrac{6}{e} = 2$

$\therefore \quad 3 = e$

hence $\quad b^2 = a^2(e^2 - 1) = 6^2(3^2 - 1) = 36(8) = 288$

Therefore, the equation of the hyperbola is:

$$\dfrac{x^2}{36} - \dfrac{y^2}{288} = 1$$

CONICS PURE MATHEMATICS

b) *Asymptotes*

When x and y both become very large, $\frac{x^2}{a^2} - \frac{y^2}{b^2} = 1$ approximates to $\frac{x^2}{a^2} - \frac{y^2}{b^2} = 0$ because 1 is so small as to be negligible compared with x^2 and y^2.

$$\frac{x^2}{a^2} - \frac{y^2}{b^2} = 0 \qquad \therefore \qquad \frac{x^2}{a^2} = \frac{y^2}{b^2}$$

$$\therefore \qquad y^2 = \frac{b^2}{a^2} x$$

$$\therefore \qquad y = \pm \frac{b}{a} x$$

Therefore, as x and y tend to infinity, the equation of the hyperbola tends to the equations of two straight lines known as asymptotes, ie

$$\boxed{y = \frac{b}{a} x \text{ and } y = -\frac{b}{a} x}$$

Example 18

Write down the equations of the asymptotes to the hyperbola $\frac{x^2}{4} - \frac{y^2}{3} = 1$.

Solution

From the equation $\qquad a^2 = 4 \quad$ and $\quad b^2 = 3$

$\therefore \qquad\qquad\qquad a = \pm 2 \quad$ and $\quad b = \pm \sqrt{3}$

Therefore, the equations of the asymptotes are:

$$y = \frac{\sqrt{3}}{2} x \qquad \text{and} \qquad y = -\frac{\sqrt{3}}{2} x$$

or $\qquad 2y = \sqrt{3} x \qquad$ and $\qquad 2y = -\sqrt{3} x$

c) *The rectangular hyperbola*

A rectangular hyperbola is one whose asymptotes are at right angles to one another.

The asymptote $\quad y = \frac{b}{a} x \qquad$ has gradient $\quad = \frac{b}{a}$

The asymptote $\quad y = -\frac{b}{a} x \qquad$ has gradient $\quad = -\frac{b}{a}$

Therefore, since they are perpendicular $\quad \frac{b}{a}\left(-\frac{b}{a}\right) = -1$

$$\therefore \qquad -\frac{b^2}{a^2} = -1$$

$$\therefore \qquad b^2 = a^2$$

$$\therefore \qquad b = a$$

362

So the equations of the asymptotes become y = x and y = - x. The equation of a rectangular hyperbola is therefore:

$$\frac{x^2}{a^2} - \frac{y^2}{a^2} = 1$$

or $\quad x^2 - y^2 = a^2$

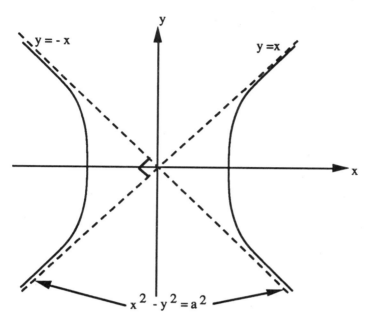

Also
$$b^2 = a^2(e^2 - 1)$$

∴ $\quad a^2 = a^2(e^2 - 1)$

∴ $\quad 1 = e^2 - 1$

∴ $\quad e^2 = 2$

∴ $\quad e = \sqrt{2}$

For a rectangular hyperbola $x^2 - y^2 = a^2$.

The equations of the asymptotes are y = x and y = - x and the eccentricity (e) of the hyperbola is always $\sqrt{2}$.

# CONICS	PURE MATHEMATICS

If the hyperbola is rotated through 45° about the origin, the asymptotes coincide with the x and y axes as shown below.

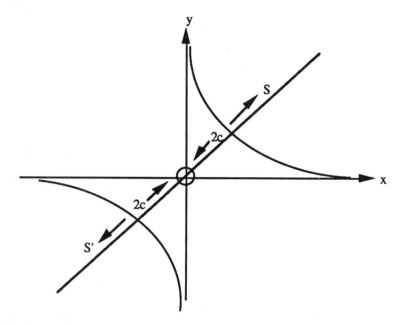

The equation of this form of the rectangular hyperbola is:

$$xy = c^2 \quad \text{where} \quad c^2 = \tfrac{1}{2}a^2$$

and the coordinates of the foci are $(\sqrt{2}c, \sqrt{2}c)$ and $(-\sqrt{2}c, -\sqrt{2}c)$.

d) *Parametric coordinates*

Since the equation of the hyperbola is of the form $(\ )^2 - (\ )^2 = 1$ a trigonometric parameter is implied $\sec^2\theta - \tan^2\theta = 1$.

$$\therefore \quad \frac{a^2\sec^2\theta}{a^2} - \frac{b^2\tan^2\theta}{b^2} = 1 \text{ giving } x^2 = a^2\sec^2\theta \text{ and } y^2 = b^2\tan^2\theta$$

so the parametric coordinates of any point P on the curve $\frac{x^2}{a^2} - \frac{y^2}{b^2} = 1$ are $(a\sec\theta, b\tan\theta)$ giving the parametric equations of:

$$x = a\sec\theta \quad y = b\tan\theta$$

Example 19

P is any point on the hyperbola $\frac{x^2}{a^2} - \frac{y^2}{b^2} = 1$ and Q is the point (a, b).

Find the locus of the point dividing PQ in the ratio 2 : 1.

Solution

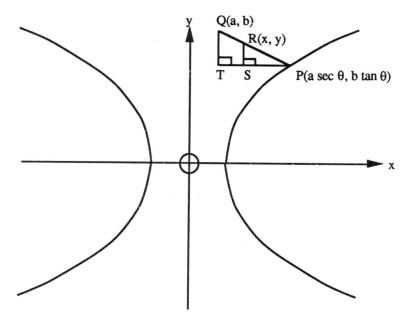

Let P be the point ($a \sec\theta$, $b \tan\theta$) on the hyperbola and R(x, y) the point dividing PQ in the ratio 2 : 1.

∴   PR : RQ = 2 : 1   ∴   PR : PQ = 2 : 3

∴ $\dfrac{PR}{PQ} = \dfrac{2}{3}$ ∴ 3PR = 2PQ

and by similar triangles $\dfrac{PS}{PT} = \dfrac{2}{3}$ ∴ 3PS = 2PT

and $\dfrac{SR}{TQ} = \dfrac{2}{3}$ ∴ 3SR = 2TQ

∴   3PS = 2PT   gives $3(a \sec\theta - x) = 2(a \sec\theta - a)$

∴ $3a \sec\theta - 3x = 2a \sec\theta - 2a$

∴ $a \sec\theta = 3x - 2a$

3SR = 2TQ gives $3(y - b \tan\theta) = 2(b - b \tan\theta)$

$3y - 3b \tan\theta = 2b - 2b \tan\theta$

∴ $3y - 2b = b \tan\theta$

Using $\sec^2\theta - \tan^2\theta = 1$ gives

$$\left(\dfrac{3x - 2a}{a}\right)^2 - \left(\dfrac{3y - 2b}{b}\right)^2 = 1$$

∴ $\dfrac{9x^2 - 12x + 4a^2}{a^2} - \left(\dfrac{9y^2 - 12by + 4b^2}{b^2}\right) = 1$

Multiplying by a^2b^2:

$$b^2(9x^2 - 12ax + 4a^2) - a^2(9y^2 - 12by + 4b^2) = a^2b^2$$

$$9b^2x^2 - 9a^2y^2 - 12ab^2x + 12a^2by - a^2b^2 = 0$$

The parametric coordinates for any pont P on the rectangular hyperbola $xy = c^2$ are easier to deal with.

They are $\left(ct, \dfrac{c}{t}\right)$ giving parametric equations of:

$$\boxed{x = ct \quad y = \dfrac{c}{t}}$$

Example 20

PQ is a chord of a rectangular hyperbola $xy = c^2$ and R is its mid-point.
If PQ is a constant length k, find the locus of R.

Solution

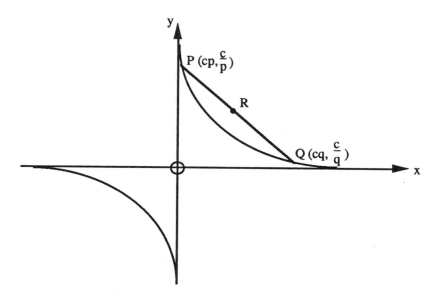

Let P be the point with parameter p and Q the point with parameter q.

Therefore, $P\left(cp, \dfrac{c}{p}\right)$ and $Q\left(cq, \dfrac{c}{q}\right)$ and R is the point (x, y).

$$PQ^2 = (cp - cq)^2 + \left(\dfrac{c}{p} - \dfrac{c}{q}\right)^2 \quad \text{but } PQ^2 = k^2$$

$\therefore \quad k^2 = c^2(p - q)^2 + c^2\left(\dfrac{1}{p} - \dfrac{1}{q}\right)^2$

$\therefore \quad k^2 = c^2(p - q)^2 + c^2\left(\dfrac{q - p}{pq}\right)^2$

$\therefore \quad k^2 = c^2(p - q)^2 + \dfrac{c^2}{p^2q^2}(q - p)^2 \quad \text{where} \quad (q - p)^2 = (p - q)^2$

PURE MATHEMATICS — CONICS

$$\therefore \quad k^2 = c^2(p-q)^2 \left[1 + \frac{1}{p^2q^2}\right]$$

Now it is necessary to find alternative expressions for $(p-q)^2$ and p^2q^2 using the coordinates of the mid-point R.

$$x = \frac{cp + cq}{2} \qquad \text{and} \qquad y = \frac{\frac{c}{p} + \frac{c}{q}}{2}$$

$$\therefore \quad x = \frac{c}{2}(p+q) \qquad \text{and} \qquad y = \frac{c}{2}\left(\frac{p+q}{pq}\right)$$

$$\therefore \quad \frac{2x}{c} = (p+q) \qquad \text{and} \qquad \frac{2y}{c} = \frac{p+q}{pq}$$

$$\therefore \quad pq = \frac{c}{2y}(p+q) = \frac{c}{2y} \cdot \frac{2x}{c}$$

$$\therefore \quad pq = \frac{x}{y}$$

and
$$(p-q)^2 = p^2 - 2pq + q^2$$
$$= (p+q)^2 - 4pq$$
$$= \left(\frac{2x}{c}\right)^2 - 4\left(\frac{x}{y}\right)$$

$$\therefore \quad (p-q)^2 = \frac{4x^2}{c^2} - \frac{4x}{y}$$

Hence $\quad k^2 = c^2(p-q)^2 \left[1 + \frac{1}{p^2q^2}\right]$

becomes $\quad k^2 = c^2\left(\frac{4x^2}{c^2} - \frac{4x}{y}\right)\left[1 + \frac{1}{\left(\frac{x}{y}\right)^2}\right]$

$$k^2 = \left(\frac{4x^2c^2}{c^2} - \frac{4xc^2}{y}\right)\left(1 + \frac{y^2}{x^2}\right)$$

$$k^2 = \left(\frac{4x^2y - 4c^2x}{y}\right)\left(\frac{x^2 + y^2}{x^2}\right)$$

$$= \frac{4x}{x^2y}(xy - c^2)(x^2 + y^2)$$

$$k^2 = \frac{4}{xy}(xy - c^2)(x^2 + y^2)$$

$$\therefore \quad xyk^2 = 4(xy - c^2)(x^2 + y^2) \text{ is the required locus equation.}$$

CONICS PURE MATHEMATICS

e) *Equations of tangent and normals*

As for the ellipse most of the work on tangents and normals is done in terms of the parameter θ for a hyperbola and in terms of t for a rectangular hyperbola.

Parametric form of the equation of the tangent to the hyperbola $\frac{x^2}{a^2} - \frac{y^2}{b^2} = 1$ at (a sec θ, b tan θ).

$$\frac{dy}{dx} = \frac{dy}{d\theta} \cdot \frac{d\theta}{dx}$$

where $x = a \sec \theta = \frac{a}{\cos \theta}$ and $y = b \tan \theta$

$$\therefore \quad \frac{dx}{d\theta} = \frac{\cos \theta \cdot (0) - a(-\sin \theta)}{\cos^2 \theta} \qquad \therefore \quad \frac{dy}{d\theta} = b \sec^2 \theta$$

$$= \frac{a \sin \theta}{\cos^2 \theta} \qquad \therefore \quad \frac{dy}{d\theta} = \frac{b}{\cos^2 \theta}$$

$$\therefore \quad \frac{dy}{dx} = \left(\frac{b}{\cos^2 \theta}\right) \left(\frac{1}{\frac{a \sin \theta}{\cos^2 \theta}}\right)$$

$$= \frac{b}{\cos^2 \theta} \times \frac{\cos^2 \theta}{a \sin \theta}$$

$$= \frac{b}{a \sin \theta}$$

$$= m$$

Using $y = mx + c$

$$y = \left(\frac{b}{a \sin \theta}\right) x + c \qquad \text{and using } x = a \sec \theta \quad \text{and } y = b \tan \theta$$

$$= \frac{a}{\cos \theta} \qquad = \frac{b \sin \theta}{\cos \theta}$$

$$\therefore \quad \frac{b \sin \theta}{\cos \theta} = \frac{b}{a \sin \theta} \left(\frac{a}{\cos \theta}\right) + c$$

$$\therefore \quad c = \frac{b \sin \theta}{\cos \theta} - \frac{b}{\sin \theta \cos \theta}$$

$$\therefore \quad c = \frac{b \sin^2 \theta - b}{\sin \theta \cos \theta} = \frac{b(\sin^2 \theta - 1)}{\sin \theta \cos \theta} = \frac{b(-\cos^2 \theta)}{\sin \theta \cos \theta} = -\frac{b \cos \theta}{\sin \theta}$$

Therefore, equation is $y = \left(\frac{b}{a \sin \theta}\right) x - \frac{b \cos \theta}{\sin \theta}$

$$\therefore \quad (a \sin \theta) y = bx - \frac{b \cos \theta}{\sin \theta} (a \sin \theta)$$

$$\therefore \quad \boxed{(a \sin \theta) y = bx - ab \cos \theta}$$

PURE MATHEMATICS — CONICS

Parametric form of the equation of the normal to $\dfrac{x^2}{a^2} - \dfrac{y^2}{b^2} = 1$ at $(a\sec\theta, b\tan\theta)$.

Gradient $\quad m' = \dfrac{-1}{m}$

$$= \dfrac{-1}{\left(\dfrac{b}{a\sin\theta}\right)}$$

$\therefore\quad m' = \dfrac{-a\sin\theta}{b}$

Using $\quad y = m'x + c$

$\quad y = \left(\dfrac{-a\sin\theta}{b}\right)x + c$ and using $x = \dfrac{a}{\cos\theta}$ and $y = \dfrac{b\sin\theta}{\cos\theta}$

$\therefore\quad \dfrac{b\sin\theta}{\cos\theta} = \dfrac{-a\sin\theta}{b}\left(\dfrac{a}{\cos\theta}\right) + c$

$\therefore\quad \dfrac{b\sin\theta}{\cos\theta} = \dfrac{-a^2\sin\theta}{b\cos\theta} + c$

$\therefore\; c = \dfrac{b\sin\theta}{\cos\theta} + \dfrac{a^2\sin\theta}{b\cos\theta} = \dfrac{b^2\sin\theta + a^2\sin\theta}{b\cos\theta} = \dfrac{\sin\theta(a^2+b^2)}{b\cos\theta}$

Therefore, equation is $\quad y = \left(\dfrac{-a\sin\theta}{b}\right)x + \dfrac{\sin\theta(a^2+b^2)}{b\cos\theta}$

$\therefore\quad \boxed{by = -(a\sin\theta)x + (a^2+b^2)\tan\theta}$

Example 21

Find the locus of the foot of the perpendicular from the origin to a tangent of the hyperbola $\dfrac{x^2}{a^2} - \dfrac{y^2}{b^2} = 1$.

Solution

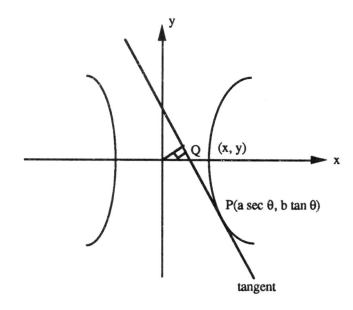

The perpendicular from the origin meets the tangent at $Q(x, y)$.

CONICS PURE MATHEMATICS

P is the point (a sec θ, b tan θ) and the equation of the tangent is (a sin θ)y = bx - ab cos θ.

Equation of the perpendicular is y = m'x, ie

$$y = \left(\frac{-a\sin\theta}{b}\right)x \qquad \therefore \qquad by = -(a\sin\theta)x$$

$$\therefore \quad -\frac{by}{ax} = \sin\theta$$

but $\cos^2\theta = 1 - \sin^2\theta$ $\qquad \therefore \qquad \cos\theta = \sqrt{1 - \left(-\frac{by}{ax}\right)^2}$

$$= \sqrt{\frac{a^2x^2 - b^2y^2}{a^2x^2}}$$

$$\therefore \quad \cos\theta = \frac{\sqrt{a^2x^2 - b^2y^2}}{ax}$$

replacing in the tangent equation

$$a \cdot \left(-\frac{by}{ax}\right)y = bx - ab \cdot \frac{\sqrt{a^2x^2 - b^2y^2}}{ax}$$

$$-\frac{by^2}{x} = bx - \frac{b\sqrt{a^2x^2 - b^2y^2}}{x} \qquad \text{Dividing by b}$$

$$\frac{\sqrt{a^2x^2 - b^2y^2}}{x} = x + \frac{y^2}{x} \qquad \text{Multiplying by x}$$

$$\sqrt{a^2x^2 - b^2y^2} = x^2 + y^2 \qquad \text{Squaring both sides}$$

$$a^2x^2 - b^2y^2 = (x^2 + y^2)^2 \qquad \text{is the required locus}$$

Now to find the parametric equations for the tangent and normal at the point $P\left(ct, \frac{c}{t}\right)$ on the rectangular hyperbola $xy = c^2$.

$$y = \frac{c^2}{x} \qquad \therefore \qquad \frac{dy}{dx} = \frac{-c^2}{x^2} = \frac{-c^2}{(ct)^2}$$

$$\therefore \qquad \frac{dy}{dx} = \frac{-1}{t^2}$$

Equation of the tangent is y = mx + k: using k instead of c to avoid confusion.

$$\therefore \qquad y = \left(\frac{-1}{t^2}\right)x + k$$

but $x = ct, y = \frac{c}{t}$ $\qquad \therefore \qquad \frac{c}{t} = \left(\frac{-1}{t^2}\right)ct + k$

$$\therefore k = \frac{c}{t} + \frac{c}{t} = \frac{2c}{t}$$

Therefore the equation of the tangent is:

$$y = \frac{-x}{t^2} + \frac{2c}{t}$$

PURE MATHEMATICS   CONICS

or $\boxed{t^2 y = -x + 2ct}$

The gradient of the normal is t^2

$$y = t^2 x + k \quad \text{but } x = ct \text{ and } y = \frac{c}{t}$$

$$\therefore \quad \frac{c}{t} = t^2(ct) + k$$

$$\therefore \quad \frac{c}{t} - ct^3 = k$$

Therefore the equation of the normal is

$$y = t^2 x + \frac{c}{t} - ct^3$$

or $\boxed{ty = t^3 x + c - ct^4}$

Example 22

The normal at any point P of the rectangular hyperbola $xy = c^2$ meets the y axis at A and the tangent meets the x axis at B. Find the coordinates of the fourth vertex Q of the rectangle APBQ in terms of t the parameter of P.

Solution

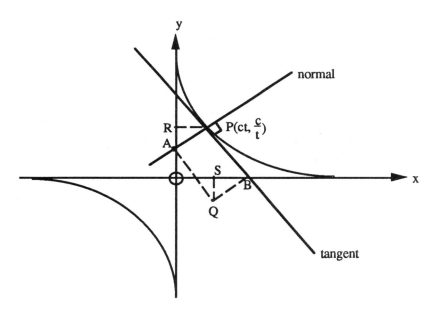

Using equation of normal $ty = t^3 x + c - ct^4$

At A (on y axis) $x = 0$ $\therefore$ $ty = t^3(0) + c - ct^4$

$$\therefore \quad y = \frac{c - ct^4}{t} \quad \text{Thus, coordinates of A are } \left(0, \frac{c - ct^4}{t}\right)$$

Using equation of tangent $t^2 y = -x + 2ct$

At B (on x axis) $y = 0$ $t^2(0) = -x + 2ct$

$$\therefore \quad x = 2ct \quad \text{Thus, coordinates of B are } (2ct, 0)$$

The coordinates of Q can be found in one of two ways: either by finding the equations of AQ and BQ and solving them for Q itself, or by using similar triangles. This second method will be used here.

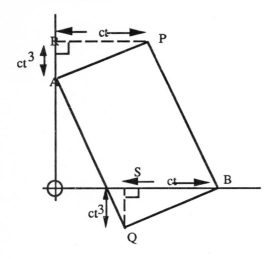

$$OR = \frac{c}{t} \text{ and } OA = \frac{c - ct^4}{t}$$

$$\therefore \quad AR = OR - OA = \frac{c}{t} - \left(\frac{c - ct^4}{t}\right)$$

$$\therefore \quad AR = \frac{ct^4}{t} = ct^3$$

$$\therefore \quad QS = AR = ct^3$$

$$\therefore \quad OS = OB - SB = 2ct - ct$$

$$\therefore \quad OS = ct$$

Therefore, the x and y coordinates of Q are given by $x = OS$ and $y = -SQ$, ie $(ct, -ct^3)$.

f) *Example ('A' level question)*

The tangent to the rectangular hyperbola $xy = c^2$ at the point $P\left(ct, \frac{c}{t}\right)$ crosses the coordinate axes at points A and B. Show that PA = PB.

The normal to the hyperbola at P meets the lines $x + y = 0$ and $x - y = 0$ at points C and D and intersects the hyperbola $x^2 - y^2 = c^2$ at the points E and F.

Show that:

(a) the points A, B, C, D are the vertices of a square;

(b) PE = PF.

Solution

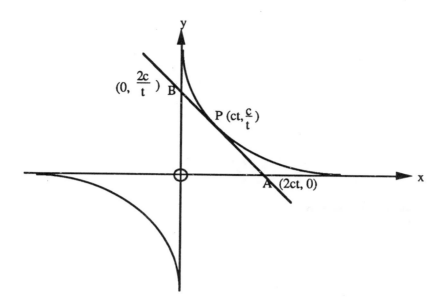

The equation of the tangent at P is

$$t^2 y = -x + 2ct$$

∴ at A $y = 0$ $t^2(0) = -x + 2ct$

∴ $x = 2ct$ ∴ A is the position $(2ct, 0)$

∴ at B $x = 0$ $t^2 y = -(0) + 2ct$

∴ $y = \dfrac{2ct}{t^2}$

   $= \dfrac{2c}{t}$ ∴ B is the position $\left(0, \dfrac{2c}{t}\right)$

Using Pythagoras $AP^2 = (2ct - ct)^2 + \left(0 - \dfrac{c}{t}\right)^2$

   $= (ct)^2 + \left(\dfrac{c}{t}\right)^2$

∴ $AP^2 = c^2 t^2 + \dfrac{c^2}{t^2}$

also $BP^2 = (0 - ct)^2 + \left(\dfrac{2c}{t} - \dfrac{c}{t}\right)^2$

   $= (-ct)^2 + \left(\dfrac{c}{t}\right)^2$

∴ $BP^2 = c^2 t^2 + \dfrac{c^2}{t^2}$

So $AP^2 = BP^2$

∴ $AP = BP$

CONICS   PURE MATHEMATICS

(a) $x + y = 0$ is $y = -x$

   $x - y = 0$ is $y = x$

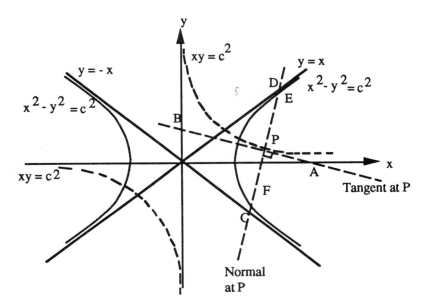

At $P\left(ct, \dfrac{c}{t}\right)$ $A(2ct, 0)$ $B\left(0, \dfrac{2c}{t}\right)$

C is the intersection of $y = -x$ and the normal

ie   of $ty = t^3x + c - ct^4$ and $y = -x$

   $t(-x) = t^3x + c - ct^4$

∴ $-tx - t^3x = c - ct^4$

∴ $-xt(1 + t^2) = c(1 - t^4)$

∴ $x = \dfrac{c(1 - t^2)(1 + t^2)}{-t(1 + t^2)}$

∴ $x = -\dfrac{c(1 - t^2)}{t}$, $y = \dfrac{c(1 - t^2)}{t}$

D is the intersection of $y = x$ and the normal

   $ty = t^3x + c - ct^4$

   $t(x) = t^3x + c - ct^4$

   $tx - t^3x = c - ct^4$

∴ $tx(1 - t^2) = c(1 - t^4)$

∴ $x = \dfrac{c(1 - t^2)(1 + t^2)}{t(1 - t^2)}$

∴ $x = \dfrac{c(1 + t^2)}{t}$, $y = \dfrac{c(1 + t^2)}{t}$

$$C\left(-\frac{c(1-t^2)}{t}, \frac{c(1-t^2)}{t}\right) \text{ and } D\left(\frac{c(1+t^2)}{t}, \frac{c(1+t^2)}{t}\right)$$

It is already known that PA = PB and that AB (part of the tangent) is at right angles to CD (part of the normal). It remains to prove PC = PD, then BDAC is a square.

$$PC^2 = \left(ct - \frac{-c(1-t^2)}{t}\right)^2 + \left(\frac{c}{t} - \frac{c(1-t^2)}{t}\right)^2$$

$$= \left(\frac{ct^2 + c(1-t^2)}{t}\right)^2 + \left(\frac{c - c(1-t^2)}{t}\right)^2$$

$$= \left(\frac{c}{t}\right)^2 + c^2 t^2$$

$$= \frac{c^2}{t^2} + c^2 t^2$$

$$PD^2 = \left(ct - \frac{c(1+t^2)}{t}\right)^2 + \left(\frac{c}{t} - \frac{c(1+t^2)}{t}\right)^2$$

$$= \left(\frac{ct^2 - c(1+t^2)}{t}\right)^2 + \left(\frac{c - c(1+t^2)}{t}\right)^2$$

$$= \left(-\frac{c}{t}\right)^2 + \left(-\frac{ct^2}{t}\right)^2$$

$\therefore \quad PD^2 = \frac{c^2}{t^2} + c^2 t^2$

$\therefore \quad PC^2 = PD^2 = PA^2 = PB^2$

Therefore, BDAC is a square because the diagonals AB and CD bisect each other at right angles.

(b) E and F are the points of intersection of the normal $ty = t^3 x + c - ct^4$ and $x^2 - y^2 = c^2$.

$\therefore$ replacing $y = \frac{t^3 x + c - ct^4}{t}$ into $x^2 - y^2 = c^2$

gives $\quad x^2 - \left(\frac{t^3 x + c - ct^4}{t}\right)^2 - c^2 = 0$

$$x^2 - \frac{(t^6 x^2 + c^2 + c^2 t^8 + 2ct^3 x - 2c^2 t^4 - 2ct^7 x)}{t^2} - c^2 = 0$$

$\therefore \quad t^2 x^2 - t^6 x^2 - c^2 - c^2 t^8 - 2ct^3 x + 2c^2 t^4 + 2ct^7 x - c^2 t^2 = 0$

$\therefore \quad (t^2 - t^6)x^2 - 2ct^3 x + 2ct^7 x - c^2 t^8 + 2c^2 t^4 - c^2 t^2 - c^2 = 0$

$\therefore \quad (t^2 - t^6)x^2 + (2ct^7 - 2ct^3)x - c^2 t^8 + 2c^2 t^4 - c^2 t^2 - c^2 = 0$

The roots of this equation (x_1 and x_2) will give the x coordinates of the points E and F on $x^2 - y^2 = c^2$. Therefore, by finding the sum of the roots ($x_1 + x_2$) the mid-point of EF can be found. Should the mid-point of EF be P then it must mean that PE = PF.

CONICS   PURE MATHEMATICS

Sum of roots $\quad (x_1 + x_2) \quad = \dfrac{-b}{a}$

$$= \dfrac{-(2ct^7 - 2ct^3)}{t^2 - t^6}$$

$$= \dfrac{-2ct(t^2 - t^6)}{(t^2 - t^6)}$$

$\therefore \quad x_1 + x_2 = 2ct$

$\therefore$ for the mid-point $\quad \dfrac{x_1 + x_2}{2} = \dfrac{2ct}{2}$

$$= ct$$

Therefore, the mid-point of EF has coordinates $\left(ct, \dfrac{c}{t}\right)$ and so is the point P.

$\therefore \quad PE = PF$

PURE MATHEMATICS — COMPLEX NUMBERS AND VECTORS

9 COMPLEX NUMBERS AND VECTORS

    9.1   Complex numbers
    9.2   Further theory of complex numbers
    9.3   Vectors

9.1 Complex numbers

a) *Real and imaginary numbers*

In all previous work it has been assumed that when any number, k, is squared the result is positive (or zero), ie $k^2 \geq 0$ - such numbers are called real numbers.

Equations such as $x^2 = -1$, therefore, have no real solution because no real number exists such that, when it is squared, the result is -1.

However, equations of this type arise quite frequently in the work and it would be useful to have some means of solving them. It is, therefore, necessary to define a new set of numbers whose squares are negative real numbers. Members of this set are called imaginary numbers and they are numbers such as $\sqrt{-1}$, $\sqrt{-10}$, $\sqrt{-26}$, etc.

The usual way of writing these numbers is:

ex. $\sqrt{-25} \;=\; \sqrt{25 \times -1}$

                   $= \sqrt{25} \times \sqrt{-1}$

$\therefore$ $\sqrt{-25} \;=\; 5i$ where $i = \sqrt{-1}$ $\therefore i^2 = -1$

so $\sqrt{-10} \;=\; \sqrt{10} \cdot i$

$$\boxed{\text{In general terms } \sqrt{-n^2} = ni \text{ where } i = \sqrt{-1}}$$

Imaginary numbers can be added and subtracted and the result will be another imaginary number.

eg $7i + 6i = 13i$

and $10i - 4i = 6i$

However, the product or quotient of two imaginary numbers will be a real number.

eg $7i \times 6i = 42i^2$ but $i^2 = -1$

                  $= 42(-1)$

$\therefore$ $7i \times 6i = -42$

and $10i \div 5i = \dfrac{10i}{5i}$

                  $= 2$

Powers of i can be simplified:

    $i^2 = -1$ $i^3 = (i^2)i \;=\; -i$

                                  $i^4 = (i^2)^2 \;=\; (-1)^2 = 1$

COMPLEX NUMBERS AND VECTORS   PURE MATHEMATICS

$$i^5 = i(i^4) = i \quad \text{etc.}$$

$$\text{and} \quad i^{-1} = \frac{1}{i} = \frac{-i^2}{i} = -i$$

b)   *Operations on complex numbers*

A complex number is the result of the addition (or subtraction) of a real number and an imaginary number eg $5 + 4i$ or $3 - 2i$, neither of which can be simplified.

> In general a complex number is of the form $a + bi$ where a and b are real numbers

(a)   Addition and subtraction

$$(a + bi) + (c + di) = (a + c) + (b + d)i$$
$$(a + bi) - (c + di) = (a - c) + (b - d)i$$

ie the real and imaginary parts are treated separately.

eg
$$(5 + 3i) + (4 + 2i) = (5 + 4) + (3 + 2)i = 9 + 5i$$
$$(5 + 3i) - (4 + 2i) = (5 - 4) + (3 - 2)i = 1 + i$$

(b)   *Multiplication*

$$(a + bi)(c + di) = ac + adi + bci + bdi^2$$
$$= ac + (ad + bc)i - bd \quad \text{as } i^2 = -1$$

$\therefore \quad (a + bi)(c + di) = ac - bd + (ad + bc)i$

eg
$$(5 + 3i)(4 + 2i) = 20 + 10i + 12i + 6i^2$$
$$= 20 + 22i - 6$$

$\therefore \quad (5 + 3i)(4 + 2i) = 14 + 22i$

A special case arises if a complex number is multiplied by its conjugate, ie the conjugate of $a + bi$ is $a - bi$.

$$(a + bi)(a - bi) = a^2 - abi + abi - b^2i^2$$
$$= a^2 + b^2 \quad \text{- a real number}$$

eg
$$(5 + 3i)(5 - 3i) = 25 - 15i + 15i - 9i^2$$
$$= 25 + 9$$
$$= 34$$

(c)   *Division*

Complex numbers cannot be divided out as they stand. It is first necessary to rationalise the denominator - this means multiplying the denominator (and numerator) by the conjugate of the denominator.

$$\frac{(a + bi)}{(c + di)} = \frac{(a + bi)(c - di)}{(c + di)(c - di)}$$

PURE MATHEMATICS   COMPLEX NUMBERS AND VECTORS

$$= \frac{ac - adi + bci - bdi^2}{c^2 - cdi + cdi - d^2i^2}$$

$$= \frac{ac + (bc - ad)i + bd}{c^2 + d^2}$$

$$\therefore \quad \frac{(a + bi)}{(c + di)} = \frac{ac + bd}{c^2 + d^2} + \frac{(bc - ad)}{c^2 + d^2}i$$

eg $\quad \dfrac{2 + 9i}{5 - 2i} = \dfrac{(2 + 9i)(5 + 2i)}{(5 - 2i)(5 + 2i)}$

$$= \frac{10 + 4i + 45i + 18i^2}{25 - 10i + 10i - 4i^2}$$

$$= \frac{10 - 18 + 49i}{25 + 4}$$

$\therefore \quad \dfrac{2 + 9i}{5 - 2i} = -\dfrac{8}{29} + \dfrac{49}{29}i$

Further points to note:

(a) A complex number such as a + bi is only equal to zero if both the real and imaginary parts equal zero, ie

$$\boxed{a + bi = 0 \quad \therefore \quad a + 0 \text{ and } b = 0}$$

(b) Two complex numbers are equal if and only if the real parts are equal and the imaginary parts are equal, ie

$$\boxed{a + bi = c + di \quad \therefore \quad a = c \text{ and } b = d}$$

Equating the real and imaginary parts of a complex equation has many useful applications as will be seen in some of the examples to come.

Example 1

Evaluate:

(a) $(2 + 7i) + (4 - 9i)$

(b) $(4 - i) - (3 + 3i)$

(c) $(2 - i)^2$

(d) $\dfrac{7 - i}{1 + 7i}$

Solution

(a) $(2 + 7i) + (4 - 9i) = 2 + 4 + 7i - 9i$

$\qquad\qquad\qquad\qquad = 6 - 2i$

(b) $(4 - i) - (3 + 3i) = 4 - i - 3 - 3i$

$\qquad\qquad\qquad\qquad = 4 - 3 - i - 3i$

$\qquad\qquad\qquad\qquad = 1 - 4i$

(c) $(2-i)^2 = (2-i)(2-i)$

$= 4 - 2i - 2i + i^2$ but $i^2 = -1$

$= 4 - 4i - 1$

$\therefore (2-i)^2 = 3 - 4i$

(d) $\dfrac{7-i}{1+7i} = \dfrac{7-i}{1+7i} \times \dfrac{(1-7i)}{(1-7i)}$ note $i - 7i$ is the conjugate of $1 + 7i$

$= \dfrac{7 - 49i - i + 7i^2}{1 - 7i + 7i - 49i^2}$ but $i^2 = -1$

$= \dfrac{7 - 50i - 7}{1 + 49}$

$= -\dfrac{50i}{50}$

$\therefore \dfrac{7-i}{1+7i} = -i$ this solution is only imaginary

Example 2

Solve the following equations:

(a) $x + yi = (3+i)(2-3i)$

(b) $x + yi = 2$

Solution

(a) $x + yi = (3+i)(2-3i)$

$= 6 - 9i + 2i - 3i^2$

$\therefore x + yi = 9 - 7i$

Therefore, equating the real and imaginary parts gives

$x = 9$ and $y = -7$

(b) $x + yi = 2$

Therefore, equating real and imaginary parts gives

$x = 2$ and $y = 0$ because $x + yi = 2 + (0)i$

PURE MATHEMATICS — COMPLEX NUMBERS AND VECTORS

Example 3

Find the square root of $15 + 8i$

Solution

Let $\sqrt{15 + 8i} = x + yi$

Squaring both sides gives

$$\left(\sqrt{15 + 8i}\right)^2 = (x + yi)^2$$

$\therefore \quad 15 + 8i = x^2 + 2xyi + y^2i^2$

$\therefore \quad 15 + 8i = x^2 - y^2 + 2xyi$

Therefore, equating real and imaginary parts gives

$$x^2 - y^2 = 15$$
$$2xy = 8$$

} solving these equations will give x and y

$\therefore \quad x = \dfrac{8}{2y}$

$\qquad \quad = \dfrac{4}{y}$ replacing in the first equation

$\left(\dfrac{4}{y}\right)^2 - y^2 = 15$

$\therefore \quad \dfrac{16}{y^2} - y^2 = 15$

$\therefore \quad 16 - y^4 = 15y^2$

$\therefore \quad 0 = y^4 + 15y^2 - 16$ a quadratic in y^2

$\therefore \quad 0 = (y^2 - 1)(y^2 + 16)$

either $y^2 = -16$ or $y^2 = 1$

$\qquad \qquad \qquad \qquad \therefore \quad y = \pm 1$

- no solutions because y is real.

and $\quad x = \dfrac{4}{y}$

$\qquad \quad = \dfrac{4}{\pm 1}$

$\qquad \quad = \pm 4$

when $x = 4$, $y = 1$ $\therefore$ $\sqrt{15 + 8i} = 4 + i$

when $x = -4$ $y = -1$ $\therefore$ $\sqrt{15 + 8i} = -4 - 1$

}

381

COMPLEX NUMBERS AND VECTORS   PURE MATHEMATICS

c) *Complex roots*

(a) *Quadratic equations*

$x^2 + 2x + 2 = 0$ is a quadratic equation. Therefore, in general:

$$x = \frac{-b \pm \sqrt{b^2 - 4ac}}{2a}$$

$$= \frac{-2 \pm \sqrt{2^2 - 4.(1).(2)}}{2.(1)}$$

$$= \frac{-2 \pm \sqrt{-4}}{2}$$

Previously the work would have stopped at this stage because x has no real solutions. However, by using complex numbers it is now possible to give solutions for this equation.

because $\sqrt{-4} = \sqrt{4i^2} = 2i$

$$\therefore x = \frac{-2 \pm 2i}{2}$$

$$\therefore x = \frac{-2 + 2i}{2} \text{ or } \frac{-2 - 2i}{2}$$

$$\therefore x = -1 + i \text{ or } -1 - i$$

Therefore the roots of $x^2 + 2x + 2 = 0$ are the complex conjugate numbers $-1 + i$ and $-1 - i$.

In general terms a quadratic equation such as $ax^2 + bx + c = 0$ will have two roots which are complex conjugate numbers if $b^2 - 4ac < 0$, ie $x = p + qi$ or $x = p - qi$ where p and q are real numbers.

By using complex numbers it is possible to extend the work on quadratic equations and quadratic expressions generally as will be seen from the next examples.

Example 4

Find the complex roots of the equation $2x^2 + 3x + 5 = 0$. If these roots are α and β confirm that $\alpha + \beta = \frac{-b}{a}$ and $\alpha\beta = \frac{c}{a}$

Solution

$$2x^2 + 3x + 5 = 0$$

$$\therefore x = \frac{-3 \pm \sqrt{3^2 - 4.2.5}}{2.2}$$

$$= \frac{-3 \pm \sqrt{9 - 40}}{4}$$

$$= \frac{-3 \pm \sqrt{-31}}{4}$$

PURE MATHEMATICS   COMPLEX NUMBERS AND VECTORS

$$= \frac{-3 \pm (\sqrt{31})i}{4}$$

Therefore the roots are:

$$x = \frac{-3 + (\sqrt{31})i}{4} \quad \text{or} \quad x = \frac{-3 - (\sqrt{31})i}{4}$$

$\therefore \quad \alpha = -\frac{3}{4} + \frac{(\sqrt{31})i}{4} \quad \text{and} \quad \beta = -\frac{3}{4} - \frac{(\sqrt{31})i}{4}$

$\alpha + \beta = -\frac{3}{4} + \frac{(\sqrt{31})i}{4} + \left(-\frac{3}{4} - \frac{(\sqrt{31})i}{4}\right)$

$\therefore \quad \alpha + \beta = -\frac{3}{2} \quad \text{and} \quad \frac{b}{a} = \frac{3}{2}$

$\therefore \quad \alpha + \beta = -\frac{b}{a}$

$\alpha\beta = \left(-\frac{3}{4} + \frac{(\sqrt{31})i}{4}\right)\left(-\frac{3}{4} - \frac{(\sqrt{31})i}{4}\right)$

$\quad = \left(-\frac{3}{4}\right)^2 + \frac{3(\sqrt{31})i}{16} - \frac{3(\sqrt{31})i}{16} - \left(\frac{(\sqrt{3}i)i}{4}\right)^2$

$\quad = \frac{9}{16} - \frac{31 i^2}{16}$

$\quad = \frac{9}{16} + \frac{31}{16}$

$\therefore \quad \alpha\beta = \frac{40}{16} = \frac{5}{2} \quad \text{and} \quad \frac{c}{a} = \frac{5}{2}$

$\therefore \quad \alpha\beta = \frac{c}{a}$

Example 5

Form the equation whose roots are $2 + i$, $2 - i$.

Solution

Since the roots are $2 + i$ and $2 - i$, then

$\quad \alpha = 2 + i \quad \text{and} \quad \beta = 2 - i$

Sum of the roots $(\alpha + \beta) = 2 + i + 2 - i$

$\quad\quad\quad\quad\quad\quad\quad\quad = 4$

Product of the roots $\alpha\beta = (2 + i)(1 - i)$

$\quad\quad\quad\quad\quad\quad\quad\quad = 4 + 2i - 2i - i^2$

$\quad\quad\quad\quad\quad\quad\quad\quad = 5$

COMPLEX NUMBERS AND VECTORS — PURE MATHEMATICS

If the equation is $ax^2 + bx + c = 0$ then $4 = -\frac{b}{a}$ and $5 = \frac{c}{a}$. Therefore, the equation becomes $x^2 - 4x + 5 = 0$.

(b) *Cubic equations*

A cubic equation such as $x^3 + ax^2 + bx + c = 0$ has three roots and equations of this type were solved in chapter 1 pp 24, 25 using the remainder and factor theorems.

Again complex numbers can be used to extend the range of such equations that can be solved, eg by inspection one root of $x^3 + 3x^2 + x - 5 = 0$ is $x = 1$. Therefore, $x - 1$ is a factor of this cubic equation and the other factor can be found using algebraic long division:

$$\begin{array}{r}
x^2 + 4x + 5 \\
x-1 \overline{\smash{\big)}\,x^3 + 3x^2 + x - 5}\\
\underline{x^3 - x^2}\\
4x^2 + x\\
\underline{4x^2 - 4x}\\
5x - 5\\
\end{array}$$

$\therefore \quad x^3 + 3x^2 + x - 5 \equiv (x - 1)(x^2 + 4x + 5) = 0$

either $x - 1 = 0 \quad$ or $\quad x^2 + 4x + 5 = 0$

$\therefore \quad x = 1 \qquad \therefore \quad x = \dfrac{-4 \pm \sqrt{4^2 - 4 \cdot (1) \cdot (5)}}{2 \cdot 1}$

$\qquad\qquad\qquad\qquad\qquad\quad = \dfrac{-4 \pm \sqrt{-4}}{2}$

$\qquad\qquad\qquad\qquad\qquad\quad = \dfrac{-4 \pm 2i}{2}$

$\qquad\qquad\qquad \therefore \quad x = -2 \pm i$

Therefore, the roots of the given cubic equation are

$x = 1, \ (-2 + i), \ (-2 - i)$

> A cubic equation has either three real roots or one real root and two conjugate complex roots.

A particular cubic equation that occurs frequently in the work is $x^3 - 1 = 0$ or $x^3 = 1$.

One obvious solution is $x = 1$. Therefore, $x - 1$ is a factor of $x^3 - 1$.

$$\begin{array}{r}
x^2 + x + 1\\
x - 1 \overline{\smash{\big)}\,x^3 - 1}\\
\underline{x^3 - x^2}\\
x^2\\
\underline{x^2 - x}\\
x - 1\\
\underline{x - 1}\\
\end{array}$$

$\therefore \quad x^3 - 1 \equiv (x - 1)(x^2 + x + 1) = 0$

PURE MATHEMATICS — COMPLEX NUMBERS AND VECTORS

either $x - 1 = 0$ or $x^2 + x + 1 = 0$

or $x = 1$ $\quad x = \dfrac{-1 \pm \sqrt{1^2 - 4 \cdot (1) \cdot (1)}}{2 \cdot 1}$

$\quad\quad\quad\quad\quad\quad\quad\quad = \dfrac{-1 \pm \sqrt{-3}}{2}$

$\quad\quad\quad\quad\quad\quad\quad\quad = \dfrac{-1 \pm (\sqrt{3})i}{2}$

Therefore, the roots of the cubic equation $x^3 - 1 = 0$ are:

$$1,\ \left(-\tfrac{1}{2} + \tfrac{(\sqrt{3})i}{2}\right),\ \left(-\tfrac{1}{2} - \tfrac{(\sqrt{3})i}{2}\right)$$

However, $x^3 = 1$ means $x = \sqrt[3]{1}$

$\therefore$ $\boxed{\sqrt[3]{1} = 1,\ \left(-\tfrac{1}{2} + \tfrac{(\sqrt{3})i}{2}\right),\ \left(-\tfrac{1}{2} - \tfrac{(\sqrt{3})i}{2}\right) \text{ are called the cube roots of unity}}$

Notes

(i) $\left(-\tfrac{1}{2} + \tfrac{(\sqrt{3})i}{2}\right)^2 = \left(-\tfrac{1}{2} + \tfrac{(\sqrt{3})i}{2}\right)\left(-\tfrac{1}{2} + \tfrac{(\sqrt{3})i}{2}\right)$

$\quad\quad\quad\quad\quad\quad\quad\quad = \left(-\tfrac{1}{2}\right)^2 - \tfrac{(\sqrt{3})i}{4} - \tfrac{(\sqrt{3})i}{4} + \left(\tfrac{(\sqrt{3})i}{2}\right)^2$

$\quad\quad\quad\quad\quad\quad\quad\quad = \tfrac{1}{4} - \tfrac{2(\sqrt{3})i}{4} + \tfrac{3i^2}{4}$

$\quad\quad\quad\quad\quad\quad\quad\quad = -\tfrac{2}{4} - \tfrac{2(\sqrt{3})i}{4}$

$\therefore \left(-\tfrac{1}{2} + \tfrac{(\sqrt{3})i}{2}\right)^2 = -\tfrac{1}{2} - \tfrac{(\sqrt{3})i}{2}$

(ii) $\left(-\tfrac{1}{2} - \tfrac{(\sqrt{3})i}{2}\right)^2 = \left(-\tfrac{1}{2} - \tfrac{(\sqrt{3})i}{2}\right)\left(-\tfrac{1}{2} - \tfrac{(\sqrt{3})i}{2}\right)$

$\quad\quad\quad\quad\quad\quad\quad\quad = \left(-\tfrac{1}{2}\right)^2 + \tfrac{(\sqrt{3})i}{4} + \tfrac{(\sqrt{3})i}{4} + \left(\tfrac{(\sqrt{3})i}{2}\right)^2$

$\quad\quad\quad\quad\quad\quad\quad\quad = \tfrac{1}{4} + \tfrac{2(\sqrt{3})i}{4} + \tfrac{3i^2}{4}$

$\quad\quad\quad\quad\quad\quad\quad\quad = -\tfrac{2}{4} + \tfrac{2(\sqrt{3})i}{4}$

$\therefore \left(-\tfrac{1}{2} - \tfrac{(\sqrt{3})i}{2}\right)^2 = -\tfrac{1}{2} + \tfrac{(\sqrt{3})i}{2}$

COMPLEX NUMBERS AND VECTORS   PURE MATHEMATICS

In other words, when either of the complex cube roots of unity is squared, it gives the other complex cube root.

> For this reason the cube roots of unity are usually written as
> $1, w, w^2$ where $1 + w + w^2 = 0$

ie

$$1 + w + w^2 = 1 + \left(-\frac{1}{2} + \frac{\sqrt{3}i}{2}\right) + \left(-\frac{1}{2} - \frac{\sqrt{3}i}{2}\right)$$

$$= 1 - \frac{1}{2} + \frac{\sqrt{3}i}{2} - \frac{1}{2} - \frac{\sqrt{3}i}{2}$$

$$\therefore \quad 1 + w + w^2 = 0$$

Example 6

By solving the equation $x^3 + 1 = 0$, find the three cube roots of -1. If one of the complex cube roots is λ express the others in terms of λ. Prove that $1 + \lambda^2 = \lambda$.

Solution

$x^3 = -1$. Therefore, $x = \sqrt[3]{-1}$ has one obvious root, $x = -1$. Therefore, $x + 1$ is a factor of $x^3 + 1$.

```
              x2 - x + 1
         _____
x + 1  |  x3            + 1
          x3  + x2
          _____
              - x2
              - x2  - x
              _____
                      x + 1
```

$\therefore \quad x^3 + 1 = (x + 1)(x^2 - x + 1) = 0$

either $\quad x + 1 = 0 \quad$ or $\quad x^2 - x + 1 = 0$

$\therefore \quad x = -1 \qquad\qquad x = \dfrac{-(-1) \pm \sqrt{(-1)^2 - 4 \cdot (1) \cdot (1)}}{2 \cdot 1}$

$$= \frac{1 \pm \sqrt{-3}}{2}$$

$$= \frac{1 \pm (\sqrt{3})i}{2}$$

$\therefore \qquad\qquad x = \dfrac{1}{2} \pm \dfrac{(\sqrt{3})i}{2}$

Therefore, the three cube roots of -1 are $-1, \left[\dfrac{1}{2} + \dfrac{(\sqrt{3})i}{2}\right], \left[\dfrac{1}{2} - \dfrac{(\sqrt{3})i}{2}\right]$

If $\lambda = \dfrac{1}{2} + \dfrac{(\sqrt{3})i}{2}$ then $\lambda^2 = \left(\dfrac{1}{2} + \dfrac{(\sqrt{3})i}{2}\right)\left(\dfrac{1}{2} + \dfrac{(\sqrt{3})i}{2}\right)$

$$= \left(\frac{1}{2}\right)^2 + \frac{(\sqrt{3})i}{4} + \frac{(\sqrt{3})i}{4} + \left(\frac{(\sqrt{3})i}{2}\right)^2$$

$$= \frac{1}{4} + \frac{2(\sqrt{3})i}{4} + \frac{3i^2}{4}$$

$$\therefore \quad \lambda^2 = -\frac{1}{2} + \frac{(\sqrt{3})i}{2}$$

$$= -\left(\frac{1}{2} - \frac{\sqrt{3}i}{2}\right)$$

Therefore, the other root $\frac{1}{2} - \frac{(\sqrt{3})i}{2} = -\lambda^2$ and vice versa.

$$1 + \lambda^2 = 1 + -\frac{1}{2} + \frac{(\sqrt{3})i}{2} = \frac{1}{2} + \frac{(\sqrt{3})i}{2}$$

$$\therefore \quad 1 + \lambda^2 = \lambda$$

d)   Argand diagrams

An Argand diagram is a means of representing a complex number graphically. Corresponding to every complex number $x + yi$, there is a point, P, (x, y) in the plane with axes Ox and Oy.

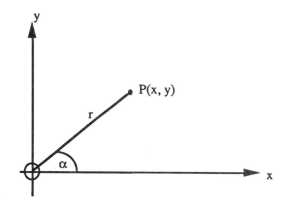

The real numbers are represented on the x axis and the imaginary numbers on the y axis.

The line OP is called a radius vector and for every complex number $x + yi$ there exists such a radius vector. This means that the magnitude and direction of OP are used to represent the complex number.

The magnitude of OP, ie the length of OP, is called the modulus of the complex number $x + yi$ and this is written as $|x + yi|$

$$\therefore \quad |x + yi| = r = \sqrt{x^2 + y^2}$$

The direction of OP is defined to be the angle that OP makes with the positive x axis, and this angle is called the argument of $x + yi$. This is written as $\text{Arg}(x + yi)$

Since $\tan \alpha = \frac{y}{x}$ where α is in radians

then $\alpha = \tan^{-1}\left(\frac{y}{x}\right)$

$\therefore \quad \arg(x + yi) = \alpha = \tan^{-1}\left(\frac{y}{x}\right)$

α is taken between the limits $-\pi < \alpha \leq \pi$.

COMPLEX NUMBERS AND VECTORS   PURE MATHEMATICS

> The modulus and argument of a complex number $x + yi$ are
> $|x + yi| = \sqrt{x^2 + y^2}$, $\arg(x + yi) = \tan^{-1} \frac{y}{x}$ respectively

This means that the point $P(x, y)$ can equally well be located using the polar coordinates (r, α), where r is the modulus of the complex number, ie $|x + yi|$, and α is the argument of $x + yi$, ie $\arg(x + yi)$.

Example 7

Represent the following numbers on an Argand diagram. Determine the modulus and argument for each number.

(a) $3 - i$

(b) $-2 + 5i$

(c) $-4 - 3i$

(d) $1 + 4i$

Solution

(a) $3 - i = x + yi$ $\therefore x = 3, \; y = -1$ $\therefore$ P is the point $(3, -1)$

$|3 - i| = r = \sqrt{3^2 + 1^2}$
$= \sqrt{10}$

$\arg(3 - i) = \alpha = \tan^{-1}\left(-\frac{1}{3}\right) = -0.32$ radians

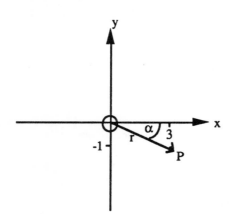

(arg(3 - i) is - 0.32 radians because the angles are measured from 0 to - π and from 0 to π, as shown in this sketch:

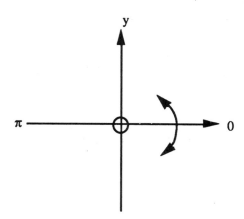

(b) $-2 + 5i = x + yi$ ∴ $x = -2, y = 5$ ∴ P is the point $(-2, 5)$

$$|-2 + 5i| = r = \sqrt{(-2)^2 + 5^2}$$
$$= \sqrt{29}$$

$$\arg(-2 + 5i) = \alpha = \tan^{-1}\frac{5}{-2}$$
$$= \pi - 1.19$$
$$= 1.95 \text{ radians}$$

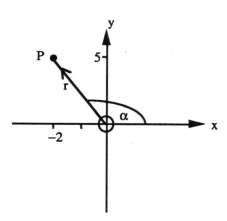

(c) $-4 - 3i = x + yi$ ∴ $x = -4, y = -3$ ∴ P is the point $(-4, -3)$

$$|-4 - 3i| = r = \sqrt{(-4)^2 + (-3)^2}$$
$$= 5$$

$$\arg(-4 - 3i) = \alpha = \tan^{-1}\frac{-3}{-4}$$
$$= -(\pi - 0.64)$$
$$= -2.50 \text{ radians}$$

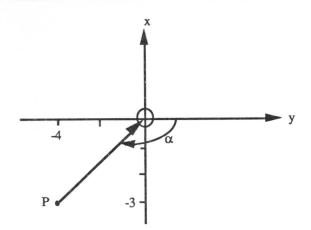

(d) $1 + 4i = x + yi$ $\therefore x = 1, y = 4$ $\therefore$ P is the point (1, 4)

$$|1 + 4i| = r = \sqrt{1^2 + 4^2}$$
$$= \sqrt{17}$$

$$\arg(1 + 4i) = \alpha = \tan^{-1}\frac{4}{1}$$
$$= 1.33 \text{ radians}$$

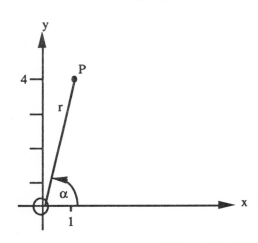

Example 8

Find the modulus and argument of:

(a)　$(2 + 10i) + (3 + 2i)$

(b)　$6 - 2i - (2 - 2i)$

(c)　$(3 + i)(4 + i)$

(d)　$\dfrac{7 - i}{3 - 4i}$

PURE MATHEMATICS   COMPLEX NUMBERS AND VECTORS

Solution

(a) $2 + 10i + (3 + 2i) = 5 + 12i$ $\therefore$ $x = 5$ and $y = 12$

$|2 + 10i + (3 + 2i)| = |5 + 12i| = \sqrt{5^2 + 12^2}$

$= \sqrt{169}$

$= 13$

$\arg(2 + 10i + (3 + 2i)) = \arg(5 + 12i) = \alpha = \tan^{-1}\dfrac{12}{5}$

$= 1.18$ radians

(If in doubt about the argument, always drawn an Argand diagram)

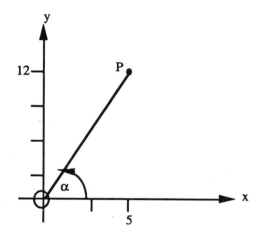

(b) $6 - 2i - (2 - 2i) = 6 - 2i - 2 + 2i$

$= 4$ $\therefore$ $x = 4$ and $y = 0$ - no imaginary part

$|6 - 2i - (2 + 2i)| = |4| = \sqrt{4^2 + 0^2}$

$= \sqrt{16}$

$= 4$

$\arg(6 - 2i - (2 + 2i)) = \arg 4 = \tan^{-1}\dfrac{0}{4}$

$= 0$

The argument is zero, of course, because P is actually on the positive y axis.

(c) $(3 + i)(4 + i) = 12 + 3i + 4i + i^2$

$= 12 + 7i - 1$

$= 11 + 7i$

$|(3 + i)(4 + 1)| = |11 + 7i| = \sqrt{11^2 + 7^2}$

$= \sqrt{170}$

COMPLEX NUMBERS AND VECTORS   PURE MATHEMATICS

$$\arg(3+i)(4+i) = \arg(11+7i) = \tan^{-1}\frac{7}{11}$$

$$= 0.57 \text{ radians.}$$

(d) $\frac{7-i}{3-4i}$: this must be rationalised using the conjugate of $3-4i$, ie $3+4i$.

$$\frac{(7-i)}{(3-4i)} \times \frac{(3+4i)}{(3+4i)} = \frac{21 + 28i - 3i - 4i^2}{9 - 12i + 12i - 16i^2}$$

$$= \frac{21 + 25i + 4}{9 - 16i^2}$$

$$= \frac{25 + 25i}{25}$$

$\therefore \quad \frac{7-i}{3-4i} = \frac{25}{25} + \frac{25i}{25} = 1 + i$

$\therefore \quad \left|\frac{7-i}{3-4i}\right| = |1+i| = \sqrt{1^2 + 1^2}$

$$= \sqrt{2}$$

$$\arg\left(\frac{7-i}{3-4i}\right) = \arg(1+i) = \tan^{-1}\frac{1}{1}$$

$$= \frac{\pi}{4} \text{ or } 0.79 \text{ radians}$$

e) *Example ('A' level question)*

(a) The complex numbers $4i$, $(2\sqrt{3} - 2i)$, $(-2\sqrt{3} - 2i)$ are represented on an Argand diagram by the points A, B and C respectively.

   (i)   Show that the triangle is equilateral.

   (ii) Show that $(4i)^2$ and $(2\sqrt{3} - 2i)(-2\sqrt{3} - 2i)$ are represented by the same point in the Argand diagram.

(b) Show that the roots of the equation $x^3 = 1$ are 1, w and w^2 where $w = -\frac{1}{2} + \frac{\sqrt{3}i}{2}$

Express the complex number $5 + 7i$ in the form $Aw + Bw^2$ where A and B are real, and give the values of A and B in surd form.

PURE MATHEMATICS   COMPLEX NUMBERS AND VECTORS

Solution

(a)   (i) $A : 4i$ $B : 2\sqrt{3} - 2i$ $C : -2\sqrt{3} - 2i$

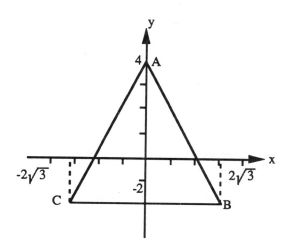

From the diagram
$$AB^2 = \sqrt{(2\sqrt{3})^2 + 6^2}$$
$$= \sqrt{12 + 36}$$
$$= \sqrt{48}$$
$$= \sqrt{16 \times 3}$$

$\therefore$ $AB = 4\sqrt{3}$

by symmetry $AC = 4\sqrt{3}$

and $BC = 2\sqrt{3} + 2\sqrt{3}$
               $= 4\sqrt{3}$

Hence: Triangle ABC is equilateral.

(ii) $(4i)^2 = 4i \cdot 4i = 16i^2$
                               $= -16$

Therefore, $(4i)^2$ is represented by the point $(-16, 0)$

$(2\sqrt{3} - 2i)(-2\sqrt{3} - 2i) = -(2\sqrt{3})^2 - 4\sqrt{3} \cdot i + 4\sqrt{3} \cdot i + (2i)^2$
$$= -12 + 4i^2$$
$$= -12 - 4$$
$$= -16$$

Therefore, $(2\sqrt{3} - 2i)(-2\sqrt{3} - 2i)$ is also represented by $(-16, 0)$.

COMPLEX NUMBERS AND VECTORS   PURE MATHEMATICS

(b)   The first part of this question is standard theory that has been included in the text (page 385) and is not repeated here.

$$5 + 7i = Aw + Bw^2$$

$$\therefore \quad 5 + 7i = A\left(-\frac{1}{2} + \frac{\sqrt{3}.i}{2}\right) + B\left(-\frac{1}{2} - \frac{\sqrt{3}.i}{2}\right)$$

$$= -\frac{A}{2} + \frac{A\sqrt{3}.i}{2} - \frac{B}{2} - \frac{B\sqrt{3}.i}{2}$$

$$= -\frac{A}{2} - \frac{B}{2} + \frac{A\sqrt{3}.i}{2} - \frac{B\sqrt{3}.i}{2}$$

$$\therefore \quad 5 + 7i = -\frac{1}{2}(A + B) + \frac{\sqrt{3}}{2}(A - B)i$$

Equating the real and imaginary parts gives:

real: $5 = -\frac{1}{2}(A + B)$ $\therefore A + B = -10$

imag: $7 = \frac{\sqrt{3}}{2}(A - B)$ $\therefore A - B = \frac{14}{\sqrt{3}}$ } +

$$2A = -10 + \frac{14}{\sqrt{3}}$$

$$\therefore A = -5 + \frac{7}{\sqrt{3}}$$

Replacing to find B $B = -10 - A$

$$B = -10 - \left(-5 + \frac{7}{\sqrt{3}}\right)$$

$$= -10 + 5 - \frac{7}{\sqrt{3}}$$

$$= -5 - \frac{7}{\sqrt{3}}$$

Therefore, A is $-5 + \frac{7}{\sqrt{3}}$ and B is $-5 - \frac{7}{\sqrt{3}}$

PURE MATHEMATICS — COMPLEX NUMBERS AND VECTORS

9.2 Further theory of complex numbers

a) *Polar coordinates*

Complex numbers are often denoted by the letter z, so consider a general complex number $z = x + yi$ which can be represented on an Argand diagram by OP where P has cartesian coordinates (x, y) and polar coordinates (r, θ).

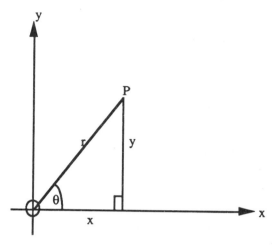

From the diagram:

$$\cos\theta = \frac{x}{r} \qquad \therefore \quad x = r\cos\theta$$

$$\sin\theta = \frac{y}{r} \qquad \therefore \quad y = r\sin\theta$$

$$\therefore \quad x + yi = r\cos\theta + r\sin\theta . i$$

$$\therefore \quad \boxed{x + yi = r(\cos\theta + i\sin\theta)}$$

Note: $\sin\theta . i$ is usually written as $i\sin\theta$ to avoid confusion.

Therefore, a complex number $z = x + yi$ can be converted into the form $r(\cos\theta + i\sin\theta)$ by finding the modulus, r, and argument, θ, of z.

Note: θ has been used instead of α for the angle because this is the more common notation when using polar coordinates.

Therefore, complex numbers can be expressed in either cartesian or polar form, and converting from one form to the other is not difficult.

Example 9

Express the following complex numbers in the form $r(\cos\theta + i\sin\theta)$:

(a) $1 - \sqrt{3}i$

(b) 2

(c) $-5i$

(d) $-2 + 2i$

Solution

(a) $|1 - \sqrt{3}i| = r = \sqrt{1^2 + (-\sqrt{3})^2}$

$\phantom{|1 - \sqrt{3}i| = r} = \sqrt{4}$

$\therefore\ r = 2$

$\arg(1 - \sqrt{3}i) = \theta = \tan^{-1} -\dfrac{\sqrt{3}}{1}$

$\phantom{\arg(1 - \sqrt{3}i) = \theta} = -\dfrac{\pi}{3} \text{ radians}$

$\therefore\ 1 - \sqrt{3}i = 2\left(\cos\left(-\dfrac{\pi}{3}\right) + i\sin\left(-\dfrac{\pi}{3}\right)\right)$

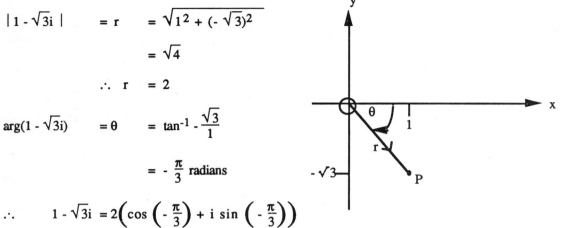

(b) $|2| = r = 2$

$\arg(2) = \theta = 0$

$\therefore\ 2 = 2(\cos 0 + i\sin 0)$

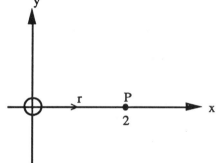

(c) $|-5i| = r = 5$

$\arg(-5i) = \theta = -\dfrac{\pi}{2}$

$\therefore\ -5i = 5\left(\cos\left(-\dfrac{\pi}{2}\right) + i\sin\left(-\dfrac{\pi}{2}\right)\right)$

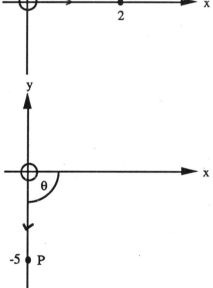

(d) $|-2 + 2i| = r = \sqrt{(-2)^2 + (2)^2}$

$= \sqrt{8} = \sqrt{4 \times 2}$

$\therefore r = 2\sqrt{2}$

$\arg(-2 + 2i) = \theta = \tan^{-1}\left(\dfrac{2}{-2}\right)$

$= \tan^{-1}(-1)$

$\therefore \theta = \dfrac{3\pi}{4}$

$\therefore -2 + 2i = 2\sqrt{2}\left(\cos\dfrac{3\pi}{4} + i\sin\dfrac{3\pi}{4}\right)$

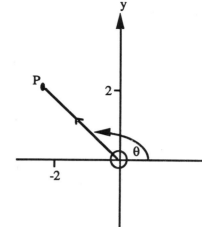

Example 10

Express the following complex numbers in the form $x + yi$:

(a) $\left(2, \dfrac{\pi}{6}\right)$

(b) $\left(3, -\dfrac{\pi}{4}\right)$

(c) $\left(1, \dfrac{2\pi}{3}\right)$

(d) $\left(4, -\dfrac{\pi}{6}\right)$

Solution

(a) $r = 2, \theta = \dfrac{\pi}{6}$ $\therefore$ $x + yi = 2\left(\cos\dfrac{\pi}{6} + i\sin\dfrac{\pi}{6}\right)$

$= 2\left(\dfrac{\sqrt{3}}{2} + i\dfrac{1}{2}\right)$

$\therefore x + yi = \sqrt{3} + i$

(b) $r = 3, \theta = -\dfrac{\pi}{4}$ $\therefore$ $x + yi = 3\left(\cos\left(-\dfrac{\pi}{4}\right) + i\sin\left(-\dfrac{\pi}{4}\right)\right)$

$= 3\left(\dfrac{1}{\sqrt{2}} - i\dfrac{1}{\sqrt{2}}\right)$

$\therefore x + yi = \dfrac{3}{\sqrt{2}} - \dfrac{3}{\sqrt{2}}i$

(c) $r = 1, \theta = \dfrac{2\pi}{3}$ $\therefore$ $x + yi = 1\left(\cos\dfrac{2\pi}{3} + i\sin\dfrac{2\pi}{3}\right)$

COMPLEX NUMBERS AND VECTORS   PURE MATHEMATICS

$$= 1\left(-\frac{1}{2} + i\frac{\sqrt{3}}{2}\right)$$

$$\therefore \quad x + yi = -\frac{1}{2} + \frac{\sqrt{3}}{2}i$$

(d) $r = 4$, $\theta = -\frac{\pi}{6}$ $\therefore$ $x + yi = 4\left(\cos\left(-\frac{\pi}{6}\right) + i\sin\left(-\frac{\pi}{6}\right)\right)$

$$= 4\left(\frac{\sqrt{3}}{2} - i\frac{1}{2}\right)$$

$$\therefore \quad x + yi = 2\sqrt{3} - 2i$$

Using polar coordinates it is possible to prove two important results concerning the product and quotient of complex numbers z_1 and z_2.

(a) *Product*

If $z_1 = r_1(\cos\theta_1 + i\sin\theta_1)$ and $z_2 = r_2(\cos\theta_2 + i\sin\theta_2)$

then $z_1 z_2 = r_1(\cos\theta_1 + i\sin\theta_1) r_2(\cos\theta_2 + i\sin\theta_2)$

$$= r_1 r_2 (\cos\theta_1 \cos\theta_2 + i\sin\theta_1 \cos\theta_2 + i\sin\theta_2 \cos\theta_1 + i^2 \sin\theta_1 \sin\theta_2)$$

$$= r_1 r_2 (\cos\theta_1 \cos\theta_2 - \sin\theta_1 \sin\theta_2 + i(\sin\theta_1 \cos\theta_2 + \sin\theta_2 \cos\theta_1))$$

$\therefore$ $z_1 z_2 = r_1 r_2 (\cos(\theta_1 + \theta_2) + i\sin(\theta_1 + \theta_2))$

$\therefore$

$z_1 z_2$ gives another complex number such that:

$|z_1 z_2| = r_1 r_2$ $\qquad \therefore \qquad |z_1 z_2| = |z_1||z_2|$

$\arg(z_1 z_2) = \theta_1 + \theta_2$ $\qquad \therefore \qquad \arg(z_1 z_2) = \arg z_1 + \arg z_2$

(b) *Quotient*

$$\frac{z_1}{z_2} = \frac{r_1(\cos\theta_1 + i\sin\theta_1)}{r_2(\cos\theta_2 + i\sin\theta_2)}$$

$$= \frac{r_1}{r_2} \frac{(\cos\theta_1 + i\sin\theta_1)(\cos\theta_2 - i\sin\theta_2)}{(\cos\theta_2 + i\sin\theta_2)(\cos\theta_2 - i\sin\theta_2)}$$

$$= \frac{r_1}{r_2} \frac{(\cos\theta_1 \cos\theta_2 + i\sin\theta_1 \cos\theta_2 - i\sin\theta_2 \cos\theta_1 - i^2 \sin\theta_1 \sin\theta_2)}{\cos^2\theta_2 + i\sin\theta_2 \cos\theta_2 - i\sin\theta_2 \cos\theta_2 - i^2 \sin\theta_2}$$

$$= \frac{r_1}{r_2} \frac{(\cos\theta_1 \cos\theta_2 + \sin\theta_1 \sin\theta_2 + i(\sin\theta_1 \cos\theta_2 - \sin\theta_2 \cos\theta_1))}{\cos^2\theta_2 + \sin^2\theta_2}$$

$$\frac{z_1}{z_2} = \frac{r_1}{r_2}(\cos(\theta_1 - \theta_2) + i\sin(\theta_1 - \theta_2))$$

∴

$\frac{z_1}{z_2}$ gives another complex number such that:

$$\left|\frac{z_1}{z_2}\right| = \frac{r_1}{r_2} \qquad \therefore \qquad \left|\frac{z_1}{z_2}\right| = \frac{|z_1|}{|z_2|}$$

$$\arg\left(\frac{z_1}{z_2}\right) = \theta_1 - \theta_2 \qquad \arg\left(\frac{z_1}{z_2}\right) = \arg z_1 - \arg z_2$$

Example 11

If $z_1 = -\sqrt{3} + i$ and $z_2 = 4 + 4i$ express $z_1 z_2$ and $\frac{z_1}{z_2}$ in the form $r(\cos\theta + i\sin\theta)$.

Solution

$$|z_1| = |-\sqrt{3} + i| \qquad \therefore \qquad r_1 = \sqrt{(-\sqrt{3})^2 + 1^2}$$
$$= 2$$

$$\arg z_1 = \arg(-\sqrt{3} + i) \qquad \therefore \qquad \theta_1 = \tan^{-1}-\frac{1}{\sqrt{3}}$$
$$= \frac{5\pi}{6}$$

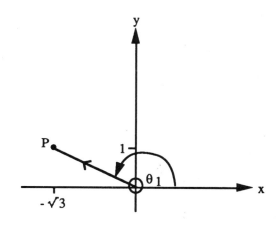

$$|z_2| = |4 + 4i| \qquad \therefore \qquad r_2 = \sqrt{4^2 + 4^2}$$
$$= \sqrt{32} = \sqrt{16 \times 2}$$
$$\therefore \quad r_2 = 4\sqrt{2}$$

$$\arg z_2 = \arg(4 + 4i) \qquad \therefore \qquad \theta_2 = \tan^{-1}\frac{4}{4} = \frac{\pi}{4}$$

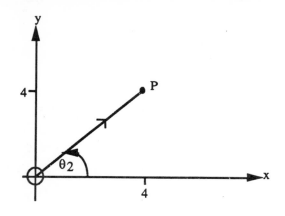

(i) $\quad z_1 z_2 = r(\cos\theta + i\sin\theta)$ where $r = r_1 r_2 = 2(4\sqrt{2})$

$$= 8\sqrt{2}$$

and $\quad \theta = \theta_1 + \theta_2 = \dfrac{5\pi}{6} + \dfrac{\pi}{4}$

$$= \dfrac{13\pi}{12}$$

$$= -\dfrac{11\pi}{12} \qquad \text{as } -\pi \leq \theta \leq \pi$$

$\therefore \quad z_1 z_2 = 8\sqrt{2}\left(\cos\left(-\dfrac{11\pi}{12}\right) + i\sin\left(-\dfrac{11\pi}{12}\right)\right)$

(ii) $\quad \dfrac{z_1}{z_2} = r(\cos\theta + i\sin\theta) \qquad$ where $r = \dfrac{r_1}{r_2} = \dfrac{2}{4\sqrt{2}}$

$$= \dfrac{1}{2\sqrt{2}}$$

and $\quad \theta = \theta_1 - \theta_2 = \dfrac{5\pi}{6} - \dfrac{\pi}{4}$

$$= \dfrac{7\pi}{12}$$

$\therefore \quad \dfrac{z_1}{z_2} = \dfrac{1}{2\sqrt{2}}\left(\cos\dfrac{7\pi}{12} + i\sin\dfrac{7\pi}{12}\right)$

b) *Loci*

When $z = x + yi$ the point $P(x, y)$ can be anywhere on the Argand diagram. But, if a special condition is imposed on z, this affects the possible positions of P, eg

$\quad |z| = 4$ means the line OP must always be at length 4

PURE MATHEMATICS   COMPLEX NUMBERS AND VECTORS

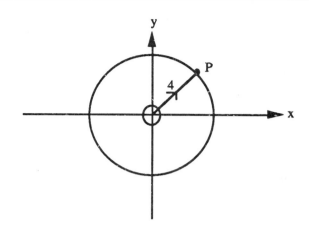

Therefore, P is restricted to any point on the circumference of a circle with centre 0 and radius 4.

The equation of this locus can be found directly from the complex equation:

$$|z| = \sqrt{x^2 + y^2} \qquad \text{but } |z| = 4$$

$$\therefore \quad \sqrt{x^2 + y^2} = 4$$

$$\therefore \quad x^2 + y^2 = 16$$

There are, therefore, two ways of obtaining the locus of a point P representing the complex number $z = x + yi$:

either   (i)   by converting the complex equation into a cartesian equation;

or   (ii)   by drawing a diagram using the information given.

Example 12

If $z_1 = 3 + i$, $z_2 = -3 - i$ and $z = x + yi$, determine the locus of the set of points $P(x, y)$ on the Argand diagram given that $|z - z_1| = |z - z_2|$

Solution

$$z - z_1 = x + yi - (3 + i) \qquad\qquad z - z_2 = x + yi - (-3 - i)$$
$$ = x - 3 + yi - i \qquad\qquad\qquad = x + yi + 3 + i$$

$$\therefore \quad z - z_1 = x - 3 + (y - 1)i \qquad \therefore \quad z - z_2 = x + 3 + (y + 1)i$$

$$|z - z_1| = \sqrt{(x - 3)^2 + (y - 1)^2} \qquad |z - z_2| = \sqrt{(x + 3)^2 + (y + 1)^2}$$

$$ = \sqrt{x^2 - 6x + 9 + y^2 - 2y + 1} \qquad = \sqrt{x^2 + 6x + 9 + y^2 + 2y + 1}$$

$$\therefore \qquad |z - z_1| = |z - z_2|$$

gives $\sqrt{x^2 - 6x + 9 + y^2 - 2y + 1} = \sqrt{x^2 + 6x + 9 + y^2 + 2y + 1}$

squaring $x^2 - 6x + 9 + y^2 - 2y + 1 = x^2 + 6x + 9 + y^2 + 2y + 1$

$$\therefore \qquad\qquad -2y - 2y = 6x + 6x$$

$$\therefore \qquad\qquad -4y = 12x$$

$$\therefore \qquad\qquad y = -3x \quad \text{is the required locus}$$

This is the equation of a straight line through (0, 0) and gradient -3.

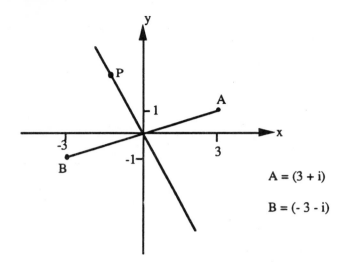

A = (3 + i)

B = (- 3 - i)

The locus $y = -3x$ is the perpendicular bisector of the line joining A and B.

Note: $|z - z_1| = \sqrt{(x - 3)^2 + (y - 1)^2}$ is the distance between any point P(x, y) and the point A(3, 1).

Similarly $|z - z_2| = \sqrt{(x + 3)^2 + (y + 1)^2}$ is the distance between any point P(x, y) and the point B(- 3, - 1).

In general terms:

$|z - z_1| = \sqrt{(x - x_1)^2 + (y - y_1)^2}$ = length of AP

where $A(x_1, y_1)$ represents $z_1 = x_1 + y_1 i$ and P(x, y) represents

$z = x + yi$ on an Argand diagram.

Example 13

If $z = x + yi$ is represented in an Argand diagram by the point P, find the locus of P when $|z| = 2|z - i + 1|$.

Solution

$|z| = \sqrt{x^2 + y^2}$ $\qquad |z - i + 1| = |x + yi - i + 1|$

$\qquad\qquad\qquad\qquad\qquad\qquad = |x + 1 + (y - 1)i|$

$\qquad\qquad\qquad\qquad\qquad\qquad = \sqrt{(x + 1)^2 + (y - 1)^2}$

Hence, if $\quad |z| = 2|z - i + 1|$

then $\quad \sqrt{x^2 + y^2} = 2\sqrt{x^2 + 2x + 1 + y^2 - 2y + 1}$

squaring $\quad x^2 + y^2 = 4(x^2 + 2x + y^2 - 2y + 2)$

$\qquad\qquad x^2 + y^2 = 4x^2 + 8x + 4y^2 - 8y + 8$

∴ $\quad 0 = 3x^2 + 8x + 3y^2 - 8y + 8$

∴ $\quad 3\left(x^2 + \dfrac{8x}{3} + y^2 - \dfrac{8x}{3}\right) = -8$

$$x^2 + \dfrac{8x}{3} + \dfrac{16}{9} + y^2 - \dfrac{8x}{3} + \dfrac{16}{9} = -\dfrac{8}{3} + \dfrac{16}{9} + \dfrac{16}{9}$$

$$\left(x + \dfrac{4}{3}\right)^2 + \left(y - \dfrac{4}{3}\right)^2 = \dfrac{8}{9}$$

This is the equation of a circle centre

$$\left(-\dfrac{4}{3},\ \dfrac{4}{3}\right) \quad \text{radius } \dfrac{\sqrt{8}}{3}$$

Loci problems involving restrictions on the argument need careful treatment and, before attempting such examples, the meaning of $\arg(z - z_1)$ will be explained where $z = x + yi$ and $z_1 = x_1 + y_1 i$.

It has already been shown that $|z - z_1|$ gives the length of the line joining A to P and it follows that $\arg(z - z_1)$ is the angle α shown in the diagram below.

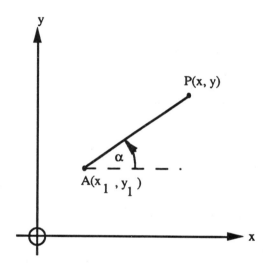

Example 14

Sketch on an Argand diagram the locus of points P(x, y) where $z = x + yi$ for which $\arg(z - 2 + 3i) = \dfrac{2\pi}{3}$

Solution

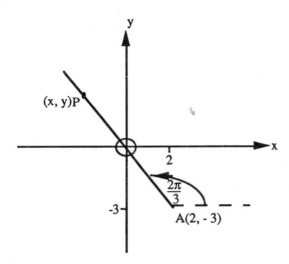

$z - 2 + 3i = z - (2 - 3i)$

∴ $z_1 = 2 - 3i$ where A is the position $(2, -3)$

and $z = x + yi$ where P is the position (x, y)

$\arg(z - z_1) = \dfrac{2\pi}{3}$

Having marked point A first, then the angle $\dfrac{2\pi}{3}$, it can be seen that P will be any point on the solid line drawn.

Example 15

If $\arg(z - 1) - \arg(z + 1) = \dfrac{\pi}{4}$ show that $P(x, y)$ representing $z = x + yi$ on an Argand diagram lies on an arc of a circle.

Solution

$z - 1 = z - (1 + 0i)$ ∴ $z_1 = 1 + 0i$ where A is the position $(1, 0)$

$z + 1 = z - (-1 + 0i)$ ∴ $z_2 = -1 + 0i$ where B is the position $(-1, 0)$

Let $\arg(z - 1) = \alpha$, say and $\arg(z + 1) = \beta$

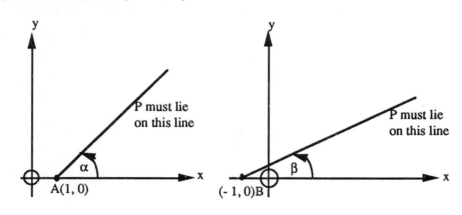

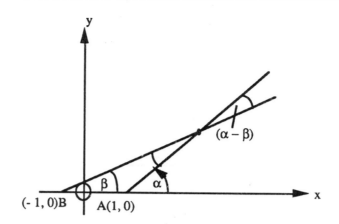

$$\arg(z-1) - \arg(z+1) = \frac{\pi}{4}$$

$$\therefore \quad \alpha - \beta = \frac{\pi}{4}$$

$$\therefore \quad A\hat{P}B = \frac{\pi}{4}$$

The line AB subtends a constant angle of $\frac{\pi}{4}$ at P so P lies on the arc of a circle cut off by the chord AB from a circle.

Example 16

Describe the locus of P, representing $z = x + yi$ in the Argand diagram if

$$\arg\left(\frac{z - (2\sqrt{3} - 2i)}{z - 4i}\right) = \frac{\pi}{2}$$

Solution

Since this is a quotient form it follows that

$$\arg\left(\frac{z - (2\sqrt{3} - 2i)}{z - 4i}\right) = \arg\left(z - (2\sqrt{3} - 2i)\right) - \arg(z - 4i)$$

$z - (2\sqrt{3} - 2i)$ $\qquad \therefore z_1 = 2\sqrt{3} - 2i \qquad$ where A is the position $(2\sqrt{3}, -2)$

$z - 4i = z - (0 + 4i) \quad \therefore z_2 = 0 + 4i \qquad$ where B is the position $(0, 4)$

Let $\arg(z - (2\sqrt{3} - 2i)) = \alpha$ and $\arg(z - 4i) = \beta$

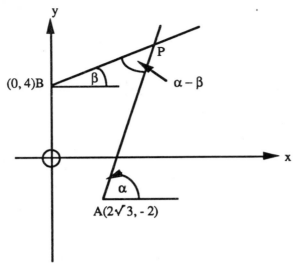

$$\therefore \quad A\hat{P}B = \alpha - \beta$$

$$\therefore \quad A\hat{P}B = \frac{\pi}{2}$$

Therefore, the locus of P is an arc of a circle and it is, in fact, a semi-circle because AB subtends a constant angle of $\frac{\pi}{2}$ at the circumference. Therefore, AB is the diameter of a circle.

---

c) *De-Moivre's theorem*

$$x + yi = r(\cos \theta + i \sin \theta)$$

However, if r is of unit length, ie r = 1, this becomes

$$z = \cos \theta + i \sin \theta$$

Considering z^2 then $|z^2| = |z\,z| = |z||z| = 1$

and $\quad \arg(z^2) = \arg z + \arg z = 2 \arg z = 2\theta$

$$\therefore \quad z^2 = \cos 2\theta + i \sin 2\theta$$

Consider now z^3 then $|z^3| = |z^2 z| = |z^2||z| = 1$

and $\quad \arg(z^3) = \arg z^2 + \arg z = 2\theta + \theta = 3\theta$

$$\therefore \quad z^3 = \cos 3\theta + i \sin 3\theta$$

Continuing this reasoning leads to the result:

$$\boxed{z^n = (\cos \theta + i \sin \theta)^n = \cos n\theta + i \sin n\theta}$$

This is known as De-Moivre's theorem and it is valid for any rational value of n - positive or negative.

$$\therefore \quad z^7 = \cos 7\theta + i \sin 7\theta$$

$$\frac{1}{z} = z^{-1} = \cos(-\theta) + i \sin(-\theta) \quad \text{but } \cos(-\theta) = \cos\theta, \text{ and } \sin(-\theta) = -\sin\theta$$

PURE MATHEMATICS   COMPLEX NUMBERS AND VECTORS

$\therefore \quad z^{-1} = \cos\theta - i\sin\theta$

$\therefore \quad z^{\frac{2}{3}} = \cos\frac{2}{3}\theta + i\sin\frac{2}{3}\theta$

This theorem can be used to derive certain trigonometric identities and to help in the simplification of complicated expressions.

Some other useful results are:

$z^n - \dfrac{1}{z^n} = z^n + z^{-n}$

$= \cos n\theta + i \sin n\theta + (\cos(-n\theta) + i \sin(-n\theta))$

$= \cos n\theta + i \sin n\theta + \cos n\theta - i \sin n\theta$

$$\boxed{\therefore \quad z^n + \frac{1}{z^n} = 2\cos n\theta \quad \text{Similarly} \quad z^n - \frac{1}{z^n} = 2i\sin n\theta}$$

Example 17

Express the following complex numbers in the form $\cos n\theta + i \sin n\theta$:

(a) $\left(\cos\dfrac{\pi}{3} + i\sin\dfrac{\pi}{3}\right)^3$

(b) $\left(\cos\dfrac{\pi}{4} + i\sin\dfrac{\pi}{4}\right)^{-2}$

Solution

(a) $\left(\cos\dfrac{\pi}{3} + i\sin\dfrac{\pi}{3}\right)^3 = \cos 3\left(\dfrac{\pi}{3}\right) + i\sin 3\left(\dfrac{\pi}{3}\right)$

$\hspace{5cm} = \cos\pi + i\sin\pi$

(b) $\left(\cos\dfrac{\pi}{4} + i\sin\dfrac{\pi}{4}\right)^{-2} = \cos -2\left(\dfrac{\pi}{4}\right) + i\sin -2\left(\dfrac{\pi}{4}\right)$

$\hspace{5cm} = \cos\left(-\dfrac{\pi}{2}\right) + i\sin\left(-\dfrac{\pi}{2}\right)$

$\hspace{5cm} = \cos\dfrac{\pi}{2} - i\sin\dfrac{\pi}{2}$

Example 18

Express each of the following complex numbers in the form $(\cos\theta + i\sin\theta)^n$:

(a) $\cos 5\theta + i\sin 5\theta$

(b) $\cos\dfrac{\theta}{2} + i\sin\dfrac{\theta}{2}$

Solution

(a) $\cos 5\theta + i \sin 5\theta = (\cos \theta + i \sin \theta)^5$

(b) $\cos \frac{\theta}{2} - i \sin \frac{\theta}{2} = \cos\left(-\frac{\theta}{2}\right) + i \sin\left(-\frac{\theta}{2}\right)$

$\qquad\qquad\quad = \cos\left(-\frac{1}{2}\theta\right) + i \sin\left(-\frac{1}{2}\theta\right)$

$\qquad\qquad\quad = (\cos \theta + i \sin \theta)^{-\frac{1}{2}}$

Example 19

Simplify the following expressions:

(a) $(\cos 3\theta + i \sin 3\theta)(\cos 6\theta - i \sin 6\theta)$

(b) $\dfrac{\cos \theta - i \sin \theta}{\cos 4\theta - i \sin 4\theta}$

(c) $\left(\cos \dfrac{\pi}{3} + i \sin \dfrac{\pi}{3}\right)^2 \left(\cos \dfrac{2\pi}{3} + i \sin \dfrac{2\pi}{3}\right)^4$

Solution

(a) $\cos 3\theta + i \sin 3\theta = z^3$

$\cos 6\theta - i \sin 6\theta = \cos(-6\theta) + i \sin(-6\theta) = z^{-6}$

$\therefore \quad z^3 \cdot z^{-6} = z^{-3}$

$\qquad\qquad = \cos(-3\theta) + i \sin(-3\theta)$

$\qquad\qquad = \cos 3\theta - i \sin 3\theta$

(b) $\cos \theta - i \sin \theta = \cos(-\theta) + i \sin(-\theta) = z^{-1}$

$\cos 4\theta - i \sin 4\theta = \cos(-4\theta) + i \sin(-4\theta) = z^{-4}$

$\qquad \dfrac{z^{-1}}{z^{-4}} = z^{-1} \cdot z^4$

$\qquad\qquad = z^3$

$\qquad\qquad = \cos 3\theta + i \sin 3\theta$

(c) $\cos \dfrac{\pi}{3} + i \sin \dfrac{\pi}{3} = (\cos \pi + i \sin \pi)^{\frac{1}{3}}$

$\therefore \left(\cos \dfrac{\pi}{3} + i \sin \dfrac{\pi}{3}\right)^2 = (\cos \pi + i \sin \pi)^{\frac{2}{3}}$

and $\cos \dfrac{2\pi}{3} + i \sin \dfrac{2\pi}{3} = (\cos \pi + i \sin \pi)^{\frac{2}{3}}$

PURE MATHEMATICS   COMPLEX NUMBERS AND VECTORS

$\therefore \quad \left(\cos\dfrac{2\pi}{3} + i\sin\dfrac{2\pi}{3}\right)^4 = (\cos\pi + i\sin\pi)^{\frac{8}{3}}$

$\therefore \quad \left(\cos\dfrac{\pi}{3} + i\sin\dfrac{\pi}{3}\right)^2 \left(\cos\dfrac{2\pi}{3} + i\sin\dfrac{2\pi}{3}\right)^4$

$\qquad\qquad\qquad = (\cos\pi + i\sin\pi)^{\frac{2}{3}} (\cos\pi + i\sin\pi)^{\frac{8}{3}}$

$\qquad\qquad\qquad = (\cos\pi + i\sin\pi)^{\frac{10}{3}}$

$\qquad\qquad\qquad = \cos\dfrac{10\pi}{3} + i\sin\dfrac{10\pi}{3}$

Example 20

Prove the following identities:

(a) $\quad \cos 4\theta = 8\cos^4\theta - 8\cos^2\theta + 1$

(b) $\quad \sin^5\theta = \dfrac{1}{16}(\sin 5\theta - 5\sin 3\theta + 10\sin\theta)$

(c) $\quad \cos^4\theta + \sin^4\theta = \dfrac{1}{4}(\cos 4\theta + 3)$

Solution

(a) $\quad \cos 4\theta + i\sin 4\theta = (\cos\theta + i\sin\theta)^4 \qquad$ - expanding

$\qquad\qquad = \cos^4\theta + 4i\cos^3\theta.\sin\theta + 6i^2\cos^2\theta.\sin^2\theta + 4i^3\cos\theta.\sin^3\theta + i^4\sin^4\theta$

$\qquad\qquad = \cos^4\theta + 4i\cos^3\theta\sin\theta - 6\cos^2\theta\sin^2\theta - 4i\cos\theta\sin^3\theta + \sin^4\theta$

Equating the real parts of each side gives:

$\qquad \cos 4\theta = \cos^4\theta - 6\cos^2\theta\sin^2\theta + \sin^4\theta$

$\qquad\qquad = \cos^4\theta - 6\cos^2\theta(1 - \cos^2\theta) + (1 - \cos^2\theta)^2$

$\qquad\qquad = \cos^4\theta - 6\cos^2\theta + 6\cos^4\theta + 1 - 2\cos^2\theta + \cos^4\theta$

$\therefore \quad \cos 4\theta = 8\cos^4\theta - 8\cos^2\theta + 1$

(b) Using $2i\sin\theta = z - \dfrac{1}{z} \qquad$ where $z = \cos\theta + i\sin\theta$

$(2i\sin\theta)^5 = \left(z - \dfrac{1}{z}\right)^5$

$2^5 i^5 \sin^5\theta = z^5 + 5z^4\left(-\dfrac{1}{z}\right) + 10z^3\left(-\dfrac{1}{z}\right)^2 + 10z^2\left(-\dfrac{1}{z}\right)^3 + 5z\left(-\dfrac{1}{z}\right)^4 + \left(-\dfrac{1}{z}\right)^5$

$\qquad\qquad = z^5 - 5z^3 + 10z - 10z^{-1} + 5z^{-3} - z^{-5}$

$32i\sin^5\theta = z^5 - z^{-5} - 5(z^3 - z^{-3}) + 10(z - z^{-1})$

$$= z^5 - \frac{1}{z^5} - 5\left(z^3 - \frac{1}{z^3}\right) + 10\left(z - \frac{1}{z}\right)$$

$$= 2i \sin 5\theta - 5(2i \sin 3\theta) + 10(2i \sin \theta)$$

$$\therefore \quad 32i \sin^5\theta = 2i \sin 5\theta - 10i \sin 3\theta + 20i \sin \theta$$

$$\sin^5\theta = \frac{2}{32}(\sin 5\theta - 5 \sin 3\theta + 10 \sin \theta)$$

$$\therefore \quad \sin^5\theta = \frac{1}{16}(\sin 5\theta - 5 \sin 3\theta + 10 \sin \theta)$$

(c) $\quad (2 \cos \theta)^4 = \left(z + \frac{1}{z}\right)^4 = z^4 + 4z^3\left(\frac{1}{z}\right) + 6z^2\left(\frac{1}{z^2}\right) + 4z\left(\frac{1}{z^3}\right) + \frac{1}{z^4}$

$$= z^4 + 4z^2 + 6 + \frac{4}{z^2} + \frac{1}{z^4}$$

$(2i \sin \theta)^4 \quad = \left(z - \frac{1}{z}\right)^4 \quad = z^4 - 4z^2 + 6 - \frac{4}{z^2} + \frac{1}{z^4}$

$$\therefore \quad (2 \cos \theta)^4 + (2i \sin \theta)^4 = 2z^4 + 12 + \frac{2}{z^4} \qquad \text{note } i^4 = (i^2)^2 = 1$$

$$\therefore \quad 16 \cos^4\theta + 16 \sin^4\theta = 2\left[\left(z^4 + \frac{1}{z^4}\right) + 6\right];$$

$$\therefore \quad \cos^4\theta + \sin^4\theta = \frac{2}{16}(2 \cos 4\theta + 6)$$

$$= \frac{4}{16}(\cos 4\theta + 3) = \frac{1}{4}(\cos 4\theta + 3)$$

---

d) *Cube roots of complex numbers*

If z is the cube root of a complex number w, then $z = \sqrt[3]{w}$. Therefore, $z^3 = w$ or $z^3 - w = 0$ and, being a cubic equation, it will have three roots, z_1, z_2, z_3.

De Moivre's theorem can now be used to find these roots, but first w must be expressed in modulus and general argument form. Therefore,

$$w = r(\cos(\theta + 2n\pi) + i \sin(\theta + 2n\pi))$$

Usually the argument of w would be θ; $\theta + 2n\pi$ is called the general argument and it allows for the fact that angles repeat themselves every 2π radians (360°). A similar idea was met when deriving general solutions of trigonometric equations:

$$z = \sqrt[3]{w} = \sqrt[3]{r(\cos(\theta + 2n\pi) + i \sin(\theta + 2n\pi))}$$

$$= \left(\sqrt[3]{r}\right)(\cos(\theta + 2n\pi) + i \sin(\theta + 2n\pi))^{\frac{1}{3}}$$

$$\therefore \quad z = \sqrt[3]{r}\left(\cos \frac{(\theta + 2n\pi)}{3} + i \sin \frac{(\theta + 2n\pi)}{3}\right)$$

By taking three consecutive values for n the three roots, z_1, z_2 and z_3 can be evaluated. Taking further values for n will only lead to repetitions in the results.

PURE MATHEMATICS   COMPLEX NUMBERS AND VECTORS

The method of actually finding these roots is best explained by an example.

Example 21

Find the cube root of 2 + 2i.

Solution

$$|2 + 2i| = r = \sqrt{2^2 + 2^2} = \sqrt{8}$$

$$\arg(2 + 2i) = \tan^{-1}\frac{2}{2} = \tan^{-1}1 = \frac{\pi}{4}$$

$$\therefore \quad z = \sqrt[3]{\sqrt{8}}\left(\cos\left(\frac{\frac{\pi}{4} + 2n\pi}{3}\right) + i\sin\left(\frac{\frac{\pi}{4} + 2n\pi}{3}\right)\right)$$

$$= \sqrt{2}\left(\cos\left(\frac{\frac{\pi}{4} + 2n\pi}{3}\right) + i\sin\left(\frac{\frac{\pi}{4} + 2n\pi}{3}\right)\right)$$

let n = 0
$$z_1 = \sqrt{2}\left(\cos\left(\frac{\frac{\pi}{4} + 0}{3}\right) + i\sin\left(\frac{\frac{\pi}{4} + 0}{3}\right)\right)$$

$$z_1 \; \sqrt{2}\left(\cos\frac{\pi}{12} + i\sin\frac{\pi}{12}\right)$$

let n = 1
$$z_2 = \sqrt{2}\left(\cos\left(\frac{\frac{\pi}{4} + 2\pi}{3}\right) + i\sin\left(\frac{\frac{\pi}{4} + 2\pi}{3}\right)\right)$$

$$= \sqrt{2}\left(\cos\frac{3\pi}{4} + i\sin\frac{3\pi}{4}\right)$$

let n = 2
$$z_3 = \sqrt{2}\left(\cos\left(\frac{\frac{\pi}{4} + 4\pi}{3}\right) + i\sin\left(\frac{\frac{\pi}{4} + 4\pi}{3}\right)\right)$$

$$= \sqrt{2}\left(\cos\frac{17\pi}{12} + i\sin\frac{17\pi}{12}\right) \qquad \text{but } \frac{17\pi}{12} = -\frac{7\pi}{12}$$

$$= \sqrt{2}\left(\cos\left(-\frac{7\pi}{12}\right) + i\sin\left(-\frac{7\pi}{12}\right)\right)$$

And on an Argand diagram these appear as follows:

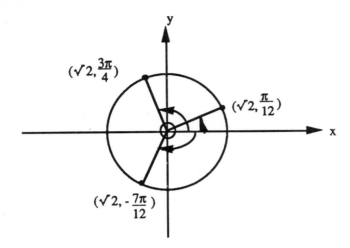

This method can obviously be extended to find any root of a number, eg the pth root.

Then $\quad z \quad = \sqrt[p]{w}$

$\quad\quad\quad\quad = \sqrt[p]{r(\cos(\theta + 2n\pi) + i\sin(\theta + 2n\pi))}$

$\therefore \quad z \quad = \sqrt[p]{r}\,(\cos(\theta + 2n\pi) + i\sin(\theta + 2n\pi))^{\frac{1}{p}}$

$\therefore \quad \boxed{z \quad = \sqrt[p]{r}\left(\cos\left(\frac{\theta + 2n\pi}{p}\right) + i\sin\left(\frac{\theta + 2n\pi}{p}\right)\right)}$

Example 22

Find the fifth roots of -1.

Solution

$\quad |-1| \quad = \sqrt{(-1)^2} = 1$

$\quad \arg(-1) \quad = \pi$

$\therefore \quad z = \sqrt[5]{1}\left(\cos\left(\frac{\pi + 2n\pi}{5}\right) + i\sin\left(\frac{\pi + 2n\pi}{5}\right)\right)$

let $\quad n = 0 \quad\quad z_1 = \cos\frac{\pi}{5} + i\sin\frac{\pi}{5}$

let $\quad n = 1 \quad\quad z_2 = \cos\frac{3\pi}{5} + i\sin\frac{3\pi}{5}$

let $\quad n = 2 \quad\quad z_3 = \cos\pi + i\sin\pi$

let $\quad n = 3 \quad\quad z_4 = \cos\frac{7\pi}{5} + i\sin\frac{7\pi}{5} \quad$ but $\frac{7\pi}{5} = -\frac{3\pi}{5}$

$\quad\quad\quad\quad\quad\quad\quad\quad = \cos\left(-\frac{3\pi}{5}\right) + i\sin\left(-\frac{3\pi}{5}\right)$

PURE MATHEMATICS — COMPLEX NUMBERS AND VECTORS

let $n = 4$ $\quad z_5 = \cos\frac{9\pi}{5} + i\sin\frac{9\pi}{5}\quad$ but $\frac{9\pi}{5} = -\frac{\pi}{5}$

$$= \cos\left(-\frac{\pi}{5}\right) + i\sin\left(-\frac{\pi}{5}\right)$$

These are then the fifth roots of -1.

e) Exponential form of a complex number

Consider:

$$e^{i\theta} = 1 + i\theta + \frac{(i\theta)^2}{2!} + \frac{(i\theta)^3}{3!} + \frac{(i\theta)^4}{4!} + \frac{(i\theta)^5}{5!} + \frac{(i\theta)^6}{6!} + \ldots$$

$$= 1 + i\theta + \frac{i^2\theta^2}{2!} + \frac{i^3\theta^3}{3!} + \frac{i^4\theta^4}{4!} + \frac{i^5\theta^5}{5!} + \frac{i^6\theta^6}{6!} + \ldots$$

$$= 1 + i\theta - \frac{\theta^2}{2!} - \frac{i\theta^3}{3!} + \frac{\theta^4}{4!} + \frac{i\theta^5}{5!} - \frac{\theta^6}{6!} + \ldots$$

$$\therefore\ e^{i\theta} = 1 - \frac{\theta^2}{2!} + \frac{\theta^4}{4!} - \frac{\theta^6}{6!} + \ldots + i\left(\theta - \frac{\theta^3}{3!} + \frac{\theta^5}{5!} - \ldots\right)$$

but the expansion of $\cos\theta = 1 - \frac{\theta^2}{2!} + \frac{\theta^4}{4!} - \frac{\theta^6}{6!} + \ldots$

and for $\sin\theta = \theta - \frac{\theta^3}{3!} + \frac{\theta^5}{5!} - \ldots$

$$\therefore\quad \boxed{e^{i\theta} = \cos\theta + i\sin\theta}$$

$$\boxed{\text{So a complex number can be written as: } z = x + yi = r(\cos\theta + i\sin\theta) = re^{i\theta}}$$

Example 23

Express these complex numbers in the form $re^{i\theta}$

(a) $1 + i$

(b) $\sqrt{3} - i$

Solution

(a) $|1 + i| = \sqrt{1^2 + 1^2} = \sqrt{2}\qquad \therefore r = 2$

$\quad\arg(1+i) = \tan^{-1}\frac{1}{1} = \frac{\pi}{4}\qquad \therefore \theta = \frac{\pi}{4}$

$$\} \; 1 + i = \sqrt{2}\,e^{\frac{i\pi}{4}}$$

(b) $|\sqrt{3} - i| = \sqrt{(\sqrt{3})^2 + (-1)^2} = \sqrt{4}\quad \therefore r = 2$

$$\} \; \sqrt{3} - i = 2\,e^{-\frac{i\pi}{6}}$$

COMPLEX NUMBERS AND VECTORS PURE MATHEMATICS

$$\arg(\sqrt{3} - i) = \tan^{-1} -\frac{1}{\sqrt{3}} = -\frac{\pi}{6} \quad \therefore \theta = -\frac{\pi}{6}$$

Example 24

Express these numbers in the form $a + bi$:

(a) $\quad 5e^{i\pi}$

(b) $\quad e^{-i\frac{\pi}{2}}$

Solution

(a) $\quad 5e^{i\pi} \quad \therefore r = 5 \quad \theta = \pi$

$\therefore \quad 5e^{i\pi} = 5(\cos \pi - i \sin \pi)$

$\qquad = -5$

(b) $\quad e^{-i\frac{\pi}{2}} \quad \therefore r = 1 \quad \theta = -\frac{\pi}{2}$

$\therefore \quad e^{-i\frac{\pi}{12}} = 1\left(\cos\left(-\frac{\pi}{2}\right) + i \sin\left(-\frac{\pi}{2}\right)\right)$

$\qquad = -i$

f) *Example ('A' level question)*

Express in the form $a + ib$:

(a) $\quad \dfrac{3 + 4i}{5 - 2i}$

(b) $\quad \left(\cos \dfrac{\pi}{6} + i \sin \dfrac{\pi}{6}\right)^5$

(c) $\quad e^{i\frac{\pi}{3}}$

Solve the equation $z^3 + 27 = 0$ and represent the roots on an Argand diagram:

The equation $z^3 + pz^2 + 40z + q = 0$ has a root $3 + i$, where p and q are real. Find the values of p and q.

Solution

(a) $\quad \dfrac{3 + 4i}{5 - 2i} = \dfrac{(3 + 4i)}{(5 - 2i)} \times \dfrac{(5 + 2i)}{(5 + 2i)}$

$\qquad = \dfrac{15 + 6i + 20i + 8i^2}{25 - 10i + 10i - 4i^2}$

$\qquad = \dfrac{7 + 26i}{25 + 4}$

$$\therefore \frac{3+4i}{5-2i} = \frac{7+26i}{29}$$

$$= \frac{7}{29} + \frac{26i}{29}$$

(b) $\left(\cos\frac{\pi}{6} + i\sin\frac{\pi}{6}\right)^5 = \cos 5\left(\frac{\pi}{6}\right) + i\sin 5\left(\frac{\pi}{6}\right)$

$$= \cos\frac{5\pi}{6} + i\sin\frac{5\pi}{6}$$

$$\therefore \left(\cos\frac{\pi}{6} + i\sin\frac{\pi}{6}\right)^5 = -\frac{\sqrt{3}}{2} + i\cdot\frac{1}{2}$$

(c) $e^{i\frac{\pi}{3}}$ $\therefore r = 1$ and $\theta = \frac{\pi}{3}$

$$\therefore e^{i\frac{\pi}{3}} = \cos\frac{\pi}{3} + i\sin\frac{\pi}{3}$$

$$= \frac{1}{2} + i\cdot\frac{\sqrt{3}}{2}$$

$z^3 = -27$ $\therefore z = \sqrt[3]{-27}$

$|-27| = \sqrt{(-27)^2} = 27$

$\arg(-27) = \pi$

$$\therefore z = \sqrt[3]{27}\left(\cos\left(\frac{\pi + 2n\pi}{3}\right) + i\sin\left(\pi + \frac{2n\pi}{3}\right)\right)$$

let $n = 0$ $z_1 = 3\left(\cos\frac{\pi}{3} + i\sin\frac{\pi}{3}\right) = 3\left(\frac{1}{2} + i\frac{\sqrt{3}}{2}\right)$

$$= \frac{3}{2} + \frac{3\sqrt{3}\cdot i}{2}$$

let $n = 1$ $z_2 = 3(\cos\pi + i\sin\pi) = 3(-1 + i0)$

$$= -3$$

let $n = 2$ $z_3 = 3\left(\cos\frac{5\pi}{3} + i\sin\frac{5\pi}{3}\right) = 3\left(\frac{1}{2} - i\frac{\sqrt{3}}{2}\right)$

$$= \frac{3}{2} - \frac{3\sqrt{3}\cdot i}{2}$$

COMPLEX NUMBERS AND VECTORS   PURE MATHEMATICS

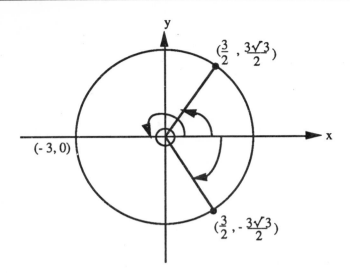

$z^3 + pz^2 + 40z + q = 0$ has one root $z = 3 + i$ replacing

$(3 + i)^3 + p(3 + i)^2 + 40(3 + i) + q = 0$

$(3^3 + 3(3^2 i) + 3(3i^2) + i^3) + p(9 + 6i + i^2) + 40(3 + i) + q = 0$

$(27 + 27i - 9 - i) + 9p + 6pi - p + 120 + 40i + q = 0 + 0i$

Real part $27 - 9 + 9p - p + 120 + q = 0$ ------------ (1)

Imaginary part $27i - i + 6pi + 40i = 0i$ ------------ (2)

∴   (1) becomes $8p + q = -138$

     (2) becomes $6p = -66$ ∴ $p = -11$

     Replacing $8(-11) + q = -138$

                   $-88 + q = -138$

∴ $q = -50$

Therefore, the values are $p = -11$, $q = -50$

9.3 Vectors

a)   *Introduction to vectors*

Scalar quantities are completely specified by a number referred to some unit of measurement, eg the temperature on a hot sunny day is 24°C or 75°F, volume of a bottle is 75 cl.

Vector quantities are completely specified by a number giving the magnitude of the quantity *and* the direction in which the quantity is acting, eg wind velocity is specified by the speed of the wind - say, 40 km/h - and the direction from which it is blowing - say, from the north-east.

Any vector may be represented by a line whose direction is that of the vector and whose length represents the magnitude of the vector.

PURE MATHEMATICS   COMPLEX NUMBERS AND VECTORS

The vector represented here is denoted by $\overline{PQ}$ and its magnitude (or modulus) is written as $|\overline{PQ}|$. Two vectors are equal if they have the same magnitude and direction.

Since the lines PQ and LM are parallel *and* equal in length $\therefore \overline{PQ} = \overline{LM}$

In general vectors do not have a fixed position in space, as such they are called 'free vectors'.

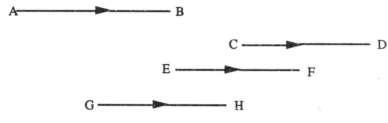

$\therefore \quad \overline{AB} = \overline{CD} = \overline{EF} = \overline{GH}$

A displacement is an example of a vector quantity. For instance, a car journey from town A to town B may be represented by the vector $\overline{AB}$. Its magnitude is the distance between A and B and its direction is that of the straight line joining A to B. Suppose a journey involves driving from A to B and then on from B to C, but on another occasion the journey involves driving straight from A to C; it can be shown diagramatically as:

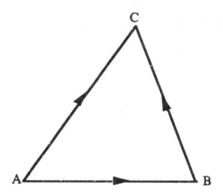

Since the result of these two journeys is the same, a vector equation can be formed: $\overline{AB} + \overline{BC} = \overline{AC}$

On another occasion the journey from A to B is followed by the return journey from B to A. The result of these two journeys is a zero displacement, ie

$$\overline{AB} + \overline{BA} = 0$$

$$\therefore \quad \overline{AB} = -\overline{BA}$$

Therefore, a vector which has the same magnitude as $\overline{AB}$ but in the opposite direction is written as $\overline{BA}$ or $-\overline{AB}$.

COMPLEX NUMBERS AND VECTORS — PURE MATHEMATICS

b)   *Operations on vectors*

It is often easier to denote a vector by a single letter, eg $\overline{AB}$ could be denoted by $\underline{a}$ and $|\overline{AB}|$ by $|\underline{a}|$. This notation will mainly be used in the following work on vectors.

Vector addition is defined by the triangle law as follows:

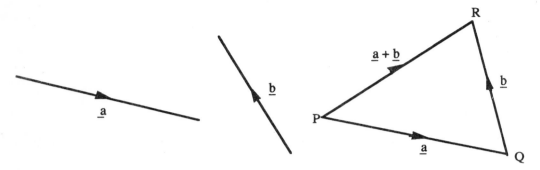

If two vectors $\underline{a}$ and $\underline{b}$ are represented by the sides $\overline{PQ}$ and $\overline{QR}$ of a triangle, then $\underline{a} + \underline{b}$ is represented by the third side $\overline{PR}$ of the same triangle ($\overline{PR} = \overline{PQ} + \overline{QR}$).

The subtraction of b from a is defined as the addition of $-\underline{b}$ to $\underline{a}$, ie $\underline{a} - \underline{b} = \underline{a} + (-\underline{b})$.

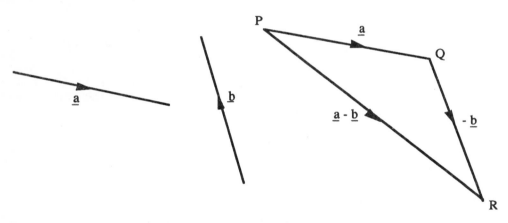

If the vectors $\underline{a}$ and $-\underline{b}$ are represented by the sides $\overline{PQ}$ and $\overline{QR}$ of a triangle, then $\underline{a} + (-\underline{b})$ is represented by the third side $\overline{PR}$ of the same triangle.

Example 25

In the given diagram $\overline{AB} = \underline{x}$, $\overline{BC} = y$, and $\overline{CD} = \underline{z}$. Find expressions for $\overline{AC}$, $\overline{BD}$ and $\overline{AD}$ in terms of $\underline{x}$, $\underline{y}$ and $\underline{z}$.

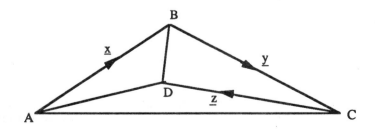

Solution

$$\overline{AC} = \overline{AB} + \overline{BC} = \underline{x} + \underline{y}$$

$$\overline{BD} = \overline{BC} + \overline{CD} = \underline{y} + \underline{z}$$

$$\overline{AD} = \overline{AB} + \overline{BD} = \underline{x} + \underline{y} + \underline{z}$$

} there are often alternative ways such as these

or $\quad \overline{AD} = \overline{AC} + \overline{CD} = \underline{x} + \underline{y} + \underline{z}$

Example 26

In the pentagon PQRST $\overline{PQ} = \underline{a}$, $\overline{QR} = \underline{b}$, $\overline{ST} = \underline{c}$ and $\overline{QT} = \underline{d}$. Find expressions for $\overline{PT}$, $\overline{RT}$, $\overline{RS}$ and $\overline{PS}$ in terms of $\underline{a}$, $\underline{b}$, $\underline{c}$ and $\underline{d}$.

Solution

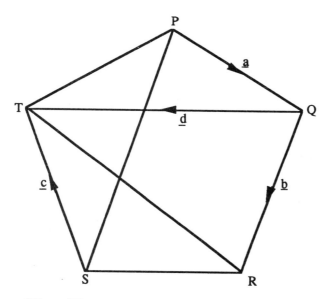

$\overline{PT} \quad = \overline{PQ} + \overline{QT} = \underline{a} + \underline{d}$

$\overline{RT} \quad = \overline{RQ} + \overline{QT} = -\underline{b} + \underline{d} \quad = \underline{d} - \underline{b}$

$\overline{RS} = \overline{RT} + \overline{TS} = \underline{d} - \underline{b} + (-\underline{c}) = \underline{d} - \underline{b} - \underline{c}$

$\overline{PS} = \overline{PT} + \overline{TS} = \underline{a} + \underline{d} + (-\underline{c}) = \underline{a} + \underline{d} - \underline{c}$

The multiplication of two vectors will be defined in two different ways:

the scalar (or dot) product (see section f)

the vector (or cross) product (see chapter 10)

In this section the multiplication of a vector by a scalar will be defined. The scalar multiple of a vector $\underline{a}$ can be defined using vector addition as follows:

COMPLEX NUMBERS AND VECTORS   PURE MATHEMATICS

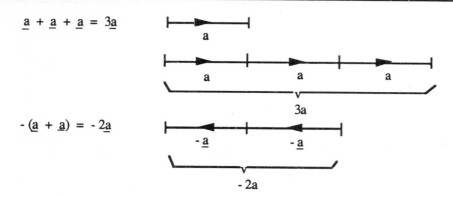

In general terms if k is a scalar and $\underline{a}$ is a vector, then the magnitude of $k\underline{a}$ is k times the magnitude of $\underline{a}$. The direction of $k\underline{a}$ is the same as that of $\underline{a}$ if k is positive (see $3\underline{a}$ above) but opposite to $\underline{a}$ if k is negative (see $-2\underline{a}$ above).

Division of a vector by a scalar k ($k \neq 0$) is defined as multiplicaton by $\frac{1}{k}$.

Example 27

PQRS is a parallelogram with $\overline{PQ} = 2\underline{a}$ and $\overline{PS} = \underline{b}$. The point T is such that $\overline{PT} = 2\underline{b}$. If the lines PR and QS intersect at X and the lines RS and QT intersect at Y, find expressions for $\overline{TR}$, $\overline{PY}$, $\overline{QY}$ and $\overline{XY}$ in terms of $\underline{a}$ and $\underline{b}$.

Solution

Firstly, a good clear diagram is essential.

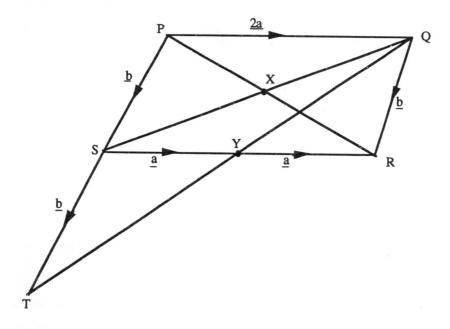

Notes on the diagram

(i) $\overline{PT} = 2\underline{b}$, $\overline{PS} = \underline{b}$

∴ $\overline{ST} = \underline{b}$

420

PURE MATHEMATICS — COMPLEX NUMBERS AND VECTORS

(ii) Since PQRS is a parallelogram (opposite sides are equal in length and parallel):

$\overline{QR} = \underline{b}$ (because $\overline{PS} = \underline{b}$)

$\overline{SR} = 2\underline{a}$ (because $\overline{PQ} = 2\underline{a}$).

(iii) Y is the midpoint of SR because triangles TSY and QRY are identical in every way (congruent).

∴ $\overline{SY} = \overline{YR} = \underline{a}$

$\overline{TR} = \overline{TS} + \overline{SR}$ using $\triangle TSR$

$= -\underline{b} + 2\underline{a} = 2\underline{a} - \underline{b}$

$\overline{PY} = \overline{PS} + \overline{SY}$ using $\triangle PYS$

$= \underline{b} + \underline{a} \qquad = \underline{a} + \underline{b}$

$\overline{QY} = \overline{QR} + \overline{RY}$ using $\triangle QRY$

$= \underline{b} + (-\underline{a}) = \underline{b} - \underline{a}$

$\overline{XY} = \tfrac{1}{2}\underline{b}$ because XY is parallel to PS and equal to half its length.

---

c) *Component vectors and unit vectors*

Resolving a vector into component vectors means expressing it as a sum of two non-parallel vectors, eg $\underline{r}$ can be resolved in the directions of vectors $\underline{a}$ and $\underline{b}$ as follows:

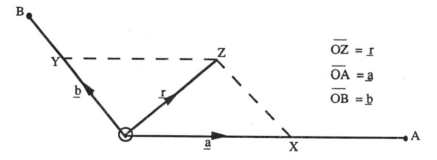

$\overline{OZ} = \underline{r}$
$\overline{OA} = \underline{a}$
$\overline{OB} = \underline{b}$

XZ is drawn parallel to OB and YZ is drawn parallel to OA, then $\overline{OX}$ is the resolved component of $\underline{r}$ in the direction of $\underline{a}$ and $\overline{OY}$ is the resolved component of $\underline{r}$ in the direction of $\underline{b}$.

It is often more practical to resolve a vector into horizontal and vertical components, and these directions will be used in all the following work.

Example 28

The horizontal and vertical components of a vector $\underline{a}$ are 7 and 4 units respectively. Draw a diagram to illustrate the components of $\underline{a}$ and $\underline{a}$ itself.

Solution

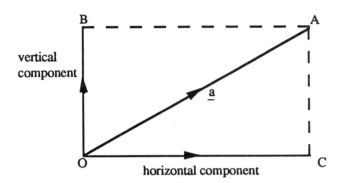

A unit vector is a vector with a magnitude of 1 unit (the actual unit being one appropriate to the problem under consideration). The standard notation that has been adopted in vector work is to let i and j represent the unit vectors in the horizontal and vertical directions respectively. This means that a in Example 28 could be expressed as a sum of its component vectors as follows:

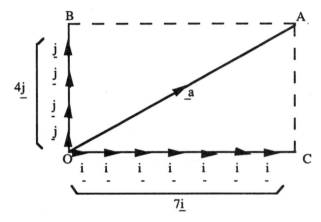

∴ $\overline{OA} = \overline{OC} + \overline{CA}$ but $\overline{CA} = \overline{OB} = 4j$

∴ $\overline{OA} = a = 7i + 4j$

Since the unit vectors i and j have been defined in the horizontal and vertical directions respectively, much vector work can be referred to a pair of rectangular axes Ox and Oy. Any vector, r, in the x, y plane can be expressed as the sum of its 'x' and 'y' components as:

$$\boxed{r = ai + bj}$$

PURE MATHEMATICS   COMPLEX NUMBERS AND VECTORS

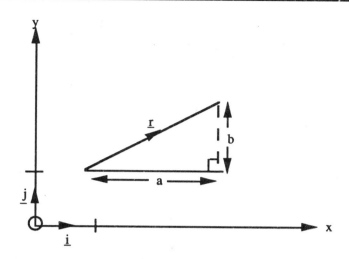

r is still a free vector as its actual position in the x, y plane is not fixed; it is only the magnitudes of its 'x' and 'y' components that are specified. It can, therefore, be drawn anywhere relative to the Ox Oy axes.

Example 29

Draw the following vectors on the same set of axes: $\underline{a} = 5\underline{i} + 3\underline{j}$, $\underline{b} = 2\underline{i} - 4\underline{j}$, $\underline{c} = -3\underline{i} - 4\underline{j}$.

Solution

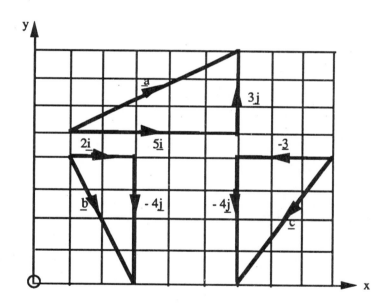

Notes to the diagram

(1)   These are not unique positions for $\underline{a}$, $\underline{b}$ and $\underline{c}$ as they are free vectors.

(2)   Consider $k\underline{i}$ if $k > 0$ component vector drawn →

                              if $k < 0$ component vector drawn ←

(3)   Consider $k\underline{j}$ if $k > 0$ component vector drawn ↑

                              if $k < 0$ component vector drawn ↓

COMPLEX NUMBERS AND VECTORS   PURE MATHEMATICS

Example 30

In quadrilateral ABCD $\overline{AB} = 2i - j$, $\overline{BC} = 3i + 4j$, $\overline{AD} = i + 5j$ and M is the midpoint of CD. Express in terms of i and j:

(a) $\overline{AC}$;

(b) $\overline{BD}$;

(c) $\overline{CD}$;

(d) $\overline{DM}$;

(e) $\overline{AM}$;

(f) $\overline{BM}$.

Solution

Firstly, draw a diagram to show the relative positions of A, B, C and D.

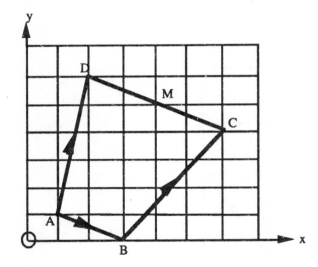

(a) $\overline{AC}$ = $\overline{AB} + \overline{BC}$

  = $(2i - j) + (3i + 4j)$

  = $5i + 3j$

(b) $\overline{BD}$ = $\overline{BA} + \overline{AD}$ ($\overline{BA} = -\overline{AB}$)

  = $-(2i - j) + (i + 5j)$

  = $-i + 6j$

(c) $\overline{CD} = \overline{CB} + \overline{BA} + \overline{AD}$ ($\overline{CB} = -\overline{BC}$)

$= -(3\underline{i} + 4\underline{j}) - (2\underline{i} - \underline{j}) + (\underline{i} + 5\underline{j})$

$= -4\underline{i} + 2\underline{j}$

(d) $\overline{DM} = \frac{1}{2}\overline{DC}$ because M is the midpoint of DC

$= -\frac{1}{2}(-4\underline{i} + 2\underline{j})$

$= 2\underline{i} - \underline{j}$

(e) $\overline{AM} = \overline{AD} + \overline{DM}$

$= (\underline{i} + 5\underline{j}) + (2\underline{i} - \underline{j})$

$= 3\underline{i} + 4\underline{j}$

(f) $\overline{BM} = \overline{BC} + \overline{CM}$ ($\overline{CM} = \overline{MD} = -\overline{DM}$)

$= (3\underline{i} + 4\underline{j}) - (2\underline{i} - \underline{j}) = \underline{i} + 5\underline{j}$

d) *Magnitude and direction of a vector*

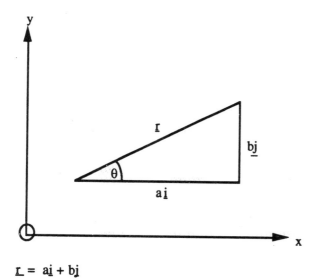

$\underline{r} = a\underline{i} + b\underline{j}$

The length of $\underline{r}$ (also known as the modulus of $\underline{r}$) can be found by applying Pythagoras theorem to the given triangle:

$$|\underline{r}| = \sqrt{a^2 + b^2}$$

If θ is the angle made by the vector with the positive direction of the x axis, then

$\tan \theta = \frac{b}{a}$

$\therefore \quad \boxed{\theta = \tan^{-1}\left(\frac{b}{a}\right)}$

COMPLEX NUMBERS AND VECTORS   PURE MATHEMATICS

Example 31

If $\underline{p} = 3\underline{i} + 2\underline{j}$ and $\underline{q} = 2\underline{i} + 4\underline{j}$ find:

(a) $|\underline{p}|$;

(b) $|\underline{q}|$;

(c) the angle $\underline{p}$ makes with the positive x axis;

(d) the angle $\underline{q}$ makes with the positive x axis;

(e) $|\underline{p} - \underline{q}|$.

Solution

(a) $|p| = \sqrt{3^2 + 2^2} = \sqrt{13}$

(b) $|q| = \sqrt{2^2 + 4^2} = \sqrt{20}$

(c) Using $\theta = \tan^{-1}\frac{b}{a}$ then $\theta = \tan^{-1}\frac{2}{3}$

$= 33.7°$ (0.59 radians)

(d) Using $\phi = \tan^{-1}\frac{b}{a}$ then $\phi = \tan^{-1}\frac{4}{2}$

$= 63.4°$ (1.11 radians)

(e) $|\underline{p} - \underline{q}|$ firstly $\underline{p} - \underline{q} = (3\underline{i} + 2\underline{j}) - (2\underline{i} + 4\underline{j})$

$= \underline{i} - 2\underline{j}$

$\therefore |\underline{p} - \underline{q}| = \sqrt{1^2 + (-2)^2}$

$= \sqrt{5}$

So far unit vectors have been defined in the horizontal and vertical directions only. However, it is quite possible to define a unit vector in the direction of a given vector, say $\underline{r}$. Since a unit vector has a magnitude of a and $\underline{r}$ has magnitude $|\underline{r}|$ then $\frac{\underline{r}}{|\underline{r}|}$ will have unit length and hence will be a unit vector, eg from Example 31:

$\underline{p} = 3\underline{i} + 2\underline{j}$

$|\underline{p}| = \sqrt{13}$

$\therefore \frac{\underline{p}}{|\underline{p}|} = \frac{3\underline{i} + 2\underline{j}}{\sqrt{13}}$

$= \frac{3}{\sqrt{13}}\underline{i} + \frac{2}{\sqrt{13}}\underline{j}$

PURE MATHEMATICS   COMPLEX NUMBERS AND VECTORS

This is clearly still a vector since $\underline{p}$ has been divided by its magnitude $|\underline{p}|$ which is a scalar quantity. Now to prove it has unit length:

$$\left|\frac{\underline{p}}{|\underline{p}|}\right| = \sqrt{\left(\frac{3}{\sqrt{13}}\right)^2 + \left(\frac{2}{\sqrt{13}}\right)^2}$$

$$= \sqrt{\frac{9}{13} + \frac{4}{13}}$$

$$= \sqrt{\frac{13}{13}}$$

$$= 1$$

∴   Unit vector ($\hat{\underline{r}}$) in the direction of $\underline{r}$ is given by

$$\boxed{\hat{\underline{r}} = \frac{\underline{r}}{|\underline{r}|}}$$

To make the notation less cumbersome in future $|\underline{r}|$ will be denoted by r.

∴ $$\boxed{\hat{\underline{r}} = \frac{\underline{r}}{r}}$$

Example 32

Find unit vectors in the directions of:

(a) $6\underline{i} + 8\underline{j}$;

(b) $\underline{i}\sqrt{2} - \underline{j}\sqrt{2}$

Solution

(a)   Let $\underline{a} = 6\underline{i} + 8\underline{j}$

∴ $|\underline{a}| = a$

$$= \sqrt{6^2 + 8^2}$$

$$= \sqrt{100}$$

$$= 10$$

∴ $\hat{\underline{a}} = \frac{\underline{a}}{a}$

$$= \frac{6\underline{i} + 8\underline{j}}{10}$$

$$= \frac{6}{10}\underline{i} + \frac{8}{10}\underline{j}$$

$$= 0.6\underline{i} + 0.8\underline{j}$$

(b) Let $\underline{b} = \sqrt{2}\underline{i} - \sqrt{2}\underline{j}$

$\therefore |\underline{b}| = b$

$= \sqrt{(\sqrt{2})^2 + (-\sqrt{2})^2}$

$= \sqrt{4}$

$= 2$

$\therefore \hat{b} = \dfrac{\underline{b}}{b}$

$= \dfrac{\sqrt{2}\underline{i} - \sqrt{2}\underline{j}}{2}$

$= \dfrac{\sqrt{2}}{2}\underline{i} - \dfrac{\sqrt{2}}{2}\underline{j}$

$= \dfrac{\sqrt{2}}{2}(\underline{i} - \underline{j})$

e) *Position vectors*

Taking a fixed point O as origin the position of any point P can be specified by giving the vector $\overline{OP}$, which is then called the position vector of P.

When working in a cartesian coordinate system, the point with coordinates (x, y) will have a possible vector $x\underline{i} + y\underline{j}$ relative to O.

Unlike free vectors which do not have a unique position in the x, y plane, position vectors are firmly fixed in one place in the plane.

Example 33

Write down, in terms of unit vectors $\underline{i}$ and $\underline{j}$, the position vectors of the points A(2, 3), B(5, -1) and C(4, 4). If

D is the fourth vertex of the parallelogram ABCD find the vector $\overline{AD}$. Hence find the position vector and the coordinates of D.

Solution

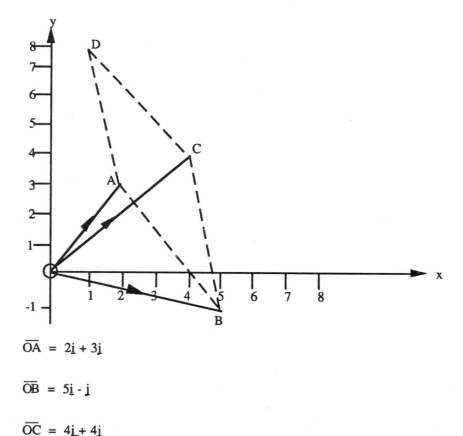

$\overline{OA} = 2\underline{i} + 3\underline{j}$

$\overline{OB} = 5\underline{i} - \underline{j}$

$\overline{OC} = 4\underline{i} + 4\underline{j}$

Since ABCD is to form a parallelogram with sides AB and DC parallel, and also sides BC and AD parallel, it follows that $\overline{AB} = \overline{DC}$ and $\overline{BC} = \overline{AD}$.

$$\therefore \quad \overline{AD} = \overline{BC} = \overline{BO} + \overline{OC}$$
$$= -(5\underline{i} - \underline{j}) + (\underline{i} + 4\underline{j})$$
$$= -\underline{i} + 5\underline{j}$$

The position vector of D will be $\overline{OD}$.

$$\therefore \quad \overline{OD} = \overline{OA} + \overline{AD}$$
$$= (2\underline{i} + 3\underline{j}) + (-\underline{i} + 5\underline{j})$$
$$= \underline{i} + 8\underline{j}$$

The coordinates of D must be (1, 8) as it has position vector $\underline{i} + 8\underline{j}$.

Example 34

The points P, Q and R have position vectors $\underline{a}$, $2\underline{a} + \underline{b}$, $4\underline{a} + 3\underline{b}$ respectively. Prove PQR are collinear and find $\dfrac{QR}{PQ}$.

Solution

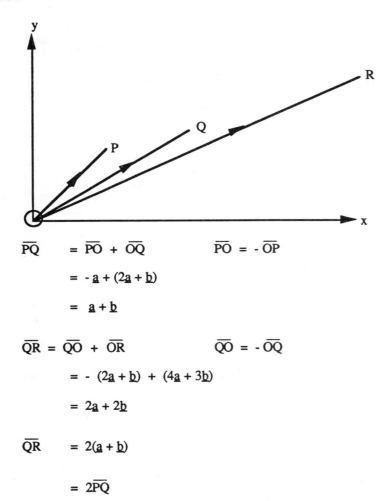

$$\overline{PQ} = \overline{PO} + \overline{OQ} \qquad \overline{PO} = -\overline{OP}$$
$$= -\underline{a} + (2\underline{a} + \underline{b})$$
$$= \underline{a} + \underline{b}$$

$$\overline{QR} = \overline{QO} + \overline{OR} \qquad \overline{QO} = -\overline{OQ}$$
$$= -(2\underline{a} + \underline{b}) + (4\underline{a} + 3\underline{b})$$
$$= 2\underline{a} + 2\underline{b}$$

$$\therefore \quad \overline{QR} = 2(\underline{a} + \underline{b})$$
$$= 2\overline{PQ}$$

This means that the vector $\overline{QR}$ is parallel to the vector $\overline{PQ}$ but twice as long. However, since the point Q is common to both these vectors QR must be a continuation of PQ and, therefore, P, Q and R lie on the same straight line.

Also $\dfrac{QR}{PQ} = 2$ since QR is twice as long as PQ.

---

f) *The scalar (or dot) product*

As has already been stated it is necessary to define carefully what is meant by the product of two vectors. This first is called the scalar product because the result of the multiplication is a scalar quantity. The second, which will be dealt with in chapter 10, is called the vector product because the result of the multiplication is a vector quantity.

PURE MATHEMATICS   COMPLEX NUMBERS AND VECTORS

If two vectors $\underline{a}$ and $\underline{b}$ are inclined to each other at an angle θ, then the scalar (or dot) product of $\underline{a}$ and $\underline{b}$ is defined as:

$$\boxed{\underline{a} \cdot \underline{b} = a \cdot b \cos \theta}$$ where $a = |\underline{a}|$ and $b = |\underline{b}|$

For example, if $\underline{a}$ has magnitude 4 units and $\underline{b}$ has magnitude 3 units and the angle between $\underline{a}$ and $\underline{b}$ is 60°, then

$$\underline{a} \cdot \underline{b} = 4 \times 3 \times \cos 60°$$
$$= 6$$

Some important results concerning parallel and perpendicular vectors follow from this definition.

If $\theta = 0°$, $\cos \theta = 1$ $\therefore \underline{a} \cdot \underline{b} = ab$ if $\underline{a}$ and $\underline{b}$ are in the same direction

If $\theta = 180°$, $\cos \theta = -1$ $\therefore \underline{a} \cdot \underline{b} = -ab$ if $\underline{a}$ and $\underline{b}$ are in opposite directions

If $\theta = 90°$, $\cos \theta = 0$ $\therefore \underline{a} \cdot \underline{b} = 0$ if $\underline{a}$ and $\underline{b}$ are perpendicular

However, if $\underline{a} \cdot \underline{b} = 0$ it does not automatically follow that $\underline{a}$ and $\underline{b}$ are perpendicular because $\underline{a} = 0$ or $\underline{b} = 0$ will make $\underline{a} \cdot \underline{b} = 0$.

Also note $\underline{a} \cdot \underline{a} = a \times a \times \cos 0 = a^2$

The scalar product can also be applied to vectors written in component form

since $\underline{i} \cdot \underline{i} = i^2 = 1$ and $\underline{j} \cdot \underline{j} = j^2 = 1$

and $\underline{i} \cdot \underline{j} = i \times j \times \cos 90°$

$$= 0$$
$$= \underline{j} \cdot \underline{i}$$

Therefore   if $a = a_1 \underline{i} + a_2 \underline{j}$

and $b = b_1 \underline{i} + b_2 \underline{j}$

then $\underline{a} \cdot \underline{b} = (a_1 \underline{i} + a_2 \underline{j}) \cdot (b_1 \underline{i} + b_2 \underline{j})$

$$= a_1 \underline{i} \cdot b_1 \underline{i} + a_1 \underline{i} \cdot b_2 \underline{j} + a_2 \underline{j} \cdot b_1 \underline{i} + a_1 \underline{i} \cdot b_2 \underline{j}$$

$$= a_1 b_1 \underline{i} \cdot \underline{i} + a_1 b_2 \underline{i} \cdot \underline{j} + a_2 b_1 \underline{j} \cdot \underline{i} + a_2 b_2 \underline{j} \cdot \underline{j}$$

$$= a_1 b_1 + a_2 b_2$$

$\therefore$ $\boxed{\underline{a} \cdot \underline{b} = a_1 b_1 + a_2 b_2}$

The scalar product can also be used to find the angle θ between two vectors $\underline{a}$ and $\underline{b}$ since:

$$\underline{a} \cdot \underline{b} = ab \cos \theta$$

$\therefore$ $\cos \theta = \dfrac{\underline{a} \cdot \underline{b}}{ab}$

$\therefore$ $\boxed{\theta = \cos^{-1}\left(\dfrac{\underline{a} \cdot \underline{b}}{ab}\right)}$

$$\boxed{\text{If } \underline{a} \text{ and } \underline{b} \text{ are perpendicular } a_1 b_1 + a_2 b_2 = 0}$$

COMPLEX NUMBERS AND VECTORS PURE MATHEMATICS

Example 35

Two vectors $\underline{a}$ and $\underline{b}$ are inclined to one another at an angle θ. Find $\underline{a} \cdot \underline{b}$ if:

(a) $a = 12, b = 7, \cos\theta = -\frac{2}{3}$

(b) $a = 3\sqrt{8}, b = 4, \theta = 135°$.

Solution

(a) $\underline{a} \cdot \underline{b} = ab\cos\theta$

$= 12 \times 7 \times \left(-\frac{2}{3}\right)$

$= -56$

(b) $\underline{a} \cdot \underline{b} = ab\cos\theta$

$= 3\sqrt{8} \times 4 \times \cos 135°$

$= 3\sqrt{8} \times 4 \times -\frac{1}{\sqrt{2}}$

$= -\frac{12\sqrt{8}}{\sqrt{2}}$

$= -24$ because $\frac{\sqrt{8}}{\sqrt{2}} = \sqrt{\frac{8}{2}} = \sqrt{4} = 2$

Example 36

Find $\overline{AB} \cdot \overline{AC}$ if A is the point (0, (2), B is the point (5, 7) and C is the point (0, 6).

Solution

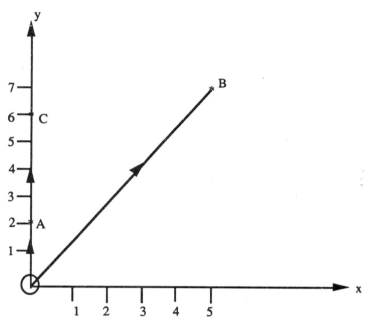

Position vector of A $\qquad \overline{OA} = 2\underline{j}$

Position vector of B $\qquad \overline{OB} = 5\underline{i} + 7\underline{j}$

Position vector of C $\qquad \overline{OC} = 6\underline{j}$

$$\overline{AB} = \overline{AO} + \overline{OB}$$
$$= -(2\underline{j}) + (5\underline{i} + 7\underline{j}) = 5\underline{i} + 5\underline{j}$$

$$\overline{AC} = \overline{AO} + \overline{OC}$$
$$= -2\underline{j} + 6\underline{j} = 4\underline{j}$$

using $\underline{a} \cdot \underline{b} = a_1 b_1 + a_2 b_2$ for vectors expressed in components

$$\overline{AB} \cdot \overline{AC} = (5 \times 0) + (5 \times 4) = 20$$

Some questions involving the scalar product require a more algebraic rather than numerical approach as will be seen from the next example.

Example 37

Given that $\underline{a}$ and $\underline{b}$ are non-zero vectors and that $\underline{a} \cdot (6\underline{a} - 2\underline{b}) = 3\underline{a} \cdot (\underline{b} + 2\underline{a})$ prove that $\underline{a}$ and $\underline{b}$ are perpendicular.

Solution

$$\begin{aligned}
\underline{a} \cdot (6\underline{a} - 2\underline{b}) &= \underline{a} \cdot 6\underline{a} - \underline{a} \cdot 2\underline{b} \\
&= 6\underline{a} \cdot \underline{a} - 2\underline{a} \cdot \underline{b} \\
&= 6a^2 - 2\underline{a} \cdot \underline{b} \\
3\underline{a} \cdot (\underline{b} + 2\underline{a}) &= 3\underline{a} \cdot \underline{b} + 3\underline{a} \cdot 2\underline{a} \\
&= 3\underline{a} \cdot \underline{b} + 6\underline{a} \cdot \underline{a} \\
&= 3\underline{a} \cdot \underline{b} + 6a^2
\end{aligned}$$

Since $\underline{a} \cdot (6\underline{a} - 2\underline{b}) = 3\underline{a} \cdot (\underline{b} + 2\underline{a})$

then $6a^2 - 2\underline{a} \cdot \underline{b} = 3\underline{a} \cdot \underline{b} + 6a^2$

∴ $6a^2 - 6a^2 = 3\underline{a} \cdot \underline{b} + 2\underline{a} \cdot \underline{b}$

∴ $5\underline{a} \cdot \underline{b} = 0$

Since $5 \neq 0$, $\underline{a} \neq 0$, $\underline{b} \neq 0$ ($\underline{a}$ and $\underline{b}$ are non-zero vectors) the only possibility is that $\underline{a}$ and $\underline{b}$ are perpendicular giving $\theta = 90°$ and $\cos \theta = 0$.

So far the vector work has been limited to two dimensional work. In the next chapter these basic ideas will be extended into three dimensions.

PURE MATHEMATICS — THREE-DIMENSIONAL GEOMETRY

10 THREE-DIMENSIONAL GEOMETRY AND VECTORS

    10.1   Three-dimensional coordinate geometry
    10.2   Further three-dimensional coordinate geometry
    10.3   Vectors in three dimensions
    10.4   Further vector work

10.1 Three-dimensional coordinate geometry

a) *Location of a point*

All the coordinate geometry so far has been concerned with two dimensions, and equations could be represented by a graph drawn in the plane of the paper - using axes Ox and Oy.

This work will now be extended into three dimensions - using axes Ox, Oy and Oz as shown below.

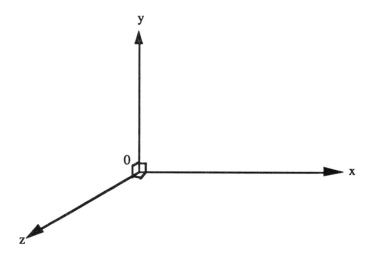

To locate a point, P, in space it needs to have three coordinates (a, b, c) which means point P is a units from O in the direction Ox, b units from O in the direction Oy and c units from O in the direction Oz.

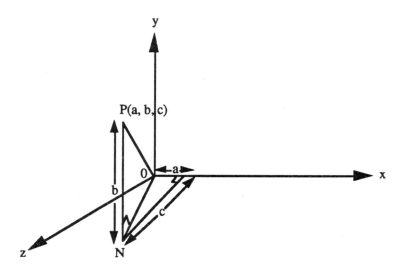

To find an expression for the length of OP:

$$ON^2 = a^2 + c^2$$

$$\therefore \quad ON = \sqrt{a^2 + c^2}$$

THREE-DIMENSIONAL GEOMETRY   PURE MATHEMATICS

$$OP^2 = ON^2 + b^2$$

$$\therefore \quad OP^2 = a^2 + c^2 + b^2$$

> The distance d of a point P(a, b, c) from the origin O is given by $\quad d = \sqrt{a^2 + b^2 + c^2}$

b)   *Direction ratios and direction cosines of a line*

The coordinates of any point P(a, b, c) fix the direction of the line OP and the ratios a : b : c are called the direction ratios of OP.

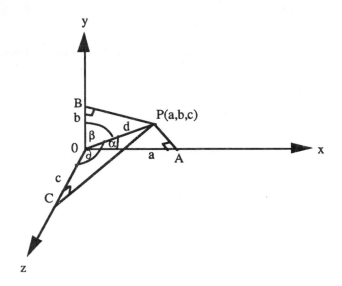

This diagram looks very confusing. However, PA, PB and PC are lines drawn from P perpendicular to the axes Ox, Oy and Oz.

$$\therefore \quad P\hat{A}O = 90°, \quad P\hat{B}O = 90°, \quad P\hat{C}O = 90°$$

α, β and ∂ are the angles between OP and Ox, Oy and Oz respectively.

$$\therefore \quad P\hat{O}A = \alpha, \quad P\hat{O}B = \beta, \quad P\hat{O}C = \partial$$

$$\therefore \quad \cos\alpha = \frac{a}{d} = l, \quad \cos\beta = \frac{b}{d} = m, \quad \cos\partial = \frac{c}{d} = n$$

Since the angles fix the direction of OP they are called the direction cosines of OP and are often given the letters l, m and n.

> Direction ratios are a : b : c and the direction cosines are:
>
> $$\frac{a}{\sqrt{a^2 + b^2 + c^2}}, \quad \frac{b}{\sqrt{a^2 + b^2 + c^2}}, \quad \frac{c}{\sqrt{a^2 + b^2 + c^2}} \quad \text{ie } l, m, n$$

$$l^2 + m^2 + n^2 = \left(\frac{a}{\sqrt{a^2 + b^2 + c^2}}\right)^2 + \left(\frac{b}{\sqrt{a^2 + b^2 + c^2}}\right)^2 + \left(\frac{c}{\sqrt{a^2 + b^2 + c^2}}\right)^2$$

$$= \frac{a^2}{a^2 + b^2 + c^2} + \frac{b^2}{a^2 + b^2 + c^2} + \frac{c^2}{a^2 + b^2 + c^2}$$

PURE MATHEMATICS   THREE-DIMENSIONAL GEOMETRY

$$\therefore \quad l^2 + m^2 + n^2 = \frac{a^2 + b^2 + c^2}{a^2 + b^2 + c^2} = 1$$

> The sum of the squares of the direction cosines is always 1, ie $l^2 + m^2 + n^2 = 1$

Since $\cos \alpha = \frac{a}{d} = l$ then $a = d \cos \alpha$ or dl

and $\cos \beta = \frac{b}{d} = m$ then $b = d \cos \beta$ or dm

and $\cos \partial = \frac{c}{d} = n$ then $c = d \cos \partial$ or dn

> The coordinates of P(a, b, c) can be written as (dl, dm, dn).

Example 1

Find the direction cosines of the line OP where P is the point (1, 8, 4).

Solution

$\therefore$ for P $a = 1, \; b = 8, \; c = 4$

$$d = \sqrt{1^2 + 8^2 + 4^2}$$

$$= \sqrt{81}$$

$\therefore \quad d = 9$

$\therefore \quad \cos \alpha = \frac{1}{9}, \quad \cos \beta = \frac{8}{9}$ and $\cos \partial = \frac{4}{9}$ are the direction cosines.

Example 2

A line OP is inclined at 60° to Ox and Oz. If OP = 3, find the coordinates of P.

Solution

If the coordinates of P are (a, b, c) and d = 3, then

$\cos \alpha = \frac{a}{d} = l$ $\therefore \quad l = \cos 60° = \frac{a}{3}$

$$3 \times \frac{1}{2} = a$$

$$\frac{3}{2} = a$$

$\cos \partial = \frac{c}{d} = n$ $\therefore \quad n = \cos 60° = \frac{c}{3}$

$$\frac{3}{2} = c$$

The angle and hence the direction cosine is not known for b but using $l^2 + m^2 + n^2 = 1$

gives $\left(\dfrac{1}{2}\right)^2 + m^2 + \left(\dfrac{1}{2}\right)^2 = 1$

$\dfrac{1}{4} + m^2 + \dfrac{1}{4} = 1$

$\therefore \qquad m^2 = \dfrac{1}{2}$

$\therefore \qquad m = \pm \dfrac{1}{\sqrt{2}}$

$\cos \beta = \dfrac{b}{d} = m \quad \therefore \quad m = \pm \dfrac{1}{\sqrt{2}} = \dfrac{b}{3}$

$\therefore \quad \pm \dfrac{3}{\sqrt{2}} = b$

Therefore, the coordinates of P are $\left(\dfrac{3}{2}, \pm \dfrac{3}{\sqrt{2}}, \dfrac{3}{2}\right)$

c) *Properties of a line joining two points*

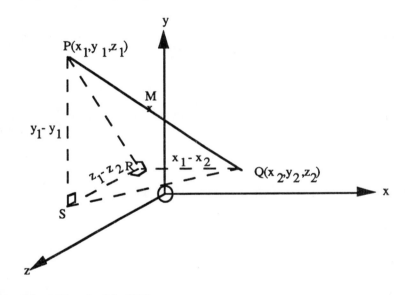

The coordinates of the mid-point M of PQ are:

$$M : \left(\dfrac{x_1 + x_2}{2}, \dfrac{y_1 + y_2}{2}, \dfrac{z_1 + z_2}{2}\right)$$

To find the length of PQ:

$PR^2 = (y_1 - y_2)^2 + (z_1 - z_2)^2$

$\therefore \quad PR = \sqrt{(y_1 - y_2)^2 + (z_1 - z_2)^2}$

$PQ^2 = RQ^2 + PR^2$

$\therefore \quad PQ^2 = (x_1 - x_2)^2 + (y_1 - y_2)^2 + (z_1 - z_2)^2$

PURE MATHEMATICS — THREE-DIMENSIONAL GEOMETRY

> The length of a line, PQ, joining two points $P(x_1, y_1, z_1)$ and $Q(x_2, y_2, z_2)$ is given by:
>
> $$\sqrt{(x_1 - x_2)^2 + (y_1 - y_2)^2 + (z_1 - z_2)^2}$$

The direction ratios of PQ are $\quad (x_2 - x_1) : (y_2 - y_1) : (z_2 - z_1)$

and the direction ratios of QP are $\quad (x_1 - x_2) : (y_1 - y_2) : (z_1 - z_2)$

However, since they are ratios rather than absolute values these two statements are the same, eg the ratios $5 : 4 : 3$ can be multiplied by -1 to give $-5 : -4 : -3$ which is numerically the same.

$\therefore \quad$ > Direction ratios of PQ (and QP) are $(x_2 - x_1) : (y_2 - y_1) : (z_2 - z_1)$

The direction cosines do depend on the direction of the line.

> Direction cosines of PQ are $\dfrac{(x_2 - x_1)}{PQ}, \dfrac{(y_2 - y_1)}{PQ}, \dfrac{(z_2 - z_1)}{PQ}$
>
> Direction cosines of QP are $\dfrac{(x_1 - x_2)}{PQ}, \dfrac{(y_1 - y_2)}{PQ}, \dfrac{(z_1 - z_2)}{PQ}$

Example 3

Find the length and direction cosines of the line LM whre L is the mid-point of AB, M is the mid-point of BC and A, B and C are the points $(3, -1, 5)$, $(7, 1, 3)$, $(-5, 9, -1)$ respectively.

Solution

L is the mid-pont of AB $\quad \therefore L\left(\dfrac{3+7}{2}, \dfrac{-1+1}{2}, \dfrac{5+3}{2}\right) \quad = (5, 0, 4)$

M is the mid-point of BC $\quad \therefore M\left(\dfrac{7+-5}{2}, \dfrac{1+9}{2}, \dfrac{3+-1}{2}\right) \quad = (1, 5, 1)$

$\therefore \quad LM^2 = (5-1)^2 + (0-5)^2 + (4-1)^2$

$\qquad \qquad = 16 + 25 + 9$

$\therefore \quad LM^2 = 50 \text{ and } LM = \sqrt{50}$

The direction cosines of LM are given by:

$\dfrac{(1-5)}{\sqrt{50}}, \dfrac{(5-0)}{\sqrt{50}}, \dfrac{(1-4)}{\sqrt{50}} \quad$ ie $\quad \dfrac{-4}{\sqrt{50}}, \dfrac{5}{\sqrt{50}}, \dfrac{-3}{\sqrt{50}}$

(Note the order: for direction cosines of LM: $\dfrac{\text{Coordinates of M - Coordinates of L}}{LM}$)

THREE-DIMENSIONAL GEOMETRY   PURE MATHEMATICS

Example 4

The points OABC form a tetrahedron where A, B and C are the points (4, 1, - 2), (3, 5, - 1), (- 1, 2, 4) respectively. Show that the line joining the mid-point of OA to the mid-point of OB is parallel to the line joining the mid-point of AC to the mid-point of CB.

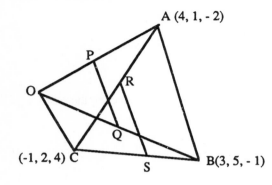

Let mid-point   of OA be P

of OB be Q

of AC be R

of CB be S

To find the mid-points: $P\left(\dfrac{0+4}{2}, \dfrac{0+1}{2}, \dfrac{0+-2}{2}\right)$ ie $\left(2, \dfrac{1}{2}, -1\right)$

$Q\left(\dfrac{0+3}{2}, \dfrac{0+5}{2}, \dfrac{0+-1}{2}\right)$ ie $\left(\dfrac{3}{2}, \dfrac{5}{2}, -\dfrac{1}{2}\right)$

$R\left(\dfrac{4+-1}{2}, \dfrac{1+2}{2}, \dfrac{-2+4}{2}\right)$ ie $\left(\dfrac{3}{2}, \dfrac{3}{2}, 1\right)$

$S\left(\dfrac{-1+3}{2}, \dfrac{2+5}{2}, \dfrac{4+-1}{2}\right)$ ie $\left(1, \dfrac{7}{2}, \dfrac{3}{2}\right)$

The direction ratios of PQ are $\left(2 - \dfrac{3}{2}\right) : \left(\dfrac{1}{2} - \dfrac{5}{2}\right) : \left(-1 - -\dfrac{1}{2}\right)$

$\dfrac{1}{2} : -\dfrac{4}{2} : -\dfrac{1}{2}$ ie 1 : - 4 : - 1

The direction ratios of RS are $\left(\dfrac{3}{2} - 1\right) : \left(\dfrac{3}{2} - \dfrac{7}{2}\right) : \left(1 - \dfrac{3}{2}\right)$

$\dfrac{1}{2} : -\dfrac{4}{2} : -\dfrac{1}{2}$ ie 1 : - 4 : - 1

Since the direction ratios fix the direction of a line in space, lines having the same direction ratios must go in the same direction and, therefore, are parallel.

---

d)   *Equations of a straight line*

A line is uniquely fixed in space if either:

(a)   it has a known direction and passes through a known point; or

(b)   it passes through two known points.

PURE MATHEMATICS — THREE-DIMENSIONAL GEOMETRY

Considering each case in turn gives:

(a) The equation of a straight line with direction ratios a : b : c which passes through the point $A(x_1, y_1, z_1)$ has equation

$$\boxed{\frac{x - x_1}{a} = \frac{y - y_1}{b} = \frac{z - z_1}{c}}$$

This is being given as a statement of fact and will be derived using vector methods later.

∴ If $\dfrac{x - x_1}{a} = \dfrac{y - y_1}{b} = \dfrac{z - z_1}{c} = \lambda$ (say)

the equation can be written in parametric form as:

$$x - x_1 = a\lambda \quad \therefore \quad x = a\lambda + x_1$$
$$y - y_1 = b\lambda \quad \therefore \quad y = b\lambda + y_1$$
$$z - z_1 = c\lambda \quad \therefore \quad z = c\lambda + z_1$$

Example 5

Find the equations of the line with direction ratios 2 : -1 : -4 which passes through $\left(3, -1, \dfrac{1}{2}\right)$

Solution

Equation will be:
$$\frac{x - 3}{2} = \frac{y - (-1)}{-1} = \frac{z - \left(\frac{1}{2}\right)}{-4}$$

∴
$$\frac{x - 3}{2} = \frac{y + 1}{-1} = \frac{z - \frac{1}{2}}{-4}$$

∴
$$\frac{x - 3}{2} = \frac{-(y + 1)}{1} = \frac{-(2z - 1)}{8}$$

Example 6

Find the direction cosines of the line whose equation is $x = 2y - 4 = 3 - 4z$.

Solution
$$x = 2y - 4 = 3 - 4z$$

becomes
$$\frac{x}{1} = \frac{2(y - 2)}{1} = \frac{-4\left(z - \frac{3}{4}\right)}{1}$$

$$\frac{x}{1} = \frac{y - 2}{\frac{1}{2}} = \frac{z - \frac{3}{4}}{-\frac{1}{4}}$$

Having written this equation in the standard form the direction ratios are $1 : \dfrac{1}{2} : -\dfrac{1}{4}$, ie 4 : 2 : -1 (multiplying by 4).

THREE-DIMENSIONAL GEOMETRY   PURE MATHEMATICS

Direction cosines in general are $\dfrac{a}{\sqrt{a^2 + b^2 + c^2}}, \dfrac{b}{\sqrt{a^2 + b^2 + c^2}}, \dfrac{c}{\sqrt{a^2 + b^2 + c^2}}$

where $a = 4, b = 2, c = -1$ $\therefore$ $\sqrt{a^2 + b^2 + c^2} = \sqrt{4^2 + 2^2 + (-1)^2}$

$$= \sqrt{21}$$

Therefore, direction cosines are $\dfrac{4}{\sqrt{21}}, \dfrac{2}{\sqrt{21}}, \dfrac{-1}{\sqrt{21}}$ respectively.

(b)   The equation of a straight line passing through two points $P(x_1, y_1, z_1)$ and $Q(x_2, y_2, z_2)$ is given by:

$$\boxed{\dfrac{x - x_1}{x_2 - x_1} = \dfrac{y - y_1}{y_2 - y_1} = \dfrac{z - z_1}{z_2 - z_1}}$$

since the direction ratios are $x_2 - x_1 : y_2 - y_1 : z_2 - z_1$.

Example 7

Find the equation of the line passing through $(3, -1, 5)$ and $(7, -2, 8)$.

Solution

$$\dfrac{x - 3}{7 - 3} = \dfrac{y - (-1)}{-2 - (-1)} = \dfrac{z - 5}{8 - 5}$$

$\therefore$ $\dfrac{x - 3}{4} = \dfrac{y + 1}{-1} + \dfrac{z - 5}{3}$

$\therefore$ $\dfrac{x - 3}{4} = \dfrac{-(y + 1)}{1} = \dfrac{z - 5}{3}$

Example 8

Write the equations $\dfrac{x - 2}{1} = \dfrac{1 - 3y}{2} = \dfrac{4 - 2z}{1}$ in parametric form. Hence or otherwise find the coordinates of the point where the line cuts:

(a)   the xy plane;

(b)   the xz plane;

(c)   the yz plane;

(d)   the plane parallel to the yz plane which cuts Ox at $(2, 0, 0)$.

Solution

$$\dfrac{x - 2}{1} = \dfrac{1 - 3y}{2} = \dfrac{4 - 2z}{1} = \lambda \text{ (say)}$$

$\therefore$ $\dfrac{x - 2}{1} = \lambda$ giving $x = \lambda + 2$

$\dfrac{1 - 3y}{2} = \lambda$ $\therefore \ 1 - 3y = 2\lambda$ $\therefore \ y = \dfrac{1 - 2\lambda}{3}$

and $\quad \dfrac{4-2z}{1} = \lambda \quad \therefore 4 - 2z = \lambda \quad \therefore z = \dfrac{4-\lambda}{2}$

(a) On the xy plane $z = 0$ $\quad \therefore 0 = \dfrac{4-\lambda}{2}$

$$\therefore \lambda = 4$$

So $x = 4 + 2 = 6$ and $y = \dfrac{1 - 2(4)}{3} = \dfrac{-7}{3}$

Therefore, the line cuts the xy plane at $\left(6,\ \dfrac{-7}{3},\ 0\right)$

(b) On the xz plane $y = 0$ $\quad \therefore -\dfrac{1-2\lambda}{3}$

$$\therefore \lambda = \dfrac{1}{2}$$

So $x = \dfrac{1}{2} + 2 = \dfrac{5}{2}$ and $z = \dfrac{4-\dfrac{1}{2}}{2} = \dfrac{7}{4}$

Therefore, the line cuts the xz plane at $\left(\dfrac{5}{2},\ 0,\ \dfrac{7}{4}\right)$

(c) On the yz plane $x = 0$ $\quad \therefore 0 = \lambda + 2$

$$\therefore \lambda = -2$$

So $y = \dfrac{1 - 2(-2)}{3} = \dfrac{5}{3}$ and $z = \dfrac{4 - (-2)}{2} = 3$

Therefore, the line cuts the yz plane at $\left(0,\ \dfrac{5}{3},\ 3\right)$

(d) Since the plane is parallel to the yz plane and cuts Ox at $(2, 0, 0)$

$$x = 2 \quad \therefore 2 = \lambda + 2$$

$$\therefore \lambda = 0$$

So the coordinates of y and z are $\dfrac{1 - 2(0)}{3} = \dfrac{1}{3},\ \dfrac{4-0}{2} = 2$

Therefore, the required point is $\left(2,\ \dfrac{1}{3},\ 2\right)$.

e) *Pairs of lines*

Two lines in space may be located so that:

(a) the lines are parallel;

(b) the lines are not parallel and intersect;

(c) the lines are not parallel and do not intersect. Such lines are called skew lines.

THREE-DIMENSIONAL GEOMETRY PURE MATHEMATICS

Taking each of these in turn:

(a) If lines L_1 and L_2 are parallel they have the same direction ratios and this can be observed directly from the equations of L_1 and L_2.

Example 9

Prove that the following pair of lines are parallel:

$$\frac{x-1}{3} = \frac{y-1}{-2} = \frac{z-2}{4}, \quad \frac{x-2}{-6} = \frac{y+1}{4} = \frac{z-3}{-8}$$

Solution

The direction ratios of the first line are $3:-2:4$ and of the second line are $-6:4:-8$ which, when divided by -2, gives $3:-2:4$, ie the same. Therefore, lines are parallel.

(b) If lines L_1 and L_2 are not parallel and do intersect, then there must exist a unique set of values for λ and μ such that

$$L_1 : \frac{x-x_1}{a_1} = \frac{y-y_1}{b_1} = \frac{z-z_1}{c_1} = \lambda$$

$$L_2 : \frac{x-x_2}{a_2} = \frac{y-y_2}{b_2} = \frac{z-z_2}{c_2} = \mu$$

are both satisfied by $x = x_3$, $y = y_3$ and $z = z_3$.

Example 10

Prove L_1 and L_2 intersect and find the point of intersection of the lines

$$L_1 : x - 1 = 1 - y = z - 3 \qquad \text{and} \quad L_2 : \frac{x-2}{2} = y - 4 = -z$$

Solution

Let $\quad x - 1 = 1 - y = z - 3 = \lambda \qquad$ and $\quad \dfrac{x-2}{2} = y - 4 = -z = \mu$

$L_1:$
$\therefore x = \lambda + 1$
$\therefore y = -\lambda + 1 \qquad$ and
$\therefore z = \lambda + 3$

$L_2:$
$\therefore x = 2\mu + 2$
$\therefore y = \mu + 4$
$\therefore z = -\mu$

At the point of intersection the x, y and z values must be equal. Using x and y alone to find λ and μ gives:

$\lambda + 1 = x = 2\mu + 2 \qquad\qquad \therefore \quad \lambda + 1 = 2\mu + 2$
$\therefore \quad -\lambda + 1 = y = \mu + 4 \qquad\qquad \therefore \quad -\lambda + 1 = \mu + 4 \quad \} +$

$\qquad\qquad\qquad\qquad\qquad\qquad\qquad\qquad 2 = 3\mu + 6$

$\qquad\qquad\qquad\qquad\qquad\qquad \therefore \quad 2 - 6 = 3\mu$

$\qquad\qquad\qquad\qquad\qquad\qquad \therefore \quad -\frac{4}{3} = \mu, \quad \lambda + 1 = 2\left(-\frac{4}{3}\right) + 2$

$\qquad\qquad\qquad\qquad\qquad\qquad \therefore \quad \lambda = -\frac{5}{3}$

PURE MATHEMATICS   THREE-DIMENSIONAL GEOMETRY

Replacing these in z for each line gives:

$$L_1: z = \lambda + 3 = -\frac{5}{3} + 3 \qquad\qquad L_2: z = -\mu = -\left(-\frac{4}{3}\right)$$

$$\therefore \quad z = \frac{4}{3} \qquad\qquad\qquad\qquad \therefore \quad z = \frac{4}{3}$$

So the λ and μ values found from equating the x and y coordinates for each line give the same value for z on L_1 and L_2. Therefore, the lines must intersect.

$$x = \lambda + 1 = -\frac{5}{3} + 1 = -\frac{2}{3} \qquad \left[\text{or } x = 2\mu + 2 = 2\left(-\frac{4}{3}\right) + 2 = -\frac{2}{3} \right]$$

and $\quad y = -\lambda + 1 = -\left(-\frac{5}{3}\right) + 1 = \frac{8}{3} \qquad \left[\text{and } y = \mu + 4 = -\frac{4}{3} + 4 = \frac{8}{3} \right]$

Therefore, the point of intersection is $\left(-\frac{2}{3}, \frac{8}{3}, \frac{4}{3}\right)$

(c)   For skew lines it is necessary to prove that they are not parallel and do not intersect.

Example 11

Prove that the lines L_1 and L_2 are skew, where

$$L_1: x - 1 = \frac{y}{3} = \frac{z-1}{4}, \qquad\qquad L_2: \frac{x-2}{4} = \frac{y-3}{-1} = z$$

Solution

$L_1: \dfrac{x-1}{1} = \dfrac{y}{3} = \dfrac{z-1}{4} \qquad\qquad \therefore$ direction ratios $1 : 3 : 4$

$L_2: \dfrac{x-2}{4} = \dfrac{y-3}{-1} = \dfrac{z}{1} \qquad\qquad \therefore$ direction ratios $4 : -1 : 1$

Lines L_1 and L_2 are not parallel. Now it is necessary to determine whether or not they intersect.

Assuming that $x - 1 = \dfrac{y}{3} = \dfrac{z-1}{4} = \lambda \quad$ and $\quad \dfrac{x-2}{4} = \dfrac{y-3}{-1} = z = \mu$

$L_1: \quad x = \lambda + 1$ $L_2: \quad x = 4\mu + 2$

$\qquad\quad y = 3\lambda \qquad$ and $\qquad\qquad\quad y = -\mu + 3$

$\qquad\quad z = 4\lambda + 1 \qquad\qquad\qquad\qquad\quad z = \mu$

Using x and y: $\lambda + 1 = 4\mu + 2 \qquad\qquad\qquad \lambda + 1 = 4\mu + 2$
(as before)
$\qquad\qquad\qquad\quad (3\lambda = -\mu + 3) \times 4 \qquad\qquad 12\lambda = -4\mu + 12 \quad \Big\} +$

$\qquad\qquad\qquad\qquad\qquad\qquad\qquad\qquad\qquad 13\lambda + 1 = 14$

$\qquad\qquad\qquad\qquad\qquad\qquad\qquad\qquad \therefore \quad 13\lambda = 13$

$\qquad\qquad\qquad\qquad\qquad\qquad\qquad\qquad \therefore \quad \lambda = 1, \; 1 + 1 = 4\mu + 2$

$\qquad\qquad\qquad\qquad\qquad\qquad\qquad\qquad \therefore \quad 2 = 4\mu + 2$

$$\therefore \quad 0 = \mu$$

Replacing these in z for each line gives:

L_1: $z = 4\lambda + 1 = 4(1) + 1$ L_2: $z = \mu = 0$

$\therefore \quad z = 5$ $\therefore z = 0$

Since these values for z are different, the lines do not intersect.

Since the lines are not parallel and do not intersect they must be skew.

Another property to be studied when considering a pair of lines is the angle between the lines.

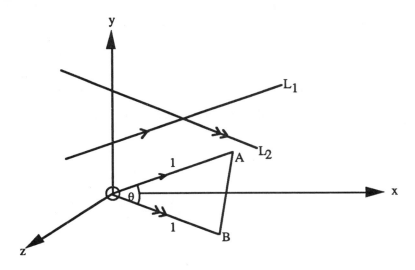

Consider two lines L_1 and L_2. If OA is drawn parallel to L_1 and OB is drawn parallel to L_2, then the angle between L_1 and L_2 is defined to be $A\hat{O}B = \theta$, say.

This means that the angle between the lines depends on their directions and not on their positions in space. Therefore, there exists an angle between skew lines as well as between intersecting lines.

If L_1 has direction cosines l_1, m_1, n_1

and L_2 has direction cosines l_2, m_2, n_2

and OA and OB are each of unit length, then A is the point (l_1, m_1, n_1) and B the point (l_2, m_2, n_2).

$$\therefore \quad AB^2 = (l_1 - l_2)^2 + (m_1 - m_2)^2 + (n_1 - n_2)^2$$

$$= l_1^2 - l_1 l_2 + l_2^2 + m_1^2 - 2m_1 m_2 + m_2^2 + n_1^2 - 2n_1 n_2 + n_2^2$$

$$= l_1^2 + m_1^2 + n_1^2 + l_2^2 + m_2^2 + n_2^2 - 2(l_1 l_2 + m_1 m_2 + n_1 n_2)$$

However, since l_1, m_1 and n_1 are direction cosines then

$$l_1^2 + m_1^2 + n_1^2 = 1 \text{ and similarly } l_2^2 + m_2^2 + n_2^2 = 1$$

$$\therefore \quad AB^2 = 2 - 2(l_1 l_2 + m_1 m_2 + n_1 n_2)$$

PURE MATHEMATICS — THREE-DIMENSIONAL GEOMETRY

Using the cosine formula:

$$AB^2 = OA^2 + OB^2 - 2\, OA \cdot OB \cdot \cos\theta$$

where $OA = OB = 1$ unit of length

then $\quad \cos\theta = \dfrac{1^2 + 1^2 - [2 - 2(l_1 l_2 + m_1 m_2 + n_1 n_2)]}{2(1)(1)}$

$\therefore \quad \cos\theta = l_1 l_2 + m_1 m_2 + n_1 n_2.$

So the angle θ between two lines L_1 and L_2 is given by:

$$\cos\theta = l_1 l_2 + m_1 m_2 + n_1 n_2$$

For perpendicular lines $\theta = 90°$ $\quad\therefore \cos 90° = 0$

$\therefore \; l_1 l_2 + m_1 m_2 + n_1 n_2 = 0$

However, for L_1: $a_1 = d_1 l_1$, $b_1 = d_1 m_1$, $c_1 = d_1 n_1$, where $a_1 : b_1 : c_1$ are the direction ratios and similarly for L_2: $a_2 = d_2 l_2$, $b_2 = d_2 m_2$, $c_2 = d_2 n_2$. Therefore, with $d_1 = d_2 = 1$, this becomes:

$$\boxed{l_1 l_2 + m_1 m_2 + n_1 n_2 = a_1 a_2 + b_1 b_2 + c_1 c_2 = 0}$$

Example 12

Find the angle between the lines

$$L_1: \frac{x-3}{2} = \frac{y+4}{-1} = \frac{z-3}{2} \quad \text{and} \quad L_2: \frac{x+1}{6} = \frac{y-2}{-3} = \frac{z-3}{-2}$$

Solution

For L_1 the direction ratios are $a_1 : a_2 : a_3 = 2 : -1 : 2$

and $\quad |a| = \sqrt{a_1^2 + a_2^2 + a_3^2} = \sqrt{2^2 + (-1)^2 + 2^2} = 3$

$\therefore$ direction cosines are $\dfrac{a_1}{a}, \dfrac{a_2}{a}, \dfrac{a_3}{a} = \dfrac{2}{3}, \dfrac{-1}{3}, \dfrac{2}{3}$

Likewise for L_2

direction cosines are $\dfrac{6}{7}, \dfrac{-3}{7}, \dfrac{-2}{7}$.

$\therefore \quad l_1 = \dfrac{2}{3}, m_1 = -\dfrac{2}{3}, n_1 = \dfrac{2}{3}, l_2 = \dfrac{6}{7}, m_2 = -\dfrac{3}{7}, n_2 = -\dfrac{2}{7}$

$\therefore \quad \cos\theta = \left(\dfrac{2}{3}\right)\left(\dfrac{6}{7}\right) + \left(-\dfrac{1}{3}\right)\left(-\dfrac{3}{7}\right) + \left(\dfrac{2}{3}\right)\left(-\dfrac{2}{7}\right)$

$\qquad\qquad = \dfrac{12}{21} + \dfrac{3}{21} - \dfrac{4}{21}$

$\qquad\qquad = \dfrac{11}{21}$

$\therefore \quad \theta = \cos^{-1}\left(\dfrac{11}{21}\right)$

$\qquad\qquad = 58.4°$

Example 13

Find the equations of the line through A(2, -1, 5) which is perpendicular to, and intersects, the line whose equations are $\dfrac{x-3}{1} = \dfrac{y-1}{2} = \dfrac{z+1}{2}$

Solution

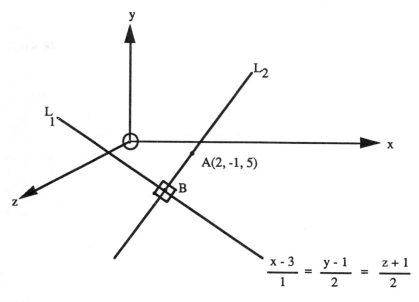

L_1 is the given line and L_2 is the required line; B is the point of intersection of the two lines. Since one point is known on L_2 it is necessary to find the direction ratios for L_2 in order to determine the equations of L_2.

Writing the equations of L_1 in terms of a parameter λ gives

$$\frac{x-3}{1} = \frac{y-1}{2} = \frac{z+1}{2} = \lambda$$

$\therefore \quad x = \lambda + 3 \text{ and } y = 2\lambda + 1 \text{ and } z = 2\lambda - 1$

In parametric terms B is the point $(\lambda + 3, 2\lambda + 1, 2\lambda - 1)$ and A is the point $(2, -1, 5)$, so the direction ratios of AB are

$(\lambda + 3) - 2 \;:\; (2\lambda + 1) - (-1) \;:\; (2\lambda - 1) - 5$

$\lambda + 1 \;:\; 2\lambda + 2 \;:\; 2\lambda - 6$

From its equation the direction ratios of L_1 are $1 : 2 : 2$

Therefore, for perpendicular lines $(\lambda + 1) \cdot 1 + (2\lambda + 2) \cdot 2 + (2\lambda - 6) \cdot 2 \quad = 0 \quad (\text{as } \theta = 90°)$

$\lambda + 1 + 4\lambda + 4 + 4\lambda - 12 \quad = 0$

$9\lambda \quad = 7$

$\therefore \quad \lambda \quad = \dfrac{7}{9}$

Therefore, the direction ratios of AB are $\dfrac{7}{9} + 1 : 2\left(\dfrac{7}{9}\right) + 2 : 2\left(\dfrac{7}{9}\right) - 6$

PURE MATHEMATICS — THREE-DIMENSIONAL GEOMETRY

ie $\quad \dfrac{16}{9} : \dfrac{32}{9} : -\dfrac{40}{9} = 16 : 32 : -40$ (multiplying by 9)

$\qquad\qquad\qquad\qquad = 2 : 4 : -5 \quad$ (dividing by 8)

So the equations of L_2 are $\quad \dfrac{x-2}{2} = \dfrac{y-(-1)}{4} = \dfrac{z-5}{-5}$

$\qquad\qquad\qquad\qquad\qquad\qquad \dfrac{x-2}{2} = \dfrac{y+1}{4} = \dfrac{z-5}{-5}$

(Alternatively, having found λ, the actual coordinates of B could have been calculated and then the equations of L_2 using:

$\dfrac{x - x_1}{x_2 - x_1} = \dfrac{y - y_1}{y_2 - y_1} = \dfrac{z - z_1}{z_2 - z_1} \quad)$

---

f) *Example 'A' level question*

The lines L_1 and L_2 have the following equations:

$\qquad L_1 : \quad \dfrac{x+1}{2} = \dfrac{y-2}{-1} = \dfrac{z-2}{1}$

$\qquad L_2 : \quad \dfrac{x+5}{1} = \dfrac{y+1}{-3} = \dfrac{z-7}{4}$

Show that the lines intersect and find the point of intersection.

Find also:

(a) the angle between L_1 and L_2;

(b) the perpendicular distance from $(1, 2, 3)$ to the line L_1

Solution

Let $\quad \dfrac{x+1}{2} = \dfrac{y-2}{-1} = \dfrac{z-2}{1} = \lambda, \qquad \dfrac{x+5}{1} = \dfrac{y+1}{-3} = \dfrac{z-7}{4} = \mu$

$\therefore \quad x = 2\lambda - 1 \qquad)\qquad\qquad\qquad (\ x = \mu - 5$
$\qquad\qquad\qquad\qquad\qquad)\qquad\qquad\qquad (\ $
$\therefore \quad y = -\lambda + 2 \qquad) \quad$ and $\quad (\ y = -3\mu - 1$
$\qquad\qquad\qquad\qquad\qquad)\qquad\qquad\qquad (\ $
$\therefore \quad z = \lambda + 2 \qquad\;\;)\qquad\qquad\qquad (\ z = 4\mu + 7$

Taking the x and y values and equating to find λ and μ gives:

$\qquad 2\lambda - 1 = \mu - 5 \quad : \quad 2\lambda - 1 = \mu - 5$
$\qquad\qquad\qquad\qquad\qquad\qquad\qquad\qquad\quad \}$ add
$\qquad (-\lambda + 2 = -3\mu - 1) \times 2 : \quad -2\lambda + 4 = -6\mu - 2$

$\qquad\qquad\qquad\qquad\qquad\qquad 3 = -5\mu - 7$

$\qquad\qquad\qquad\therefore \qquad 10 = -5\mu \qquad \therefore \mu = -2$

$\qquad\qquad\qquad$ and $\quad 2\lambda - 1 = (-2) - 5$

$\qquad\qquad\qquad\qquad\qquad\qquad 2\lambda = -6 \qquad \therefore \lambda = -3$

449

Now to see if these values of λ and μ give the same value for z.

$$z = \lambda + 2 \qquad\qquad z = 4\mu + 7$$

∴ $\quad z = -3 + 2 = -1 \qquad\qquad z = 4(-2) + 7 = -1$

Since the z value is the same for both, the lines L_1 and L_2 do intersect at $x = 2(-3) - 1 = -7$, $y = -(-3) + 2 = 5$, $z = -1$, ie $(-7, 5, -1)$.

(a) The angle between L_1 and L_2 is found using:

$$\cos\theta = l_1 l_2 + m_1 m_2 + n_1 n_2$$

for L_1 the direction ratios are $2 : -1 : 1$ and $|a| = \sqrt{2^2 + (-1)^2 + 1^2} = \sqrt{6}$

Therefore, the direction cosines are $\dfrac{2}{\sqrt{6}}, \dfrac{-1}{\sqrt{6}}, \dfrac{1}{\sqrt{6}}$

for L_2 the direction ratios are $1 : -3 : 4$ and $|b| = \sqrt{1^2 + (-3)^2 + 4^2} = \sqrt{26}$

Therefore, the direction cosines are $\dfrac{1}{\sqrt{26}}, \dfrac{-3}{\sqrt{26}}, \dfrac{4}{\sqrt{26}}$

∴ $\quad \cos\theta = \dfrac{2}{\sqrt{6}}\left(\dfrac{1}{\sqrt{26}}\right) + \dfrac{-1}{\sqrt{6}}\left(\dfrac{-3}{\sqrt{26}}\right) + \dfrac{1}{\sqrt{6}}\left(\dfrac{4}{\sqrt{26}}\right)$

$$= \dfrac{2}{\sqrt{156}} + \dfrac{3}{\sqrt{156}} + \dfrac{4}{\sqrt{156}}$$

$$\cos\theta = \dfrac{9}{\sqrt{156}}$$

∴ $\quad \theta = \cos^{-1}\left(\dfrac{9}{\sqrt{156}}\right)$

$$= 43.9°$$

(b) To find the perpendicular distance from the point $(1, 2, 3)$ - A - to the line L_1 it is necessary to find the point of intersection (B) of L_1 and the perpendicular to L_1 through $(1, 2, 3)$.

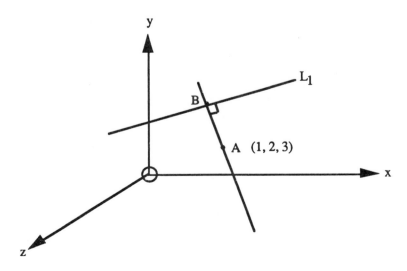

Since B lies on L_1 it will have parametric coordinates $(2\lambda - 1, -\lambda + 2, \lambda + 2)$ and so the direction ratios of AB are

$$(2\lambda - 1) - 1 : (-\lambda + 2) - 2 : (\lambda + 2) - 3$$

ie $\quad 2\lambda - 2 : -\lambda : \lambda - 1$

and the direction ratios for L_1 are $2 : -1 : 1$

$\therefore \quad (2\lambda - 2)2 + (-\lambda)(-1) + (\lambda - 1)1 \quad = 0$ for perpendicular lines

$$4\lambda - 4 + \lambda + \lambda - 1 = 0$$

$$6\lambda = 5$$

$\therefore \quad \lambda = \dfrac{5}{6}$

So the cordinates of B are

$$\left(2\left(\dfrac{5}{6}\right) - 1, \ -\dfrac{5}{6} + 2, \ \dfrac{5}{6} + 2\right)$$

ie $\quad \left(\dfrac{4}{6}, \dfrac{7}{6}, \dfrac{17}{6}\right)$

Therefore, the distance AB can be found from the distance formula:

$$AB = \sqrt{\left(\dfrac{4}{6} - 1\right)^2 + \left(\dfrac{7}{6} - 2\right)^2 + \left(\dfrac{17}{6} - 3\right)^2}$$

$$= \sqrt{\left(-\dfrac{2}{6}\right)^2 + \left(-\dfrac{5}{6}\right)^2 + \left(-\dfrac{1}{6}\right)^2}$$

$$= \sqrt{\dfrac{4}{36} + \dfrac{25}{36} + \dfrac{1}{36}} \quad = \sqrt{\dfrac{30}{36}}$$

Therefore, required distance $= \sqrt{\dfrac{5}{6}}$

THREE-DIMENSIONAL GEOMETRY — PURE MATHEMATICS

10.2 Further three-dimensional coordinate geometry

a) *The equation of a plane*

A plane is located in space if a point on the plane is known and if the plane is perpendicular to a particular direction.

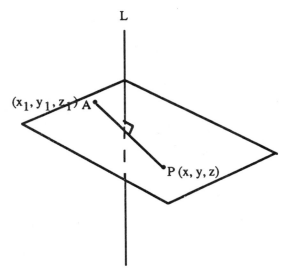

If L is a line with direction ratios $a : b : c$ normal to the plane, then any line in a plane will be perpendicular to L.

Therefore, if A is a known point with coordinates (x_1, y_1, z_1) and P is any other point in the plane with coordinates (x, y, z), the direction ratios of AP will be $(x - x_1) : (y - y_1) : (z - z_1)$ and

$$a(x - x_1) + b(y - y_1) + c(z - z_1) = 0 \text{ for perpendicular lines}$$

$$\therefore \quad ax - ax_1 + by - by_1 + cz - cz_1 = 0$$

$$\therefore \quad ax + by + cz = ax_1 + bx_1 + x_1$$

The equation of a plane is of the form

$$ax + by + cz = D$$

where $a : b : c$ are the direction ratios of any normal to the plane

---

This is basically the only way of deriving the equation of a plane. However, as will be seen from the followng examples, planes can be defined in a number of different ways.

Example 14

Find the equation of the plane containing the point A(3, -1, 0) and normal to a line with direction ratios $2 : -1 : 3$.

PURE MATHEMATICS — THREE-DIMENSIONAL GEOMETRY

Solution

Therefore, if P is a general point (x, y, z) in the plane, the direction ratios of AP are $(x - 3) : (y - (-1)) : (z - 0)$, ie

$$(x - 3) : (y + 1) : z$$

$\therefore \quad 2(x - 3) + (-1)(y + 1) + 3z = 0$

$\therefore \quad 2x - 6 - y - 1 + 3z = 0$

$\therefore \quad 2x - y + 3z = 7$

Alternatively, this can be found by replacing directly in

$$ax + by + cz = ax_1 + by_1 + cz_1$$
$$2x - y + 3z = 2(3) + -1(-1) + 3(0)$$
$$2x - y + 3z = 7$$

Example 15

Find the equation of the plane passing through (2, -1, 4) (3, 2, -6) (4, 1, 5)

Solution

Here is an alternative way of locating a plane in space - by knowing three points in the plane itself. The only way to find the equation of this plane is to use the standard form of the equation: $ax + by + cz = D$ and, since the given points lie in the plane, their coordinates must satisfy the equation.

(2, -1, 4)	$a(2) + b(-1) + c(4) = D$ ...	$2a - b + 4c = D$	----------	(1)
(3, 2, -6)	$a(3) + b(2) + c(-6) = D$	$3a + 2b - 6c = D$	----------	(2)
(4, 1, 5)	$a(4) + b(1) + c(5) = D$	$4a + b + 5c = D$	----------	(3)

These equations must be solved for a, b and c in terms of D:

(1) x 2 $4a - 2b + 8c = 2D$
} add $7a + 2c = 3D$ (4)
(2) $3a + 2b - 6c = D$

(2) $3a + 2b - 6c = D$
} subtract $-5a - 16c = -D$ (5)
(3) x 2 $8a + 2b + 10c = 2D$

(4) x 8 $56a + 16c = 24D$
} add $51a = 23D$
(5) $-5a - 16c = -D$

$$\therefore \quad a = \frac{23D}{51}$$

Replacing a in (4) $7\left(\frac{23D}{51}\right) + 2c = 3D$

$$\therefore \quad 2c = 3D - \frac{161D}{51}$$

THREE-DIMENSIONAL GEOMETRY PURE MATHEMATICS

$$\therefore \quad 2c = -\frac{8D}{51}$$

$$\therefore \quad c = -\frac{4D}{51}$$

Replacing a and c in (3)

$$4\left(\frac{23D}{51}\right) + b + 5\left(-\frac{4D}{51}\right) = D$$

$$\frac{92D}{51} + b - \frac{20D}{51} = D$$

$$\therefore \quad b = D - \frac{72D}{51} = -\frac{21D}{51}$$

Therefore, the equation of the plane is:

$$\frac{23D}{51}x - \frac{21D}{51}y - \frac{4D}{51}z = D \quad \text{multiply by } \frac{51}{D}$$

$$\therefore \quad 23x - 21y - 4z = 51$$

Example 16

Find the equation of the plane containing the lines

$$L_1 : \frac{x-3}{5} = \frac{y+1}{2} = \frac{z-3}{1}, \quad \text{and} \quad L_2 : \frac{x-3}{2} = \frac{y+1}{4} = \frac{z-3}{3}$$

Solution

Two lines can be contained in the same plane if they intersect or if they are parallel.

As has been seen in the last example a plane is defined by three non-collinear points, ie the three points must not be on the same straight line. Therefore, it is necessary to find two points on one line and one point on the other line.

Using the parametric form of the equations:

$$L_1 : \frac{x-3}{5} = \frac{y+1}{2} = \frac{z-3}{1} = \lambda \qquad L_2 : \frac{x-3}{2} = \frac{y+1}{4} = \frac{z-3}{3} = \mu$$

Each value of the parameter λ or μ corresponds to one and only one point on the line.

$\therefore \quad L_1 \quad x = 5\lambda + 3 \qquad \text{and} \qquad L_2 \quad x = 2\mu + 3$

$\qquad\qquad y = 2\lambda - 1 \qquad\qquad\qquad\qquad y = 4\mu - 1$

$\qquad\qquad z = \lambda + 3 \qquad\qquad\qquad\qquad\; z = 3\mu + 3$

$\therefore \quad \lambda = 0$ gives the point $(3, -1, 3)$ $\qquad\qquad \mu = 0$ gives the point $(3, -1, 3)$

$\therefore \quad \lambda = -1$ gives the point $(-2, -3, 2)$ $\qquad \mu = -1$ gives the point $(1, -5, 0)$

So the plane passes through $(3, -1, 3)$ $(-2, -3, 2)$ and $(1, -5, 0)$.

(*Note*: $\mu = 0$ gave $(3, -1, 3)$ which is also on L_1 and so cannot be used again. Therefore, it was necessary to find a different point on L_2.)

PURE MATHEMATICS — THREE-DIMENSIONAL GEOMETRY

The method of finding the equation of the plane would then be identical to the last example, ie

$3a - b + 3c = D$

$-2a - 3b + 2c = D$ } solved to give $2x - 13y + 16z = 67$

$a - 5b = D$

Note that any plane through the origin has equation $ax + by + cz = 0$ because it is satisfied by the point $(0, 0, 0)$. Therefore $D = 0$.

---

b) *Distance of a plane from the origin*

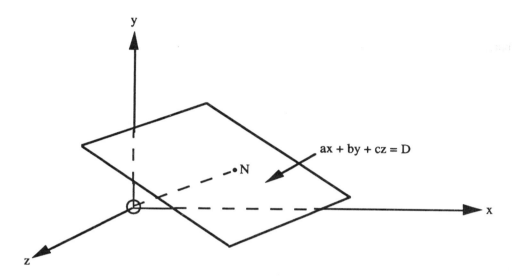

Consider a plane with equation $ax + by + cz = D$ where the perpendicular from O meets the plane at N. Therefore, ON is normal to the plane.

From the equation of the plane, ON must have direction ratios $a : b : c$ and direction cosines.

$$l = \frac{a}{\sqrt{a^2 + b^2 + c^2}} \quad m = \frac{b}{\sqrt{a^2 + b^2 + c^2}} \quad n = \frac{c}{\sqrt{a^2 + b^2 + c^2}}$$

Therefore, if the length of ON is d, the coordinates of ON are:

$$\left(\frac{a.d}{\sqrt{a^2 + b^2 + c^2}}, \; \frac{b.d}{\sqrt{a^2 + b^2 + c^2}}, \; \frac{c.d}{\sqrt{a^2 + b^2 + c^2}} \right)$$

and since N is a point on the plane, these coordinates satisfy the equation of the plane.

$\therefore \quad a\left(\dfrac{a.d}{\sqrt{a^2 + b^2 + c^2}} \right) + b\left(\dfrac{b.d}{\sqrt{a^2 = b^2 + c^2}} \right) + c\left(\dfrac{c.d}{\sqrt{a^2 + b^2 + c^2}} \right) = D$

$\therefore \quad \dfrac{a^2 d + b^2 d + c^2 d}{\sqrt{a^2 + b^2 + c^2}} = D$

$\therefore \quad d = \dfrac{D\sqrt{a^2 + b^2 + c^2}}{a^2 + b^2 + c^2}$

$\therefore \quad d = \dfrac{D}{\sqrt{a^2 + b^2 + c^2}}$

THREE-DIMENSIONAL GEOMETRY — PURE MATHEMATICS

> The perpendicular distance of the plane from O is given by: $d = \dfrac{D}{\sqrt{a^2 + b^2 + c^2}}$

Example 17

Find the distance of the plane $2x - y + 5z = 2$ from the origin.

Solution

From the equation of the plane $a = 2$, $b = -1$, $c = 5$, $D = 2$.

$$\therefore \quad d = \dfrac{2}{\sqrt{2^2 + (-1)^2 + 5^2}}$$

$$= \dfrac{2}{\sqrt{30}}$$

Required distance $= \dfrac{2}{\sqrt{30}}$

c) Distance of a point from a plane

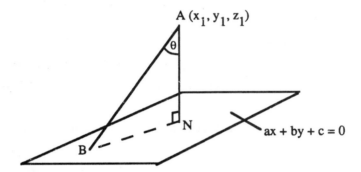

The required distance is the perpendicular distance of A from the plane, ie the distance AN. If B is any other point on the plane $AN = AB \cos \theta$. To explain this in general terms leads to a rather cumbersome formula. Therefore, the method will be explained by reference to a numerical example.

Let $A(-1, 3, 2)$ \qquad Equation of plane: $x - 3y + z = 1$

Choosing B as any point on the plane, let $y = 0$, $z = 0$, then $x = 1$. Therefore, B is $(1, 0, 0)$.

To find AN it is necessary to find AB and $\cos \theta$

$$AB^2 = (-1 - 1)^2 + (3 - 0)^2 + (2 - 0)^2$$

$$AB = \sqrt{17}$$

$$\cos \theta = l_1 l_2 + m_1 m_2 + n_1 n_2$$

where l_1, m_1, n_1 are the direction cosines for AB and l_2, m_2, n_2 for AN.

AB : direction ratios $(-1 - 1) : (3 - 0) : (2 - 0) = -2 : 3 : 2$

$$\therefore \quad \text{direction cosines} = \dfrac{-2}{\sqrt{17}}, \dfrac{3}{\sqrt{17}}, \dfrac{2}{\sqrt{17}}$$

AN : direction ratios 1 : - 3 : 1 from equation of plane

$$\therefore \quad \text{direction cosines} = \frac{1}{\sqrt{11}}, \frac{-3}{\sqrt{11}}, \frac{1}{\sqrt{11}}$$

$$\therefore \quad \cos\theta = \frac{-2}{\sqrt{17}}\left(\frac{1}{\sqrt{11}}\right) + \frac{3}{\sqrt{17}}\left(\frac{-3}{\sqrt{11}}\right) + \frac{2}{\sqrt{17}}\left(\frac{1}{\sqrt{11}}\right)$$

$$= \frac{-2 - 9 + 2}{\sqrt{17}\sqrt{11}}$$

$$\therefore \quad \cos\theta = \frac{-9}{\sqrt{17}.\sqrt{11}}$$

The negative sign arises from the direction ratios and can be ignored because - 2 : 3 : 2 is the same as 2 : - 3 : - 2 which would change the sign away.

$$\therefore \quad AN = AB \cos\theta$$

$$= \sqrt{17} \cdot \frac{9}{\sqrt{17}\sqrt{11}}$$

$$= \frac{9}{\sqrt{11}}$$

---

d) *Example 'A' level question*

The coordinates of A, B and C are:

        A(3, 0, 0)   B(0, 4, 0)   C(0, 0, 5)

(a) Find the equations of:

    (i) the line BC;

    (ii) the plane ABC.

(b) Show that angle $A\hat{B}C$ is approximately 60°.

(c) Find the volume of the tetrahedron ABCD where D is the point (7, 8, 9).

THREE-DIMENSIONAL GEOMETRY   PURE MATHEMATICS

Solution

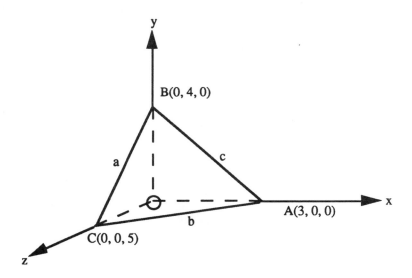

(a) (i) *Equation of BC*

Using $\dfrac{x - x_1}{x_2 - x_1} = \dfrac{y - y_1}{y_2 - y_1} = \dfrac{z - z_1}{z_2 - z_1}$

where (x_1, y_1, z_1) is $(0, 4, 0)$ and (x_2, y_2, z_2) is $(0, 0, 5)$.

$$\dfrac{x - 0}{0 - 0} = \dfrac{y - 4}{0 - 4} = \dfrac{z - 0}{5 - 0}$$

$$\dfrac{x}{1} = \dfrac{y - 4}{-4} = \dfrac{z}{5}$$

(ii) *Equation of plane ABC*

Using $ax + by + cz = D$

A: $(3, 0, 0)$ $\therefore a(3) + b(0) + c(0) = D$ $\therefore a = \dfrac{D}{3}$

B: $(0, 4, 0)$ $\therefore a(0) + b(4) + c(0) = D$ $\therefore b = \dfrac{D}{4}$

C: $(0, 0, 5)$ $\therefore a(0) + b(0) + c(5) = D$ $\therefore c = \dfrac{D}{5}$

$\therefore \quad \dfrac{D}{3} x + \dfrac{D}{4} y + \dfrac{D}{5} z = D$

or $\quad 20x + 15y + 12z = 60$

(b) The angle $A\hat{B}C$ can be found by applying the cosine formula to $\triangle ABC$.

$\therefore \quad \cos A\hat{B}C = \dfrac{a^2 + c^2 - b^2}{2ac}$

$AB^2 = c^2 = (0 - 3)^2 + (4 - 0)^2 + (0 - 0)^2$

$\qquad\qquad\quad = 25$

$$\therefore \quad c = 5$$

$$AC^2 = b^2 = (3-0)^2 + (0-0)^2 + (0-5)^2$$
$$= 34$$

$$\therefore \quad b = \sqrt{34}$$

$$BC^2 = a^2 = (0-0)^2 + (4-0)^2 + (0-5)^2$$
$$= 41$$

$$\therefore \quad a = \sqrt{41}$$

$$\cos A\hat{B}C = \frac{41 + 25 - 34}{2\sqrt{41}\,(5)}$$

$$= \frac{32}{10\sqrt{41}}$$

$$= \frac{32}{64.03}$$

Since $\cos A\hat{B}C$ is approximately equal to 0.5 $A\hat{B}C$ is nearly 60°.

(c) Volume of tetrahedron $= \frac{1}{3}$ (base area) height

base area $= $ area ABC $= \frac{1}{2} ac \sin B$

$$= \frac{1}{2}\left(\sqrt{41}\right) . (5) \sin 60°$$

$\therefore$ base area $= 13.9$ square units

height $=$ AN $=$ perpendicular distance of (7, 8, 9) from plane ABC

Using A(3, 0, 0) and D(7, 8, 9) then DN = AD cos θ

$$AD = \sqrt{(3-7)^2 + (0-8)^2 + (0-9)^2}$$

$$AD = \sqrt{161}$$

For AD: direction ratios (7 - 3) : (8 - 0) : (9 - 0) = 4 : 8 : 9

∴ direction cosines are $\dfrac{4}{\sqrt{161}}, \dfrac{8}{\sqrt{161}}, \dfrac{9}{\sqrt{161}}$

For DN: direction ratios 20 : 15 : 12 from equation of plane

∴ direction cosines $= \dfrac{20}{\sqrt{769}}, \dfrac{15}{\sqrt{769}}, \dfrac{12}{\sqrt{769}}$

$\cos\theta = l_1 l_2 + m_1 m_2 + n_1 n_2$

$\cos\theta = \dfrac{4}{\sqrt{161}}\left(\dfrac{20}{\sqrt{769}}\right) + \dfrac{8}{\sqrt{161}}\left(\dfrac{15}{\sqrt{769}}\right) + \dfrac{9}{\sqrt{161}}\left(\dfrac{12}{\sqrt{769}}\right)$

$= \dfrac{80 + 120 + 108}{\sqrt{161}\,\sqrt{769}}$

$= \dfrac{308}{\sqrt{161}\,\sqrt{769}}$

DN $= $ AD $\cos\theta = \sqrt{161} \cdot \dfrac{308}{\sqrt{161}\,\sqrt{769}}$

$= \dfrac{308}{\sqrt{769}}$

$= 11.1$ units

∴ Volume $= \dfrac{1}{3}(13.9)(11.1)$

$= 51.43$ units^3

10.3 Vectors in three dimensions

a) *From 2 to 3 dimensions*

The 2 dimensional work of chapter 9 can easily be extended to 3 dimensional space. The unit vector acting in the direction of Oz is denoted by **k** and so any vector **r** can be expressed as the sum of its 'x', 'y' and 'z' components as:

$$\boxed{\mathbf{r} = a\mathbf{i} + b\mathbf{j} + c\mathbf{k}}$$

and the length is:

$$\boxed{|\mathbf{r}| = \sqrt{a^2 + b^2 + c^2}}$$

The position vector of a point P(x, y, z) relative to an origin 0 is $x\mathbf{i} + y\mathbf{j} + z\mathbf{k}$.

The scalar product of two vectors **a** and **b** is still defined as $\mathbf{a} \cdot \mathbf{b} = ab\cos\theta$ where θ is the angle between **a** and **b**.

If **a** $= a_1\mathbf{i} + a_2\mathbf{j} + a_3\mathbf{k}$

and **b** $= b_1\mathbf{i} + b_2\mathbf{j} + b_3\mathbf{k}$

then **a** . **b** $= (a_1\mathbf{i} + a_2\mathbf{j} + a_3\mathbf{k}) \cdot (b_1\mathbf{i} + b_2\mathbf{j} + b_3\mathbf{k})$

PURE MATHEMATICS — THREE-DIMENSIONAL GEOMETRY

$$= a_1b_1 + a_2b_2 + a_3b_3$$

If $\underline{a}$ and $\underline{b}$ are perpendicular $\quad \underline{a} \cdot \underline{b} = 0$

$\therefore \quad a_1b_1 + a_2b_2 + a_3b_3 = 0$

Example 18

If $\underline{a} = 2\underline{i} - 4\underline{j} + 2\underline{k}$ and $\underline{b} = 3\underline{i} + 4\underline{j} - 5\underline{k}$, find:

(a) the magnitudes of $\underline{a}$, $\underline{b}$, $5\underline{a} + 2\underline{b}$;

(b) the unit vector in the direction $\underline{a} + 2\underline{b}$.

Solution

(a)
$$|\underline{a}| = a$$
$$= \sqrt{2^2 + (-4)^2 + 2^2}$$
$$= \sqrt{24}$$

$$|\underline{b}| = b$$
$$= \sqrt{3^2 + 4^2 + (-5)^2}$$
$$= \sqrt{50}$$

$$5\underline{a} + 2\underline{b} = 5(2\underline{i} - 4\underline{j} + 2\underline{k}) + 2(3\underline{i} + 4\underline{j} - 5\underline{k})$$
$$= 10\underline{i} - 20\underline{j} + 10\underline{k} + 6\underline{i} + 8\underline{j} - 10\underline{k}$$
$$= 16\underline{i} - 12\underline{j} + 0\underline{k}$$
$$= 16\underline{i} - 12\underline{j}$$

$$\therefore |5\underline{a} + 2\underline{b}| = \sqrt{16^2 + (-12)^2}$$
$$= \sqrt{400}$$
$$= 20$$

(b)
$$\underline{a} + 2\underline{b} = 2\underline{i} - 4\underline{j} + 2\underline{k} + 2(3\underline{i} + 4\underline{j} - 5\underline{k})$$
$$= 2\underline{i} - 4\underline{j} + 2\underline{k} + 6\underline{i} + 8\underline{j} - 10\underline{k}$$
$$= 8\underline{i} + 4\underline{j} - 8\underline{k}$$

This is not a unit vector since its magnitude

$$= |\underline{a} + 2\underline{b}| = \sqrt{8^2 + 4^2 + (-8)^2}$$
$$= \sqrt{144}$$
$$= 12$$

However, a unit vector $\hat{r}$ in the direction of a given vector r was defined as $\hat{r} = \frac{r}{r}$ where $r = |r|$. This definition is as true for 3 dimensions as it was for 2 dimensions.

Therefore, the required unit vector $= \dfrac{8i + 4j - 8k}{12}$

$= \dfrac{8}{12}i + \dfrac{4}{12}j - \dfrac{8}{12}k$

$= \dfrac{2}{3}i + \dfrac{1}{3}j - \dfrac{2}{3}k$

Example 19

Given that $a = 6i + (p - 10)j + (3p - 5)k$ and $|a| = 11$, find the possible values of p.

Solution

$a = |a| = \sqrt{6^2 + (p - 10)^2 + (3p - 5)^2}$

but $|a| = 11$

∴ $\sqrt{6^2 + (p - 10)^2 + (3p - 5)^2} = 11$ squaring both sides

$36 + (p - 10)^2 + (3p - 5)^2 = 121$

∴ $36 + p^2 - 20p + 100 + 9p^2 - 30p + 25 = 121$

$10p^2 - 50p + 161 - 121 = 0$

$10p^2 - 50p + 40 = 0$ dividing by 10

$p^2 - 5p + 4 = 0$

$(p - 4)(p - 1) = 0$

∴ either $p - 4 = 0$ or $p - 1 = 0$

         $p = 4$ or $p = 1$

The possible values of p are 4 and 1.

b)   *Direction ratios and direction cosines*

A point P with coordinates (a, b, c) has direction ratios a : b : c and direction cosines

$l = \dfrac{a}{\sqrt{a^2 + b^2 + c^2}}$, $m = \dfrac{b}{\sqrt{a^2 + b^2 + c^2}}$, $n = \dfrac{c}{\sqrt{a^2 + b^2 + c^2}}$ (see page 435).

The position vector r of P would be $r = ai + bj + ck$. The direction ratios would still be a : b : c and the direction cosines would be written as $l = \dfrac{a}{r}$, $m = \dfrac{b}{r}$, $n = \dfrac{c}{r}$ where $l^2 + m^2 + n^2 = 1$ and $r = \sqrt{a^2 + b^2 + c^2}$

The free vector $v = ai + bj + ck$ has the same direction as r and the same magnitude and hence the same direction ratios a : b : c and the same direction cosines $l = \dfrac{a}{v}$, $m = \dfrac{b}{v}$, $n = \dfrac{c}{v}$ where $v = r$.

PURE MATHEMATICS — THREE-DIMENSIONAL GEOMETRY

It follows that parallel vectors will have equal direction ratios. If the vectors are in the same direction they will also have equal direction cosines. However, if the vectors are in opposite directions, the direction cosines will be equal in magnitude but opposite in sign.

From this it should be realised that the direction ratios and direction cosines can be calculated for any vector - irrespective of whether it is a free or a position vector.

A unit vector $\hat{r}$ in the direction of a given vector r is defined as $\hat{r} = \dfrac{r}{r}$. For $r = a\underline{i} + b\underline{j} + c\underline{k}$,

then $\hat{r} = \dfrac{a\underline{i} + b\underline{j} + c\underline{k}}{r} = \dfrac{a}{r}\underline{i} + \dfrac{b}{r}\underline{j} + \dfrac{c}{r}\underline{k}$

$\therefore \quad \boxed{\hat{r} = l\underline{i} + m\underline{j} + n\underline{k}}$

Example 20

Find the angles at which the vector $4\underline{i} + 8\underline{j} + \underline{k}$ is inclined to each of the coordinate axes.

Solution

Let $\quad r = 4\underline{i} + 8\underline{j} + \underline{k}$

$\therefore \quad r = \sqrt{4^2 + 8^2 + 1^2}$

$\quad = \sqrt{16 + 64 + 1}$

$\quad = 9$

the direction ratios are $4 : 8 : 1$ and the direction cosines are

$l = \dfrac{4}{9},\ m = \dfrac{8}{9},\ n = \dfrac{1}{9}$

r is inclined to the x axis at an angle α where $\cos \alpha = \dfrac{4}{9} = 0.444$. Therefore $\alpha = 63.6°$.

It is inclined to the y axis at an angle β where $\cos \beta = \dfrac{8}{9} = 0.888$. Therefore $\beta = 27.3°$.

And it is inclined to the z axis at an angle δ where $\cos \delta = \dfrac{1}{9} = 0.111$. Therefore $\delta = 83.6°$.

Example 21

Find $\underline{v}$ if $\underline{v}$ is parallel to the vector $8\underline{i} + \underline{j} + 4\underline{k}$ and is equal in magnitude to the vector $\underline{i} - 2\underline{j} + 2\underline{k}$.

Solution

Let $\quad r = 8\underline{i} + \underline{j} + 4\underline{k}$

$\therefore \quad r = \sqrt{8^2 + 1^2 + 4^2}$

$\quad r = 9$

THREE-DIMENSIONAL GEOMETRY PURE MATHEMATICS

Therefore, the unit vector $\hat{r}$, in the direction of $\underline{r}$ is:

$$\hat{r} = \frac{8\underline{i} + \underline{j} + 4\underline{k}}{9}$$

$$= \frac{8}{9}\underline{i} + \frac{1}{9}\underline{j} + \frac{4}{9}\underline{k}$$

However, $\underline{v}$ must have magnitude $\sqrt{1^2 + (-2)^2 + (-2)^2} = 3$

$$\therefore \quad \underline{v} = 3 \times \hat{r}$$

$$= 3\left(\frac{8}{9}\underline{i} + \frac{1}{9}\underline{j} + \frac{4}{9}\underline{k}\right)$$

$$= \frac{8}{3}\underline{i} + \frac{1}{3}\underline{j} + \frac{4}{3}\underline{k}$$

c) *Properties of a vector joining two points*

From the previous work it should be realised that vector methods are very similar to the cartesian methods used in sections 10.1 and 10.2. They should be regarded as an alternative approach to the same work, leading to the same end results.

If A is the point (x_1, y_1, z_1) and B is the point (x_2, y_2, z_2), then their position vectors will be

$\overline{OA} = \underline{a} = x_1\underline{i} + y_1\underline{j} + z_1\underline{k}$ and $\overline{OB} = \underline{b} = x_2\underline{i} + y_2\underline{j} + z_2\underline{k}$

Therefore, the vector joining AB: $\overline{AB} = \overline{AO} + \overline{OB}$

$$= -\underline{a} + \underline{b}$$

$$= (x_2 - x_1)\underline{i} + (y_2 - y_1)\underline{j} + (z_2 - z_1)\underline{k}$$

$$AB = |\overline{AB}|$$

$$= \sqrt{(x_2 - x_1)^2 + (y_2 - y_1)^2 + (z_2 - z_1)^2}$$

The direction ratios of $\overline{AB}$ are $(x_2 - x_1) : (y_2 - y_1) : (z_2 - z_1)$

and the direction cosines are $\frac{(x_2 - x_1)}{AB}, \frac{(y_2 - y_1)}{AB}, \frac{(z_2 - z_1)}{AB}$

If M is the midpoint of AB it will have coordinates $\left(\frac{x_1 + x_2}{2}, \frac{y_1 + y_2}{2}, \frac{z_1 + z_2}{2}\right)$

and position vector $\overline{OM} = \frac{1}{2}(x_1 + x_2)\underline{i} + \frac{1}{2}(y_1 + y_2)\underline{j} + \frac{1}{2}(z_1 + z_2)\underline{k}$

All these results were derived on page 437 et seq without the use of vectors.

Example 22

A, B and C are the points with position vectors $\underline{i} + \underline{j} - \underline{k}$, $\underline{j} + \underline{k}$ and $2\underline{i} + \underline{j}$.

Find the direction cosines in $\overline{BC}$ and $\overline{AB}$.

Solution

$$\vec{BC} = \vec{BO} + \vec{OC}$$
$$= -\vec{OB} + \vec{OC}$$
$$= -(\underline{j} + \underline{k}) + (2\underline{i} + \underline{j})$$
$$= 2\underline{i} + \underline{k}$$

$$BC = \sqrt{2^2 + (-1)^2}$$
$$= \sqrt{5}$$

The direction cosines are $\dfrac{2}{\sqrt{5}}, \dfrac{0}{\sqrt{5}}, \dfrac{-1}{\sqrt{5}}$, ie $\dfrac{2}{\sqrt{5}}, 0, \dfrac{-1}{\sqrt{5}}$

$$\vec{AB} = \vec{AO} + \vec{OB}$$
$$= -\vec{OA} + \vec{OB}$$
$$= -(\underline{i} + \underline{j} - \underline{k}) + (\underline{j} + \underline{k})$$
$$= -\underline{i} + 2\underline{k}$$

$$AB = \sqrt{(-1)^2 + 2^2}$$
$$= \sqrt{5}$$

The direction cosines are $\dfrac{-1}{\sqrt{5}}, \dfrac{0}{\sqrt{5}}, \dfrac{2}{\sqrt{5}}$, ie $\dfrac{-1}{\sqrt{5}}, 0, \dfrac{2}{\sqrt{5}}$

---

d) *Equations of a straight line*

As has already been stated a line is uniquely defined if:

(a) it has a known direction and passes through a known fixed point; or

(b) it passes through two known fixed points.

THREE-DIMENSIONAL GEOMETRY   PURE MATHEMATICS

(a)   Consider a line, L, parallel to a vector $\underline{b}$ and which passes through a fixed point $A(x_1, y_1, z_1)$ with position vector $\underline{a}$.

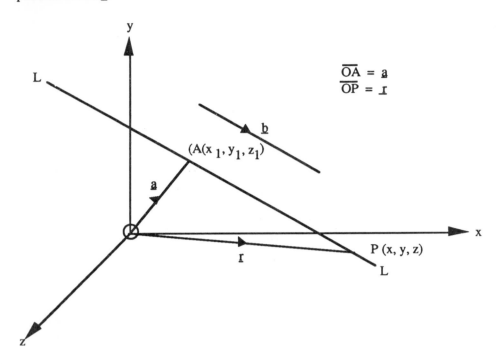

$P(x, y, z)$ is any point on the line L. Therefore the vector $\overline{AP}$ will be a scalar multiple, say λ, of $\underline{b}$ since $\overline{AP}$ and $\underline{b}$ are parallel, but of different magnitudes.

$\therefore \quad \overline{AP} = \lambda \underline{b} \qquad$ but $\overline{AP} = \overline{AO} + \overline{OP}$

$$= -\underline{a} + \underline{r}$$

$\therefore \quad -\underline{a} + \underline{r} = \lambda \underline{b}$

$\therefore \quad \boxed{\underline{r} = \underline{a} + \lambda \underline{b}}$

For each value of λ (a parameter) this vector equation gives the position vector $\underline{r}$ of a single point on the line.

It can now be seen from where the equation $\dfrac{x - x_1}{a} = \dfrac{y - y_1}{b} = \dfrac{z - z_1}{c} = \lambda$ is derived.

$\underline{a}$ is the vector $x_1\underline{i} + y_1\underline{j} + z_1\underline{k}$, $\underline{b}$ the vector $a\underline{i} + b\underline{j} + c\underline{k}$ and $\underline{r}$ will be $x\underline{i} + y\underline{j} + z\underline{k}$.

Therefore, $x\underline{i} + y\underline{j} + z\underline{k} = (x_1\underline{i} + y_1\underline{j} + z_1\underline{k}) + \lambda(a\underline{i} + b\underline{j} + c\underline{k})$

$\qquad\qquad\qquad x = x_1 + \lambda a \quad \therefore \quad \dfrac{x - x_1}{a} = \lambda$

comparing
components $\{\qquad y = y_1 + \lambda b \quad \therefore \quad \dfrac{y - y_1}{b} = \lambda$

$\qquad\qquad\qquad z = z_1 + \lambda c \quad \therefore \quad \dfrac{z - z_1}{c} = \lambda$

PURE MATHEMATICS   THREE-DIMENSIONAL GEOMETRY

So $\quad \dfrac{x - x_1}{a} = \dfrac{y - y_1}{b} = \dfrac{z - z_1}{c} = \lambda \qquad$ again

Note: The direction ratios for $\underline{r} = \underline{a} + \lambda \underline{b}$ come from $\underline{b}$ because the line is parallel to $\underline{b}$ and so has the same direction ratios $a : b : c$.

Example 23

(a) The cartesian equations of a line are $\dfrac{x - 5}{3} = \dfrac{y + 4}{7} = \dfrac{z - 6}{2}$.

Find the vector equation.

(b) The vector equation of a line is $\underline{r} = \underline{i} - 3\underline{j} + 2\underline{k} + \lambda(5\underline{i} + 2\underline{j} - \underline{k})$. Express the equation in parametric form and find the coordinates of the point where the line crosses the xy plane.

Solution

(a) $\quad \dfrac{x - 5}{3} = \dfrac{y + 4}{7} = \dfrac{z - 6}{2}$

$\quad \dfrac{x - 5}{3} = \dfrac{y - (-4)}{7} = \dfrac{z - 6}{2}$

Therefore, A is the point $(5, -4, 6)$ and the direction ratios of the line are $3 : 7 : 2$.

Therefore, the position vector of a is $5\underline{i} - 4\underline{j} + 6\underline{k}$

and $\quad$ b is $3\underline{i} + 7\underline{j} + 2\underline{k}$

$\therefore \quad \underline{r} = 5\underline{i} - 4\underline{j} + 6\underline{k} + \lambda(3\underline{i} + 7\underline{j} + 2\underline{k})$

(b) $\underline{r} = \underline{i} - 3\underline{j} + 2\underline{k} + \lambda(5\underline{i} + 2\underline{j} - \underline{k})$

taking $\underline{r} = x\underline{i} + y\underline{j} + z\underline{k}$ and comparing components

gives $\quad \begin{aligned} x &= 1 + 5\lambda \\ y &= -3 + 2\lambda \\ z &= 2 - \lambda \end{aligned} \}$ parametric form

On the xy plane $z = 0$, thus $0 = 2 - \lambda$, and $\lambda = 2$ giving $x = 1 + 5(2) = 11$, $y = -3 + 2(2) = 1$, $z = 0$. Therefore, the line crosses the xy plane at $(11, 1, 0)$.

Consider a line L which passes through two fixed points $A(x_1, y_1, z_1)$ with position vector $\underline{a}$ and $B(x_2, y_2, z_2)$ with position vector $\underline{b}$.

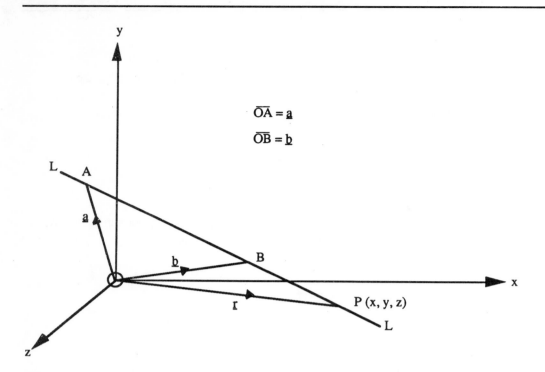

$\overline{OA} = \underline{a}$

$\overline{OB} = \underline{b}$

$P(x, y, z)$ is any point on this line L. Therefore the vector $\overline{AP}$ will be a scalar multiple, say λ, of $\overline{AB}$.

$\therefore \quad \overline{AP} = \lambda \overline{AB} \quad$ but $\quad \overline{AB} = \overline{AO} + \overline{OB} = -\underline{a} + \underline{b}$

and $\quad \overline{AP} = \overline{AO} + \overline{OP} = -\underline{a} + \underline{r}$

$\therefore \quad -\underline{a} + \underline{r} = \lambda(-\underline{a} + \underline{b})$

$\therefore \quad \boxed{\underline{r} = \underline{a} + \lambda(\underline{b} - \underline{a})}$

For each value of the parameter λ this vector equation gives the position vector $\underline{r}$ of a single point on the line. This can be converted into the equivalent cartesian equation by replacing

$\underline{r} = x\underline{i} + y\underline{j} + z\underline{k}, \quad \underline{a} = x_1\underline{i} + y_1\underline{j} + z_1\underline{k}, \quad \underline{b} = x_2\underline{i} + y_2\underline{j} + z_2\underline{k}$

$\therefore \quad x\underline{i} + y\underline{j} + z\underline{k} = x_1\underline{i} + y_1\underline{j} + z_1\underline{k} + \lambda\left[(x_2 - x_1)\underline{i} + (y_2 - y_1)\underline{j} + (z_2 - z_1)\underline{k}\right]$

comparing components in the $\underline{i}, \underline{j}, \underline{k}$ directions

$x = x_1 + \lambda(x_2 - x_1) \quad \therefore \quad \dfrac{x - x_1}{x_2 - x_1} = \lambda$

$y = y_1 + \lambda(y_2 - y_1) \quad \therefore \quad \dfrac{y - y_1}{y_2 - y_1} = \lambda$

$z = z_1 + \lambda(z_2 - z_1) \quad \therefore \quad \dfrac{z - z_1}{z_2 - z_1} = \lambda$

So $\quad \dfrac{x - x_1}{x_2 - x_1} = \dfrac{y - y_1}{y_2 - y_1} = \dfrac{z - z_1}{z_2 - z_1} = \lambda \quad$ again

Note: The direction ratios for $\underline{r} = \underline{a} + \lambda(\underline{b} - \underline{a})$ come from the $(\underline{b} - \underline{a})$ part of the equation and are $(x_2 - x_1) : (y_2 - y_1) : (z_2 - z_1)$.

PURE MATHEMATICS — THREE-DIMENSIONAL GEOMETRY

Example 24

(a) Show that the line through the points with position vectors $\underline{i} + \underline{j} - 3\underline{k}$ and $4\underline{i} + 7\underline{j} + \underline{k}$ is parallel to the line $\underline{r} = \underline{i} - \underline{k} + \lambda \left(\frac{3}{2}\underline{i} + 3\underline{j} + 2\underline{k}\right)$.

(b) Find the coordinates of the point where the line through A(3, 4, 1) and B(5, 1, 6) crosses the xy plane.

Solution

(a) Let $\quad \underline{a} = \underline{i} + \underline{j} - 3\underline{k} \quad$ and $\quad \underline{b} = 4\underline{i} + 7\underline{j} + \underline{k}$

then $\quad \underline{b} - \underline{a} = 4\underline{i} + 7\underline{j} + \underline{k} - (\underline{i} + \underline{j} - 3\underline{k})$

$$= 3\underline{i} + 6\underline{j} + 4\underline{k}$$

Therefore equation of the line through the given points is:

$$\underline{r} = \underline{i} + \underline{j} - 3\underline{k} + \lambda(3\underline{i} + 6\underline{j} + 4\underline{k})$$

and has direction ratios 3 : 6 : 4.

The given line $\underline{r} = \underline{i} - \underline{k} + \lambda \left(\frac{3}{2}\underline{i} + 3\underline{j} + 2\underline{k}\right)$ has direction ratios $\frac{3}{2}$: 3 : 2 which are (multiplying by 2) 3 : 6 : 4.

The lines are parallel as they have the same direction ratios.

(b) It is first necessary to find the equation of the line through A and B.

Let $\quad \underline{a} = 3\underline{i} + 4\underline{j} + \underline{k} \quad$ and $\quad \underline{b} = 5\underline{i} + \underline{j} + 6\underline{k}$

then $\quad \underline{b} - \underline{a} = 5\underline{i} + \underline{j} + 6\underline{k} - (3\underline{i} + 4\underline{j} + \underline{k})$

$$= 2\underline{i} - 3\underline{j} + 5\underline{k}$$

Therefore equation of the line through the given points is:

$$\underline{r} = 3\underline{i} + 4\underline{j} + \underline{k} + \lambda(2\underline{i} - 3\underline{j} + 5\underline{k})$$

On the xy plane z = 0, so working with the parametric equations:

$$x = 3 + 2\lambda \qquad y = 4 - 3\lambda \qquad z = 1 + 5\lambda$$

$\therefore$ z = 0 gives $1 + 5\lambda = 0$. $\therefore \lambda = -\frac{1}{5}$ and $x = 3 + 2\left(-\frac{1}{5}\right) = \frac{13}{5}$ and $y = 4 - 3\left(-\frac{1}{5}\right) = \frac{23}{5}$.

The line crosses the xy plane at $\left(\frac{13}{5}, \frac{23}{5}, 0\right)$

---

e) *Pairs of lines*

As has already been stated a pair of lines can be:

(a) parallel;

(b) not parallel and intersecting;

(c) skew (not parallel and not intersecting).

THREE-DIMENSIONAL GEOMETRY PURE MATHEMATICS

Vector questions concerning pairs of lines are solved in exactly the same way as they were in section e) page 442 et seq, ie considering the direction ratios and using the parametric forms of the equations.

Example 25

The lines L_1 and L_2 are given by vector equations.

$$L_1: \quad \underline{r} = \underline{j} - \underline{k} + t(2\underline{i} - \underline{j} + 2\underline{k})$$

$$L_2: \quad \underline{r} = p\underline{i} + 3\underline{j} + s(2\underline{i} + 2\underline{j} - \underline{k})$$

where t and s are parameters. Given that L_1 and L_2 intersect, find the value of p.

Find the distance from the point with coordinates $(p, 3, 0)$ to the point of intersection of L_1 and L_2.

Solution

This is an 'A' level question and probably appears harder than the other 'textbook' examples given so far. Do not be put off by the use of t and s instead of λ and μ: they are just parameters whatever letter is used.

$L_1: \quad \underline{r} = \underline{j} - \underline{k} + t(2\underline{i} - \underline{j} + 2\underline{k}) \qquad L_2: \quad \underline{r} = p\underline{i} + 3\underline{j} + s(2\underline{i} + 2\underline{j} - \underline{k})$

$\therefore \quad x = 0 + 2t \qquad\qquad\qquad\qquad\qquad \therefore \quad x = p + 2s$

$\therefore \quad y = 1 - t \qquad \{ \text{working with parameters} \} \qquad \therefore \quad y = 3 + 2s$

$\therefore \quad z = -1 + 2t \qquad\qquad\qquad\qquad\qquad \therefore \quad z = 0 - s$

At the point of intersection the x, y and z values must be the same.

$\therefore \quad 2t = p + 2s \qquad \text{-------- (1)}$

$\therefore \quad 1 - t = 3 + 2s \qquad \text{-------- (2)}$

$\therefore \quad -1 + 2t = -s \qquad \text{-------- (3)}$

$\}$ Using equations (2) and (3) as they do not involve p

Equation (2) $1 - 3 = t + 2s \qquad \therefore \quad -2 = t + 2s$

(3) $-1 = -2t - s \qquad \therefore \quad -2 = -4t - 2s$ $\}$ add

$\therefore \quad -4 = -3t$

$\therefore \quad t = \dfrac{4}{3}$

replacing $-1 = -2\left(\dfrac{4}{3}\right) - s$

$\therefore \quad s = -\dfrac{5}{3}$

Since L_1 and L_2 intersect it follows that

$L_1: \quad x = 0 + 2t = 0 + 2\left(\dfrac{4}{3}\right) = \dfrac{8}{3} \qquad L_2: \quad x = p + 2s = p + 2\left(-\dfrac{5}{3}\right) = p - \dfrac{10}{3}$

$\therefore \quad \dfrac{8}{3} = p - \dfrac{10}{3}$

PURE MATHEMATICS — THREE-DIMENSIONAL GEOMETRY

$\therefore \quad \dfrac{8}{3} + \dfrac{10}{3} = p$

$\quad\quad p = 6$

So the point of intersection of L_1 and L_2 is $x = \dfrac{8}{3}$, $y = 1 - t = 1 - \left(\dfrac{4}{3}\right) = -\dfrac{1}{3}$, $z = -s = -\left(-\dfrac{5}{3}\right) = \dfrac{5}{3}$, ie $\left(\dfrac{8}{3}, -\dfrac{1}{3}, \dfrac{5}{3}\right)$.

The distance, d, between the point $(p, 3, 0)$, ie $(6, 3, 0)$ and the point of intersection $\left(\dfrac{8}{3}, -\dfrac{1}{3}, \dfrac{5}{3}\right)$ is

$$d = \sqrt{\left(6 - \dfrac{8}{3}\right)^2 + \left(3 - \left(-\dfrac{1}{3}\right)\right)^2 + \left(0 - \dfrac{5}{3}\right)^2}$$

$$= \sqrt{\left(\dfrac{10}{3}\right)^2 + \left(\dfrac{10}{3}\right)^2 + \left(-\dfrac{5}{3}\right)^2}$$

$$= \sqrt{\dfrac{100}{9} + \dfrac{100}{9} + \dfrac{25}{9}}$$

$$= \sqrt{\dfrac{225}{9}}$$

$$= \sqrt{25}$$

$$= 5$$

So the required distance is 5 units.

Finding the angle between two lines was studied on page 445 et seq and the expression $\cos \theta = l_1 l_2 + m_1 m_2 + n_1 n_2$ was derived. This same expression can be derived more easily using the scalar product.

If $\quad \underline{a} = a_1 \underline{i} + a_2 \underline{j} + a_3 \underline{k}$

and $\quad \underline{b} = b_1 \underline{i} + b_2 \underline{j} + b_3 \underline{k}$

then $\quad \underline{a} \cdot \underline{b} = (a_1 \underline{i} + a_2 \underline{j} + a_3 \underline{k}) \cdot (b_1 \underline{i} + b_2 \underline{j} + b_3 \underline{k})$

$\quad\quad\quad\quad = a_1 b_1 + a_2 b_2 + a_3 b_3$

but $\quad \underline{a} \cdot \underline{b} = ab \cos \theta \quad$ where $a = |\underline{a}|$ and $b = |\underline{b}|$

$\therefore \quad \cos \theta = \dfrac{\underline{a} \cdot \underline{b}}{ab}$

$\quad\quad\quad\quad = \dfrac{a_1 b_1 + a_2 b_2 + a_3 b_3}{ab}$

$\quad\quad\quad\quad = \dfrac{a_1 b_1}{ab} + \dfrac{a_2 b_2}{ab} + \dfrac{a_3 b_3}{ab}$

THREE-DIMENSIONAL GEOMETRY   PURE MATHEMATICS

But $a_1 : a_2 : a_3$ are the direction ratios for $\underline{a}$

and $b_1 : b_2 : b_3$ are the direction ratios for $\underline{b}$

$\therefore$ $\dfrac{a_1}{a}, \dfrac{a_2}{a}, \dfrac{a_3}{a}$ are the direction cosines for $\underline{a}$

and $\dfrac{b_1}{b}, \dfrac{b_2}{b}, \dfrac{b_3}{b}$ are the direction cosines for $\underline{b}$

$\therefore$ $\cos\theta = l_1 l_2 + m_1 m_2 + n_1 n_2$

where l_1, m_1 and n_1 are the direction cosines for $\underline{a}$ and l_2, m_2 and n_2 are the direction cosines for $\underline{b}$.

For perpendicular lines $l_1 l_2 + m_1 m_2 + n_1 n_2 = 0$

$$a_1 b_1 + a_2 b_2 + a_3 b_3 = 0$$

Example 26

Find the angle between the lines

$$L_1: \quad \underline{r} = \underline{i} - 2\underline{j} + 3\underline{k} + \lambda(2\underline{i} - 3\underline{j} + 6\underline{k})$$

and $L_2: \quad \underline{r} = 2\underline{i} - 7\underline{j} + 10\underline{k} + \mu(\underline{i} + 2\underline{j} + 2\underline{k})$

Solution

The direction ratios for L_1 are $2 : -3 : 6$ and $\sqrt{2^2 + (-3)^2 + 6^2} = \sqrt{49} = 7$.

Therefore the direction cosines are $\dfrac{2}{7}, \dfrac{-3}{7}, \dfrac{6}{7}$ (l_1, m_1, n_1).

The direction ratios for L_2 are $1 : 2 : 2$ and $\sqrt{1^2 + 2^2 + 2^2} = \sqrt{9} = 3$.

Therefore the direction cosines are $\dfrac{1}{3}, \dfrac{2}{3}, \dfrac{2}{3}$ (l_2, m_2, n_2).

Using $\cos\theta = l_1 l_2 + m_1 m_2 + n_1 n_2$

$$= \dfrac{2}{7} \times \dfrac{1}{3} + \left(\dfrac{-3}{7}\right) \times \dfrac{2}{3} + \dfrac{6}{7} \times \dfrac{2}{3}$$

$$= \dfrac{2}{21} - \dfrac{6}{21} + \dfrac{12}{21}$$

$$= \dfrac{8}{21}$$

$$= 0.381$$

$\therefore$ $\theta = \cos^{-1}(0.381)$

$$= 67.6°$$

PURE MATHEMATICS — THREE-DIMENSIONAL GEOMETRY

Example 27

Find a unit vector perpendicular to $\overline{AB} = \underline{i} + 2\underline{j} + 3\underline{k}$ and $\overline{AC} = 4\underline{i} - \underline{j} + 2\underline{k}$.

Solution

Assume the vector perpendicular to $\overline{AB}$ and $\overline{AC}$ is $\underline{p} = a\underline{i} + b\underline{j} + c\underline{k}$

then for $\overline{AB}$ $\quad (a\underline{i} + b\underline{j} + c\underline{k}) \cdot (\underline{i} + 2\underline{j} + 3\underline{k}) = 0$

$\therefore \qquad a + 2b + 3c = 0 \qquad \underline{\qquad}$ (1)

and for $\overline{AC}$ $\quad (a\underline{i} + b\underline{j} + c\underline{k}) \cdot (4\underline{i} - \underline{j} + 2\underline{k}) = 0$

$\therefore \qquad 4a - b + 2c = 0 \qquad \underline{\qquad}$ (2)

Having two equations in three unknowns means that the values of a and b, for example, can be found in terms of c.

(1) $\qquad a + 2b = -3c$

(2) x 2 $\qquad 8a - 2b = -4c$ $\qquad$ } add $\qquad 9a = -7c$

$\qquad \therefore \quad a = \dfrac{-7c}{9}$

Replacing in (1) $\left(\dfrac{-7c}{9}\right) + 2b = -3c$

$\therefore \qquad 2b = -3c + \dfrac{7c}{9}$

$\qquad\qquad = \dfrac{-20c}{9}$

$\therefore \qquad b = \dfrac{-10c}{9}$

So the vector $\quad \underline{p} = \dfrac{-7c}{9}\underline{i} - \dfrac{10c}{9}\underline{j} + c\underline{k}$

$\qquad\qquad = \dfrac{-7c\underline{i} - 10c\underline{j} + 9c\underline{k}}{9}$

$\therefore \qquad \underline{p} = \dfrac{c}{9}(-7\underline{i} - 10\underline{j} + 9\underline{k})$

This vector is perpendicular to $\overline{AB}$ and $\overline{AC}$ but is not the unit vector $\hat{\underline{p}}$ where $\hat{\underline{p}} = \dfrac{\underline{p}}{p}$ where $p = |\underline{p}|$

$$p = |\underline{p}| = \sqrt{\left(\dfrac{-7c}{9}\right)^2 + \left(\dfrac{-10c}{9}\right)^2 \left(\dfrac{9c}{9}\right)^2}$$

$$= \sqrt{\dfrac{49c^2}{81} + \dfrac{100c^2}{81} + \dfrac{81c^2}{81}}$$

473

THREE-DIMENSIONAL GEOMETRY   PURE MATHEMATICS

$$= \sqrt{\frac{230c^2}{81}}$$

$\therefore \quad p = \frac{\sqrt{230}}{9} \cdot c$

$\therefore \quad \hat{p} = \dfrac{\frac{c}{9}(-7\underline{i} - 10\underline{j} + 9\underline{k})}{\frac{\sqrt{230}}{9} \cdot c}$

So the required unit vector is $\dfrac{-7\underline{i} - 10\underline{j} + 9\underline{k}}{\sqrt{230}}$

---

f) *Example 'A' level question*

The lines L_1 and L_2 have equations

$L_1: \quad \underline{r} = 2\underline{i} + 2\underline{j} + 2\underline{k} + t(2\underline{i} + \underline{j} - \underline{k})$

$L_2: \quad \underline{r} = 3\underline{j} - \underline{k} + s(-2\underline{i} + \underline{j} + \underline{k})$

where t and s are parameters. Show L_1 and L_2 do not intersect. The point P on L_1 has parameter p and the point Q on L_2 has parameter q. Find $\overline{PQ}$ as a vector in terms of p and q. Given that L_1 and L_2 are both perpendicular to PQ find p and q.

Solution

If L_1 and L_2 are parallel they will not intersect; for L_1 the direction ratios are $2 : 1 : -1$, and for L_2 the direction ratios are $-2 : 1 : 1$.

Therefore, L_1 and L_2 are not parallel. However, it is still necessary to prove that they do not intersect. Using the parametric forms for the equations ($\underline{r} = x\underline{i} + y\underline{j} + z\underline{k}$)

$L_1 \quad x = 2 + 2t \qquad\qquad L_2 \quad x = -2s$

$\quad y = 2 + t \qquad\qquad \quad y = 3 + s$

$\quad z = 2 - t \qquad\qquad \quad z = -1 + s$

Using x and y: $\quad 2 + 2t = -2s \quad \therefore \quad 2 = -2t - 2s$ ⎱ add

$\quad 2 + t = 3 + s \quad \therefore \quad -2 = -2t + 2s$ ⎰

$ 0 = -4t \quad \therefore t = 0$

Replacing $2 + 2(0) = -2s \quad \therefore s = -1$

Replacing $t = 0$ and $s = -1$ in z for each line

$L_1: \quad z = 2 - t$

$ = 2 - 0$

$ = 2$

474

PURE MATHEMATICS — THREE-DIMENSIONAL GEOMETRY

L_2 : $z = -1 + s$

$\qquad\qquad = -1 + -1$

$\qquad\qquad = -2$

Since these values for z are different the lines do not intersect.

At P $\quad t = p \quad \therefore \underline{r} = \overline{OP} = 2\underline{i} + 2\underline{j} + 2\underline{k} + p(2\underline{i} + \underline{j} - \underline{k})$

At Q $\quad s = q \quad \therefore \underline{r} = \overline{OQ} = 3\underline{j} - \underline{k} + q(-2\underline{i} + \underline{j} + \underline{k})$

$\qquad \overline{PQ} = \overline{PO} + \overline{OQ}$

$\qquad\qquad = -[(2\underline{i} + 2\underline{j} + 2\underline{k} + p(2\underline{i} + \underline{j} - \underline{k})] + [3\underline{j} - \underline{k} + q(-2\underline{i} + \underline{j} + \underline{k})]$

$\qquad\qquad = (-2 - 2p - 2q)\underline{i} + (-2 - p + 3 + q)\underline{j} + (-2 + p - 1 + q)\underline{k}$

$\therefore \qquad \overline{PQ} = (-2 - 2p - 2q)\underline{i} + (1 - p + q)\underline{j} + (-3 + p + q)\underline{k}$

L_1 and L_2 are perpendicular to PQ. Therefore, the dot product can be used.

For L_1 the direction ratios are $2 : 1 : -1$ and for L_2 - $2 : 1 : 1$.

PQ and L_1 $\quad 2(-2 - 2p - 2q) + 1(1 - p + q) + (-1)(-3 + p + q) = 0$

$\qquad\qquad\qquad -4 - 4p - 4q + 1 - p + q + 3 - p - q = 0$

$\qquad\qquad\qquad\qquad\qquad -6p - 4q = 0 \quad\text{---------} (1)$

PQ and L_2 $\quad -2(-2 - 2p - 2q) + 1(1 - p + q) + 1(-3 + p + q) = 0$

$\qquad\qquad\qquad 4 + 4p + 4q + 1 - p + q - 3 + p + q = 0$

$\qquad\qquad\qquad\qquad\qquad 2 + 4p + 6q = 0 \quad\text{---------} (2)$

(1) $\quad -6p - 4q = 0 \quad$ x 2 $\qquad -12p - 8q = 0$

$\qquad\qquad\qquad\qquad\qquad\qquad\qquad\qquad\qquad\quad\}$ add

(2) $\quad 4p + 6q = -2 \quad$ x 2 $\qquad 12p + 18q = -6$

$\qquad\qquad\qquad\qquad\qquad\qquad\qquad 10q = -6 \quad \therefore q = \dfrac{-3}{5}$

$\qquad\qquad\qquad \text{Replacing } -6p - 4\left(\dfrac{-3}{5}\right) = 0$

$\qquad\qquad\qquad\qquad\qquad -6p + \dfrac{12}{5} = 0 \quad \therefore p = \dfrac{2}{5}$

The values for p and q are $\dfrac{2}{5}$ and $\dfrac{-3}{5}$.

THREE-DIMENSIONAL GEOMETRY   PURE MATHEMATICS

10.4 Further vector work

a) *The vector (or cross) product*

The vector product $\underline{a} \times \underline{b}$ of two vectors $\underline{a}$ and $\underline{b}$ which are inclined at an angle θ is defined as a vector of magnitude $ab \sin \theta$ in a direction perpendicular to the plane containing $\underline{a}$ and $\underline{b}$.

$\therefore \quad \boxed{\underline{a} \times \underline{b} = ab \sin \theta \; \hat{\underline{n}}}$

where $\underline{n}$ is a unit vector perpendicular to the plane containing $\underline{a}$ and $\underline{b}$.

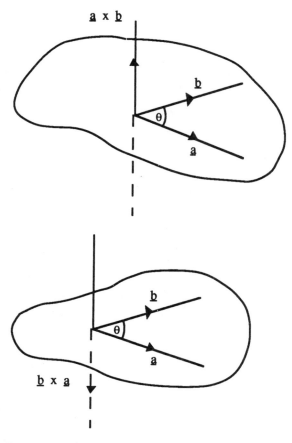

Note: The direction of the vectors $\underline{a} \times \underline{b} = - \underline{b} \times \underline{a}$

$\xrightarrow{\quad\quad \underline{a} \quad\quad}$ For parallel vectors

$\xrightarrow{\quad\quad\quad \underline{b} \quad\quad\quad}$ $|\underline{a} \times \underline{b}| = ab \sin \theta = ab \sin 0° = 0$

$\therefore \quad \boxed{\text{For parallel vectors } \underline{a} \times \underline{b} = O(\hat{\underline{n}}) = 0}$

PURE MATHEMATICS　　THREE-DIMENSIONAL GEOMETRY

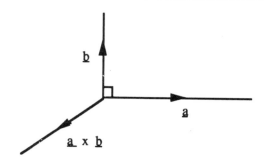

For perpendicular vectors

$|a \times b| = ab \sin \theta = ab \sin 90° = ab$

∴　　For perpendicular vectors　$\underline{a} \times \underline{b} = ab\,\underline{\hat{n}}$

This last result is particularly important as regards $\underline{i}$, $\underline{j}$ and $\underline{k}$.

$$\underline{i} \times \underline{j} = \underline{k} \qquad \therefore \qquad \underline{j} \times \underline{i} = -\underline{k}$$
$$\underline{j} \times \underline{k} = \underline{i} \qquad \therefore \qquad \underline{k} \times \underline{j} = -\underline{i}$$
$$\underline{k} \times \underline{i} = \underline{j} \qquad \therefore \qquad \underline{i} \times \underline{k} = -\underline{j}$$
$$\text{Also } \underline{i} \times \underline{i} = \underline{j} \times \underline{j} = \underline{k} \times \underline{k} = 0$$

The vectors $\underline{a}$ and $\underline{b}$ can be expressed as the sum of their component vectors, ie

$$\underline{a} = a_1\underline{i} + a_2\underline{j} + a_3\underline{k}, \qquad \underline{b} = b_1\underline{i} + b_2\underline{j} + b_3\underline{k}$$

∴
$$\underline{a} \times \underline{b} = (a_1\underline{i} + a_2\underline{j} + a_3\underline{k}) \times (b_1\underline{i} + b_2\underline{j} + b_3\underline{k})$$

$$= a_1b_1(\underline{i} \times \underline{i}) + a_1b_2(\underline{i} \times \underline{j}) + a_1b_3(\underline{i} \times \underline{k}) + a_2b_1(\underline{j} \times \underline{i})$$
$$+ a_2b_2(\underline{j} \times \underline{j}) + a_2b_3(\underline{j} \times \underline{k}) + a_3b_1(\underline{k} \times \underline{i}) + a_3b_2(\underline{k} \times \underline{j}) + a_3b_3(\underline{j} \times \underline{k})$$

$$= a_1b_2\underline{k} + a_1b_3(-\underline{j}) + a_2b_1(-\underline{k}) + a_2b_3\underline{i} + a_3b_1\underline{j} + a_3b_2(-\underline{i})$$

∴　　$\underline{a} \times \underline{b} = (a_2b_3 - a_3b_2)\underline{i} + (a_3b_1 - a_1b_3)\underline{j} + (a_1b_2 - a_2b_1)\underline{k}$

Not an easy formula to memorise although there is a certain pattern to it.

Example 28

If $\underline{a} = \underline{i} + \underline{j}$ and $\underline{b} = 2\underline{i} + \underline{k}$, find the sine of the angle between $\underline{a}$ and $\underline{b}$ and also the unit vector perpendicular to $\underline{a}$ and $\underline{b}$.

THREE-DIMENSIONAL GEOMETRY   PURE MATHEMATICS

Solution

To find $\sin \theta$, for example, the vector product can be used.

$$|\underline{a} \times \underline{b}| = ab \sin \theta \qquad \therefore \sin \theta = \frac{|\underline{a} \times \underline{b}|}{ab}$$

$$\begin{aligned}
\underline{a} \times \underline{b} &= (\underline{i} + \underline{j}) \times (2\underline{i} + \underline{k}) \\
&= 2\underline{i} \times \underline{i} + \underline{i} \times \underline{k} + 2\underline{j} \times \underline{i} + \underline{j} \times \underline{k} \\
&= -\underline{j} + 2(-\underline{k}) + \underline{i} \\
&= \underline{i} - \underline{j} - 2\underline{k}
\end{aligned}$$

$$|\underline{a} \times \underline{b}| = \sqrt{1^2 + (-1)^2 + (-2)^2} \qquad = \sqrt{6}$$

$$a = |\underline{a}| = \sqrt{1^2 + 1^2} \qquad = \sqrt{2}$$

$$b = |\underline{b}| = \sqrt{2^2 + 1^2} \qquad = \sqrt{5}$$

Replacing in the formula gives:

$$\sin \theta = \frac{\sqrt{6}}{\sqrt{2}\sqrt{5}}$$

$$= \sqrt{\frac{6}{10}}$$

$$= \sqrt{\frac{3}{5}}$$

$\underline{a} \times \underline{b} = \underline{i} - \underline{j} - 2\underline{k}$ and $\underline{b} \times \underline{a} = -(\underline{i} - \underline{j} - 2\underline{k})$ are both vectors perpendicular to the plane containing $\underline{a}$ and $\underline{b}$ but they are not unit vectors since $|\underline{a} \times \underline{b}| = |-\underline{a} \times \underline{b}| = \sqrt{6}$.

Therefore the required unit vectors are $\dfrac{\pm(\underline{i} - \underline{j} - 2\underline{k})}{\sqrt{6}}$

Example 29

A, B and C are the points (0, 1, 2), (3, 2, 1) and (1, -1, 0) respectively. Find the unit vector which is perpendicular to the plane ABC.

Solution

A is (0, 1, 2)	$\therefore$	$\overline{OA}$	$= 0\underline{i} + \underline{j} + 2\underline{k} = \underline{j} + 2\underline{k}$
B is (3, 2, 1)	$\therefore$	$\overline{OB}$	$= 3\underline{i} + 2\underline{j} + \underline{k}$
C is (1, -1, 0)	$\therefore$	$\overline{OC}$	$= \underline{i} + (-1)\underline{j} + 0\underline{k} \qquad = \underline{i} - \underline{j}$

PURE MATHEMATICS   THREE-DIMENSIONAL GEOMETRY

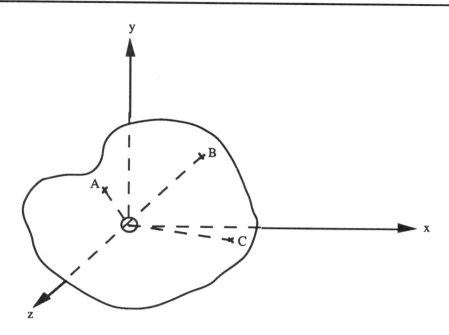

$\overline{OA}$, $\overline{OB}$ and $\overline{OC}$ are the position vectors of A, B and C. It is always possible to draw a plane through three points (see page 452) but to find the unit vector perpendicular to ABC it is necessary to find two vectors *in* the plane and use the vector product. There are three possible vectors in the plane, $\overline{AB}$, $\overline{BC}$, $\overline{CA}$ (and $\overline{BA}$, $\overline{CB}$, $\overline{AC}$ - six vectors really). Using any two of these three will do, say, $\overline{AB}$ and $\overline{BC}$.

$$\overline{AB} = \overline{AO} + \overline{OB} = -(\underline{j} + 2\underline{k}) + (3\underline{i} + 2\underline{j} + \underline{k})$$
$$= 3\underline{i} + \underline{j} - \underline{k}$$

$$\overline{BC} = \overline{BO} + \overline{OC} = -(3\underline{i} + 2\underline{j} + \underline{k}) + (\underline{i} - \underline{j})$$
$$= -2\underline{i} - 3\underline{j} - \underline{k}$$
$$= -(2\underline{i} + 3\underline{j} + \underline{k})$$

$$\overline{AB} \times \overline{BC} = (3\underline{i} + \underline{j} - \underline{k}) \times [-(2\underline{i} + 3\underline{j} + \underline{k})]$$
$$= -[6\underline{i}\times\underline{i} + 9\underline{i}\times\underline{j} + 3\underline{i}\times\underline{k} + 2\underline{i}\times\underline{i} + 3\underline{j}\times\underline{j} + \underline{j}\times\underline{k}$$
$$\quad - 2\underline{i}\times\underline{k} - 3\underline{j}\times\underline{k} - \underline{k}\times\underline{k}]$$
$$= -[9\underline{k} + 3(-\underline{j}) + 2(-\underline{k}) + \underline{i} - 2\underline{j} - 3(-\underline{i})]$$

$\therefore \quad \overline{AB} \times \overline{BC} = -[4\underline{i} - 5\underline{j} + 7\underline{k}]$

and $\quad \overline{BC} \times \overline{AB} = 4\underline{i} - 5\underline{j} + 7\underline{k}$

So the vectors $\pm(4\underline{i} - 5\underline{j} + 7\underline{k})$ are perpendicular to the plane ABC but they are not unit vectors, so

$$|\overline{AB} \times \overline{BC}| = |\overline{BC} \times \overline{AB}| = \sqrt{4^2 + (-5)^2 + 7^2}$$
$$= \sqrt{90}$$

$$= \sqrt{9 \times 10}$$
$$= 3\sqrt{10}$$

So the required unit vectors are:

$$\frac{\pm(4\mathbf{i} - 5\mathbf{j} + 7\mathbf{k})}{3\sqrt{10}}$$

The vector product has many applications including the calculation of areas, volumes and distances.

(a)   The area of a parallelogram:

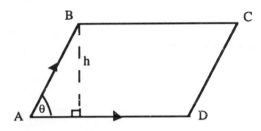

Area = base x height

= AD x h   but h = AB sin θ

= AD x AB sin θ

∴ | Area of parallelogram = $|\overrightarrow{AD} \times \overrightarrow{AB}|$ |

(b)   The area of a triangle:

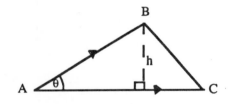

Area = $\frac{1}{2}$ base x height

= $\frac{1}{2}$ AC x h

= $\frac{1}{2}$ AC x AB sin θ

∴ | Area of triangle = $\frac{1}{2} |\overrightarrow{AC} \times \overrightarrow{AB}|$ |

(c) Volume of a tetrahedron:

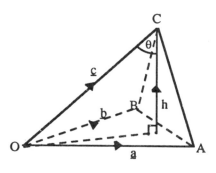

Volume $= \frac{1}{3}$ area x height

$= \frac{1}{3}\left(\frac{1}{2}|\underline{a} \times \underline{b}|\right)h$ but $h = OC \cos \theta = c \cos \theta$

$= \frac{1}{3}\left(\frac{1}{2}|\underline{a} \times \underline{b}|\right) c \cos \theta$

$= \frac{1}{6}|\underline{a} \times \underline{b}| c \cos \theta$

but θ is the angle between $\underline{a} \times \underline{b}$ and $\underline{c}$. $\therefore |\underline{a} \times \underline{b}| c \cos \theta$ is the scalar product of $\underline{a} \times \underline{b}$ and $\underline{c}$.

$\therefore$ $\boxed{\text{Volume of tetrahedron} = \frac{1}{6}(\underline{a} \times \underline{b} \cdot \underline{c})}$

(d) Distance, d, of a point, P, from a line, L:

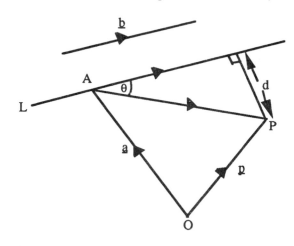

L has equation $\underline{r} = \underline{a} + \lambda \underline{b}$

P has position vector $\underline{p}$ and A has position vector $\underline{a}$

$\therefore \overline{AP} = \overline{AO} + \overline{OP} = \underline{p} - \underline{a}$

$\sin \theta = \frac{d}{AP}$ $\therefore d = AP \sin \theta$

but $|\underline{b} \times \overline{AP}| = b(AP) \sin \theta$

$\therefore \frac{|\underline{b} \times \overline{AP}|}{b} = \frac{b(AP \sin \theta)}{b} = d$

$\therefore$ $\boxed{\text{The distance of a point from a line} = \frac{|\underline{b} \times (\underline{p} - \underline{a})|}{b}}$

THREE-DIMENSIONAL GEOMETRY PURE MATHEMATICS

Example 30

ABCD is a tetrahedron and ABCD are the points (0, 1, 0), (0, 0, 4), (1, 1, 1), (- 1, 3, 2). Find the volume of the tetrahedron.

Solution

Volume $= \frac{1}{6}(\underline{a} \times \underline{b} \cdot \underline{c})$ where $\underline{a} = \overline{AB}$, $\underline{b} = \overline{AC}$ and $\underline{c} = \overline{AD}$

($\underline{a}$, $\underline{b}$ and $\underline{c}$ could have been defined differently, eg $\underline{a} = \overline{BC}$, $\underline{b} = \overline{BA}$, $\underline{c} = \overline{BD}$, but the same answer would have been calculated).

The position vectors of A, B, C, D are:

$\overline{OA} = \underline{j}$, $\overline{OB} = 4\underline{k}$, $\overline{OC} = \underline{i} + \underline{j} + \underline{k}$, $\overline{OD} = -\underline{i} + 3\underline{j} + 2\underline{k}$

$\underline{a} = \overline{AB} = \overline{AO} + \overline{OB} = -\underline{j} + 4\underline{k}$

$\underline{b} = \overline{AC} = \overline{AO} + \overline{OC} = -\underline{j} + \underline{i} + \underline{j} + \underline{k} = \underline{i} + \underline{k}$

$\underline{c} = \overline{AD} = \overline{AO} + \overline{OD} = -\underline{j} + (-\underline{i} + 3\underline{j} + 2\underline{k}) = -\underline{i} + 2\underline{j} + 2\underline{k}$

$\underline{a} \times \underline{b} = (-\underline{j} + 4\underline{k}) \times (\underline{i} \times \underline{k})$

$= (-\underline{j} \times \underline{i}) - (\underline{j} \times \underline{k}) + (4\underline{k} \times \underline{i}) + (4\underline{k} \times \underline{k})$ } see page 476

$= \underline{k} - \underline{i} + 4\underline{j}$

$= -\underline{i} + 4\underline{j} + \underline{k}$

∴ $(\underline{a} \times \underline{b} \cdot \underline{c}) = (-\underline{i} + 4\underline{j} + \underline{k}) \cdot (-\underline{i} + 2\underline{j} + 2\underline{k})$

$= (-1)(-1) + 4(2) + 1(2)$

$= 11$

∴ Volume $= \frac{1}{6}(\underline{a} \times \underline{b} \cdot \underline{c})$

$= \frac{1}{6}(11)$

$= \frac{11}{6}$

PURE MATHEMATICS — THREE-DIMENSIONAL GEOMETRY

Example 31

Find the perpendicular distance from $(1, 0, 2)$ to $\underline{r} = \underline{i} + \underline{j} + \lambda(\underline{i} - \underline{j} + \underline{k})$

Solution

$$\text{Distance} = \frac{|\underline{b} \times (\underline{p} - \underline{a})|}{b} \qquad \text{where } \underline{p} = 1\underline{i} + 0\underline{j} + 2\underline{k} = \underline{i} + 2\underline{k}$$

$$\underline{a} = \underline{i} + \underline{j}$$

$$\underline{b} = \underline{i} - \underline{j} + \underline{k}$$

$\therefore \quad \underline{p} - \underline{a} = \underline{i} + 2\underline{k} - (\underline{i} + \underline{j})$

$\qquad\qquad\quad = -\underline{j} + 2\underline{k}$

$\underline{b} \times (\underline{p} - \underline{a}) = (\underline{i} - \underline{j} + \underline{k}) \times (-\underline{j} + 2\underline{k})$

$\qquad\qquad = (-\underline{i} \times \underline{j}) + (2\underline{i} \times \underline{k}) + (\underline{j} \times \underline{j}) - (2\underline{j} \times \underline{k}) - (\underline{k} \times \underline{j}) + (2\underline{k} \times \underline{k})$

$\qquad\qquad = -\underline{k} + 2(-\underline{j}) - 2(\underline{i}) - (-\underline{i}) \qquad\qquad \}\ \text{see page 476}$

$\qquad\qquad = -\underline{i} - 2\underline{j} - \underline{k}$

$|\underline{b} \times (\underline{p} - \underline{a})| = \sqrt{(-1)^2 + (-2)^2 + (-1)^2}$

$\qquad\qquad = \sqrt{6}$

$b = |\underline{b}|$

$\quad = \sqrt{1^2 + (-1)^2 + 1^2}$

$\quad = \sqrt{3}$

$\therefore \quad \text{Distance} = \dfrac{\sqrt{6}}{\sqrt{3}} = \sqrt{2}$

b)   *Equation of a plane*

The equation of a plane was derived on page 451. Vector methods will now be used to derive an alternative form.

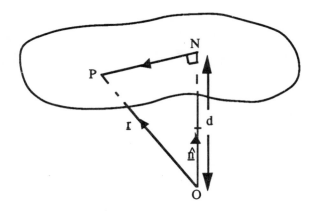

Point P has position vector $\underline{r}$

ON is the perpendicular distance, d, from the origin to the plane (unit vector $\hat{\underline{n}}$).

$\therefore \quad \overline{ON} = d\hat{\underline{n}}$ and $\overline{ON} \cdot \overline{NP} = 0$

but $\quad \overline{NP} = \overline{NO} + \overline{OP}$

$\qquad \qquad = -d\hat{\underline{n}} + \underline{r}$

$\therefore \quad \overline{ON} \cdot \overline{NP} = d\hat{\underline{n}} \cdot (\underline{r} - d\hat{\underline{n}}) = 0$

$\therefore \quad d\hat{\underline{n}} \cdot \underline{r} - d^2 \hat{\underline{n}} \cdot \hat{\underline{n}} = 0 \quad$ but $\hat{\underline{n}} \cdot \hat{\underline{n}} = 1$

$\therefore \quad d\hat{\underline{n}} \cdot \underline{r} = d^2$

$\therefore \quad \hat{\underline{n}} \cdot \underline{r} = d$

> The standard form of the vector equation of a plane is $\underline{r} \cdot \hat{\underline{n}} = d$

Also:

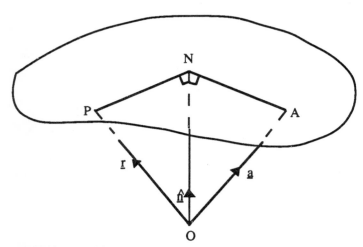

Point A has position vector $\underline{a}$

From the diagram $d = \underline{a} \cdot \hat{\underline{n}}$ but $d = \underline{r} \cdot \hat{\underline{n}}$

∴ $\boxed{\text{The equation of the plane is } \underline{r} \cdot \hat{\underline{n}} = \underline{a} \cdot \hat{\underline{n}}}$

The standard (or perpendicular) form $\underline{r} \cdot \hat{\underline{n}} = d$ of the vector equation of a plane can be converted into a more convenient form by multiplying both sides by a constant so that it becomes

$\boxed{\underline{r} \cdot \underline{n} = D}$ where $\underline{n}$ is any vector perpendicular to the plane.

To convert to the standard form both sides have to be divided by $n = |\underline{n}|$

thus $\quad \dfrac{\underline{r} \cdot \underline{n}}{n} = \dfrac{D}{n} \quad$ but $\quad \dfrac{\underline{n}}{n} = \hat{\underline{n}}$

∴ $\quad \underline{r} \cdot \hat{\underline{n}} = d \quad$ where $d = \dfrac{D}{n}$

The cartesian equation can be derived from the vector equation.

If $\quad \underline{r} = x\underline{i} + y\underline{j} + z\underline{k} \;$ and $\; \underline{n} = a\underline{i} + b\underline{j} + c\underline{k}$

then $\quad \underline{r} \cdot \underline{n} = (x\underline{i} + y\underline{j} + z\underline{k}) \cdot (a\underline{i} + b\underline{j} + c\underline{k}) = D$

∴ $\quad\quad\quad\quad\quad ax + by + cz = D$

where $a : b : c$ are the direction ratios of any normal to the plane.

Example 32

A plane passes through the point $(2, -1, -5)$ and is perpendicular to the vector $3\underline{i} - 2\underline{j} + 4\underline{k}$. Find the equation of the plane and its distance from the origin.

Solution

Using $\quad \underline{r} \cdot \underline{n} = \underline{a} \cdot \underline{n} \quad\quad$ where $\underline{n} = 3\underline{i} - 2\underline{j} + 4\underline{k}$

$\quad\quad\quad\quad\quad\quad\quad\quad\quad\quad\quad$ and $\underline{a} = 2\underline{i} - \underline{j} - 5\underline{k}$

$\quad\quad \underline{r} \cdot (3\underline{i} - 2\underline{j} + 4\underline{k}) = (2\underline{i} - \underline{j} - 5\underline{k}) \cdot (3\underline{i} - 2\underline{j} + 4\underline{k})$

∴ $\quad \underline{r} \cdot (3\underline{i} - 2\underline{j} + 4\underline{k}) = 2(3) - 1(-2) - 5(4)$

∴ $\quad \underline{r} \cdot (3\underline{i} - 2\underline{j} + 4\underline{k}) = -12$ is the equation of the plane

To find the distance, d, from the origin the equation has to be converted to the standard form $\underline{r} \cdot \hat{\underline{n}} = d$ by dividing by $n = |\underline{n}|$.

∴ $\quad \underline{n} = \sqrt{3^2 + (-2)^2 + 4^2} = \sqrt{29}$

∴ $\quad \underline{r} \cdot \dfrac{(3\underline{i} - 2\underline{j} + 4\underline{k})}{\sqrt{29}} = \dfrac{-12}{\sqrt{29}}$ and the required distance is $\dfrac{12}{\sqrt{29}}$

THREE-DIMENSIONAL GEOMETRY PURE MATHEMATICS

Example 33

Find the vector equation of the line through the point (2, 1, 1) which is perpendicular to the plane $\mathbf{r} \cdot (\mathbf{i} + 2\mathbf{j} - 3\mathbf{k}) = 6$.

Solution

Since the line is perpendicular to the plane it is parallel to $\mathbf{i} + 2\mathbf{j} - 3\mathbf{k}$.

Using $\quad \mathbf{r} = \mathbf{a} + \lambda \mathbf{b} \quad$ with $\quad \mathbf{a} = 2\mathbf{i} + \mathbf{j} + \mathbf{k}$

and $\quad \mathbf{b} = \mathbf{i} + 2\mathbf{j} - 3\mathbf{k}$

then $\quad \mathbf{r} = 2\mathbf{i} + \mathbf{j} + \mathbf{k} + \lambda(\mathbf{i} + 2\mathbf{j} - 3\mathbf{k})$

Example 34

Find the point of intersection of the line $\mathbf{r} = (\mathbf{i} + \mathbf{j} - 2\mathbf{k}) + \lambda(\mathbf{i} - \mathbf{j} + \mathbf{k})$

and the plane $\mathbf{r} \cdot (\mathbf{i} + 2\mathbf{j} - \mathbf{k}) = 2$.

Solution

Rearranging the equation of the line gives $\mathbf{r} = (1 + \lambda)\mathbf{i} + (1 - \lambda)\mathbf{j} + (-2 + \lambda)\mathbf{k}$

Substituting in the equation of that plane:

$$[(1 + \lambda)\mathbf{i} + (1 - \lambda)\mathbf{j} + (-2 + \lambda)\mathbf{k}] \cdot (\mathbf{i} + 2\mathbf{j} - \mathbf{k}) = 2$$

$$(1 + \lambda)(1) + (1 - \lambda)(2) + (-2 + \lambda)(-1) = 2$$

$$1 + \lambda + 2 - 2\lambda + 2 - \lambda = 2$$

$$-2\lambda + 5 = 2$$

$$\therefore \quad -2\lambda = -3$$

$$\therefore \quad \lambda = \frac{3}{2}$$

At the point of intersection $\lambda = \frac{3}{2}$. Therefore the coordinates are $x = (1 + \lambda) = \frac{5}{2}$,

$y = 1 - \lambda = -\frac{1}{2}$, $z = -2 + \lambda = -\frac{1}{2}$, ie $\left(\frac{5}{2}, -\frac{1}{2}, -\frac{1}{2}\right)$.

---

c) *Angles, planes and lines*

The angle between two planes is the angle between any two normals to the planes and can be found directly from the equations.

If the planes have equations $\mathbf{r} \cdot \hat{\mathbf{n}}_1 = d_1$ and $\mathbf{r} \cdot \hat{\mathbf{n}}_2 = d_2$ and θ is the angle between them, then $\hat{\mathbf{n}}_1 \cdot \hat{\mathbf{n}}_2 = n_1 n_2 \cos\theta$ but $n_1 = 1$ and $n_2 = 1$

$$\therefore \quad \boxed{\hat{\mathbf{n}}_1 \cdot \hat{\mathbf{n}}_2 = \cos\theta}$$

PURE MATHEMATICS   THREE-DIMENSIONAL GEOMETRY

> Also if $\hat{\underline{n}}_1 \cdot \hat{\underline{n}}_2 = 0$ the planes are perpendicular and if $\hat{\underline{n}}_1 = \hat{\underline{n}}_2$ the planes are parallel

The angle between a line and a plane can be found as follows:

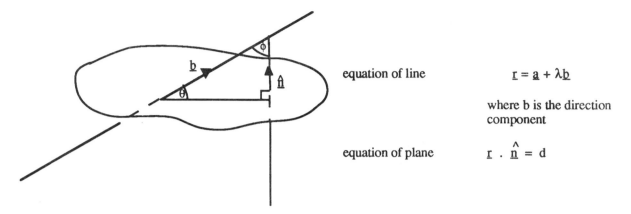

equation of line $\underline{r} = \underline{a} + \lambda \underline{b}$

where b is the direction component

equation of plane $\underline{r} \cdot \hat{\underline{n}} = d$

Using the scalar product $\underline{b} \cdot \hat{\underline{n}} = b(1) \cos \phi$

$$\therefore \quad \cos \phi = \frac{\underline{b} \cdot \hat{\underline{n}}}{b}$$

but ϕ is the angle between the normal and the line, θ is the angle between the plane and the line. Therefore, $\theta = 90 - \phi$.

$$\therefore \quad \sin \theta = \sin(90 - \phi)$$
$$= \cos \phi$$

The angle between a line and a plane is given by:

$$\boxed{\sin \theta = \frac{\underline{b} \cdot \hat{\underline{n}}}{b}}$$

Example 35

(a) Find the cosine of the angle between the two planes $\underline{r} \cdot (\underline{i} - \underline{j} + 3\underline{k}) = 3$ and $\underline{r} \cdot (2\underline{i} - \underline{j} + 2\underline{k}) = 5$.

(b) Find the sine of the angle between the line $\underline{r} = \underline{i} - \underline{j} + \lambda(\underline{i} + \underline{j} + \underline{k})$ and the plane $\underline{r} \cdot (\underline{i} - 2\underline{j} + 2\underline{k}) = 4$.

Solution

(a) $\underline{r} \cdot (\underline{i} - \underline{j} + 3\underline{k}) = 3 \quad \therefore \quad \hat{\underline{n}} = \dfrac{(\underline{i} - \underline{j} + 3\underline{k})}{\sqrt{11}}$

$\underline{r} \cdot (2\underline{i} - \underline{j} + 2\underline{k}) = 5 \quad \therefore \quad \hat{\underline{n}}_2 = \dfrac{(2\underline{i} - \underline{j} + 2\underline{k})}{\sqrt{9}}$

$\therefore \quad \cos \theta = \hat{\underline{n}}_1 \cdot \hat{\underline{n}}_2 = \dfrac{(\underline{i} - \underline{j} + 3\underline{k})}{\sqrt{11}} \cdot \dfrac{(2\underline{i} - \underline{j} + 2\underline{k})}{3}$

$= \dfrac{1(2) + (-1)(-1) + 3(2)}{3\sqrt{11}}$

$$= \frac{9}{3\sqrt{11}}$$

$$\therefore \quad \cos\theta = \frac{3}{\sqrt{11}}$$

(b) From the equation of the line $\underline{b} = \underline{i} + \underline{j} + \underline{k}$ and $b = \sqrt{3}$, and from the equation of the plane $\hat{\underline{n}} = \frac{\underline{i} - 2\underline{j} + 2\underline{k}}{\sqrt{9}}$

$$\therefore \quad \sin\theta = \frac{\underline{b} \cdot \hat{\underline{n}}}{b} = \frac{(\underline{i} + \underline{j} + \underline{k}) \cdot \frac{(\underline{i} - 2\underline{j} + 2\underline{k})}{3}}{\sqrt{3}}$$

$$= \frac{1(1) + 1(-2) + 1(2)}{3\sqrt{3}}$$

$$= \frac{1}{3\sqrt{3}}$$

$$\therefore \quad \sin\theta = \frac{1}{3\sqrt{3}}$$

d) *Distance of a point from a plane*

To find the distance of a point A, position vector $\underline{a}$, from a plane whose equation is in standard form $\underline{r} \cdot \hat{\underline{n}} = d$, imagine a second plane parallel to the first plane passing through A. The equation of this plane would be:

$$\underline{r} \cdot \hat{\underline{n}} = \underline{a} \cdot \hat{\underline{n}}$$

The first plane is at a distance d from the origin, the second (and parallel) plane is at a distance $\underline{a} \cdot \hat{\underline{n}}$ from the origin. Therefore, the distance between these planes is $\underline{a} \cdot \hat{\underline{n}} - d$ (if a negative value results, then A is between the original plane and the origin) but the distance between the planes is the same as the distance of the point from the plane.

$$\therefore \quad \boxed{\text{Required distance} = \underline{a} \cdot \hat{\underline{n}} - d}$$

Example 36

Find the distance of the point $(1, 3, 2)$ from the plane $\underline{r} \cdot (7\underline{i} + 4\underline{j} + 4\underline{k}) = 9$.

Solution

$\underline{r} \cdot (7\underline{i} + 4\underline{j} + 4\underline{k}) = 9$ must be made into its standard form by dividing by n,
ie $n = |\underline{n}| = \sqrt{7^2 + 4^2 + 4^2} = \sqrt{81} = 9$.

$$\therefore \quad \underline{r} \cdot \frac{(7\underline{i} + 4\underline{j} + 4\underline{k})}{9} = \frac{9}{9}$$

ie $\quad \underline{r} \cdot \hat{\underline{n}} = d$

So this plane is 1 unit from the origin.

Imagine another plane through (1, 3, 2) ie position vector $\underline{i} + 3\underline{j} + 2\underline{k}$ - with equation

and $\quad \underline{r} \cdot \hat{\underline{n}} = \underline{a} \cdot \hat{\underline{n}}$

$$\underline{r} \cdot \frac{(7\underline{i} + 4\underline{j} + 4\underline{k})}{9} = (\underline{i} + 3\underline{j} + 2\underline{k}) \cdot \frac{(7\underline{i} + 4\underline{j} + 4\underline{k})}{9}$$

$$= \frac{1(7) + 3(4) + 2(4)}{9}$$

$$= \frac{27}{9}$$

$$= 3$$

So the imaginary plane through (1, 3, 2) is 3 units from the origin. Therefore, the point is 3 - 1 = 2 units from the plane.

e) *Intersection of two planes*

Unless two planes are parallel they will intersect in a line

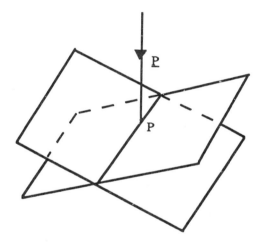

Assuming that the equations of the planes are $\underline{r} \cdot \hat{\underline{n}}_1 = d_1$ and $\underline{r} \cdot \hat{\underline{n}}_2 = d_2$, then the position vector $\underline{p}$ of any point, P, on the line of intersection must satisfy both equations. Therefore, $\underline{p} \cdot \hat{\underline{n}}_1 = d_1$ and $\underline{p} \cdot \hat{\underline{n}}_2 = d_2$.

Therefore, for any value of k (not just k = 1)

$$\underline{p} \cdot \hat{\underline{n}}_1 - k\underline{p} \cdot \hat{\underline{n}}_2 = d_1 - kd_2$$

$\therefore \quad \underline{p} \cdot (\hat{\underline{n}}_1 - k\hat{\underline{n}}_2) = d_1 - kd_2$

But the equation $\underline{r} \cdot (\hat{\underline{n}}_1 - k\hat{\underline{n}}_2) = d_1 - kd_2$ represents a third plane which is such that if any vector r satisfies the equations of the first two planes it also satisfies the equation of this third plane. So different values of k give a 'family' of planes, all of which pass through the line of intersection of $\underline{r} \cdot \hat{\underline{n}}_1 = d_1$ and $\underline{r} \cdot \hat{\underline{n}}_2 = d_2$.

To find the equation of the line of intersection of two planes the method of the next example should be used.

THREE-DIMENSIONAL GEOMETRY   PURE MATHEMATICS

Example 37

Find the vector equation of the line of intersection of the planes
$\underline{r} \cdot (\underline{i} - 2\underline{j} + \underline{k}) = 3$ and $\underline{r} \cdot (3\underline{i} + \underline{j} - 2\underline{k}) = 4$.

Solution

It is easiest to convert these equations to their cartesian form with $\underline{r} = x\underline{i} + y\underline{j} + z\underline{k}$.

$\therefore \quad \underline{r} \cdot (\underline{i} - 2\underline{j} + \underline{k}) = 3 \quad$ is $x(1) + y(-2) + z(1) = 3 \quad$ ---------- (1)

and $\quad \underline{r} \cdot (3\underline{i} + \underline{j} - 2\underline{k}) = 4 \quad$ is $x(3) + y(1) + z(-2) = 4 \quad$ ---------- (2)

Eliminate z:

$\therefore \quad$ Equation $\quad (1) \times 2 \qquad 2x - 4y + 2z = 6$
$\qquad \qquad \qquad \qquad \qquad \qquad \qquad \qquad \qquad \qquad \qquad \quad$ } add
$\qquad$ Equation $\quad (2) \qquad \qquad \quad 3x + y - 2z = 4$

$\qquad \qquad \qquad \therefore \qquad 5x - 3y = 10$

$\qquad \qquad \qquad \therefore \qquad 5x = 10 + 3y$

$\qquad \qquad \qquad \therefore \qquad x = \dfrac{10 + 3y}{5}$

Eliminate y
Equation (1) $\qquad \qquad x - 2y + z = 3$
$\qquad \qquad \qquad \qquad \qquad \qquad \qquad \qquad \qquad$ } add
Equation (2) $\times$ 2 $\qquad \quad 6x + 2y - 4z = 8$

$\qquad \qquad \qquad \qquad \qquad 7x - 3z = 11$

$\qquad \qquad \qquad \therefore \qquad 7x = 11 + 3z$

$\qquad \qquad \qquad \therefore \qquad x = \dfrac{11 + 3z}{7}$

So the solutions are:

$$x = \frac{10 + 3y}{5} = \frac{11 + 3z}{7} = \text{say } \lambda$$

This is the equation of a straight line in cartesian form. To convert it into vector form a little re-arranging is necessary:

$x = \lambda, \quad \dfrac{10 + 3y}{5} = \lambda \quad \therefore \; 10 + 3y = 5\lambda \quad \therefore \; y = \dfrac{-10 + 5\lambda}{3}$

$\qquad \qquad \dfrac{11 + 3z}{7} = \lambda \quad \therefore \; 11 + 3z = 7\lambda \quad \therefore \; z = \dfrac{-11 + 7\lambda}{3}$

$\therefore \qquad x = \lambda, \; y = \dfrac{-10 + 5\lambda}{3}, \; z = \dfrac{-11 + 7\lambda}{3}$

$\therefore \qquad \underline{r} = \lambda\underline{i} + \left(\dfrac{-10 + 5\lambda}{3}\right)\underline{j} + \left(\dfrac{-11 + 7\lambda}{3}\right)\underline{k}$

$\therefore \qquad \underline{r} = -\dfrac{10}{3}\underline{j} - \dfrac{11}{3}\underline{k} + \lambda\left(\underline{i} + \dfrac{5}{3}\underline{j} + \dfrac{7}{3}\underline{k}\right)$

PURE MATHEMATICS — THREE-DIMENSIONAL GEOMETRY

$$\underline{r} = -\frac{1}{3}(10\underline{j} + 11\underline{k}) + \frac{\lambda}{3}(3\underline{i} + 5\underline{j} + 7\underline{k})$$

replacing λ with, say, 3μ gives

$$\underline{r} = -\frac{1}{3}(10\underline{j} + 11\underline{k}) + \mu(3\underline{i} + 5\underline{j} + 7\underline{k})$$

---

f) *Example 'A' level question*

The coordinates of the points A, B and C are (1, 2, 1), (2, -1, 0), (3, 1, 2) respectively. The line L and the plane have vector equations:

  L: $\underline{r} = 3\underline{j} + 5\underline{k} + t(2\underline{i} + \underline{j} - 2\underline{k})$

  Plane: $\underline{r} \cdot (2\underline{i} - 2\underline{j} + \underline{k}) = -1$

Show:

(a)   that the plane contains A and L;

(b)   that BC is parallel to the plane and perpendicular to L;

(c)   that BC is equal to the perpendicular distance from A to L.

Solution

(a)   If the plane contains A the position vector of A, $\overline{OA}$, should satisfy the equation of the plane.

  $\overline{OA} = \underline{i} + 2\underline{j} + \underline{k}$ replacing for $\underline{r}$ in the equation for the plane gives

  $(\underline{i} + 2\underline{j} + \underline{k}) \cdot (2\underline{i} - 2\underline{j} + \underline{k}) = 1(2) + 2(-2) + 1(1)$

  $= -1$

so $\underline{a}$ satisfies the equation of the plane and is contained in the plane.

If the plane contains L the position vector of any point on L should satisfy the equation of the plane.

  $\underline{r} = 3\underline{j} + 5\underline{k} + t(2\underline{i} + \underline{j} - 2\underline{k})$ rearranging

  $\therefore \quad \underline{r} = 2t\underline{i} + (3 + t)\underline{j} + (5 - 2t)\underline{k}$ replacing in the equation of the plane

  $(2t\underline{i} + (3 + t)\underline{j} + (5 - 2t)\underline{k}) \cdot (2\underline{i} - 2\underline{j} + \underline{k}) = 2t(2) + (3 + t)(-2) + (5 - 2t)(1)$

  $= 4t - 6 - 2t + 5 - 2t$

  $= -1$

So the position vector of any point on the line satisfies the equation of the plane and is contained in the plane.

(b) $\overline{BC} = \overline{BO} + \overline{OC}$ where $\overline{OB} = 2\underline{i} - \underline{j}$

and $\overline{OC} = 3\underline{i} + \underline{j} + 2\underline{k}$

$\therefore \quad \overline{BC} = -(2\underline{i} - \underline{j}) + (3\underline{i} + \underline{j} + 2\underline{k})$

$= \underline{i} + 2\underline{j} + 2\underline{k}$

If BC is parallel to the plane it will be perpendicular to the normal to the plane, ie to $2\underline{i} - 2\underline{j} + \underline{k} = \underline{n}$.

$\therefore \quad \overline{BC} \cdot \underline{n} = (\underline{i} + 2\underline{j} + 2\underline{k}) \cdot (2\underline{i} - 2\underline{j} + \underline{k})$

$= 1(2) + 2(-2) + 2(1)$

$= 0$

Therefore, BC is perpendicular to $\underline{n}$ and so it is parallel to the plane.

If BC is perpendicular to L it will be perpendicular to any vector parallel to
L: $\underline{r} = 3\underline{j} + 5\underline{k} + t(2\underline{i} + \underline{j} - 2\underline{k})$ in particular it will be perpendicular to $2\underline{i} + \underline{j} - 2\underline{k}$.

Therefore, $(\underline{i} + 2\underline{j} + 2\underline{k}) \cdot (2\underline{i} + \underline{j} - 2\underline{k}) = 1(2) + 2(1) + 2(-2)$

$= 0$

Therefore, BC is perpendicular to L.

(c) $BC = \sqrt{(2-3)^2 + (-1-1)^2 + (0-2)^2}$

$= \sqrt{9}$

$= 3$

To find the perpendicular distance from A to the line L the formula $\dfrac{|\underline{b} \times (\underline{p} - \underline{a})|}{b}$ will be used, where $\underline{a}$ and $\underline{b}$ come from L: $\underline{r} = \underline{a} + \lambda \underline{b}$ and $\underline{p}$ is the position vector of $A(\overline{OA})$.

$\underline{r} = 3\underline{j} + 5\underline{k} + t(2\underline{i} + \underline{j} - 2\underline{k})$

where $\underline{a} = 3\underline{j} + 5\underline{k}$

and $\underline{b} = 2\underline{i} + \underline{j} - 2\underline{k}$ and $b = |\underline{b}| = \sqrt{2^2 + 1^2 + (-2)^2} = 3$

A is the position $(1, 2, 1)$

$\therefore \quad \underline{p} = \underline{i} + 2\underline{j} + \underline{k}$

$\therefore \quad \underline{p} - \underline{a} = \underline{i} + 2\underline{j} + \underline{k} - (3\underline{j} + 5\underline{k})$

$= \underline{i} - \underline{j} - 4\underline{k}$

$\underline{b} \times (\underline{p} - \underline{a}) = (2\underline{i} + \underline{j} - 2\underline{k}) \times (\underline{i} - \underline{j} - 4\underline{k})$

$$= (2\underline{i} \times \underline{i}) - (2\underline{i} \times \underline{j}) - (8\underline{i} \times \underline{k}) + (\underline{j} \times \underline{i}) - (\underline{j} \times \underline{j}) - (4\underline{j} \times \underline{k})$$
$$- (2\underline{k} \times \underline{i}) + (2\underline{k} \times \underline{j}) + (8\underline{k} \times \underline{k})$$

} see page 476

$$= -2\underline{k} - 8(-\underline{j}) + (-\underline{k}) - 4\underline{i} - 2\underline{j} + 2(-\underline{i})$$

$$= -6\underline{i} + 6\underline{j} - 3\underline{k}$$

$$|\underline{b} \times (\underline{p} - \underline{a})| = \sqrt{(-6)^2 + 6^2 + (-3)^2}$$

$$= \sqrt{81}$$

$$= 9$$

$\therefore\quad$ Distance $= \dfrac{|\underline{b} \times (\underline{p} - \underline{a})|}{b}$

$$= \frac{9}{3}$$

$$= 3$$

$$= BC$$

PURE MATHEMATICS — FURTHER CALCULUS

Another special case of integrating by parts arises when integrating products such as:

$$\int e^x \cos x \, dx \quad \text{let} \quad u = e^x \quad \text{and} \quad \frac{dv}{dx} = \cos x$$

$$\therefore \quad \frac{du}{dx} = e^x \quad \text{and} \quad v = \sin x$$

$$\therefore \quad \int e^x \cos x \, dx = e^x \sin x - \int e^x \sin x \, dx \quad \text{- no better (1)}$$

Try again $\int e^x \cos x \, dx \quad$ let $\quad u = \cos x \quad$ and $\quad \dfrac{dv}{dx} = e^x$

$$\therefore \quad \frac{du}{dx} = -\sin x \quad \text{and} \quad v = e^x$$

$$\therefore \quad \int e^x \cos x \, dx = e^x \cos x - \int e^x (-\sin x) \, dx$$

$$= e^x \cos x + \int e^x \sin x \, dx \quad \text{- no better (2)}$$

However (1) add (2) gives:

$$\int e^x \cos x \, dx + \int e^x \cos x \, dx = e^x \sin x - \int e^x \sin x \, dx + e^x \cos x + \int e^x \sin x \, dx$$

$$\therefore \quad 2\int e^x \cos x \, dx = e^x(\sin x + \cos x) + c$$

$$\therefore \quad \int e^x \cos x \, dx = \frac{e^x}{2}(\sin x + \cos x) + c$$

f) *Integration by reduction formulae*

Integrating $\cos^2 x$ is relatively easy because:

$$\cos^2 x = \frac{1}{2}(\cos 2x + 1) \qquad \therefore \quad \int \cos^2 x \, dx = \int \frac{1}{2}(\cos 2x + 1) \, dx$$

However, integrating $\cos^{20} x$ would not be easy - in fact at the moment it would be impossible - but there is a method (based on integration by parts) which reduces the power of the function, eg $\int \cos^n x \, dx$ where n is a positive integer.

First $\cos^n x$ can be written as $\cos x \cdot \cos^{n-1} x$.

Let $\quad I_n = \int \cos^n x \, dx = \int \cos x \cdot \cos^{n-1} x \, dx \quad$ let $\quad u = \cos^{n-1} x \quad$ and $\quad \dfrac{dv}{dx} = \cos x$

$$\therefore \quad \frac{du}{dx} = (n-1)\cos^{n-2} x (-\sin x) \quad \text{and} \quad v = \sin x$$

$$I_n = \cos^{n-1} x \cdot (\sin x) - \int (n-1) \cos^{n-2} x (-\sin x) \sin x \, dx$$

$$= \sin x \cos^{n-1} x + \int (n-1) \sin^2 x \cos^{n-2} x \, dx$$

$$= \sin x \cos^{n-1} x + (n-1) \int (1 - \cos^2 x) \cos^{n-2} x \, dx$$

$$= \sin x \cos^{n-1} x + (n-1) \int \cos^{n-2} x - \cos^n x \, dx$$

$$= \sin x \cos^{n-1} x + (n-1) \int \cos^{n-2} x \, dx - (n-1) \int \cos^n x \, dx$$

but $\quad \int \cos^n x \, dx = I_n \quad \therefore \quad \int \cos^{n-2} x \, dx = I_{n-2}$

FURTHER CALCULUS — PURE MATHEMATICS

Hence $I_n = \sin x \cos^{n-1} x + (n-1)I_{n-2} - (n-1)I_n$

$\therefore I_n + (n-1)I_n = \sin x \cos^{n-1} x + (n-1)I_{n-2}$

$nI_n = \sin x \cos^{n-1} x + (n-1)I_{n-2}$

So $I_n = \dfrac{\sin x \cos^{n-1} x}{n} + \dfrac{(n-1)}{n} I_{n-2}$

This reduction formula - as it is called - can be used to integrate $\cos^5 x$, say.

$\therefore \int \cos^5 x \, dx = I_5 = \dfrac{\sin x \cos^4 x}{5} + \dfrac{4}{5} I_3$

$I_3 = \dfrac{\sin x \cos^2 x}{3} + \dfrac{2}{3} I_1$

$I_1 = \int \cos x \, dx = \sin x$

$\therefore \int \cos^5 x \, dx = \dfrac{\sin x \cos^4 x}{5} + \dfrac{4}{5}\left(\dfrac{\sin x \cos^2 x}{3} + \dfrac{2}{3} \sin x\right) + c$

$= \dfrac{\sin x \cos^4 x}{5} + \dfrac{4 \sin x \cos^2 x}{15} + \dfrac{8}{15} \sin x + c$

Example 14

Find a reduction formula to integrate $x^n e^x$ and use it to integrate $x^4 e^x$.

Solution

$\int x^n e^x \, dx = I_n$ let $u = x^n$ and $\dfrac{dv}{dx} = e^x$

$\therefore \dfrac{du}{dx} = nx^{n-1}$ and $v = e^x$

$\therefore I_n = x^n e^x - \int nx^{n-1} e^x \, dx$

$\therefore I_n = x^n e^x - n\int x^{n-1} e^x \, dx$ but $\int x^{n-1} e^x \, dx = I_{n-1}$

$\therefore I_n = x^n e^x - n I_{n-1}$

$\therefore \int x^4 e^x \, dx = I_4$

$\therefore I_4 = x^4 e^x - 4 I_3$

$I_3 = x^3 e^x - 3 I_2$

$I_2 = x^2 e^x - 2 I_1$

$I_1 = x e^x - 1 \cdot I_0$

$I_0 = \int x^0 e^x \, dx = \int e^x \, dx = e^x$

$\therefore \int x^4 e^x \, dx = x^4 e^x - 4[x^3 e^x - 3(x^2 e^x - 2\{x e^x - 1 \, e^x\})] + c$

$= x^4 e^x - 4x^3 e^x + 12x^2 e^x - 24x \, e^x + 24 e^x$

PURE MATHEMATICS   FURTHER CALCULUS

Example 15

Given that $I_n = \int_0^{\pi/2} x^n \sin x \, dx$ prove that for $n \geq 2$

$$I_n = n\left(\frac{\pi}{2}\right)^{n-1} - n(n-1)I_{n-2}$$

Find I_3 in the form $p\pi^2 - q$ where p and q are constants to be determined.

Solution

$\int_0^{\pi/2} x^n \sin x \, dx - I_n \quad$ let $\quad u = x^n \quad$ and $\quad \dfrac{dv}{dx} = \sin x$

$\qquad\qquad\qquad\qquad\qquad \therefore \quad \dfrac{du}{dx} = nx^{n-1} \qquad v = -\cos x$

$I_n = \left[x^n(-\cos x)\right]_0^{\pi/2} - \int_0^{\pi/2} n x^{n-1}(-\cos x) \, dx$

$\quad\;\; = \left[-x^n \cos x\right]_0^{\pi/2} + n \int_0^{\pi/2} x^{n-1} \cos x \, dx$

$\quad\;\; = \left[-\left(\dfrac{\pi}{2}\right)^n \cos \dfrac{\pi}{2} - (-0^n \cos 0)\right] + n \int_0^{\pi/2} x^{n-1} \cos x \, dx$

$I_n = n \int_0^{\pi/2} x^{n-1} \cos x \, dx \quad$ let $\quad u = x^{n-1} \quad$ and $\quad \dfrac{dv}{dx} = \cos x$

$\qquad\qquad\qquad\qquad\qquad \therefore \quad \dfrac{du}{dx} = (n-1)x^{n-2} \qquad v = \sin x$

$\therefore \quad I_n = n\left(\left[x^{n-1} \sin x\right]_0^{\pi/2} - \int_0^{\pi/2} (n-1)x^{n-2} \sin x \, dx\right)$

$\qquad\;\; = n\left[\left(\dfrac{\pi}{2}\right)^{n-1} \sin \dfrac{\pi}{2} - 0^{n-1} \sin 0\right] - n(n-1)\int_0^{\pi/2} x^{n-2} \sin x \, dx$

$\therefore \quad I_n = n\left(\dfrac{\pi}{2}\right)^{n-1} - n(n-1)I_{n-2} \qquad$ as required

FURTHER CALCULUS — PURE MATHEMATICS

Hence $I_3 = 3\left(\dfrac{\pi}{2}\right)^2 - 3(2)I_1$

$ = \dfrac{3\pi^2}{4} - 6I_1$

and $I_1 = 1\left(\dfrac{\pi}{2}\right)^0 - 1 \cdot (0)I_{-1} = 1$

$\therefore \quad I_3 = \dfrac{3\pi^2}{4} - 6 \qquad \therefore \quad p = \dfrac{3}{4},\ q = 6$

---

g) *Example ('A' level question)*

(a) If $I_n = \int_0^1 x^n e^x\, dx$ where n is a positive integer, show that

$$I_{n+1} = e - (n+1)I_n$$

Hence evaluate $\int_0^{0.2} t^3 e^{5t}\, dt$ leaving e in answer

(b) Given $y = a^{7x}$ where a is a constant, find:

(i) $\dfrac{dy}{dx}$

(ii) $\int y\, dx$

Solution

(a) $I_n = \int_0^1 x^n e^x\, dx \qquad$ let $\quad u = e^x \quad$ and $\quad \dfrac{dv}{dx} = x^n$

$\therefore \quad \dfrac{du}{dx} = e^x \quad$ and $\quad v = \dfrac{x^{n+1}}{n+1}$

$I_n = \left[\dfrac{e^x \cdot x^{n+1}}{n+1}\right]_0^1 - \int_0^1 \dfrac{e^x x^{n+1}}{n+1}\, dx$

$ = \left[\dfrac{e^x \cdot x^{n+1}}{n+1}\right]_0^1 - \dfrac{1}{n+1}\int_0^1 e^x x^{n+1}\, dx$

$ = \left[\dfrac{e^1 1^{n+1}}{n+1} - \dfrac{e^0 0^{n+1}}{n+1}\right] - \dfrac{1}{n+1}I_{n+1}$

$I_n = \dfrac{e}{n+1} - \dfrac{1}{n+1}I_{n+1}$

$(n+1)I_n = e - I_{n+1}$

$\therefore \quad I_{n+1} = e - (n+1)I_n \quad$ as required

PURE MATHEMATICS — FURTHER CALCULUS

Hence: $I_3 = \int_0^1 x^3 e^x \, dx$

If this is compared with $\int_0^{0.2} t^3 e^{5t} \, dt$ it means that x has been replaced by 5t with some adjustments of constants as follows:

Let $x = 5t$ ∴ $\dfrac{dx}{dt} = 5$ and $dx = 5.dt$

and for the limits: if $t = \cdot 2$ then $5t = 1$

Thus $I_3 = \int_0^1 x^3 e^x \, dx$ becomes $\int_0^{0.2} (5t)^3 e^{5t} (5dt)$

ie $I_3 = 625 \int_0^{0.2} t^3 e^{5t} \, dt$

But from above:

$I_3 = e - 3I_2 = e - 3[e - 2I_1] = e - 3[e - 2(e - I_0)]$

and $I_0 = \int_0^1 x^0 e^x \, dx = (e^x)\Big|_0^1 = e - 1$

∴ $I_3 = e - 3[e - 2(e - (e - 1))] = -2e + 6$

∴ $I_3 = 625 \int_0^{0.2} t^3 e^{5t} \, dt = 6 - 2e$

Hence $\int_0^{0.2} t^3 e^{5t} \, dt = \dfrac{6 - 2e}{625}$

(b) (i) $y = a^{7x}$

To find $\dfrac{dy}{dx}$ take logarithms:

$\log_e y = \log_e(a^{7x})$

∴ $\log_e y = (7x) \log_e a$

∴ $\dfrac{d}{dx}(\log_e y) = 7 \log_e a$

∴ $\dfrac{1}{y} \dfrac{dy}{dx} = 7 \log_e a$

∴ $\dfrac{dy}{dx} = y \cdot 7 \log_e a = 7 a^{7x} \log_e a$

(ii) $\int y \, dx = \int a^{7x} \, dx$

From (i) $\int 7a^{7x} \log_e a \, dx = y + c$ because integration is the reverse of differentiation.

$\therefore \quad 7 \log_e a \int a^{7x} \, dx = a^{7x} + c$

$\therefore \quad \int a^{7x} \, dx = \dfrac{a^{7x}}{7 \log_e a} + c$

11.2 Applications of calculus

a) *Length of an arc*

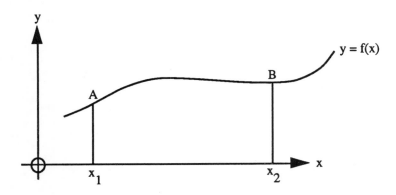

The length of the arc AB of the curve $y = f(x)$ can be found using the formula:

$$\text{length of arc, } s = \int_{x_1}^{x_2} \sqrt{1 + \left(\frac{dy}{dx}\right)^2} \, dx$$

This formula has not been derived as it can be quoted in all examples in which the arc length is required.

Example 16

Find the length of the arc of the parabola $y = x^2$ bounded by the line $y - 2 = 0$.

Solution

First it is necessary to find the coordinates of the points of intersection of the line and the curve, ie

$y = x^2 \quad$ and $y - 2 = 0 \quad \therefore \quad y = 2$

$\therefore \quad 2 = x^2$

$\therefore \quad \pm \sqrt{2} = x$

So the points of intersection are P($\sqrt{2}$, 2) and Q(-$\sqrt{2}$, 2)

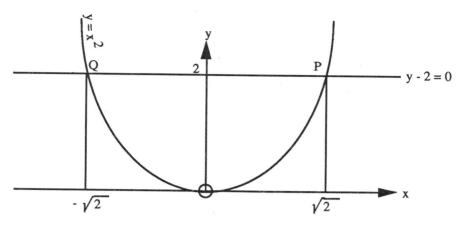

and the required arc length is POQ.

$$y = x^2$$

∴ $\frac{dy}{dx} = 2x$ so using $s = \int \sqrt{1 + \left(\frac{dy}{dx}\right)^2}\, dx$

∴ $s = \int_{-\sqrt{2}}^{\sqrt{2}} \sqrt{1 + (2x)^2}\, dx$

$= \int_{-\sqrt{2}}^{\sqrt{2}} \sqrt{1 + 4x^2}\, dx$

Since the curve is symmetrical about the y axis, ie arc PO = arc QO, this can be written as:

$$s = 2 \cdot \int_0^{\sqrt{2}} \sqrt{1 + 4x^2}\, dx$$

To integrate: let $x = \frac{1}{2} \sinh u$ ∴ $\frac{dx}{du} = \frac{1}{2} \cosh u$ ∴ $dx = \frac{1}{2} \cosh u\, du$

∴ $s = 2 \cdot \int \sqrt{1 + 4\left(\frac{1}{2} \sinh u\right)^2} \left(\frac{1}{2} \cosh u\, du\right)$

$= 2 \cdot \int \sqrt{1 + 4\left(\frac{1}{4} \sinh^2 u\right)} \left(\frac{1}{2} \cosh u\, du\right)$ using $\cosh^2 u - \sinh^2 u = 1$

$= 2 \cdot \int \cosh u \cdot \frac{1}{2} \cosh u\, du$

$= 2 \cdot \frac{1}{2} \int \cosh^2 u\, du$ using $\cosh 2u = 2 \cosh^2 u - 1$

$= \int \frac{1}{2}(\cosh 2u + 1)\, du$

FURTHER CALCULUS — PURE MATHEMATICS

$$\therefore \quad s = \frac{1}{2}\left(\frac{\sinh 2u}{2} + u\right)$$

but $\sinh 2u = 2\sinh u \cdot \cosh u$

$$= 2(2x)\sqrt{1 + (2x)^2}$$

$$= 4x\sqrt{1 + 4x^2}$$

Replacing to give s in terms of x:

$$s = \frac{1}{2}\left[\frac{4x\sqrt{1 + 4x^2}}{2} + \sinh^{-1} 2x\right]_0^{\sqrt{2}}$$

$$= \frac{1}{2}\left[2\sqrt{2}\sqrt{1 + 4(2)} + \sinh^{-1}(2\sqrt{2})\right] - \frac{1}{2}\left[2(0)\sqrt{1 + 4(0)} + \sinh^{-1} 2(0)\right]$$

$$= \frac{1}{2}\left(2\sqrt{2} \cdot 3 + \sinh^{-1}(2\sqrt{2}) - 0\right)$$

$$\therefore \quad s = 3\sqrt{2} + \frac{1}{2}\sinh^{-1}(2\sqrt{2})$$

If the equation of a curve is given in terms of a parameter, say t, ie $x = f(t)$ and $y = g(t)$, the working can still be done in terms of x and y. However, this may not always be easy and, anyway, it would seem better to work in terms of the parameter, t, as this would certainly be quicker and probably be simpler.

The formula for finding the arc length, s, of a curve with parametric equations is:

$$\boxed{\,s = \int_{t_1}^{t_2} \left(\sqrt{\left(\frac{dx}{dt}\right)^2 + \left(\frac{dy}{dt}\right)^2}\right) dt\,}$$

where t_1 and t_2 are the boundary values of the parameter t.

Example 17

A curve is given parametrically by the equations $x = a(t - \sin t)$ and $y = a(1 - \cos t)$ where a is a constant.

Find the length of the arc between the points where $t = 0$ and $t = \frac{\pi}{2}$.

Solution

$$x = a(t - \sin t) \qquad\qquad y = a(1 - \cos t)$$

$$\therefore \quad \frac{dx}{dt} = a(1 - \cos t) \qquad\qquad \therefore \quad \frac{dy}{dt} = a \sin t$$

$$\therefore \quad \left(\frac{dx}{dt}\right)^2 = a^2(1 - \cos t)^2 \qquad\qquad \therefore \quad \left(\frac{dy}{dt}\right)^2 = a^2 \sin^2 t$$

$$= a^2(1 - 2\cos t + \cos^2 t)$$

$$\therefore \quad s = \int_0^{\frac{\pi}{2}} \left(\sqrt{a^2(1 - 2\cos t + \cos^2 t) + a^2 \sin^2 t} \right) dt$$

$$= \int_0^{\frac{\pi}{2}} \left(\sqrt{a^2 - 2a^2 \cos t + a^2 \cos^2 t + a^2 \sin^2 t} \right) dt$$

$$= \int_0^{\frac{\pi}{2}} \left(\sqrt{a^2 - 2a^2 \cos t + a^2(\cos^2 t + \sin^2 t)} \right) dt$$

$$= \int_0^{\frac{\pi}{2}} \left(\sqrt{a^2 - 2a^2 \cos t + a^2} \right) dt$$

$$= \int_0^{\frac{\pi}{2}} \left(\sqrt{2a^2 - 2a^2 \cos t} \right) dt$$

$$= \int_0^{\frac{\pi}{2}} \left(\sqrt{2a^2(1 - \cos t)} \right) dt$$

This can be integrated using the double angle formula $\cos 2A = 1 - 2\sin^2 A$ where $A = \frac{t}{2}$, ie

$$\cos 2\left(\frac{t}{2}\right) = 1 - 2\sin^2\left(\frac{t}{2}\right)$$

$$\therefore \quad s = \int_0^{\frac{\pi}{2}} \left(\sqrt{2a^2\left(1 - \left(1 - 2\sin^2\left(\frac{t}{2}\right)\right)\right)} \right) dt$$

$$= \int_0^{\frac{\pi}{2}} \left(\sqrt{2a^2\left(1 - 1 + 2\sin^2\right)\left(\frac{t}{2}\right)} \right) dt$$

$$= \int_0^{\frac{\pi}{2}} \left(\sqrt{4a^2 \sin^2\left(\frac{t}{2}\right)} \right) dt$$

$$= \int_0^{\frac{\pi}{2}} 2a \sin\left(\frac{t}{2}\right) dt$$

$$= \left[\frac{-2a \cos\left(\frac{t}{2}\right)}{\frac{1}{2}}\right]_0^{\frac{\pi}{2}}$$

$$= \left[-4a \cos \frac{\pi}{4} - (-4a \cos 0)\right]$$

$$= -4a \cdot \frac{1}{\sqrt{2}} + 4a$$

$$\therefore \quad s = 4a\left(1 - \frac{1}{\sqrt{2}}\right)$$

b)   *Area of a surface of revolution*

When a section of a curve is rotated completely about the x axis a three-dimensional surface is formed. The area of this surface can be determined using the formula:

$$\boxed{\text{Area of a surface of revolution, } A = \int 2\pi y \, ds}$$ where s is the length of the arc

This can be written in an alternative form:

$$s = \int \sqrt{1 + \left(\frac{dy}{dx}\right)^2} \, dx$$

$$\therefore \quad \frac{ds}{dx} = \sqrt{1 + \left(\frac{dy}{dx}\right)^2}$$

$$\therefore \quad ds = \sqrt{1 + \left(\frac{dy}{dx}\right)^2} \, dx$$

$$\therefore \quad \boxed{A = \int 2\pi y \sqrt{1 + \left(\frac{dy}{dx}\right)^2} \, dx}$$

As with arc length no attempt has been made to derive this formula, and it will be quoted in any examples requiring evaluation of the surface area of a solid of revolution.

Example 18

Find the area of the surface generated when the arc of the parabola $y^2 = 4x$ between $(0, 0)$ and $(4, 4)$ is rotated once about the x axis.

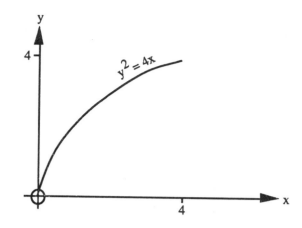

$$y^2 = 4x$$

$$\therefore \quad y = \sqrt{4x} = 2x^{\frac{1}{2}}$$

$$\therefore \quad \frac{dy}{dx} = 2 \cdot \frac{1}{2} \cdot x^{-\frac{1}{2}} = \frac{1}{\sqrt{x}}$$

$$\therefore \quad 1 + \left(\frac{dy}{dx}\right)^2 = 1 + \left(\frac{1}{\sqrt{x}}\right)^2 = 1 + \frac{1}{x}; \quad \text{using } A = 2\pi \int y \sqrt{1 + \left(\frac{dy}{dx}\right)^2} \, dx$$

$$\therefore \quad A = 2\pi \int_0^4 2\sqrt{x} \sqrt{1 + \frac{1}{x}} \, dx \quad = 2\pi \int_0^4 2\sqrt{x} \sqrt{\frac{x+1}{x}} \, dx$$

$$= 2\pi \int_0^4 2\sqrt{x} \frac{\sqrt{x+1}}{\sqrt{x}} \, dx \quad = 4\pi \int_0^4 \sqrt{x+1} \, dx$$

let $u^2 = x + 1 \quad \therefore \quad 2u \frac{du}{dx} = 1 \quad \therefore \quad 2u \, du = dx$

$$\therefore \quad A = 4\pi \int u \cdot (2u \, du)$$

$$= 4\pi \int 2u^2 \, du$$

$$\therefore \quad A = 8 \cdot \pi \frac{u^3}{3} + C$$

Replacing to give A in terms of x again:

$$A = \left[\frac{8\pi(x+1)^{\frac{3}{2}}}{3} \right]_0^4$$

FURTHER CALCULUS   PURE MATHEMATICS

$$= \left[\frac{8\pi(4+1)^{\frac{3}{2}}}{3}\right] - \left[\frac{8\pi(0+1)^{\frac{3}{2}}}{3}\right]$$

$$= \frac{8\pi(5)^{\frac{3}{2}}}{3} - \frac{8\pi(1)^{\frac{3}{2}}}{3}$$

$$= \frac{8\pi}{3} 5\sqrt{5} - \frac{8\pi}{3}$$

$$\therefore \quad A = \frac{8\pi}{3}(5\sqrt{5} - 1)$$

The integral $\int \sqrt{x+1}\, dx$ can be done directly and without a change of variable (substitution) if preferred.

In parametric terms the formula becomes:

Area $= \int 2\pi y\, ds$ but $\frac{ds}{dt} = \sqrt{\left(\frac{dx}{dt}\right)^2 + \left(\frac{dy}{dt}\right)^2}$

$$\therefore \quad \boxed{\text{Area} = \int 2\pi y \sqrt{\left(\frac{dx}{dt}\right)^2 + \left(\frac{dy}{dt}\right)^2}\, dt} \quad \text{where } x = f(t) \text{ and } y = g(t)$$

Example 19

Continuing with the data of Example 17.

Now find the area of the curved surface formed by rotating this arc about the x axis through 2π radians.

Solution

$$x = a(t - \sin t) \qquad y = a(1 - \cos t)$$

$$\sqrt{\left(\frac{dx}{dt}\right)^2 + \left(\frac{dy}{dt}\right)^2} = \sqrt{4a^2 \sin^2 \frac{t}{2}}$$

$$= 2a \sin \frac{t}{2} \qquad \text{cf pages 519, 520}$$

$$\therefore \quad A = \int_0^{\frac{\pi}{2}} 2\pi a(1 - \cos t)\left(2a \sin \frac{t}{2}\right) dt$$

$$= 4\pi a^2 \int_0^{\frac{\pi}{2}} (1 - \cos t) \sin\left(\frac{t}{2}\right) dt$$

$$= 4\pi a^2 \int_0^{\frac{\pi}{2}} \left[1 - \left(1 - 2\sin^2\left(\frac{t}{2}\right)\right)\right] \sin\left(\frac{t}{2}\right) dt$$

PURE MATHEMATICS — FURTHER CALCULUS

$$\therefore \quad A = 8\pi a^2 \int_0^{\frac{\pi}{2}} \sin^3\left(\frac{t}{2}\right) dt$$

To integrate use the change of variable:

$$\text{Let: } u = \cos\left(\frac{t}{2}\right) \quad \frac{du}{dt} = -\frac{1}{2}\sin\left(\frac{t}{2}\right) \quad \therefore \quad \frac{-2\,du}{\sin\left(\frac{t}{2}\right)} = dt$$

$$\therefore \quad A = 8\pi a^2 \int \sin^3\left(\frac{t}{2}\right) \cdot \frac{-2\,du}{\sin\left(\frac{t}{2}\right)}$$

$$= -16\pi a^2 \int \sin^2\left(\frac{t}{2}\right) du \quad \text{but } \sin^2\left(\frac{t}{2}\right) = 1 - \cos^2\left(\frac{t}{2}\right) = 1 - u^2$$

$$= -16\pi a^2 \int (1 - u^2) du$$

$$= -16\pi a^2 \left(u - \frac{u^3}{3}\right)$$

$$= -16\pi a^2 \left[\cos\left(\frac{t}{2}\right) - \frac{1}{3}\cos^3\left(\frac{t}{2}\right)\right]_0^{\frac{\pi}{2}}$$

$$= -16\pi a^2 \left[\cos\left(\frac{\pi}{4}\right) - \frac{1}{3}\cos^3\left(\frac{\pi}{4}\right)\right] - \left[-16\pi a^2 \left[\cos 0 - \frac{1}{3}\cos^3 0\right]\right]$$

$$= -16\pi a^2 \left(\frac{1}{\sqrt{2}} - \frac{1}{3}\left(\frac{1}{2\sqrt{2}}\right)\right) + 16\pi a^2 \left(1 - \frac{1}{3}\right)$$

$$= \frac{32\pi a^2}{3} - 16\pi a^2 \left(\frac{6-1}{6\sqrt{2}}\right)$$

$$\therefore \quad \text{Area} = \frac{32\pi a^2}{3} - \frac{40\pi a^2}{3\sqrt{2}}$$

c) *Centroids*

The centroid of an object is its geometric centre. For an object made of uniform material the centroid coincides with the point at which the body can be supported in a perfectly balanced state.

FURTHER CALCULUS — PURE MATHEMATICS

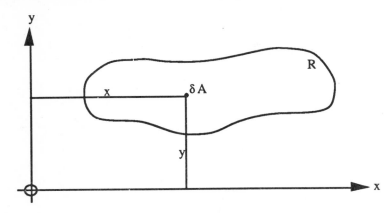

In this diagram R is a region in the x, y plane and δA is a small increment of area located at the point (x, y). Then the x-moment of this area (ie the moment about the y axis) is defined as $x.\delta A$. Similarly, the y-moment of this area (ie the moment about the x axis) is defined as $y.\delta A$.

For the whole region R, the x and y moments are defined as the sums of the corresponding moments of small elements taken over the whole of R,

eg   the x-moment $= \Sigma x \delta A$

                             $= \int x \, \delta A$

Example 20

Calculate the moment about the y axis of the area enclosed by the curve $y = \dfrac{1}{1 + x^2}$ and the lines $x = 0$, $x = 1$, $y = 0$.

Solution

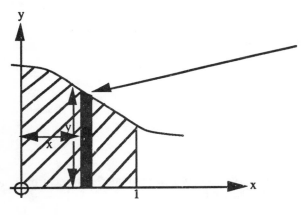

The shaded region is made up of strips such as this with area δA; the moment of the strip about the y axis is $x \, \delta A$.

The total moment about the y axis is defined as the x-moment $= \Sigma x \delta A = \int_0^1 x \, dA$.

However, it is necessary to find an expression for dA in terms of x and y.

If $A = \int_0^1 y \, dx$, then it follows $dA = y \, dx = \dfrac{1}{1 + x^2} \, dx$

$\therefore$ x-moment $= \int_0^1 x . \dfrac{1}{1 + x^2} \, dx$

$$= \left[\tfrac{1}{2} \log_e (1 + x^2)\right]_0^1$$

$$= \tfrac{1}{2}(\log_e 2 - \log_e 1)$$

$$= \tfrac{1}{2} \log_e 2$$

FURTHER CALCULUS — PURE MATHEMATICS

The moment about the x axis (ie the y moment) of this *same* strip would be $\frac{1}{2}y\,\delta A$ because it can be treated as though the area of this strip is concentrated at the midpoint of the strip.

Therefore, the total moment about the x-axis is defined as the

$$\text{y-moment} = \Sigma \tfrac{1}{2} y\,\delta A = \int \tfrac{1}{2} y\,dA$$

but $dA = y\,dx$

$$= \int \tfrac{1}{2} y(y\,dx)$$

$$= \int \tfrac{1}{2} y^2\,dx$$

The centroid of a region R lying in the x, y plane is the point $(\bar{x}, \bar{y})$ where

$$\boxed{\bar{x} = \frac{\text{x-moment of R}}{\text{area of R}} \qquad \bar{y} = \frac{\text{y-moment of R}}{\text{area of R}}}$$

Example 21

Find the coordinates of the centroid of the region bounded by the curve $y = e^x$ the coordinate axes and the line $x = 1$.

Solution

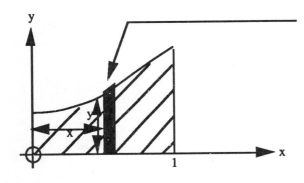

For this strip:

x-moment $= x\,\delta A$

y-moment $= \tfrac{1}{2} y\,\delta A$

The area of this region $= \int_0^1 y\,dx$

$$= \int_0^1 e^x\,dx$$

$$= [e^x]_0^1$$

$$= e^1 - e^0$$

$$= e - 1$$

Finding the x and y-moments gives:

for x: Moment $= \int x\,dA = \int_0^1 xy\,dx = \int_0^1 x \cdot e^x\,dx$ integrating by parts (see Example 12)

PURE MATHEMATICS — FURTHER CALCULUS

$$= [xe^x - ex]_0^1 \quad = (1(e) - e) - (0(1) - 1) \quad = 0 - (-1) = 1$$

for y:
$$\int_0^1 \frac{1}{2} y^2 \, dx = \int_0^1 \frac{1}{2}(ex)^2 \, dx$$

$$= \int_0^1 \frac{1}{2} e^{2x} \, dx = \left[\frac{e^{2x}}{4}\right]_0^1 = \left(\frac{e^2}{4} - \frac{e^0}{4}\right)$$

$$= \frac{1}{4}(e^2 - 1) = \frac{1}{4}(e-1)(e+1)$$

So the centroid is at the position $(\bar{x}, \bar{y})$ where $\bar{x} = \dfrac{1}{e-1}$ and $\bar{y} = \dfrac{\frac{1}{4}(e-1)(e+1)}{(e-1)} = \dfrac{1}{4}(e+1)$

Moments can also be defined for volumes of revolution in just the same way.

x-moment = $\Sigma x \delta V$ where δV is an element of volume at a distance x from the y axis.

$\therefore$ total x-moment = $\int x \, dV$ but $V = \int \pi y^2 \, dx$

$\therefore \quad dV = \pi y^2 \, dx$

$$= \int x \cdot (\pi y^2 \, dx)$$
$$= \int \pi x y^2 \, dx$$

The centroid of a volume of revolution is defined for, say, $\bar{x}$ as

$$\frac{\text{x-moment of } V}{\text{volume of } V}$$

If a curve is rotated about the x axis then by symmetry $\bar{y} = 0$; likewise if it is rotated about the y axis $\bar{x} = 0$.

Example 22

The region defined in Example 21 is rotated about the x axis to form a solid of revolution. Find the coordinates of the centroid of this solid.

Solution

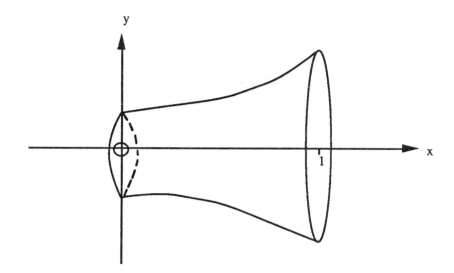

The volume of this solid $= \int_0^1 \pi y^2 \, dx$

$= \int_0^1 \pi (e^x)^2 \, dx$

$= \int_0^1 \pi e^{2x} \, dx$

$= \left[\dfrac{\pi e^{2x}}{2}\right]_0^1$

$= \left(\dfrac{\pi e^2}{2} - \dfrac{\pi}{2}\right)$

$= \dfrac{\pi}{2}(e^2 - 1)$

x-moment $= \int_0^1 \pi x y^2 \, dx$

$= \pi \int_0^1 x e^{2x} \, dx \quad$ integrating by parts

$= \pi\left[\dfrac{xe^{2x}}{2} - \dfrac{e^{2x}}{4}\right]_0^1$

$= \pi\left(\dfrac{1(e^2)}{2} - \dfrac{e^2}{4}\right) - \pi\left(\dfrac{0(1)}{2} - \dfrac{1}{4}\right)$

$= \pi\left(\dfrac{e^2}{4}\right) - \pi\left(-\dfrac{1}{4}\right)$

PURE MATHEMATICS — FURTHER CALCULUS

$$= \frac{\pi}{4}(e^2 + 1)$$

$$\therefore \quad \bar{x} = \frac{\frac{\pi}{4}(e^2 + 1)}{\frac{\pi}{2}(e^2 - 1)}$$

$$= \frac{(e^2 + 1)}{2(e^2 - 1)}$$

and $\bar{y} = 0$ by symmetry

d) *Example ('A' level question)*

Sketch the curve $y = \cosh x$.

Find the mean value of $\cosh x$ for $0 \leq x \leq \log_e 2$.

Find the perimeter of the region defined by the inequalities $0 \leq x \leq \log_e 2$, $0 \leq y \leq \cosh x$.

This region is rotated completely about the x axis. Show that the total surface area of the solid formed is $\frac{1}{2}\pi(7 + \log_e 4)$.

Solution

Graph:

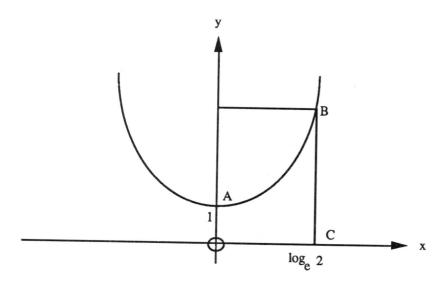

$$\text{Mean value} = \frac{1}{b-a}\int_a^b f(x)\, dx$$

ie $\quad$ mean value $= \dfrac{1}{\log_e 2 - 0}\displaystyle\int_0^{\log_e 2}\cosh x\, dx$

$$= \frac{1}{\log_e 2}[\sinh x]_0^{\log_e 2}$$

FURTHER CALCULUS — PURE MATHEMATICS

$$= \frac{1}{\log_e 2} \left[\frac{1}{2}(e^x - e^{-x}) \right]_0^{\log_e 2}$$

$$= \frac{1}{\log_e 2} \left[\frac{1}{2}(e^{\log_e 2} - e^{-\log_e 2}) - \frac{1}{2}(e^0 - e^0) \right]$$

$$= \frac{1}{\log_e 2} \left[\frac{1}{2} \left(2 - \frac{1}{2} \right) \right]$$

$\therefore$ mean value $= \dfrac{1}{\log_e 2} \left(\dfrac{3}{4} \right)$

$$= \frac{3}{4 \log_e 2}$$

The perimeter of the region defined by OABC is required:

- OA = 1 unit
- AB = arc length s
- BC = $\cosh x \,(\log_e 2) = \frac{1}{2}\left[e^{\log_e 2} + e^{-\log_e 2} \right] = \frac{1}{2}\left[2 + \frac{1}{2} \right] = \frac{5}{4}$
- CO = $\log_e 2$

To find AB - using $\quad s = \displaystyle\int_0^{\log_e 2} \sqrt{1 + \left(\frac{dy}{dx}\right)^2}\, dx$

let $\quad y = \cosh x$

$\therefore \quad \dfrac{dy}{dx} = \sinh x$

$\therefore \quad \sqrt{1 + \left(\dfrac{dy}{dx}\right)^2} = \sqrt{1 + \sinh^2 x}$

$$= \cosh x$$

$\therefore \quad s = \displaystyle\int_0^{\log_e 2} \cosh x \, dx$

$$= [\sinh x]_0^{\log_e 2}$$

$$= \sinh(\log_e 2) - \sinh(0)$$

$\therefore$ arc AB $= s = \dfrac{3}{4}$ unit

Total perimeter $= OA + AB + BC + CO$

$$= 1 + \frac{3}{4} + \frac{5}{4} + \log_e 2$$

$$= 3 + \log_e 2$$

PURE MATHEMATICS — FURTHER CALCULUS

Area:
$$A = 2\pi \int y \sqrt{1 + \left(\frac{dy}{dx}\right)^2}\, dx$$

$$= 2\pi \int_0^{\log_e 2} \cosh x \cdot \cosh x \, dx$$

$$= 2\pi \int_0^{\log_e 2} \cosh^2 x \, dx \qquad \text{but } \cosh^2 x = \tfrac{1}{2}(\cosh 2x + 1)$$

$$= 2\pi \int_0^{\log_e 2} \tfrac{1}{2}(\cosh 2x + 1)\, dx$$

$$= \pi \left[\frac{\sinh 2x}{2} + x\right]_0^{\log_e 2}$$

$$= \pi \left[\frac{\sinh 2\log_e 2}{2} + \log_e 2\right] - \left[\frac{\sinh 2(0)}{2} + 0\right]$$

$$= \pi \left[\tfrac{1}{2}\sinh \log_e 2^2 + \log_e 2\right] - 0$$

$$= \frac{\pi}{2}\left[\tfrac{1}{2}\left(e^{\log_e 4} - e^{-\log_e 4}\right) + 2\log_e 2\right]$$

$$= \frac{\pi}{2}\left[\tfrac{1}{2}\left(4 - \tfrac{1}{4}\right) + \log_e 2^2\right]$$

$$A = \frac{\pi}{2}\left(\frac{15}{8} + \log_e 4\right)$$

The total surface area will also include the areas of both the circular ends.

at 0, radius = 1 $\therefore$ Area = πr^2 = $\pi(1)^2 = \pi$

at C, radius = $\frac{5}{4}$ $\therefore$ Area = $\pi\left(\frac{5}{4}\right)^2 = \frac{25\pi}{16}$

$\therefore$ Total surface area
$$= \frac{\pi}{2}\left(\frac{15}{8} + \log_e 4\right) + \pi + \frac{25\pi}{16}$$

$$= \frac{\pi}{2}\left(\frac{15}{8} + \log_e 4 + 2 + \frac{25}{8}\right)$$

$$= \frac{\pi}{2}\left(\frac{15 + 16 + 25}{8} + \log_e 4\right)$$

$$= \frac{\pi}{2}\left(\frac{56}{8} + \log_e 4\right)$$

$$= \frac{\pi}{2}(7 + \log_e 4)$$

FURTHER CALCULUS PURE MATHEMATICS

11.3 Differential equations

a) *Differential equations - introduction*

An equation containing differential coefficients such as $\frac{dy}{dx}$, $\frac{d^2y}{dx^2}$, etc is called a differential equation, eg $\frac{dy}{dx} = 3$ has a general solution - by integration -

$$y = 3x + c$$

If it is also known that $y = 5$ when $x = 1$, ie

$$5 = 3(1) + c$$
$$5 = 3 + c$$
$$\therefore \quad c = 2$$

So the particular solution is $y = 3x + 2$.

A differential equation defines some property common to a family of curves. The general solution, involving constants, is the equation of any member of the family. The particular solution is the equation of one member of the family.

The order of a differential equation is determined by the highest differential coefficient in the equation.

b) *Differential equations - variables separable*

Certain equations can be solved directly by integrating both sides - once they have been written in a suitable form, eg

$$\frac{dy}{dx} = xy \text{ can be written as } \frac{1}{y}\frac{dy}{dx} = x$$

$$\therefore \quad \frac{1}{y}dy = x\,dx$$

Now both sides can be integrated, the left hand side with respect to y (ie dy) and the right hand side with respect to x (ie dx).

This method of solving a differential equation is known as 'separating the variables', because all the y's are on one side of the equation and all the x's on the other side.

Example 23

Find the general solutions of the following differential equations:

(a) $\quad e^{-x}\frac{dy}{dx} = y^2 - 1$

(b) $\quad \sqrt{x^2 + 1}\,\frac{dy}{dx} = \frac{x}{y}$

Solution

(a) $\qquad e^{-x}\frac{dy}{dx} = y^2 - 1$

$\qquad \therefore \quad \frac{1}{y^2 - 1}\frac{dy}{dx} = e^x$

$$\therefore \quad \frac{1}{y^2 - 1} \, dy = e^x \, dx$$

Before integration $\frac{1}{y^2 - 1}$ must be expressed in partial fractions:

$$\frac{1}{y^2 - 1} = \frac{1}{(y-1)(y+1)}$$

$$\frac{1}{y^2 - 1} = \frac{A}{y-1} + \frac{B}{y+1}$$

$$1 = A(y+1) + B(y-1)$$

let $y = 1$ \qquad $1 = A(1+1) + B(0)$ \qquad $\therefore A = \frac{1}{2}$

let $y = -1$ \qquad $1 = A(0) + B(-1-1)$ \qquad $\therefore B = -\frac{1}{2}$

$$\therefore \quad \frac{1}{y^2 - 1} = \frac{1}{2(y-1)} - \frac{1}{2(y+1)}$$

$$\int \frac{1}{2(y-1)} - \frac{1}{2(y+1)} \, dy = \int e^x \, dx$$

$$\therefore \quad \frac{1}{2} \log_e (y-1) - \frac{1}{2} \log_e (y+1) = e^x + c$$

$$\frac{1}{2} \log_e \left(\frac{y-1}{y+1} \right) = e^x + c$$

$$\log_e \left(\frac{y-1}{y+1} \right)^{\frac{1}{2}} = e^x + c$$

$$\log_e \sqrt{\frac{y-1}{y+1}} = e^x + c \text{ is the general solution}$$

(b) $\qquad \sqrt{x^2 + 1} \, \frac{dy}{dx} = \frac{x}{y}$

$\therefore \qquad y \frac{dy}{dx} = \frac{x}{\sqrt{x^2 + 1}}$

$\therefore \qquad y \, dy = \frac{x}{\sqrt{x^2 + 1}} \, dx$

Using a substitution to integrate $\frac{x}{\sqrt{x^2 + 1}} \, dx$ \qquad let \qquad $u^2 = x^2 + 1$

$$\therefore \quad 2u \frac{du}{dx} = 2x$$

$$\frac{u}{x} du = dx$$

FURTHER CALCULUS — PURE MATHEMATICS

$$\therefore \quad \int y\, dy = \int \frac{x}{u} \frac{u}{x}\, du$$

$$\therefore \quad \int y\, dy = \int du$$

$$\therefore \quad \frac{y^2}{2} = u + c$$

$$\therefore \quad y^2 = 2\sqrt{x^2+1} + c \text{ is the general solution}$$

Example 24

Given that $(1 + \sin^2 x)\frac{dy}{dx} = e^{-2y} \sin 2x$ and $y = 1$ when $x = 0$, find the value of y when $x = \frac{\pi}{2}$.

Solution

$$(1 + \sin^2 x)\frac{dy}{dx} = e^{-2y} \sin 2x$$

$$e^{2y}\frac{dy}{dx} = \frac{\sin 2x}{1 + \sin^2 x}$$

$$e^{2y}\, dy = \frac{\sin 2x}{1 + \sin^2 x}\, dx \qquad \text{Note: } \frac{d}{dx}(1 + \sin^2 x) = 2 \sin x \cos x = \sin 2x$$

$$\therefore \quad \int e^{2y}\, dy = \int \frac{\sin 2x}{1 + \sin^2 x}\, dx$$

$$\frac{e^{2y}}{2} = \log_e(1 + \sin^2 x) + c$$

when $y = 1$, $x = 0$
$$\frac{e^2}{2} = \log_e(1 + \sin^2 0) + c$$

$$\frac{e^2}{2} = \log_e 1 + c \qquad \text{but} \qquad \log_e 1 = 0$$

$$\therefore \quad \frac{e^2}{2} = c$$

$$\frac{e^{2y}}{2} = \log_e(1 + \sin^2 x) + \frac{e^2}{2}$$

$$\therefore \quad e^{2y} = 2 \log_e(1 + \sin^2 x) + e^2$$

when $x = \frac{\pi}{2}$
$$e^{2y} = 2 \log_e\left(1 + \sin^2 \frac{\pi}{2}\right) + e^2$$

$$e^{2y} = 2 \log_e 2 + e^2$$

$$\therefore \quad 2y = \log_e(2 \log_e 2 + e^2)$$

$$\therefore \quad y = \frac{1}{2} \log_e(2 \log_e 2 + e^2)$$

Example 25

Find y in terms of x if $\frac{dy}{dx} = (1 + y)^2 \sin^2 x \cos x$, and $y = 2$ when $x = 0$.

Solution

$$\frac{dy}{dx} = (1 + y)^2 \sin^2 x \cos x$$

$$\therefore \quad \frac{1}{(1 + y)^2} dy = \sin^2 x \cos x \, dx \qquad \text{Note: } \frac{d}{dx}(\sin^3 x) = 3 \sin^2 x \cos x$$

$$\therefore \quad \int (1 + y)^{-2} dy = \frac{1}{3} \int 3 \sin^2 x \cos x \, dx$$

$$\frac{(1 + y)^{-1}}{-1} = \frac{1}{3} \sin^3 x + c$$

$$\frac{-1}{(1 + y)} = \frac{1}{3} \sin^3 x + c$$

$y = 2, \; x = 0$: $\quad \frac{-1}{(1 + 2)} = \frac{1}{3} \sin^3 0 + c$

$$\therefore \quad -\frac{1}{3} = c$$

$$\therefore \quad \frac{-1}{1 + y} = \frac{1}{3} \sin^3 x - \frac{1}{3}$$

$$\therefore \quad \frac{-3}{1 + y} = \sin^3 x - 1$$

$$\therefore \quad -3 = (1 + y)(\sin^3 x - 1)$$

$$-3 = \sin^3 x - 1 + y \sin^3 x - y$$

$$\therefore \quad -3 - \sin^3 x + 1 = y(\sin^3 x - 1)$$

$$-2 - \sin^3 x = y(\sin^3 x - 1)$$

$$\frac{-2 - \sin^3 x}{\sin^3 x - 1} = y$$

---

c) *First order - exact*

Consider the equation $2xy \frac{dy}{dx} + y^2 = e^{2x}$; the variables cannot be separated this time. However, the left hand side:

$$2xy \frac{dy}{dx} + y^2 = \frac{d}{dx}(xy^2)$$

$$\therefore \quad \frac{d}{dx}(xy^2) = e^{2x}$$

and now both sides can be integrated with respect to x:

$$\int \frac{d}{dx}(xy^2) \, dx = \int e^{2x} \, dx$$

$$\therefore \quad xy^2 = \frac{e^{2x}}{2} + c$$

Example 26

Solve the following exact equations:

(a) $\quad \dfrac{t^2}{x} \dfrac{dx}{dt} + 2t \log_e x = 3 \cos t$

(b) $\quad x^2 \cos u \dfrac{du}{dx} + 2x \sin u = \dfrac{1}{x}$

Solution

(a) $\quad \dfrac{t^2}{x} \dfrac{dx}{dt} + 2t \log_e x = \dfrac{d}{dt}(t^2 \log_e x)$

$\therefore \quad \dfrac{d}{dt}(t^2 \log_e x) = 3 \cos t$

Integrating both sides with respect to t:

$$\int \frac{d}{dt}(t^2 \log_e x) \, dt = \int 3 \cos t \, dt$$

$\therefore \quad t^2 \log_e x = 3 \sin t + c$

(b) $\quad x^2 \cos u \dfrac{du}{dx} + 2x \sin u = \dfrac{d}{dx}(x^2 \sin u)$

$\therefore \quad \dfrac{d}{dx}(x^2 \sin u) = \dfrac{1}{x}$

Integrating both sides with respect to x:

$$\int \frac{d}{dx}(x^2 \sin u) \, dx = \int \frac{1}{x} \, dx$$

$\therefore \quad x^2 \sin u = \log_e x + c$

d) *First order - using an integrating factor*

There are some equations which are not exact as they stand but can be made so by multiplying each side by an integrating factor, eg the equation $xy \dfrac{dy}{dx} + y^2 = 3x$ cannot have the variables separated and is not exact as it stands.

The left hand side is not an exact differential. Try:

$$\frac{d}{dx}(x^2y) = x^2\frac{dy}{dx} + 2xy \quad - \text{no}$$

$$\frac{d}{dx}(x^2y^2) = x^2 2y\frac{dy}{dx} + 2xy^2 \quad - \text{no}$$

$$= 2x\left(xy\frac{dy}{dx} + y^2\right) \quad - \text{yes}$$

Therefore, multiplying the left hand side by 2x gives an exact differential: 2x is known as the integrating factor. The right hand side must also be multiplied by 2x, of course.

$$xy\frac{dy}{dx} + y^2 = 3x$$

$$\therefore \quad 2x\left(xy\frac{dy}{dx} + y^2\right) = 2x(3x)$$

$$\therefore \quad 2x^2y\frac{dy}{dx} + 2xy^2 = 6x^2$$

$$\therefore \quad \frac{d}{dx}(x^2y^2) = 6x^2$$

Integrating both side with respect to x:

$$\int \frac{d}{dx}(x^2y^2)\,dx = \int 6x^2\,dx$$

$$\therefore \quad x^2y^2 = \frac{6x^3}{3} + c$$

$$\therefore \quad x^2y^2 = 2x^3 + c$$

Example 27

Find the integrating factor required to make these equations exact and then solve them:

(a) $\quad xe^y \frac{dy}{dx} + 2e^y = x$

(b) $\quad 2x^2y \frac{dy}{dx} + xy^2 = 1$

Solution

(a) $\quad xe^y \frac{dy}{dx} + 2e^y = x$

$$\frac{d}{dx}(x^2 e^y) = x^2 e^y \frac{dy}{dx} + 2xe^y$$

$$= x\left(xe^y \frac{dy}{dx} + 2e^y\right)$$

Therefore the integrating factor is x:

FURTHER CALCULUS — PURE MATHEMATICS

$$\therefore \quad x\left(xe^y \frac{dy}{dx} + 2e^y\right) = x(x)$$

$$\therefore \quad x^2 e^y \frac{dy}{dx} + 2xe^y = x^2$$

$$\frac{d}{dx}(x^2 e^y) = x^2$$

Integrating both sides with respect to x:

$$\int \frac{d}{dx}(x^2 e^y)\, dx = \int x^2\, dx$$

$$x^2 e^y = \frac{x^3}{3} + c$$

(b) $\quad 2x^2 y \frac{dy}{dx} + xy^2 = 1$

$$\frac{d}{dx}(xy^2) = x\cdot 2y\frac{dy}{dx} + y^2$$

$$= \frac{1}{x}(2x^2 y + xy^2)$$

Therefore the integrating factor is $\frac{1}{x}$

$$\frac{1}{x}\left(2x^2 y \frac{dy}{dx} + xy^2\right) = \frac{1}{x}(1)$$

$$2xy \frac{dy}{dx} + y^2 = \frac{1}{x}$$

$$\frac{d}{dx}(xy^2) = \frac{1}{x}$$

Integrating both sides with respect to x:

$$\int \frac{d}{dx}(xy^2)\, dx = \int \frac{1}{x}\, dx$$

$$xy^2 = \log_e x + c$$

A special type of integrating factor occurs when dealing with linear equations.

ie   equations of the type

$$\frac{dy}{dx} + Py = Q \quad \text{where P and Q are functions of x or constants}$$

The integrating factor is of the form $e^{\int P\, dx}$ - no proof is offered but it will be shown that this form of integrating factor works, eg take the equation.

$$\frac{dy}{dx} + 3y = e^{2x}$$

PURE MATHEMATICS — FURTHER CALCULUS

This is a linear equation with $P = 3$ and $Q = e^{2x}$ and so the integrating factor will be

$$e^{\int 3\,dx} = e^{3x}$$

$$\therefore \quad e^{3x}\left(\frac{dy}{dx} + 3y\right) = e^{3x}(e^{2x})$$

$$e^{3x}\frac{dy}{dx} + 3e^{3x}y = e^{5x} \quad \text{and} \quad \frac{d}{dx}(e^{3x}y) = e^{3x}\frac{dy}{dx} + 3e^{3x}y$$

$$\therefore \quad \frac{d}{dx}(e^{3x}y) = e^{5x}$$

$$\therefore \quad \int \frac{dy}{dx}(e^{3x}y)\,dx = \int e^{5x}\,dx$$

$$e^{3x}y = \frac{e^{5x}}{5} + c \quad \text{is the general solution of the differential equation } \frac{dy}{dx} + 3y = e^{2x}$$

Example 28

Solve the following linear equations:

(a) $\quad \dfrac{dy}{dx} + y\cot x = \cos x$

(b) $\quad \dfrac{dy}{dx} + y + 3 = x$

Solution

(a) $\quad \dfrac{dy}{dx} + y\cot x = x \quad \therefore$ Integrating factor $= e^{\int \cot x\,dx}$ where $P = \cot x$

$$= e^{\int \frac{\cos x}{\sin x}\,dx} = e^{\log_e \sin x} = \sin x$$

$$\therefore \quad \sin x\left(\frac{dy}{dx} + y\cot x\right) = \sin x\,(\cos x)$$

$$\therefore \quad \sin x\,\frac{dy}{dx} + y\sin x \cdot \frac{\cos x}{\sin x} = \frac{\sin 2x}{2}$$

$$\therefore \quad \sin x\,\frac{dy}{dx} + y\cos x = \frac{\sin 2x}{2}$$

$$\therefore \quad \frac{d}{dx}(y\sin x) = \frac{\sin 2x}{2}$$

$$\therefore \quad \int \frac{d}{dx}(y\sin x)\,dx = \int \frac{\sin 2x}{2}\,dx$$

$$\therefore \quad y\sin x = -\frac{\cos 2x}{4} + c$$

(b)
$$\frac{dy}{dx} + y = x - 3 \qquad \text{Integrating factor} = e^{\int 1\,dx} = e^x$$

$$e^x\left(\frac{dy}{dx} + y\right) = e^x(x - 3)$$

$$e^x \frac{dy}{dx} + e^x y = xe^x - 3e^x$$

$$\therefore \quad \frac{d}{dx}(e^x y) = xe^x - 3e^x$$

$$\therefore \quad \int \frac{d}{dx}(e^x y)\,dx = \int xe^x\,dx - \int 3e^x\,dx$$

$$e^x y = \int xe^x\,dx - 3e^x$$

Using integration by parts for $\int xe^x\,dx$ \quad let \quad $u = x$ \quad and \quad $\frac{dv}{dx} = e^x$

$$\therefore \quad \frac{du}{dx} = 1 \quad \text{and} \quad v = e^x$$

$$\therefore \quad \int xe^x\,dx = xe^x - \int e^x\,dx$$

$$= xe^x - e^x$$

$$\therefore \quad e^x y = (xe^x - e^x) - 3e^x + c$$

$$\therefore \quad e^x y = xe^x - 4e^x + c$$

Example 29

Solve the equation $\frac{dy}{dx} + \frac{2y}{x} = \frac{1}{x - 1}$ given that $y = 2$ when $x = 2$.

Solution

$$\frac{dy}{dx} + \frac{2}{x} \cdot y = \frac{1}{x - 1} \qquad \text{Integrating factor} = e^{\int \frac{2}{x}\,dx} = e^{2\log_e x} = e^{\log_e x^2} = x^2$$

$$x^2\left(\frac{dy}{dx} + \frac{2}{x} y\right) = x^2\left(\frac{1}{x - 1}\right)$$

$$x^2 \frac{dy}{dx} + 2xy = \frac{x^2}{x - 1} \qquad \text{and} \qquad \begin{array}{r} x + 1 \\ x - 1 \overline{\smash{\big)}\, x^2} \\ \underline{x^2 - x} \\ x \\ \underline{x - 1} \\ 1 \end{array}$$

$$\frac{d}{dx}(x^2 y) = x + 1 + \frac{1}{x - 1}$$

$$\therefore \quad \int \frac{d}{dx}(x^2 y)\,dx = \int \left(x + 1 + \frac{1}{x - 1}\right) dx$$

$$\therefore \quad x^2y = \frac{x^2}{2} + x + \log_e(x-1) + c$$

$x = 2, y = 2$
$$2^2 \cdot 2 = \frac{2^2}{2} + 2 + \log_e(2-1) + c$$
$$8 = 2 + 2 + 0 + c$$
$$4 = c$$

$$\therefore \quad x^2y = \frac{x^2}{2} + x + \log_e(x-1) + 4$$

Example 30

Solve the equation $\dfrac{dy}{dx} = \dfrac{2x - y}{x(x+1)}$ given $y = 0$ when $x = 0$

Solution

$$\frac{dy}{dx} = \frac{2x}{x(x+1)} - \frac{y}{x(x+1)}$$

$$\therefore \quad \frac{dy}{dx} + \frac{1}{x(x+1)} \cdot y = \frac{2}{(x+1)}$$

This is a linear equation. $\quad \therefore$ Integrating factor $= e^{\int \frac{1}{x(x+1)} dx}$

Using partial fractions

$$\int \frac{1}{x(x+1)} dx = \int \left[\frac{1}{x} - \frac{1}{x+1} \right] dx$$

$$= \log_e(x) - \log_e(x+1)$$

$$= \log_e\left(\frac{x}{x+1}\right)$$

$$\therefore \quad e^{\int \frac{1}{x(x+1)} dx} = e^{\log_e\left(\frac{x}{x+1}\right)} = \frac{x}{x+1}$$

$$\frac{x}{x+1}\left(\frac{dy}{dx} + \frac{1}{x(x+1)} \cdot y\right) = \frac{x}{x+1}\left(\frac{2}{x+1}\right)$$

$$\therefore \quad \frac{x}{x+1}\frac{dy}{dx} + \frac{1}{(x+1)^2} \cdot y = \frac{2x}{(x+1)^2}$$

$$\therefore \quad \frac{d}{dx}\left(\frac{x}{x+1} \cdot y\right) = \frac{2x}{(x+1)^2}$$

FURTHER CALCULUS — PURE MATHEMATICS

To check the left hand side, differentiate:

$$\frac{d}{dx}\left(\frac{x}{x+1}y\right) = \frac{x}{x+1}\frac{dy}{dx} + \left(\frac{(x+1)\cdot 1 - x(1)}{(x+1)^2}\right)y$$

$$= \frac{x}{x+1}\frac{dy}{dx} + \frac{1}{(x+1)^2}\cdot y$$

$\therefore \quad \int \frac{dy}{dx}\left(\frac{x}{x+1}y\right)dx = \int \frac{2x}{(x+1)^2}dx$

$\therefore \quad \frac{xy}{x+1} = \int\left(\frac{2}{x+1} - \frac{2}{(x+1)^2}\right)dx \quad$ by partial fractions

$\therefore \quad \frac{xy}{x+1} = 2\log_e(x+1) + \frac{2}{x+1} + c$

$x=0, y=0 \quad \frac{0\cdot 0}{0+1} = 2\log_e(0+1) + \frac{2}{0+1} + c$

$\therefore \quad 0 = 0 + 2 + c \quad\quad \therefore c = -2$

$\therefore \quad \frac{xy}{x+1} = 2\log_e(x+1) + \frac{2}{x+1} - 2$

(The actual derivation of the partial fractions has been omitted from this example as it has been explained in detail on a number of occasions.)

e) *Second order equations*

These are equations of the type $a\frac{d^2y}{dx^2} + b\frac{dy}{dx} + cy = f(x)$ where a, b and c are constants and f(x) is a simple function of x.

The general solution is made up of two distinct parts, ie

y = complementary function (CF) + particular integral (PI)

Taking each of these in turn:

(a) *Complementary function*

This is the solution of the equation $a\frac{d^2y}{dx^2} + b\frac{dy}{dx} + cy = 0$

which is solved by considering the quadratic equation '$au^2 + bu + c = 0$' known as the auxiliary quadratic equation.

(*Note*: u^2 replaces $\frac{d^2y}{dx^2}$, u replaces $\frac{dy}{dx}$ in the differential equation)

Although this quadratic equation has been formed in a very unusual way, the usual rules governing the roots of such an equation still apply, ie

$b^2 - 4ac > 0 \quad$ roots α and β are real and different

$b^2 - 4ac = 0 \quad$ roots α and β are real and equal, ie $\alpha = \beta$

$b^2 - 4ac < 0 \quad$ roots α and β are complex conjugates, $p \pm qi$

PURE MATHEMATICS — FURTHER CALCULUS

Furthermore, knowing the nature of the roots of the auxiliary equation means that the complementary function can be quoted according to the following rules:

Consider the auxiliary equation $au^2 + bu + c = 0$. Then

$b^2 - 4ac > 0$ complementary function $y = Ae^{\alpha x} + Be^{\beta x}$

$b^2 - 4ac = 0$ complementary function $y = e^{\alpha x}(A + Bx)$

$b^2 - 4ac < 0$ complementary function $y = Ae^{px}\cos(qx + \theta)$

No attempt is being made to derive these; they are given as a statement of fact and as such must be learnt.

Example 31

Find the complementary functions for each of the following differential equations:

(a) $\dfrac{d^2y}{dx^2} - 6\dfrac{dy}{dx} + 5y = 3$

(b) $\dfrac{d^2y}{dx^2} - 2\dfrac{dy}{dx} + y = e^{2x}$

(c) $\dfrac{d^2y}{dx^2} + \dfrac{dy}{dx} + y = 1 + x$

Solution

Note: In each case only the complementary function is being determined, not the general solution.

(a) Consider $\dfrac{d^2y}{dx^2} - 6\dfrac{dy}{dx} + 5y = 0$ ∴ auxiliary equation

becomes $u^2 - 6u + 5 = 0$ ($b^2 - 4ac > 0$)

$(u - 5)(u - 1) = 0$

Therefore, either $u - 5 = 0$ or $u - 1 = 0$, so $u = 5$ or 1.

The roots of the auxiliary equation are 5 or 1. Therefore, $\alpha = 5$ and $\beta = 1$ and CF $y = Ae^{5x} + Be^x$.

(b) Consider $\dfrac{d^2y}{dx^2} - 2\dfrac{dy}{dx} + y = 0$ ∴ auxiliary equation

becomes $u^2 - 2u + 1 = 0$ ($b^2 - 4ac = 0$)

$(u - 1)(u - 1) = 0$ ie $(u - 1)^2 = 0$

The roots of the auxiliary equation are equal. Therefore, $\alpha = \beta = 1$ and CF $y = e^x(A + Bx)$

FURTHER CALCULUS — PURE MATHEMATICS

(c) Consider $\dfrac{d^2y}{dx^2} + \dfrac{dy}{dx} + y = 0$ ∴ auxiliary equation

becomes $u^2 + u + 1 = 0$ $(b^2 - 4ac < 0)$

$$\therefore u = \frac{-1 \pm \sqrt{1-4}}{2}$$

$$= \frac{-1 \pm \sqrt{-3}}{2}$$

$$\therefore u = \frac{-1 \pm \sqrt{3(i^2)}}{2}$$

$$= \frac{-1 \pm \sqrt{3}\, i}{2}$$

The roots of the auxiliary equation are the complex conjugates $\dfrac{-1 + \sqrt{3}\,i}{2}, \dfrac{-1 - \sqrt{3}\,i}{2}$.

Therefore, $p = -\dfrac{1}{2}$, $q = \dfrac{\sqrt{3}}{2}$

$$\therefore \quad CF\ y = Ae^{-\frac{1}{2}x} \cos\left(\frac{\sqrt{3}}{2}x + \theta\right)$$

(b) *Particular integral*

The particular integral is another solution of the differential equation and it depends on the nature of f(x). It is found by substituting trial solutions into the differential equation until the correct one is found. This sounds rather hazardous but in general the nature of the trial solution is not difficult to determine.

f(x)	Trial solution
e^{kx}	try λe^{kx}
a constant	try λ
$px + q$	try $\lambda x + \mu$
$px^2 + qx + r$	try $\lambda x^2 + \mu x + \delta$
$p \sin x$ $p \cos x$ $p \sin x + q \sin x$	try $\lambda \sin x + \mu \cos x$

where k, p, q, r and k, λ, μ, δ are all constants.

PURE MATHEMATICS — FURTHER CALCULUS

Example 32

Find the general solution for each of the following differential equations (from Example 31 where the complementary functions were determined):

(a) $\dfrac{d^2y}{dx^2} - 6\dfrac{dy}{dx} + 5y = 3$

(b) $\dfrac{d^2y}{dx^2} - 2\dfrac{dy}{dx} + y = e^{2x}$

(c) $\dfrac{d^2y}{dx^2} + \dfrac{dy}{dx} + y = 1 + x$

Solution

(a) To find the particular integral try $y = \lambda$ (since $f(x)$ = constant).

$\therefore \quad y = \lambda$
$\dfrac{dy}{dx} = 0$
and $\dfrac{d^2y}{dx^2} = 0$

replacing in the original equation $\dfrac{d^2y}{dx^2} - 6\dfrac{dy}{dx} + 5y = 3$ becomes

$0 - 6(0) + 5\lambda = 3 \therefore \lambda = \dfrac{3}{5}$

Therefore, PI $y = \dfrac{3}{5}$ and CF $y = Ae^{5x} + Be^{x}$ giving

a general solution $y = \dfrac{3}{5} + Ae^{5x} + Be^{x}$

(b) To find the particular integral try $y = \lambda e^{2x}$ (since $f(x) = e^{2x}$).

$y = \lambda e^{2x}$
$\dfrac{dy}{dx} = 2\lambda e^{2x}$
$\dfrac{d^2y}{dx^2} = 4\lambda e^{2x}$

replacing in the original equation $\dfrac{d^2y}{dx^2} - 2\dfrac{dy}{dx} + y = e^{2x}$

$\therefore \quad 4\lambda e^{2x} - 2(2\lambda e^{2x}) + \lambda e^{2x} = e^{2x}$

$\therefore \quad \lambda e^{2x} = e^{2x}$

$\therefore \quad \lambda = 1$

Therefore, PI $y = e^{2x}$ and CF $y = e^{x}(A + Bx)$ giving a general solution $y = e^{2x} + e^{x}(A + Bx)$.

(c) To find the particular integral try $y = \lambda x + \mu$ (since $f(x) = 1 + x$).

$y = \lambda x + \mu$
$\dfrac{dy}{dx} = \lambda$
$\dfrac{d^2y}{dx^2} = 0$

replacing in the original equation $\dfrac{d^2y}{dx^2} + \dfrac{dy}{dx} + y = 1 + x$

$\therefore \quad 0 + \lambda + (\lambda x + \mu) = 1 + x$

$$\lambda x + (\lambda + \mu) = 1 + x$$

comparing like terms:

$$\lambda x = x \quad \therefore \lambda = 1$$

$$\lambda + \mu = 1 \quad \therefore \mu = 0$$

Therefore PI $y = x$ and CF $y = Ae^{-\frac{1}{2}x} \cos\left(\frac{\sqrt{3}}{2}x + \theta\right)$, giving a general solution

$$y = x + Ae^{-\frac{1}{2}x} \cos\left(\frac{\sqrt{3}}{2}x + \theta\right).$$

All these solutions have contained arbitrary constants (A, B, θ) which are contained in the complementary function. In order to evaluate these, some extra facts need to be given in the question.

Example 33

Find the general solution of the differential equation

$$\frac{d^2y}{dx^2} + 8\frac{dy}{dx} + 25y = 48\cos x - 16\sin x.$$

Find the particular solution for which $y = 8$, $\frac{dy}{dx} = -27$ when $x = 0$.

Solution

To determine the CF put $\frac{d^2y}{dx^2} + 8\frac{dy}{dx} + 25y = 0$. Therefore, the auxiliary equation becomes:

$$u^2 + 8u + 25 = 0 \qquad (b^2 - 4ac < 0)$$

$$\therefore \quad u = \frac{-8 \pm \sqrt{64 - 100}}{2}$$

$$= \frac{-8 \pm \sqrt{-36}}{2}$$

$$= \frac{-8 \pm \sqrt{36i^2}}{2}$$

$$= \frac{-8 \pm 6i}{2}$$

$$= -4 \pm 3i$$

The roots of the auxiliary equation are the complex conjugate $-4 + 3i$, $-4 - 3i$. Therefore, $p = -4$, $q = 3$.

$$\therefore \quad \text{CF} \quad y = Ae^{-4x}\cos(3x + \theta)$$

PURE MATHEMATICS — FURTHER CALCULUS

To determine the PI try $y = \lambda \sin x + \mu \cos x$

$$y = \lambda \sin x + \mu \cos x$$
$$\frac{dy}{dx} = \lambda \cos x - \mu \sin x$$
$$\frac{d^2y}{dx^2} = -\lambda \sin x - \mu \cos x$$

replacing in $\dfrac{d^2y}{dx^2} + 8\dfrac{dy}{dx} + 25y = 48\cos x - 16\sin x$

$\therefore\ (-\lambda \sin x - \mu \cos x) + 8(\lambda \cos x - \mu \sin x) + 25(\lambda \sin x + \mu \cos x) = 48\cos x - 16\sin x$

$\therefore\ -\lambda \sin x - \mu \cos x + 8\lambda \cos x - 8\mu \sin x + 25\lambda \sin x + 25\mu \cos x = 48\cos x - 16\sin x$

$\therefore\ (24\lambda - 8\mu)\sin x + (8\lambda + 24\mu)\cos x = 48\cos x - 16\sin x$

Comparing coefficients $\quad 24\lambda - 8\mu = -16 \quad \therefore \quad 3\lambda - \mu = -2 \quad$ ---------- (1)

and $\quad 8\lambda + 24\mu = 48 \quad \therefore \quad \lambda + 3\mu = 6 \quad$ ---------- (2)

$\therefore\ $ (1) x 3 : $\quad 9\lambda - 3\mu = -6$

(2) : $\quad \lambda + 3\mu = 6 \quad \Big\}$ add

$\quad\quad\quad 10\lambda = 0 \quad \therefore \quad \lambda = 0 \ \text{and}\ \mu = 2$

The PI is $y = 2\cos x$ and the general solution is

$y = 2\cos x + Ae^{-4x}\cos(3x + \theta)$

The particular solution comes from evaluating A and θ using $y = 8$, $\dfrac{dy}{dx} = -27$ when $x = 0$.

$$8 = 2\cos(0) + Ae^{-4(0)}\cos(3(0) + \theta)$$

$\therefore\ 8 = 2 + A\cos\theta$

$\therefore\ 6 = A\cos\theta$

$\therefore\ \cos\theta = \dfrac{6}{A} \quad$ ---------- (1)

$\dfrac{dy}{dx} = -2\sin x + (-4)Ae^{-4x}\cos(3x + \theta) + Ae^{-4x}(3)(-\sin(3x + \theta))$

$\quad\quad = -2\sin x - 4Ae^{-4x}\cos(3x + \theta) - 3Ae^{-4x}\sin(3x + \theta)$

$\therefore\ -27 = -2\sin(0) - 4Ae^0\cos(3(0) + \theta) - 3Ae^0\sin(3(0) + \theta)$

$\therefore\ -27 = -4A\cos\theta - 3A\sin\theta \quad$ ---------- (2) $\quad$ Replacing from (1)

$\therefore\ -27 = -4A\left(\dfrac{6}{A}\right) - 3A\sin\theta$

$\therefore\ -27 = -24 - 3A\sin\theta$

$\therefore\ -3 = -3A\sin\theta$

FURTHER CALCULUS — PURE MATHEMATICS

$\therefore \quad \sin\theta = \dfrac{-3}{-3A}$

$\qquad\qquad = \dfrac{1}{A}$

Squaring and adding:

$$\sin^2\theta + \cos^2\theta = \left(\dfrac{1}{A}\right)^2 + \left(\dfrac{6}{A}\right)^2 = 1$$

$\therefore \quad \dfrac{1}{A^2} + \dfrac{36}{A^2} = 1$

$\therefore \quad 1 + 36 = A^2$

$\therefore \quad 37 = A^2$

$\therefore \quad \sqrt{37} = A$

Also $\tan\theta = \dfrac{\sin\theta}{\cos\theta}$

$\qquad\qquad = \dfrac{\frac{1}{A}}{\frac{6}{A}}$

$\qquad\qquad = \dfrac{1}{6}$

Therefore the particular solution is:

$$y = 2\cos x + \sqrt{37}\, e^{-4x}\cos(3x + \theta) \qquad \text{where } \theta = \tan^{-1}\dfrac{1}{6}$$

---

f) *Example ('A' level question)*

(a) Solve the differential equation:

$$\dfrac{dy}{dx} + y\cot x = \sec^2 x \qquad \text{given that } y = \dfrac{2}{\sqrt{3}} \text{ when } x = \dfrac{\pi}{6}$$

(b) Find the possible values of m if $y = e^{mx}$ satisfies the differential equation:

$$3\dfrac{d^2y}{dx^2} - 4\dfrac{dy}{dx} + y = 0$$

(c) Solve the equation:

$$(1 + e^y)\dfrac{dy}{dx} = e^{2y}\cos^2 x \qquad \text{given } y = 0 \text{ when } x = 0$$

Solution

(a) $\dfrac{dy}{dx} + y\cot x = \sec^2 x \qquad \therefore \text{Integrating factor} = e^{\int \cot x\, dx} = e^{\int \frac{\cos x}{\sin x}\, dx}$

PURE MATHEMATICS — FURTHER CALCULUS

$$\therefore \text{Integrating factor} = e^{\log_e \sin x} = \sin x$$

$$\sin x \left(\frac{dy}{dx} + y \cot x \right) = \sin x (\sec^2 x)$$

$$\sin x \frac{dy}{dx} + y \sin x \cdot \frac{\cos x}{\sin x} = \sin x \sec^2 x$$

$$\sin x \frac{dy}{dx} + y \cos x = \sin x \sec^2 x$$

$$\therefore \quad \frac{d}{dx}(y \sin x) = \frac{\sin x}{\cos^2 x}$$

$$\therefore \quad \int \frac{d}{dx}(y \sin x)\,dx = \int \frac{\sin x}{\cos^2 x}\,dx$$

For $\int \frac{\sin x}{\cos^2 x}\,dx$: change variable

let $u = \cos x \quad \therefore \quad \frac{du}{dx} = -\sin x \quad \therefore \quad dx = -\frac{du}{\sin x}$

$$\therefore \quad \int \frac{\sin x}{\cos^2 x}\,dx = \int \left(\frac{\sin x}{u^2}\right) \cdot \left(\frac{-du}{\sin x}\right)$$

$$\therefore \quad \int \frac{d}{dx}(y \sin x)\,dx = \int -\frac{1}{u^2}\,du$$

$$\therefore \quad y \sin x = \frac{1}{u} + c$$

Replace variable: $\quad y \sin x = \frac{1}{\cos x} + c$

$$y \sin x = \sec x + c$$

But $y = \frac{2}{\sqrt{3}}, x = \frac{\pi}{6} \quad \therefore \quad \frac{2}{\sqrt{3}} \sin \frac{\pi}{6} = \sec \frac{\pi}{6} + c$

$$\frac{2}{\sqrt{3}} \left(\frac{1}{2}\right) = \frac{1}{\left(\frac{\sqrt{3}}{2}\right)} + c$$

$$\frac{1}{\sqrt{3}} = \frac{2}{\sqrt{3}} + c \quad \therefore \quad c = -\frac{1}{\sqrt{3}}$$

$$\therefore \quad y \sin x = \sec x - \frac{1}{\sqrt{3}}$$

Note: It is possible to do a change of variable on just the right hand side.

FURTHER CALCULUS PURE MATHEMATICS

(b) $\quad y = e^{mx} \quad \therefore \quad \dfrac{dy}{dx} = m\,e^{mx} \quad \dfrac{d^2y}{dx^2} = m^2 e^{mx}$

$\therefore \quad 3\dfrac{d^2y}{dx^2} - 4\dfrac{dy}{dx} + y = 0 \quad \text{becomes}$

$3(m^2 e^{mx}) - 4(m e^{mx}) + e^{mx} = 0$

$e^{mx}(3m^2 - 4m + 1) = 0$

$e^{mx}(3m - 1)(m - 1) = 0$

either $e^{mx} = 0$ or $3m - 1 = 0$ or $m - 1 = 0$

- impossible $\therefore 3m = 1$ $\therefore m = 1$

$\therefore$ no solution $\therefore m = \dfrac{1}{3}$

Therefore, the solutions are $m = \dfrac{1}{3}$ or 1.

(c) $\quad (1 + e^y)\dfrac{dy}{dx} = e^{2y}\cos^2 x \quad$ - variables separable

$\dfrac{(1 + e^y)}{e^{2y}} \dfrac{dy}{dx} = \cos^2 x$

$\therefore \quad \left(\dfrac{1}{e^{2y}} + \dfrac{e^y}{e^{2y}}\right) dy = \cos^2 x\, dx$

$\therefore \quad \int (e^{-2y} + e^{-y})\, dy = \int \cos^2 x\, dx \quad \text{but } \cos^2 x = \dfrac{\cos 2x + 1}{2}$

$\dfrac{e^{-2y}}{-2} + \dfrac{e^{-y}}{-1} = \int \left(\dfrac{\cos 2x}{2} + \dfrac{1}{2}\right) dx$

$-\dfrac{1}{2e^{2y}} - \dfrac{1}{e^y} = \dfrac{\sin 2x}{4} + \dfrac{1}{2}x + c$

$x = 0,\ y = 0$

$-\dfrac{1}{2e^0} - \dfrac{1}{e^0} = \dfrac{\sin 0}{4} + \dfrac{1}{2}(0) + c$

$-\dfrac{1}{2} - 1 = 0 + 0 + c \quad\quad \therefore \quad c = -\dfrac{3}{2}$

$-\dfrac{1}{2}e^{-2y} - e^{-y} = \dfrac{\sin 2x}{4} + \dfrac{1}{2}x - \dfrac{3}{2} \quad$ (multiplying by 4 gives)

$-2e^{-2y} - 4e^{-y} = \sin 2x + 2x - 6 \quad$ (multiplying by e^{2y} gives)

$-2 - 4e^y = e^{2y}(\sin 2x + 2x - 6)$

or $\quad 0 = e^{2y}(\sin 2x + 2x - 6) + 4e^y + 2$

PURE MATHEMATICS — SET THEORY AND PROBABILITY

12 SET THEORY AND PROBABILITY

12.1 Set theory
12.2 Functions
12.3 Permutations and combinations
12.4 Probability

12.1 Set theory

a) *Sets*

A set is a well-defined collection of objects called elements or members of the set. The elements of a set may be named in a list or may be given by a descriptions, eg

the set of days of the week = {Saturday, Sunday, Monday, Tuesday, Wednesday, Thursday, Friday}

A set may be finite, ie containing a limited number of elements as in the example just given, or it may be infinite as in the set of whole numbers.

Example 1

List the following sets:

(a) A = the set of even numbers less than 10;

(b) B = the set of vowels of the alphabet;

(c) C - the set of colours of the rainbow;

(d) D = the set of positive numbers less than zero.

Solution

(a) A = {2, 4, 6, 8};

(b) B = {a, e, i, o, u};

(c) C = {red, orange, yellow, green, blue, indigo, violet};

(d) D = { } - this is known as the empty or null set as no positive numbes are less than O. An alternative symbol for the empty set is ϕ.

b) *Language of sets*

A number of symbols are used in this work to enable statements to be made as briefly and precisely as possible. Their meanings will now be explained:

(a) $x \, \varepsilon \, S$ means x belongs to, or is an element of, set S; and $x \notin S$ means x does not belong to, or is not an element of, set S, eg

Tuesday ε {days of the week}

April $\notin$ {days of the week}

(b) $A \subset B$ - means set A is contained in set B, ie every element of set A belongs to set B and A is called a subset of B, eg

A = {a, b}   B = {a, b, c}   then $A \subset B$

Note B is not a subset of A and B contains an element that is not in A.

SET THEORY AND PROBABILITY   PURE MATHEMATICS

(c) ξ - is called the universal set and it contains all the objects under consideration in a particular problem, eg

in a geometrical problem the universal set might be a set of points or a set of lines or a set of shapes or a set of solids, depending on the type of problem.

(d)   { } or ϕ - is called the empty or null set (see example 1).

(e)   A' - is called the complement of Set A and it contains all the elements of ζ which are not elements of A, eg

if ζ = {positive whole numbers} and A = {even numbers}, then A' = odd numbers.

Note: $\zeta' = \phi$ and $\phi' = \zeta$.

(f)   {x : x > 7} - this is an alternative way of defining a set and it reads as 'the set x such that x is greater than 7'.

Example 2

State whether the following statements are true or false:

(a)   4 ε {even integers};

(b)   {a} ε {a, b, c};

(c)   51 $\notin$ {prime numbers}

(d) $\phi \subset$ {3, 7, 11, 12};

(e)   {p : p is prime and p $\neq$ 2} $\subseteq$ {x : x is an odd integer}.

Solution

(a)   True. 4 is a number of the set of even integers.

(b)   False. This is a nonsense statement - it reads as '{a} is a member of {a, b, c}. The correct symbol would have been $\subset$ as {a} is a subset of {a, b, c}, ie {a} $\subset$ {a, b, c}.

(c)   True. 51 does not belong to the set of prime numbers as it is divisible by 3 and 17.

(d)   True. The empty set is regarded as a subset of every set.

(e)   True. The first set contains all the prime numbers except 2. Therefore, it contains only odd numbers and so is a subset of the second set.

---

c)   *Number systems*

The following number systems will be considered:

       N - the set of natural numbers (positive integers)
       Z - the set of all integers (positive, negative and zero)
       Q - the set of rational numbers
       R - the set of real numbers
       C - the set of complex numbers

where $N \subset Z \subset Q \subset R \subset C$.

PURE MATHEMATICS — SET THEORY AND PROBABILITY

(a) *N - the set of natural numbers*

Therefore, $N = \{1, 2, 3, ...\}$ and if two members of this set are added or multiplied the result will always be another member of N, eg $5 + 6 = 11$, $11 \, \varepsilon \, N$ and $5 \times 6 = 30$, $30 \, \varepsilon \, N$. However, this is not necessarily true for subtraction or division, eg

$$5 - 6 = -1, \; -1 \notin N, \quad 5 \div 6 = \frac{5}{6}, \; \frac{5}{6} \notin N$$

This means that certain equations cannot be solved in N, eg $x + 6 = 5$ and $5x = 6$.

(b) *Z - the set of all integers*

Therefore, $Z = \{0, \pm 1, \pm 2, \pm 3, ...\}$ and if two members of this set are added, multiplied or subtracted the result will always be another member of Z, eg

$$(-2) + 7 = 5, \; 5 \, \varepsilon \, Z; \quad (-2) \times 7 = -14, \; -14 \, \varepsilon \, Z$$

and $(-2) - 7 = -9, \; -9 \, \varepsilon \, Z$.

However, this is still not true for division, eg

$$-2 \div 7 = -\frac{2}{7}, \; -\frac{2}{7} \notin Z$$

So some equations are still insoluble in Z, eg $5x = 6$.

(c) *Q - the set of rational numbers*

Therefore, $Q = \{\frac{m}{n} : m \, \varepsilon \, Z, \, n \, \varepsilon \, Z, \, n \neq 0\}$, so Q is the set of all numbers that can be expressed in a quotient form, eg

$$\frac{2}{5}, \; \frac{7}{11}, \; \frac{9}{22}$$

This includes integers since 10 can be written as $\frac{10}{1}$ and -12 can be written as $-\frac{12}{1}$.

Now if two members of this set are added, multiplied, subtracted or divided the result will always be another member of Q, eg

$$\frac{2}{3} + \frac{1}{4} = \frac{11}{12}, \; \frac{11}{12} \, \varepsilon \, Q, \quad \frac{2}{3} \times \frac{1}{4} = \frac{2}{12} = \frac{1}{6}, \; \frac{1}{6} \, \varepsilon \, Q$$

$$\frac{2}{3} - \frac{1}{4} = \frac{5}{12}, \; \frac{5}{12} \, \varepsilon \, Q, \quad \frac{2}{3} \div \frac{1}{4} = \frac{2}{3} \times \frac{4}{1} = \frac{8}{3}, \; \frac{8}{3} \, \varepsilon \, Q$$

All linear equations can be solved in Q but not necessarily quadratic or higher order equations, eg $x^2 = 2$, $x = \pm \sqrt{2}$, $\pm \sqrt{2} \notin Q$.

(d) *R - the set of real numbers*

This is the set which is used for most of the work. (Think how many questions involving a variable, say x, have included the statement 'where x is real'). Therefore, set R contains all the rational numbers (as defined in Q) but in addition it contains irrational numbers such as $\sqrt{2}$, the mathematical constants e and π, etc which cannot be expressed in the form $\frac{m}{n}$. (Note: $\pi = \frac{22}{7}$ is only an approximation and is not the exact value of π.) Adding, subtracting, multiplying or dividing any two members of R will always result in another member of R. However, some equations are still insoluble in R, eg

$$x^2 = -7 \qquad \therefore \; x = \pm\sqrt{-7}, \; \pm\sqrt{-7} \notin R$$

SET THEORY AND PROBABILITY — PURE MATHEMATICS

(e) **C - *the set of complex numbers***

Therefore, C is an extension of the real number system so that equations such as $x^2 = -7$ can (at last) be solved, and it is defined as:

$$C = \{a + ib : a \in R \text{ and } b \in R\} \quad i^2 = -1$$

Any two members of C can be added, subtracted, multiplied or divided and the result will always be another member of C (see the section on complex numbers). All equations (at this level) can be solved in C.

Example 3

To which of the sets N, Z, Q, R do the following numbers belong:

(a) $\dfrac{3}{2}$;

(b) $\sqrt{7}$;

(c) -13 ;

(d) π ;

(e) $\log_{10} 2$;

(f) 0 ;

(g) 2.2 ;

(h) $\log_4 2$.

Solution

(a) $\dfrac{3}{2}$ (a rational number) $\in$ Q, R

(b) $\sqrt{7}$ (an irrational number) $\in$ R

(c) -13 (a negative integer) $\in$ Z, Q, R

(d) π (an irrational number) $\in$ R

(e) $\log_{10} 2$ (an irrational number) $\in$ R

(f) 0 (zero integer) $\in$ Z, Q, R

(g) $2.2 = 2\dfrac{1}{5}$ (a rational number) $\in$ Q, R

(h) $\log_4 2 = $ say x

$\therefore \quad 2 = 4x$, but $2 = \sqrt{4} = 4^{\frac{1}{2}}$

$\therefore \quad 2 = 4^{\frac{1}{2}}$, and $x = \dfrac{1}{2}$ (a rational number) $\in$ Q, R

PURE MATHEMATICS — SET THEORY AND PROBABILITY

Example 4

To which of the sets N, Z, Q, R, C do the roots of the following equations belong:

(a) $x^2 - 3x + 2 = 0$;

(b) $2x^2 + x + 5 = 0$;

(c) $\tan x = 1$ (x in radians).

Solution

(a) $x^2 - 3x + 2 = 0$

$\therefore \quad (x - 2)(x - 1) = 0$

$\therefore \quad$ either $x = 2$ or 1 so $x \in$ N, Z, Q, R, C

(b) $2x^2 + x + 5 = 0$

$$x = \frac{-1 \pm \sqrt{1^2 - 4(2)(5)}}{2(2)}$$

$$= \frac{-1 \pm \sqrt{-39}}{4}$$

$$= \frac{-1 \pm \sqrt{39 i^2}}{4}$$

$$= \frac{-1 \pm (\sqrt{39})i}{4} \quad \text{so } x \in C$$

(c) $\tan x = 1$

$\therefore \quad x = \frac{\pi}{4}, \frac{5\pi}{4}$, etc. so $x \in$ R, C

Note: $\frac{\pi}{4}, \frac{5\pi}{4}$ may look like rational numbers but remember π itself is an irrational number. Therefore, $\frac{\pi}{4}$ and $\frac{5\pi}{4}$ must be irrational numbers.

---

d) *Union of intersection of sets*

The union of two sets, A and B, is written as $A \cup B$ and is the set containing all elements which belong to A or B (or both), eg

if $\quad A = \{1, 2, 3, 4\}$

and $\quad B = \{1, 3, 5\}$

then $\quad A \cup B = \{1, 2, 3, 4, 5\}$

The intersection of two sets is written as $A \cap B$ and is the set containing all elements which belong to both A and B, eg

$\quad A \cap B = \{1, 3\}$

SET THEORY AND PROBABILITY   PURE MATHEMATICS

Venn diagrams can be used to show relationships between sets. The universal set is shown as a rectangle and other sets are shown in circles (or shapes) inside the rectangle, eg

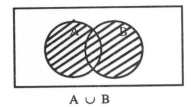
A ∪ B

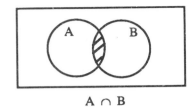
A ∩ B

If two sets X and Y are such that $X \subset Y$, then $X \cap Y = X$ and $X \cup Y = Y$.

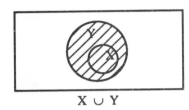
X ∪ Y

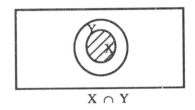
X ∩ Y

If two sets P and Q are such that they have no common elements then $P \cap Q = \phi$. Such sets are said to be disjoint.

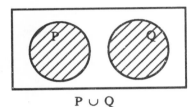
P ∪ Q

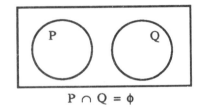
P ∩ Q = φ

Venn diagrams can be used to show the number of elements in a set (or sets).

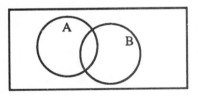

n (A) = number of elements in set A

n (B) = number of elements in set B

In the sum n (A) + n (B) the elements of A ∩ B have been counted twice (once in n (A) and again in n (B)).

∴   n (A ∪ B) = n (A) + n (B) - n (A ∩ B)

Example 5

Find P ∪ Q and P ∩ Q for the following:

(a)   P = {x : - 2 < x < 4}, Q = {x : 0 < x < 7} x ε R

(b)   P = {x : 2 ≤ x ≤ 9}, Q = {x : x ≤ 5} x ε R

PURE MATHEMATICS — SET THEORY AND PROBABILITY

Solution

(a) $P \cup Q = \{x : -2 < x < 7\}\ x \in R$

$P \cap Q = \{x : 0 < x < 4\}\ x \in R$

(b) $P \cup Q = \{x : x \leq 9\}\ x \in R$

$P \cap Q = \{x : 2 \leq x \leq 5\}\ x \in R$

Example 6

Draw Venn diagrams to show:

(a) $A' \cap B$;

(b) $(A \cup B) \cap C'$;

(c) $A' \cap B' \cap C'$.

Solution

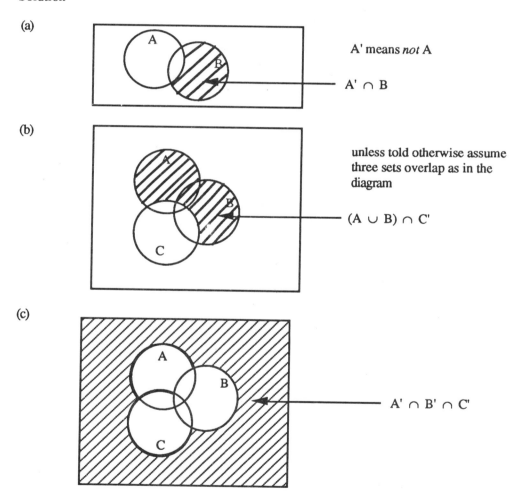

A' means *not* A

unless told otherwise assume three sets overlap as in the diagram

Example 7

In a survey of 100 children it was found that 47 have at least one brother, 58 have at least one sister and 32 children have both. How many children have no brothers or sisters?

Solution

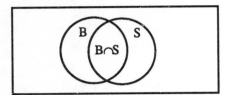

B = {at least one brother}

S = {at least one sister}

∴ n (B) = 47 and n (S) = 58

n (B ∩ S) = 32

∴ using "n (A ∪ B) = n (A) + n (B) - n (A ∩ B)"

$$n(B \cup S) = n(B) + n(S) - n(B \cap S)$$
$$= 47 + 58 - 32$$
$$= 105 - 32$$
$$= 73$$

∴ 73 children have at least one brother and/or sister

∴ 100 - 73 = 72 children have no brothers or sisters

In section 12.4 f page 591 this set notation will be related to probability theory.

PURE MATHEMATICS — SET THEORY AND PROBABILITY

12.2 Functions

a) *Functions as mappings*

A mapping (or function f) is a rule which assigns to an object x an image y. This can be written as $f : x \to y$ which means 'f maps x to y', eg

Consider a function f which maps each weekday onto its initial letter. This can be represented in a diagram as:

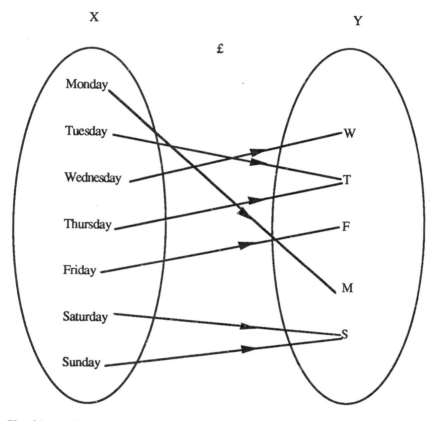

where X = {days of the week}

Y = {all letters of the alphabet}

Therefore, $f : X \to Y$ and set X is called the domain of the function, set Y is called the codomain and the set {W, T, F, M, S} is called the range of the function. So the range contains all the images of the elements in X and it is denoted by f(X).

Every element of the range belongs to the codomain. Therefore, $F(X) < Y$.

Also F is the image of Friday but T is the image of Tuesday and Thursday. This type of function is called a 'many to one mapping' because some elements of the range are the images of more than one element of the domain.

Now to take a more mathematical example. Let X = {1, 2, 3, 4} and Y = {1, 2, 3, 4, 5, 6, 7, 8} and each element (x) of X is mapped onto its double (2x) in Y.

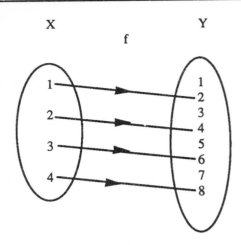

$\therefore$ $X = \{1, 2, 3, 4\}$ is the domain of the function

$Y = \{1, 2, 3, 4, 5, 6, 7, 8\}$ is the codomain of the function

$f(X) = \{2, 4, 6, 8\}$ is the range of the function

$\therefore$ $f(X) < Y$

This type of function is a 'one to one mapping' because every element of the range is the image of only one element of the domain.

The domain and codomain are usually taken to be R - the set of real numbers - and functions are defined by statements such as $f : x \to 2x - 5$. This reads as 'the function f which maps x to $2x - 5$' and furthermore this can be written using the function notation introduced at the beginning of the manual as $f(x) = 2x - 5$. So $f(x)$ is used to denote the image of x and is often called y. The image of $x = 4$, for instance, would be $f(4) = 2(4) - 5 = 3$.

Example 8

The domain of the function $g : x \to 4x - 3$ is $\{x : 0 \leq x \leq 2\}$ $x \, \varepsilon \, R$. Find the range of g.

Solution

$g : x \to 4x - 3$ $\therefore$ $g(x) = 4x - 3$ since $x \, \varepsilon \, R$

It can take any real value subject to $0 \leq x \leq 2$ (not only the whole values 0, 1, 2 but values such as 1.65, 0.32, etc.). Considering the limits of the domain $x = 0$ and $x = 2$, then $f(0) = 4(0) - 3 = -3$, and $f(2) = 4(2) - 3 = 5$. Therefore, the range of g is $\{x : -3 \leq x \leq 5\}$

Example 9

The domain of the function $h : x \to 5 - 2x^2$ is R. Find the range of h.

Solution

$h : x \to 5 - 2x^2$ $\therefore$ $h(x) = 5 - 2x^2$ since $x \, \varepsilon \, R$ the limits of the domain are $\pm \infty$

However, replacing either of these values in $f(x)$ will result in $-\infty$. This is because squaring $\pm \infty$ will give $+\infty$ each time and 5 minus $+\infty$ is going to be $-\infty$. Trying the mid value of R, ie $x = 0$ gives $f(0) = 5 - 2(0) = 5$, so the range of h is $\{x : -\infty \leq x \leq 5\}$.

Finding the range of a function is rather a trial and error process.

b) *Composite functions*

If $f: x \to x^2$ and $g: x \to x + 1$ where $x \in R$, then fg is a composite function and it means 'do g then f'.

$\therefore \quad g(x) = x + 1$

so $\quad fg(x) = f(x + 1)$

$\qquad\qquad\quad = (x + 1)^2$

$\qquad\qquad\quad = x^2 + 2x + 1$

$\therefore \quad fg: x \to x^2 + 2x + 1, x \in R$

gf is also a composite function and it means 'do f then g'. This gives

$\qquad f(x) = x^2$

then $\quad gf(x) = g(x^2)$

$\qquad\qquad\quad = x^2 + 1$

$\therefore \quad gf: x \to x^2 + 1, x \in R$

Note: fg is not the same as gf - the order of the functions is very important.

Example 10

Given $f: x \to x - 3$ and $g: x \to 2x$, express as single mappings:

(a) fg;

(b) gf;

(c) ff;

(d) gg;

(e) fgf.

Solution

(a) fg: $\quad g(x) = 2x$

$\qquad \therefore \quad fg(x) = f(2x)$

$\qquad\qquad\qquad = (2x) - 3$

$\qquad\qquad\qquad = 2x - 3$

(b) gf: $\quad f(x) = x - 3$

$\qquad \therefore \quad gf(x) = g(x - 3)$

$\qquad\qquad\qquad = 2(x - 3)$

$\qquad\qquad\qquad = 2x - 6$

(c) ff: $f(x) = x - 3$

∴ $ff(x) = f(x - 3)$

$= (x - 3) - 3$

$= x - 6$

(d) gg: $g(x) = 2x$

∴ $gg(x) = g(2x)$

$= 2(2x)$

$= 4x$

(e) fgf: $f(x) = x - 3$

∴ $gf(x) = g(x - 3)$

$= 2(x - 3)$

$= 2x - 6$

∴ $fgf(x) = f(2x - 6)$

$= (2x - 6) - 3$

$= 2x - 9$

These solutions have been written as functions but the example asked for mappings. This means that the functions should be written using the mapping notation thus:

(a) $fg : x \rightarrow 2x - 3$

(b) $gf : x \rightarrow 2x - 6$

(c) $ff : x \rightarrow x - 6$

(d) $gg : x \rightarrow 4x$

(e) $fgf : x \rightarrow 2x - 9$

Example 11

Given $f(x) = \frac{x}{2}$, $g(x) = 2x - 4$ and $h(x) = x + 3$, obtain expressions for:

(a) hfg;

(b) fgh;

(c) ghf.

Solution

(a) hfg: $g(x) = 2x - 4$

$\therefore fg(x) = f(2x - 4)$

$= \dfrac{2x - 4}{2}$

$= x - 2$

$\therefore hfg(x) = h(x - 2)$

$= (x - 2) + 3$

$= x + 1$

(b) fgh: $h(x) = x + 3$

$\therefore gh(x) = g(x + 3)$

$= 2(x + 3) - 4$

$= 2x + 2$

$\therefore fgh(x) = f(2x + 2)$

$= \dfrac{2x + 2}{2}$

$= x + 1$

(c) ghf: $f(x) = \dfrac{x}{2}$

$\therefore hf(x) = h\left(\dfrac{x}{2}\right)$

$= \dfrac{x}{2} + 3$

$= \dfrac{x + 6}{2}$

$\therefore ghf(x) = g\left(\dfrac{x + 6}{2}\right)$

$= 2\left(\dfrac{x + 6}{2}\right) - 4$

$= x + 2$

The answers have been left as functions since the question is given in terms of the function notation.

c) *Inverse functions*

If a relationship between x and y can be expressed in the two forms $y = f(x)$ and $x = g(y)$, then the function g is the inverse of the function f and vice versa. The usual notation for the inverse function is f^{-1}, ie

$f : x \rightarrow y$ $\qquad \therefore \qquad y = f(x)$ and $x = f^{-1}(y)$

Note: f and f^{-1} refer to the function and not to the variable. Therefore, both f and f^{-1} can be written as functions of x, eg

if $f : x \to 2x$, $x \in R$, ie $f(x) = 2x$, then the inverse function is

$$f^{-1}(x) = \frac{x}{2}$$

Showing this diagrammatically:

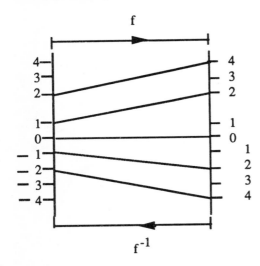

Not all functions have an inverse and it is essential that the mapping $f : x \to y$ is a one to one mapping as in the examples above. Also the domain of the inverse function is the range of the original function.

The simplest method of obtaining inverse functions is to express the original function as a formula giving y in terms of x and then rearrange it to give x in terms of y, eg

if $\quad f : x \to 2x \quad$ let $y = 2x$, then $x = \frac{y}{2}$

$\therefore \quad f^{-1} : y \to \frac{y}{2} \quad$ but this can equally well be written in term of x as the choice of letter is arbitrary.

Therefore, the inverse function can be written as:

$$f^{-1} : y \to \frac{y}{2} \text{ or } f^{-1} : x \to \frac{x}{2} \text{ or } f^{-1}(x) = \frac{x}{2}$$

and the last two forms will be used in most work.

Example 12

Given $f : x \to \dfrac{2x - 1}{2x + 2}$, find the inverse function $f^{-1}(x)$.

Solution

$$f : x \to \frac{2x - 1}{2x + 2} \qquad \therefore f(x) = \frac{2x - 1}{2x + 2}$$

$$\text{let} \qquad y = \frac{2x - 1}{2x + 2}$$

$\therefore \qquad (2x + 2)y = 2x - 1$

PURE MATHEMATICS   SET THEORY AND PROBABILITY

$\therefore$ $2xy + 2y = 2x - 1$

$\therefore$ $2y + 1 = 2x - 2xy$

$\therefore$ $2y + 1 = 2x(1 - y)$

$\therefore$ $\dfrac{2y + 1}{2(1 - y)} = x$

So $f^{-1} : y \to \dfrac{2y + 1}{1(1 - y)}$ but changing the choice of variable to x as usual

$f^{-1} : x \to \dfrac{2x + 1}{2(1 - x)}$ $\therefore f^{-1}(x) = \dfrac{2x + 1}{2(1 - x)}$

Just to give some examples of functions for which no inverse exists, eg

$f : x \to x^3 - x^2 + 1$ let $y = x^3 - x^2 + 1$

This cannot be rearranged to give x in terms of y, and therefore there is no inverse.

$f : x \to x^2$ $\therefore$ $1 \to 1^2$, ie $1 \to 1$

but $-1 \to (-1)^2$, ie $-1 \to 1$

This is not a one to one mapping and so no inverse exists unless the domain is restricted to the set of positive real numbes and zero - R+, 0, then:

$f \quad : x \to x^2$ and $x \in$ R+, 0   let $y = x^2$ $\therefore x = \sqrt{y}$

$f^{-1} \quad : x \to \sqrt{x}$ so $f^{-1}(x) = \sqrt{x}$

Inverses exist for composite functions and $(fg)^{-1} = g^{-1}f^{-1}$. Note the order is reversed.

To explain the reason for reversing the order consider the functions: put on your shoes and do up the laces. To cancel these out, the order would be: undo the laces and take off your shoes.

Example 13

If $f(x) = \dfrac{x}{2}$ and $g(x) = x - 5$, find:

(a) $(fg)^{-1}(x)$;

(b) $g^{-1}f^{-1}(x)$.

Solution

(a)   To find $fg(x)$ $g(x) = x - 5$

        $\therefore$ $fg(x) = f(x - 5)$

                                   $= \dfrac{x - 5}{2}$

        To find $(fg)^{-1}(x)$ let $y = \dfrac{x - 5}{2}$

        $\therefore$ $2y + 5 = x$

(replacing y with x as usual)

$$(fg)^{-1}(x) = 2x + 5$$

(b) To find $g^{-1}f^{-1}(x)$ $f(x) = \dfrac{x}{2}$

let $y = \dfrac{x}{2}$

∴ $2y = x$

replacing y with x gives $f^{-1}(x) = 2x$.

To find $g^{-1}f^{-1}(x)$ $g^{-1}f^{-1}(x) = g^{-1}(2x)$

let $2y = x - 5$

∴ $2y + 5 = x$

replacing y with x gives $g^{-1}f^{-1}(x) = 2x + 5$ - again

d) *Graphical representation*

Both a function and its inverse (if it exists) can be shown graphically.

Example 14

Draw on the same axes the graphs of the functions $f(x) = x^2$ and $f^{-1}(x)$ for $x \geq 0$.

Solution

Let $y = x^2$ ∴ $\sqrt{y} = x$

∴ $f^{-1}(x) = \sqrt{x}$

So the graphs that are required are $y = x^2$ and $y = \sqrt{x}$.

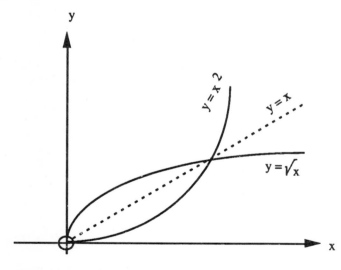

Note: The two graphs are symmetrical about $y = x$.

Example 15

Draw on the same axes the graphs of the function $f(x) = 10x$ and $f^{-1}(x)$.

Solution

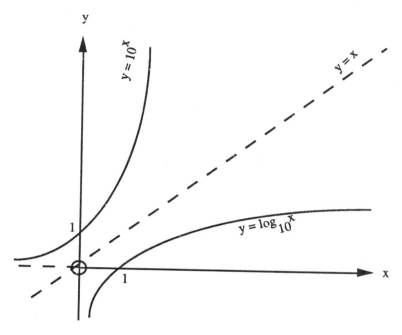

Note: The two graphs are again symmetrical about $y = x$.

When a function and its inverse are both expressed in terms of x (as in Examples 14 and 15) the resulting graph is symmetrical about the line $y = x$, ie the function and its inverse are reflections in the line $y = x$, ie if the point (a, b) lies on one of the curves the point (b, a) will lie on the other curve.

e)   *Transformations*

A mapping of a set of points in the (x, y) plane into itself is called a transformation. Transformations can be very complicated. However, the only ones that are required are $y = af(x)$, $y = f(x) + a$, $y = f(x - a)$, $y = f(ax)$ where a is a constant.

Taking each of these in turn and considering a worked example, the effect of the transformation will be seen.

(a) $y = af(x)$

Comparing $y = af(x)$ with the original function $y = f(x)$ it can be seen that the x values are unaltered but the y values are multiplied by the factor a, ie (x, y) is mapped onto (x, ay).

Example 16

What is the effect of the transformation $y = 4f(x)$ on the function $f(x) = x^2 - 3$?

Solution

Let $y = f(x)$ ∴ $y = x^2 - 3$

and $y' = 4f(x)$ ∴ $y' = 4(x^2 - 3)$

let	x = 0	∴ y = -3 and y' = -12
∴	(0, -3)	is mapped on to (0, -12)
let	x = 5	∴ y = 22 and y' = 88
∴	(5, 22)	is mapped on to (5, 88)

— arbitrary choice of value for x
— any value could have been used

In general terms (x, y) is mapped on to (x, 4y).

(b) $y = f(x) + a$

Comparing $y = f(x) + a$ with the original function $y = f(x)$, it can be seen that the x values are unaltered but the y values are increased by a, ie (x, y) is mapped on to (x, y + a).

Example 17

What is the effect of the transformation $y = f(x) + 4$ on the function $f(x) = x^2 - 3$?

Solution

let $y = f(x)$ ∴ $y = x^2 - 3$

and $y' = f(x) + 4$ ∴ $y' = (x^2 - 3) + 4 = x^2 + 1$

let   x = 0   ∴   y = -3 and y' = 1

∴   (0, -3) is mapped on to (0, 1)

let   x = 5   ∴   y = 22 and y' = 26

∴   (5, 22) is mapped on to (5, 26)

In general terms (x, y) is mapped on to (x, y + 4).

(c) $y = f(x - a)$

Comparing $y = f(x - a)$ with the original $y = f(x)$, if the x values in $f(x - a)$ are increased by a the y values will remain unaltered.

Example 18

What is the effect of the transformation $y = f(x - 4)$ on the function $f(x) = x^2 - 3$?

Solution

let $y = f(x)$ ∴ $y = x^2 - 3$

and $y' = f(x - 4)$ ∴ $y' = (x - 4)^2 - 3$

when   x = 0, y = -3   but   if x = 4, y' = -3

∴   (0, -3) is mapped on to (4, -3)

when   x = 5, y = 22   but   if x = 9, y' = 22

∴   (5, 22) is mapped on to (9, 22)

In general terms (x, y) is mapped on to (x + a, y).

(d) $y = f(ax)$

Comparing $y = f(ax)$ with the original $y = f(x)$, if the x values in f(ax) are divided by a the y values will remain unaltered.

Example 19

What is the effect of the transformation $y = f(4x)$ on the function $f(x) = x^2 - 3$?

Solution

let $y = f(x)$ $\therefore$ $y = x^2 - 3$

and $y' = f(4x)$ $\therefore$ $y' = (4x)^2 - 3$

when $x = 1, y = -2$ but if $x = \frac{1}{4}, y' = -2$

$\therefore$ $(1, -2)$ is mapped on to $\left(\frac{1}{4}, -2\right)$

($x = 0$ was not used this time as $4x = 0$)

when $x = 5, y = 22$ but if $x = \frac{5}{4}, y' = 22$

$\therefore$ $(5, 22)$ is mapped on to $\left(\frac{5}{4}, 22\right)$

In general terms (x, y) is mapped on to $\left(\frac{x}{4}, y\right)$.

12.3 Permutations and combinations

a)   *Arrangements*

This section is concerned with problems involving arrangements and selections of items such as letters, numbers, books, etc.

For example, the letters A, B and C can be arranged in six different ways:

     ABC - ACB - BAC - BCA - CAB - CBA

because any of the three letters can occupy the first place, either of the remaining two letters the second place, leaving one letter for the third place. Therefore, number of arrangements = 3 x 2 x 1 = 6.

The following examples will illustrate the method of approach to these types of problems; more definite rules will be given later.

Example 20

In how many ways can four letters of the word BRIDGE be arranged in a row, if no letter is repeated?

Solution

Any of the six letters can be used in the first place, leaving five for the second place, four for the third place and three for the last place.
Number of arrangements = 6 x 5 x 4 x 3 = 360.

SET THEORY AND PROBABILITY — PURE MATHEMATICS

Example 21

A man who works a five day week can travel to work on foot, by cycle or by bus. In how many ways can he arrange a week's travelling to work?

Solution

There are no restrictions on the ways the man can travel to work, so on Monday he can travel on foot, by cycle or by bus, ie in any of the three ways. The same is true for the other days of the week. Number of arrangements = 3 x 3 x 3 x 3 x 3 = 243.

Example 22

How many five figure odd numbers can be made from he digits 1, 2, 3, 4, 5 if no digit is repeated?

Solution

Since the numbers must be odd they must end in 1 or 3 or 5.

Assuming the number ends in 1, that leaves a choice of four numbers (2, 3, 4 or 5) for the first digit, three for the second digit, two for the third digit, one for the fourth digit and one for the fifth digit. Number of arrangements = 4 x 3 x 2 x 1 x 1 = 24.

Assuming the number ends in 3, the number of arrangements must be the same as above, ie
4 x 3 x 2 x 1 x 1 = 24.

And, assuming the number ends in 5, the same reasoning as above will apply, giving
4 x 3 x 2 x 1 x 1 = 24 arrangements.

Therefore, total number of arrangements = 24 + 24 + 24 = 72.

Example 23

There are 15 books in three different sizes - five of each. In how many ways can they be arranged on a shelf if books of the same size must be kept together?

Solution

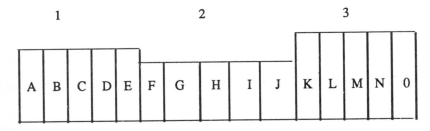

Books ABCDE form Group 1, FGHIJ form Group 2, KLMNO form Group 3.

As the books stand on the shelf in the diagram Group 1 books can be arranged in 5 x 4 x 3 x 2 x 1 = 120 ways amongst themselves because any of the five (ABCD or E) can occupy the first place, leaving a choice of four for the second place, three for the third place, two for the fourth place and one for the fifth place.

Each group can be similarly rearranged amongst itself; so for Group 2, number of arrangements = 120 and for Group 3, number of arrangements = 120.

Also any one of the 120 arrangements of Group 1 books can go with any one of the 120 arrangements of Group 2 books can go with any one of the 120 arrangements of Group 3 books.

Therefore, number of arrangements = 120 x 120 x 120 = 1,728,000.

BUT the three groups of books themselves can also be arranged in six ways, ie 123 or 132 or 213 or 231 or 312 or 321.

Therefore, total number of arrangements = 1,728,000 x 6 = 10,368,000.

b) *Factorial notation*

Factorials have already been introduced in earlier work on the binomial theorem. They have been included again at this stage as they will be used extensively in the following sections.

eg 6! (pronounced six factorial) means 6 x 5 x 4 x 3 x 2 x 1 and equals 720.

They are so important that many calculators have the factorial function included - usually as x!

The following examples will show how the factorial notation can be used and interpreted.

Example 24

Evaluate:

(a) 5!

(b) $\dfrac{12!}{9!}$

(c) $\dfrac{10!}{2!3!5!}$

Solution

(a) 5! = 5 x 4 x 3 x 2 x 1

= 120

(b) $\dfrac{12!}{9!} = \dfrac{12 \times 11 \times 10 \times 9 \times 8 \times 7 \times 6 \times 5 \times 4 \times 3 \times 2 \times 1}{9 \times 8 \times 7 \times 6 \times 5 \times 4 \times 3 \times 2 \times 1}$

= 12 x 11 x 10

= 1,320

or $\dfrac{12!}{9!} = \dfrac{497,001,600}{362,880}$

= 1,320 by calculator

(c) $\dfrac{10!}{2!3!5!} = \dfrac{10 \times 9 \times 8 \times 7 \times 6 \times 5 \times 4 \times 3 \times 2 \times 1}{(2 \times 1) \times (3 \times 2 \times 1) \times (5 \times 4 \times 3 \times 2 \times 1)}$

$= \dfrac{10 \times 9 \times 8 \times 7}{2}$

= 2,520

or $\dfrac{10!}{2!3!5!} = \dfrac{3,628,000}{(2)(6)(120)}$

= 2,520 by calculator

SET THEORY AND PROBABILITY — PURE MATHEMATICS

Example 25

Express in factorial notation:

(a) $6 \times 5 \times 4$

(b) $12 \times 11 \times 10 \times 9$

(c) $\dfrac{7 \times 6 \times 5}{3 \times 2 \times 1}$

Solution

(a) $6 \times 5 \times 4$ is not a factorial as written so some rearrangement is needed:

$$6 \times 5 \times 4 = \frac{6 \times 5 \times 4 \times 3 \times 2 \times 1}{3 \times 2 \times 1}$$

$$= \frac{6!}{3!}$$

(b) $12 \times 11 \times 10 \times 9$ - again some extra work needed first:

$$12 \times 11 \times 10 \times 9 = \frac{12 \times 11 \times 10 \times 9 \times 8 \times 7 \times 6 \times 5 \times 4 \times 3 \times 2 \times 1}{8 \times 7 \times 6 \times 5 \times 4 \times 3 \times 2 \times 1}$$

$$= \frac{12!}{8!}$$

(c) $\dfrac{7 \times 6 \times 5}{3 \times 2 \times 1} = \dfrac{7 \times 6 \times 5}{3!}$

$$= \frac{7 \times 6 \times 5 \times 4 \times 3 \times 2 \times 1}{3! \times 4 \times 3 \times 2 \times 1}$$

$$= \frac{7!}{3!4!}$$

Example 26

Express in factorial notation:

(a) $n(n-1)(n-2)$

(b) $(n+2)(n+1)n$

(c) $\dfrac{2n(2n-1)}{2 \times 1}$

Solution

A more algebraic example but using the same methods.

(a) $n(n-1)(n-2) = \dfrac{n(n-1)(n-2)(n-3)(n-4) \ldots \times 3 \times 2 \times 1}{(n-3)(n-4) \ldots 3 \times 2 \times 1}$

$$= \frac{n!}{(n-3)!}$$

PURE MATHEMATICS — SET THEORY AND PROBABILITY

(b) $\quad (n+2)(n+1)n = \dfrac{(n+2)(n+1)n(n-1)(n-2) \ldots \times 3 \times 2 \times 1}{(n-1)(n-2) \ldots \times 3 \times 2 \times 1}$

$\qquad\qquad\qquad\qquad = \dfrac{(n+2)!}{(n-1)!}$

(c) $\quad \dfrac{2n(2n-1)}{2 \times 1} = \dfrac{2n(2n-1)}{2!}$

$\qquad\qquad\qquad = \dfrac{2n(2n-1)(2n-2)(2n-3) \ldots \times 3 \times 2 \times 1}{2!(2n-2)(2n-3) \ldots \times 3 \times 2 \times 1}$

$\qquad\qquad\qquad = \dfrac{(2n)!}{2!(2n-2)!}$

c) *Permutations*

Having dealt with arrangements in general terms it is now necessary to give more definite rules.

A permutation is an arrangement of a number of objects in a definite order.

For example, there are six permutations of the letters ABC because they can be arranged in six different ways: ABC - ACB - BAC - BCA - CAB - CBA.

There are 24 permutations of the letters CATS because they can be arranged in 24 different ways, ie

```
CATS  CAST  CTAS  CTSA  CSAT  CSTA
ACTS  ACST  ATCS  ATSC  ASCT  ASTC
TCAS  TCSA  TACS  TASC  TSCA  TSAC
SCAT  SCTA  SATC  SACT  STAC  STCA
```

So 3 objects can be arranged in $3 \times 2 \times 1$ $\qquad = 3!$ ways

4 objects can be arranged in $4 \times 3 \times 2 \times 1$ $\qquad = 4!$ ways

and n objects can be arranged in $n(n-1)(n-2) \ldots 2 \times 1$ $\; = n!$ ways

> So the number of permutations of n unlike objects is $^nP_n = n!$

It is not necessary to use all the objects and sometimes only a few of them are used.

For example, there are 60 permutations of any 3 cards chosen from 5 unlike cards - Ace, King, Queen, Jack, 10 - because the first card can be chosen in 5 ways, the second in 4 ways and the third in 3 ways.

Number of permutations $= 5 \times 4 \times 3$

$\qquad\qquad\qquad\qquad = \dfrac{5 \times 4 \times 3 \times 2 \times 1}{2 \times 1}$

$\qquad\qquad\qquad\qquad = \dfrac{5!}{2!}$

> So the number of permutations of r objects chosen from n unlike objects is $^nP_r = \dfrac{n!}{(n-r)!}$

Example 27

A code word consists of 3 letters followed by 2 digits. How many code words can be made if no letter or digit is repeated in any code word?

Solution

There are 26 letters to choose from, of which only 3 are required.

Therefore, $\quad {}^{26}P_3 = \dfrac{26!}{(26-3)!}$

$\qquad\qquad\qquad = \dfrac{26!}{23!}$

$\qquad\qquad\qquad = 15{,}600$

There are 10 digits to choose from - 0, 1, 2, 3, 4, 5, 6, 7, 8, 9 - of which only 2 are required.

Therefore, $\quad {}^{10}P_2 = \dfrac{10!}{(10-2)!}$

$\qquad\qquad\qquad = \dfrac{10!}{8!}$

$\qquad\qquad\qquad = 90$

Any one of the 15,600 permutations of letters can go with any one of the 90 permutations of numbers.

Therefore, total number of permutations $= 15{,}600 \times 90$

$\qquad\qquad\qquad\qquad\qquad\qquad\quad = 1{,}404{,}000$

The letters and numbers cannot be reversed and must stay in the order letters - numbers.

Example 28

7 boys and 2 girls are to sit together on a bench. In how many ways can they arrange themselves so that the girls do not sit next to each other?

Solution

It is usually easier to work out the number of ways that the children can be seated so that the girls do sit next to each other, and then subtract this from the total number of ways in which 9 children can be seated to obtain the number of ways in which the girls will not be together.

Total number of ways in which 9 children can be seated is ${}^9P_9 = 9!$

$\qquad\qquad\qquad\qquad\qquad\qquad\qquad\qquad\qquad = 362{,}880$

The number of ways in which the children can be seated so that the two girls are together is ${}^8P_8 = 8!$

$\qquad\qquad\qquad\qquad\qquad\qquad\qquad\qquad\qquad = 40{,}320$

because the two girls are treated as one person giving 8 'objects' to be arranged. However, the 2 girls can swop places and still be next to one another so the number of ways in which the two girls are together is
$2 \times 40{,}320 = 80{,}640$.

PURE MATHEMATICS — SET THEORY AND PROBABILITY

Therefore, number of ways in which the girls are not together is

362,880 - 80,640 = 282,240

There are two special cases where the number of permutations has to be worked out even more carefully. These are:

(a) If the objects are placed in a circle, eg if 5 people are sitting at a round table and then all move one place round the table, the arrangement of people is still the same, ie the position of the people relative to the table is unimportant: it is the position of the people relative to one another that matters.

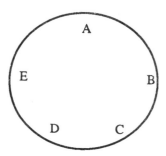 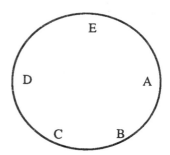

Therefore, one person is always regarded as being fixed and the number of permutations of 5 people sitting at a round table is

$$^4P_4 = 4! = 24$$

(b) If the objects are *not* all different, ie some are alike - for example, if the letters of the word FOOD are rearranged, instead of the usual 4! permutations, there are $\frac{4!}{2!}$ ie 12 permutations.

To show this it is first necessary to distinguish between the two O's in some way, ie O_1 and O_2 say.

FO_1O_2D FO_1DO_2 FO_2O_1D FO_2DO_1 FDO_1O_2 FDO_2O_1

O_1FO_2D O_1FDO_2 O_1O_2FD O_1O_2DF O_1DFO_2 O_1DO_2F

O_2FO_1D O_2FDO_1 O_2O_1FD O_2O_1DF O_2DFO_1 O_2DO_1F

DFO_1O_2 DFO_2O_1 DO_1FO_2 DO_1O_2F DO_2O_1F DO_2FO_1

It would not really be possible to distinguish between the two O's, and so the above display of 24 permutations actually only contains 12 different ones - each of which is duplicated because of the effect of having two letters the same, ie

$FO_1O_2D = FO_2O_1D = FOOD$

$FO_1DO_2 = FO_2DO_1 = FOOD$ ... etc.

So the number of permutations of n objects when p are alike of one kind and q are alike of another kind is

$$^nP_n = \frac{n!}{p!q!}$$

So the rules for calculating the number of permutations have been given with examples. However, it must be stressed that each problem needs to be treated carefully as additional restrictions are often introduced.

SET THEORY AND PROBABILITY — PURE MATHEMATICS

Example 29

Find how many different integers can be formed from the digits 1, 2, 3, 4, 5, 6, when each digit can be used at most once in each integer.

Solution

Number of digits		Number of permutations	
Using one digit	$^6P_1 = \dfrac{6!}{(6-1)!} = \dfrac{6!}{5!} =$	6	
Using two digits	$^6P_2 = \dfrac{6!}{(6-2)!} = \dfrac{6!}{4!} =$	30	+
Using three digits	$^6P_3 = \dfrac{6!}{(6-3)!} = \dfrac{6!}{3!} =$	120	+
Using four digits	$^6P_4 = \dfrac{6!}{(6-4)!} = \dfrac{6!}{2!} =$	360	+
Using five digits	$^6P_5 = \dfrac{6!}{(6-5)!} = \dfrac{6!}{1!} =$	720	+
Using six digits	$^6P_6 = \dfrac{6!}{(6-6)!} = \dfrac{6!}{0!} =$	720	(remember $0! = 1$)

$\therefore$ Total number of integers formed $= 1{,}965$

Example 30

Given that n is such that the number of permutations of n different things taken 5 at a time is three times the number of permutations of n + 1 different things taken 4 at a time, find n.

Solution

Number of permutations of n different things taken 5 at a time is

$$^nP_5 = \frac{n!}{(n-5)!} \qquad \text{------------ (1)}$$

Number of permutations of n + 1 different things taken 4 at a time is

$$^{n+1}P_4 = \frac{(n+1)!}{[(n+1)-4]!} \qquad \text{------------ (2)}$$

But it is given that: (1) = 3 x (2)

$\therefore \quad \dfrac{n!}{(n-5)!} = 3\dfrac{(n+1)!}{(n+3)!}$

$\dfrac{n(n-1)(n-2)(n-3)(n-4)(n-5)!}{(n-5)!} = \dfrac{3 \cdot (n+1)n(n-1)(n-2)(n-3)!}{(n-3)!}$

$\therefore \quad n(n-1)(n-2)(n-3)(n-4) = 3(n+1)n(n-1)(n-2)$

$\therefore \quad (n-3)(n-4) = 3(n+1)$

$\qquad n^2 - 3n - 4n + 12 = 3n + 3$

PURE MATHEMATICS — SET THEORY AND PROBABILITY

$$n^2 - 10n + 9 = 0$$
$$(n-1)(n-9) = 0$$

∴ Either $n - 1 = 0$ or $n - 9 = 0$
$n = 1$ $\qquad n = 9$

$n = 1$ is impossible.   Therefore, $n = 9$ is the required answer.

---

d) *Combinations*

Combinations differ from permutations in that the actual order of the objects is disregarded.

For example, there are 6 different permutations of the letters ABC but they are all the same combinations.

Similarly there are 24 different permutations of the letters CATS but they are all the same combination.

So the number of combinations of n unlike objects taken together is always 1.

However, the number of combinations of r objects chosen from n unlike objects

$$^nC_r = \frac{^nP_r}{r!} = \frac{n!}{(n-r)!r!}$$

The reason that the number of combinations is obtained from the number of permutations ÷ r! is that r objects can be arranged amongst themselves in r! different ways making r! different permutations but only one combination each time. The next example may help to clarify the link between permutations and combinations.

Example 31

In how many ways can 4 letters be chosen from 6 different letters?

Solution

Say the 6 different letters are ABCDEF, then the number of combinations

$$^6C_4 = \frac{6!}{(6-4)(!4!)}$$

$$= \frac{6!}{2!4!}$$

$$= 15$$

ie
```
ABCD   ABCE   ABCF
ABDE   ABDF   ABEF
ACDE   ACDF   ACEF
ADEF   BCDE   BCDF
BCEF   BDEF   CDEF
```

Each group of 4 letters could be arranged amongst itself in $^4P_4 = 4! = 24$ ways so the total number of permutations would be 15 x 24 = 360.

SET THEORY AND PROBABILITY — PURE MATHEMATICS

or, using $^6P_4 = \dfrac{6!}{(6-4)!}$

$\qquad\qquad\quad = \dfrac{6!}{2!}$

$\qquad\qquad\quad = \dfrac{720}{2}$

$\qquad\qquad\quad = 360$

$\therefore \quad \dfrac{^6P_4}{4!} = \dfrac{360}{4!}$

$\qquad\qquad = \dfrac{360}{24}$

$\qquad\qquad = 15$ - the number of combinations.

Example 32

A committee of 11 people is to be chosen from 6 men and 8 women. Find the number of ways in which the committee can be selected if it is to contain:

(a) exactly 4 men;

(b) at least 4 men;

(c) one particular man and one particular woman.

Solution

This is a combinations question since it is the number of ways in which the committee can be formed that is of interest and not the order in which the people are chosen:

(a) 'exactly 4 men' must mean 4 out of 6 men and 7 out of 8 women are chosen. These can be selected in:

$\qquad ^6C_4 \times {}^8C_7 = \dfrac{6!}{(6-4)!4!} \times \dfrac{8!}{(8-7)!7!}$

$\qquad\qquad\qquad\; = \dfrac{6!}{2!4!} \times \dfrac{8!}{1!7!}$

$\qquad\qquad\qquad\; = 120$ ways

(b) 'at least 4 men' means 4 men and 7 women (as in (a)) or 5 men and 6 women or 6 men and 5 women.

5 men and 6 women can be selected in

$\qquad ^6C_5 \times {}^8C_6 = \dfrac{6!}{(6-5)!5!} \times \dfrac{8!}{(8-6)!6!}$

$\qquad\qquad\qquad\; = \dfrac{6!}{1!5!} \times \dfrac{8!}{2!6!}$

$\qquad\qquad\qquad\; = 168$ ways

PURE MATHEMATICS — SET THEORY AND PROBABILITY

6 men and 5 women can be selected in

$$^6C_6 \times {}^8C_5 = \frac{6!}{(6-6)!6!} \times \frac{8!}{(8-5)!5!}$$

$$= \frac{6!}{0!6!} \times \frac{8!}{3!5!}$$

$$= 56 \text{ ways}$$

Therefore, total number of ways $= 120 + 168 + 56 = 344$ ways

(c) 'one particular man and one particular woman' means these two with any 9 of the remaining 12 people. These can be selected in:

$$^2C_2 \times {}^{12}C_9 = \frac{2!}{(2-2)!2!} \times \frac{12!}{(12-9)!9!}$$

$$= \frac{2!}{0!2!} \times \frac{12!}{3!9!}$$

$$= 220 \text{ ways}$$

Example 33

A group of 7 letters is to be found from the first 13 letters A, B, C, ... L, M of the alphabet. Of the possible number of ways of forming the group find how many:

(a) contain at least 2 but not more than 4 of the letters A, B, C, D, E;

(b) contain exactly 4 consecutive letters.

Solution

ABCDEFGHIJKLM

(a) Dividing the group of letters into two gives:

Group 1: ABCDE — 5 letters
Group 2: FGHIJKLM — 8 letters

So the number of possible combinations of 7 letters is:

Group 1	*Group 2*		
2 (of 5)	and 5 (of 8)	$^5C_2 \times {}^8C_5 = \frac{5!}{(5-2)!2!} \times \frac{8!}{(8-5)!5!}$	$= 560$
or			+
3 (of 5)	and 4 (of 8)	$^5C_3 \times {}^8C_4 = \frac{5!}{(5-3)!3!} \times \frac{8!}{(8-4)!4!}$	$= 700$
or			+
4 (of 5)	and 3 (of 8)	$^5C_4 \times {}^8C_3 = \frac{5!}{(5-4)!4!} \times \frac{8!}{(8-3)!3!}$	$= 280$

$\therefore$ Total number of combinations $= \underline{1,540}$ ways

(b) Consider first the consecutive letters ABCD, then these count as 4 of the 7 letters and so 3 more are required from the remaining 8 letters FGHIJKLM. Note E cannot be included as this would mean that some of the combinations would contain five or more consecutive letters.

So the number of possible combinations $= {}^4C_4 \times {}^8C_3$

$$= \frac{4!}{(4-4)!4!} \times \frac{8!}{(8-3)!3!}$$

$$= 56 \text{ ways}$$

Next consider the consecutive letters BCDE. These again count as 4 of the 7 letters and so 3 more are required from GHIJKLM. Note A and F have to be excluded this time or it would be possible to obtain combinations with five or more consecutive letters.

So the number of possible combinations $= {}^4C_4 \times {}^7C_3$

$$= \frac{4!}{(4-4)!4!} \frac{7!}{(7-3)!3!}$$

$$= 35 \text{ ways}$$

For the consecutive letters CDEF, DEFG, EFGH, FGHI, GHIJ, HIJK, IJKL there will also be 35 possible combinations each time but for JKLM there will be 56 combinations again (as for ABCD).

Therefore, total $= 56 + (8 \times 35) + 56$

$$= 392 \text{ ways}$$

e) *Example ('A' level question)*

Find the number of:

(a) combinations; and

(b) permutations

of 4 letters which can be taken from the word ELLIPSE.

Solution

Calculating the number of permutations and combinations is complicated because the word ELLIPSE contains repeat letters, ie 2 L's and 2 E's. Therefore, the working will be taken a stage at a time to allow for all possibilities.

PURE MATHEMATICS — SET THEORY AND PROBABILITY

Letters used	(a) No. of combinations	(b) No. of permutations	
LLEE	1	$\frac{4!}{2!2!}$	= 6
LLE and 1 of IPS	3	$\frac{4!}{2!} \times 3$	= 36
LL and 2 of IPS	3	$\frac{4!}{2!} \times 3$	= 36
LE and 2 of IPS	3	$4! \times 3$	= 72
L and 3 of IPS	1	$4!$	= 24
E and 3 of IPS	1	$4!$	= 24
EE and 2 of IPS	3	$\frac{4!}{2!} \times 3$	= 36
EEL and 1 of IPS	3	$\frac{4!}{2!} \times 3$	= 36
	18		**270**

There are other ways of arriving at these answers but this method has been used as it breaks the working down into logical steps.

12.4 Probability

a) *Definition*

When an ordinary unbiased dice is rolled there are 6 possible outcomes, ie that the number uppermost will be 1 or 2 or 3 or 4 or 5 or 6.

The probability that the number 5 is uppermost is $\frac{1}{6}$ because 1 out of the 6 outcomes is the number 5.

The probability that an even number is uppermost is $\frac{3}{6}$ $\left(\text{or } \frac{1}{2}\right)$ because 3 out of the 6 outcomes are even numbers.

The probability that a number greater than 6 is uppermost is $\frac{0}{6}$ (or 0) because 0 of the outcomes is a number greater than 6 - it is impossible.

The probability that a number less than 7 is uppermost is $\frac{6}{6}$ (or 1) because all 6 of the outcomes are numbers less than 7 - it is certain.

The above example illustrates how probabilities are worked out.

SET THEORY AND PROBABILITY   PURE MATHEMATICS

The actual definition of probability is:

Probability that an event occurs = $\dfrac{\text{Number of favourable outcomes}}{\text{Total number of possible outcomes}}$

Probability is measured on a scale from 0 to 1 where 0 represents impossibility and 1 represents certainty.

If p = probability that an event hapens
and q = probability that the same event does not happen
then p + q = 1   ∴ p = 1 - q or q = 1 - p

Example 34

Calculate the probability that when two dice are thrown the total score (ie the sum of the two numbers) is:

(a)   6

(b)   11

(c)   2

(d)   not 7.

Solution

When two dice are thrown there are 36 possible outcomes. They are:

1, 1	2, 1	3, 1	4, 1	5, 1	6, 1
1, 2	2, 2	3, 2	4, 2	5, 2	6, 2
1, 3	2, 3	3, 3	4, 3	5, 3	6, 3
1, 4	2, 4	3, 4	4, 4	5, 4	6, 4
1, 5	2, 5	3, 5	4, 5	5, 5	6, 5
1, 6	2, 6	3, 6	4, 6	5, 6	6, 6

So the total number of possible outcomes is 36, and these will be used to answer all the parts of this question.

(a)   Probability of 6, ie Pr(6):

　　　　1 + 5 = 6　　　2 + 4 = 6　　　3 + 3 = 6　　　4 + 2 = 6　　　5 + 1 = 6

　　　So the number of favourable outcomes　　= 5

　　　　　　　∴　Pr(6)　= $\dfrac{5}{36}$

(b)   Probability of 11, ie Pr(11):

　　　　5 + 6 = 11 and 6 + 5 = 11

　　　So the number of favourable outcomes　　= 2

　　　　　　　∴　Pr(11)　= $\dfrac{2}{36}$ = $\dfrac{1}{18}$

(c)   Probability of 2, ie Pr(2):

$1 + 1 = 2$ is the only favourable outcome

$$\therefore \quad Pr(2) = \frac{1}{36}$$

(d)   Probability of not 7, ie Pr(not 7):

It is often easier to work out the probability of 7, ie Pr(7) and then take it away from 1.

$1 + 6 = 7 \quad 2 + 5 = 7 \quad 3 + 4 = 7 \quad 4 + 3 = 7 \quad 5 + 2 = 7 \quad 6 + 1 = 7$

So $\quad Pr(7) = \frac{6}{36} \quad \therefore Pr(\text{not } 7) = 1 - \frac{6}{36} = \frac{30}{36} = \frac{5}{6}$

- this is using '$p + q = 1$' and '$q = 1 - p$'

Example 35

If the letters of TOGETHER are arranged at random, calculate the probability that the T's are apart.

Solution

This question requires the use of permutations in order to calculate the required probability.

The number of permutations using all 8 letters is $^8P_8 = \frac{8!}{2!2!} = 10{,}080$

Note: 8! has been divided by 2! twice because of the repeated letters, ie 2 T's and 2 E's.

The number of permutations in which the T's are together is $\frac{7!}{2!} = 2{,}520$.

Note: 7! not 8! this time because the 2 T's are counted as one 'letter' and this has been divided by 2! because of the repeated letter E.

The probability that the T's are together is given by $\frac{2{,}520}{10{,}080} = \frac{1}{4}$ because in 2,520 out of a total of 10,080 permutations the Ts are together.

$\therefore \quad Pr(\text{T's together}) = \frac{1}{4}$

so $\quad Pr(\text{T's apart}) = 1 - \frac{1}{4}$

$$= \frac{3}{4}$$

b)   *Addition law*

There are two main laws that govern the way in which probabilities are calculated; the addition law is the first of these two.

If A and B are mutually exclusive events then the probability that A or B occurs is given by:

$$\boxed{Pr(A \text{ or } B) = Pr(A) + Pr(B)}$$

where mutually exclusive events are events that cannot happen at the same time.

SET THEORY AND PROBABILITY   PURE MATHEMATICS

Example 36

A bag contains 4 red, 6 blue and 10 black balls. Calculate the probability of choosing a red or a black ball from the bag when one ball is removed.

Solution

$$4 \text{ red, } 6 \text{ blue, } 10 \text{ black} \quad \therefore \quad \text{Total} = 20 \text{ balls}$$

$$\text{Pr(red)} = \frac{4}{20} \quad \text{Pr(blue)} = \frac{6}{20} \quad \text{Pr(black)} = \frac{10}{20}$$

These are clearly mutually exclusive events since if the ball is red it cannot be black and vice versa.

$$\therefore \quad \text{Pr(red or black)} = \text{Pr(red)} + \text{Pr(black)}$$

$$= \frac{4}{20} + \frac{10}{20}$$

$$= \frac{14}{20} \text{ or } \frac{7}{10}$$

If the events are not mutually exclusive - ie they can occur at the same time - the law becomes:

$$\boxed{\text{Pr(A or B)} = \text{Pr(A)} + \text{Pr(B)} - \text{Pr(A and B)}}$$

Example 37

Calculate the probability of selecting a heart or a queen when 1 card is removed from a pack of playing cards.

Solution

These events are not mutually exclusive because a card can be a heart and a queen, ie the queen of hearts.

Total number of cards = 52

$$\therefore \quad \text{Pr(heart)} = \frac{13}{52} \quad \text{Pr(queen)} = \frac{4}{52} \quad \text{Pr(queen of hearts)} = \frac{1}{52}$$

$$\therefore \quad \text{Pr(heart or a queen)} = \text{Pr(heart)} + \text{Pr(queen)} - \text{Pr(queen of hearts)}$$

$$= \frac{13}{52} + \frac{4}{52} - \frac{1}{52} \quad = \frac{16}{52} \text{ or } \frac{4}{13}$$

The Pr(queen of hearts) is subtracted because that particular card has been counted twice, ie it has been included as a 'heart' and it has been included as a 'queen', so it is subtracted once from the working to remove the effect of the double counting.

c)   *Multiplication law*

This is the second main law of probability.

If A and B are independent events, then the probability of A and B occuring is given by:

$$\boxed{\text{Pr(A and B)} = \text{Pr(A)} \times \text{Pr(B)}}$$

where independent events are such that the occurrence of one does not affect the probability of the other happening.

PURE MATHEMATICS — SET THEORY AND PROBABILITY

Example 38

If three dice are thrown in succesion, calculate the probability of obtaining 1, 2 and 3 in that order.

Solution

They are independent events because the successive throws of the dice do not affect one another.

$$Pr(1) = \frac{1}{6} \qquad Pr(2) = \frac{1}{6} \qquad Pr(3) = \frac{1}{6}$$

$$\begin{aligned} Pr(1 \text{ and } 2 \text{ and } 3) &= Pr(1) \times Pr(2) \times Pr(3) \\ &= \frac{1}{6} \times \frac{1}{6} \times \frac{1}{6} \\ &= \frac{1}{216} \end{aligned}$$

If the events are not independent - ie the occurrence of the first event affects the probability of the second event happening - the law becomes:

$$\boxed{Pr(A \text{ and } B) = Pr(A) \times Pr(B/A)}$$

where $Pr(B/A)$ is the probability of B happening given that A has happened.

Example 39

A bag contains 3 black, 4 red and 13 blue marbles. Calculate the probability that if three are selected without replacement they will be red, blue and black in that order.

Solution

3 black, 4 red, 13 blue   Total = 20 marbles.

Since the first marble is *not* replaced before the second one is taken and the second one is *not* replaced before the third one is taken, the second and third probabilities are dependent ont he earlier events, ie

$$Pr(\text{first marble red}) = \frac{4}{20}$$

$$Pr(\text{second marble blue/red has been taken}) = \frac{13}{19}$$

$$Pr(\text{third marble black/red and blue have been taken}) = \frac{3}{18}$$

$$\begin{aligned} Pr(\text{red and blue and black}) &= Pr(\text{red}) \times Pr(\text{blue/red}) \times Pr(\text{black/red and blue}) \\ &= \frac{4}{20} \times \frac{13}{19} \times \frac{3}{18} \\ &= \frac{13}{570} \end{aligned}$$

Another useful point to remember is the 'at least one' rule. When several events are happening at the same time then:

$$\boxed{Pr(\text{at least one}) = 1 - Pr(\text{none})}$$

SET THEORY AND PROBABILITY PURE MATHEMATICS

This is because either none of the events occurs or at least one of them does:

∴   Pr(at least one) + Pr(none) = 1 : hence above result

Example 40

Calculate the probability of obtaining at least one 6 when 3 dice are thrown together.

Solution

These events will be independent.

So $\Pr(6) = \frac{1}{6}$ ... $\Pr(\text{not } 6) = 1 - \frac{1}{6} = \frac{5}{6}$

$$\Pr(\text{not 6 and not 6 and not 6}) = \Pr(\text{not 6}) \, \Pr(\text{not 6}) \, \Pr(\text{not 6})$$

$$= \frac{5}{6} \times \frac{5}{6} \times \frac{5}{6}$$

$$= \frac{125}{216}$$

∴   Pr(at least one 6) = 1 - Pr(no sixes)

$$= 1 - \frac{125}{216}$$

$$= \frac{91}{216}$$

These examples have all been fairly straightforward. However, the next two examples will show how several laws may have to be applied at the same time to answer a more complex problem.

Example 41

Two cards are drawn at random and without replacement from an ordinary pack of playing cards. Find the probability that:

(a)   at least one of the cards will be an ace or a face card (a king, queen or jack);

(b)   one of the cards will be an ace and the other a face card.

Solution

(a)   The probability that at least one of the cards will be an ace or a face card is easiest to calculate using:

Pr(at least one) = 1 - Pr(none)

A pack of cards contains 4 aces, 4 kings, 4 queens, 4 jacks. Total = 16.

$$\Pr(\text{ace or face card}) = \frac{16}{52}$$

∴ $\Pr(\text{not an ace or a face card}) = 1 - \frac{16}{52}$

$$= \frac{36}{52}$$

PURE MATHEMATICS — SET THEORY AND PROBABILITY

$\therefore$ Pr(neither of two cards is an ace or a face card) $= \dfrac{36}{52} \cdot \dfrac{35}{51}$

$= \dfrac{105}{221}$

$\therefore$ Pr(at least one) $= 1 - \dfrac{105}{221}$

$= \dfrac{116}{221}$

(b) The probability that one of the cards will be an ace and the other a face card is calculated as follows:

Pr(one ace and one face card) $=$ Pr(ace then face) or Pr(face then ace)

Pr(ace) $= \dfrac{4}{52}$ Pr(face) $= \dfrac{12}{52}$

$\therefore$ Pr(one ace and one face card) $= \dfrac{4}{52} \cdot \dfrac{12}{51} + \dfrac{12}{52} \cdot \dfrac{4}{51}$

$= \dfrac{4}{221} + \dfrac{4}{221}$

$= \dfrac{8}{221}$

Example 42

In an archery contest the probabilities that the archers A, B and C will hit the target with a single shot are independent and are $\dfrac{2}{3}$, $\dfrac{3}{4}$ and $\dfrac{4}{5}$ respectively.

If they each shoot an arrow simultaneously, find the probability that the target will be hit by:

(a) 3 arrows

(b) 0 arrows

(c) only 2 arrows

Solution

A: P(hits) $= \dfrac{2}{3}$

$\therefore$ P(misses) $= 1 - \dfrac{2}{3} = \dfrac{1}{3}$

B: P(hits) $= \dfrac{3}{4}$

$\therefore$ P(misses) $= 1 - \dfrac{3}{4} = \dfrac{1}{4}$

C: P(hits) $= \dfrac{4}{5}$

$\therefore$ P(misses) $= 1 - \dfrac{4}{5} = \dfrac{1}{5}$

(a) Pr(hit by 3 arrows) = Pr(A hits and B hits and C hits)

= Pr(A hits) Pr(B hits) Pr(C hits)

$= \frac{2}{3} \times \frac{3}{4} \times \frac{4}{5}$

$= \frac{2}{5}$

(b) Pr(hit by 0 arrows) = Pr(A misses and B misses and C misses)

= Pr(A misses) P(B misses) Pr(C misses)

$= \frac{1}{3} \times \frac{1}{4} \times \frac{1}{5}$

$= \frac{1}{60}$

(c) Pr(hit by 2 arrows):

This time there are three separate possibilities:

	A	B	C	Probability
1	hit	hit	miss	$\frac{2}{3} \times \frac{3}{4} \times \frac{1}{5} = \frac{6}{60}$
2	hit	miss	hit	$\frac{2}{3} \times \frac{1}{4} \times \frac{4}{5} = \frac{8}{60}$
3	miss	hit	hit	$\frac{1}{3} \times \frac{3}{4} \times \frac{4}{5} = \frac{12}{60}$

Pr(hit by 2 arrows) = Pr(1 or 2 or 3)

= Pr(1) + Pr(2) + Pr(3)

$= \frac{6}{60} + \frac{8}{60} + \frac{12}{60}$

$= \frac{26}{60} = \frac{13}{30}$

d) *Conditional probability*

When considering the law governing the calculation of probability for dependent events it was stated that:

Pr(A and B) = Pr(A) x Pr(B/A)

This same law can be used to calculate the conditional probability Pr(B/A), ie the probability that B happens given that A has already happened, ie

$Pr(B/A) = \frac{Pr(A \text{ and } B)}{Pr(A)}$

The next example will show this more clearly.

PURE MATHEMATICS — SET THEORY AND PROBABILITY

Example 43

A bag contains 5 white balls and 7 black balls. Three balls are drawn, one at a time without replacement. Calculate the probability that the three balls are black given that all three are the same colour.

Solution

Therefore, the conditional probability is:

$$\text{Pr(all three black/all three same colour)} = \frac{\text{Pr(all the same colour and black)}}{\text{Pr(all the same colour)}}$$

Pr(all the same colour and black) = Pr(black and black and black)

$$= \text{Pr(bl) Pr(bl) Pr(bl)}$$

$$= \frac{7}{12} \times \frac{6}{11} \times \frac{5}{10}$$

$$= \frac{210}{1{,}320}$$

Pr(all the same colour) = Pr(black, black, black or white, white, white)

$$= \text{Pr(bl) Pr(bl) Pr(bl)} + \text{Pr(wh) Pr(wh) Pr(wh)}$$

$$= \frac{210}{1{,}320} + \frac{5}{12} \times \frac{4}{11} \times \frac{3}{10}$$

$$= \frac{210}{1{,}320} + \frac{60}{1{,}320}$$

$$= \frac{270}{1{,}320}$$

$$\text{Pr(black/all the same colour)} = \frac{\frac{210}{1{,}320}}{\frac{270}{1{,}320}} = \frac{210}{270} = \frac{7}{9}$$

---

e) *Example ('A' level question)*

The set A consists of all those integers between 0 and 1,000 which can be formed in either of the following ways:

I — by using only digits chosen from 2, 4, 6, 8 and without a repeated digit

II — by using only digits chosen from 1, 3, 5, 7, 9, where each digit may be repeated as often as required.

Find the number of members in set A and state how many of these will be even.

One integer is to be selected at random from set A. Find the probability that the integer selected will be:

(a) even;

(b) less than 100;

(c) odd, given that the integer selected is less than 100.

Two integers are selected at random and without replacement from set A. Find the probability that the sum of the two selected integers will be even.

SET THEORY AND PROBABILITY   PURE MATHEMATICS

Solution

Since set A consists of integers between 0 and 1,000 only 1, 2 and 3 digit integers will be considered.

I   4 digits to choose from and all the integers will be even as they will end in 2 or 4 or 6 or 8.

Number of digits	Number of integers
1	$^4P_1 = \dfrac{4!}{(4-1)!} = 4$
2	$^4P_2 = \dfrac{4!}{(4-2)!} = 12$
3	$^4P_3 = \dfrac{4!}{(4-3)!} = 24$

∴   Total number of integers = 40

II   5 digits to choose from, all of which can be repeated, if necessary. All the integers so formed will be odd as they will end in 1 or 3 or 5 or 7 or 9.

Number of digits	Number of integers
1	$5 = 5$
2	$5 \times 5 = 5^2 = 25$
3	$5 \times 5 \times 5 = 5^3 = 125$

∴   Total number of integers = 155.

Therefore, set A contains 40 + 155 = 195 members, of which 40 are even.

(a) $\Pr(\text{even}) = \dfrac{40}{195}$

$= \dfrac{8}{39}$

(b)   Pr(less than 100): integer less than 100 will contain 1 or 2 digit so 4 + 12 + 5 + 25 = 46 such integers can be formed.

$\Pr(\text{less than 100}) = \dfrac{46}{195}$

(c)   Pr(odd/less than 100): a conditional probability, using:

$\Pr(\text{odd/less than 100}) = \dfrac{\Pr(\text{odd and less than 100})}{\Pr(\text{less than 100})}$

∴ $\Pr(\text{odd and less than 100}) = \dfrac{30}{195}$

$\Pr(\text{less than 100}) = \dfrac{46}{195}$

∴ Pr(odd/less than 100) = $\dfrac{\frac{30}{195}}{\frac{46}{195}}$

= $\dfrac{30}{46}$

= $\dfrac{15}{23}$

If the sum of the two integers is even then either both the integers are even or both are odd, because even + even = even, odd + odd = even.

∴ Pr(even and even or odd and odd) = Pr(even) Pr(even) + Pr(odd) Pr(odd)

= $\dfrac{40}{195} \cdot \dfrac{39}{194} + \dfrac{155}{195} \cdot \dfrac{154}{194}$

= $\dfrac{1,560}{37,830} + \dfrac{23,870}{37,830}$

= $\dfrac{25,430}{37,830}$

= 0.67

f) *Probability using set theory*

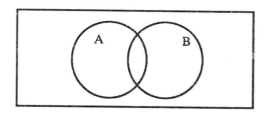

In section 12.1 d) page 555 the relationship connecting the number of elements in sets A and B was given as:
$n(A \cup B) = n(A) + n(B) - n(A \cap B)$.

Dividing this by the total number in the universal set, ie $n(\Sigma)$ it becomes:

$\dfrac{n(A \cup B)}{n(\Sigma)} = \dfrac{n(A)}{n(\Sigma)} + \dfrac{n(B)}{n(\Sigma)} - \dfrac{n(A \cap B)}{n(\Sigma)}$

and this becomes $\boxed{Pr(A \cup B) = Pr(A) + Pr(B) - Pr(A \cap B)}$

This is an alternative form for the addition law that was given earlier as
Pr(A or B) = Pr(A) + Pr(B) - Pr(A and B).

If events A and B are mutually exclusive then $A \cap B = \phi$ and the addition law simplifies to:

$\boxed{Pr(A \cup B) = Pr(A) + Pr(B)}$

Set theory can also be applied to conditional probabilities, such as Pr(B/A) which is the probability that B occurs given that A has occurred.

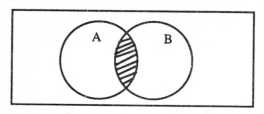

From the Venn diagram it can be seen that the subset of A in which B occurs is $A \cap B$.

$$\therefore \quad Pr(B/A) = \frac{n(A \cap B)}{n(A)}$$

$$= \frac{n(A \cap B)/n(\Sigma)}{n(A)/n(\Sigma)}$$

$$= \frac{Pr(A \cap B)}{Pr(A)}$$

$\therefore$
$$\boxed{Pr(A \cap B) = Pr(B/A) \cdot Pr(A)}$$

and vice versa
$$\boxed{Pr(A \cap B) = Pr(A/B) \cdot Pr(B)}$$

If events A and B are independent (rather than conditional), then the probability of B is unaffected by the occurrence of A.

$$\therefore \quad Pr(B/A) = Pr(B)$$

$\therefore$
$$\boxed{Pr(A \cap B) = Pr(A) \cdot Pr(B)}$$

It must be stressed that these are just alternative forms of the formulae already derived. They are not new work.

ORDER FORM - 'A' LEVEL PUBLICATIONS

Please indicate on the form below the titles and quantity of each which you require.
Then complete the despatch details overleaf where postal charges are set out.

HLT PUBLICATIONS

TEXTBOOKS	Cost £	Quantity	Total £
Accounting	10.95		
Business Studies	10.95		
Economics	9.95		
Applied Mathematics	9.95		
Pure Mathematics	9.95		
Statistics	9.95		
General Principles of English Law	10.95		
Constitutional Law	9.95		
Government and Politics	9.95		
Sociology	9.95		
WORKBOOKS			
Accounting	9.95		
Business Studies	9.95		
Economics	9.95		
		Total Cost £	

HLT PUBLICATIONS

All HLT Publications have two important qualities. First, they are written by specialists, all of whom have direct practical experience of teaching the syllabus. Second, all Textbooks are reviewed and updated each year to reflect new developments and changing trends.

They are used widely by students at polytechnics and colleges throughout the United Kingdom and overseas.

A comprehensive range of titles is covered by the following classifications:

- TEXTBOOKS
- CASEBOOKS
- SUGGESTED SOLUTIONS
- REVISION WORKBOOKS

The books listed overleaf can be ordered through your local bookshops or obtained direct from the publisher using this order form. Telephone, Fax, or Telex orders will also be accepted. Quote your Access or Visa card numbers for priority orders. To order direct from publisher please enter cost of titles you require, fill in despatch details and send it with your remittance to The HLT Group Ltd.

Please complete Order Form overleaf

DETAILS FOR DESPATCH OF PUBLICATIONS

Please insert your full name below

Please insert below the style in which you would like the correspondence from the Publisher addressed to you

TITLE Mr, Miss etc.	INITIALS	SURNAME/FAMILY NAME

Address to which study material is to be sent (please ensure someone will be present to accept delivery of your Publications).

POSTAGE & PACKING

You are welcome to purchase study material from the Publisher at 200 Greyhound Road, London W14 9RY, during normal working hours.

If you wish to order by post this may be done direct from the Publisher. Postal charges are as follows:

UK   - Orders over £30: no charge. Orders below £30: £2.00. Single paper (last exam only): 40p
OVERSEAS - See table below

The Publisher cannot accept responsibility in respect of postal delays or losses in the postal systems. All cheques must be cleared before material is despatched.

SUMMARY OF ORDER

Date of order: _____

Cost of publications ordered:
UNITED KINGDOM:

	TEXTS		Suggested Solutions (last Exam only)
	One	Each Extra	
	£1.00	£0.50	£1.00
			£1.00
			£1.50
			£1.50
			£1.50
			£1.50

Add postage and packing:

OVERSEAS:

	One	Each Extra
Eire	£3.00	£0.50
European Community	£7.50	£1.00
East Europe & North America	£8.50	£1.00
South East Asia	£10.00	£1.50
Australia / New Zealand	£12.00	£3.50
Other Countries (Africa, India etc)	£11.00	£3.00

Total cost of order: £ _____

Please ensure that you enclose a cheque or draft payable to The HLT Group Ltd for the above amount, or charge to

☐ ☐ VISA

Card Number: ☐☐☐☐ ☐☐☐☐ ☐☐☐☐ ☐☐☐☐

Expiry Date _____   Signature _____